# PLASMA PHYSICS

# PHYSIQUE DES PLASMAS

Université de Grenoble

Summer School of Theoretical Physics

Les Houches

Recent Publications

1959 La théorie des gaz neutres et ionisés. Hermann, Wiley
1960 Relations de disperison et particules élémentaires. Hermann, Wiley
1961 La physique des basses températures. Low-Temperature Physics. Gordon and Breach, Presses Universitaires
1962 Géophysique extérieure. Geophysics: The Earth's Environment. Gordon and Breach
1963 Relativité, groupes et topologie. Relativity, Groups and Topology. Gordon and Breach
1964 Optique et électronique quantiques. Quantum Optics and Electronics. Gordon and Breach
1965 Physique des hautes énergies. High Energy Physics. Gordon and Breach
1966 Hautes énergies en astrophysique. High Energy Astrophysics. Gordon and Breach
1967 Problème à *N* corps. Many-Body Problems. Gordon and Breach
1968 Physique Nucléaire. Nuclear Physics. Gordon and Breach
1969 Aspects physiques de quelques problèmes biologiques. Physical Problems in Biological Systems. Gordon and Breach
1970 Mécanique statistique et théorie quantique des champs. Statistical Mechanics and Quantum Field Theory. Gordon and Breach
1971 Physique des Particules. Particle Physics. Gordon and Breach
1972 Physique des Plasmas. Plasma Physics. Gordon and Breach
1972 Les Astres Occlus. Black Holes. Gordon and Breach

In preparation

1973 Fluid Dynamics
1973 Molecular Fluids

Université de Grenoble
Summer School of Theoretical Physics
Les Houches 1972
Supported by NATO and the
Commissariat à l'Energie Atomique

# PLASMA PHYSICS
# PHYSIQUE DES PLASMAS

edited by C. DeWitt
*Faculté des Sciences, Grenoble*
*University of Texas, Austin*

and

J. Peyraud
*Observatoire de Nice*

GORDON AND BREACH SCIENCE PUBLISHERS
New York London Paris

Gordon and Breach, Science Publishers, Inc.
One Park Avenue
New York, N.Y. 10016

*Editorial office for the United Kindom*

Gordon and Breach, Science Publishers Ltd.
42 William IV Street
London, W.C. 2.

*Editorial office for France*

Gordon & Breach
7-9, rue Emile Dubois
Paris 75014

Library of Congress catalog card number 73-87733. ISBN 0 677 15740 1.
 Printed in the United Kingdom.

CONTENTS

## THE AUTHORS

A. BERS
*Research Laboratory of Electronics, Massachusetts Institute of Technology, Cambridge*

G. BEKEFI
*Research Laboratory of Electronics, Massachusetts Institute of Technology, Cambridge*

J.L..DELCROIX
*Laboratoire de Physique des Plasmas, Université Paris XI, Orsay*

G. KALMAN
*Department of Physics, Boston College, Chestnut Hill*

G. LAVAL
*Ecole Polytechnique, Paris*

D. MONTGOMERY
*Department of Physics and Astronomy, University of Iowa*

# PREFACE

This volume gathers the lecture notes of the Session of the Ecole d'été de Physique Théorique devoted to Plasma Physics which has been held in Les Houches in July 1972.

After a rapid evolution due to research on controlled thermonuclear fusion, Plasma Physics seems to have reached a first plateau. Consequently, we have, on the one hand, synthesized the fundamental aspects of Plasma Physics, and on the other hand treated some topics at the frontier of the subject which are likely to have interesting developments.

The first part includes linear and non-linear phenomena, and radiative problems; namely:

Global description of linear waves and instability criteria which are rigorous and of easy application, a study of the main non-linear phenomena, a quasi-linear theory and the random phase approximation.

Radiation problems in Plasmas.

The second part, of a prospective nature, covers three main subjects:

The fluctuations dissipation theorem and its application to the study of dense plasmas; an approximate equation is thus provided for the correlation function which does not depend on the power expansion in the plasma parameter.

Highly magnetized plasmas where the approximation of the guide center makes it possible to give a simple and complete description of the plasmas and the microscopic phenomena (diffusion coefficients, kinetic equation).

The physical chemistry of plasmas which is currently in full evolution.

We hope that this volume will be useful to both experimenters and theorists who are interested in Plasma Physics and will give them both new ideas or a useful source of reference.

The session of the Les Houches Summer School and this volume were made possible through the financial grants from the Ministère de l'Education Nationale, the NATO Scientific Affairs Division (Advanced Study Institute Program) and the Commissariat à l'Energie Atomique under the stewardship of its Service de Physique Théorique.

Jean Peyraud

Cécile DeWitt

# PREFACE

Ce livre réunit les notes de cours de la session de l'Ecole d'été de Physique théorique consacrée à la Physique des Plasmas qui a eu lieu aux Houches en juillet 1972.

Après une évolution rapide due aux recherches sur la fusion thermonucléaire contrôlée, la physique des plasmas semble avoir atteint un premier stade. Nous avons donc, d'une part traité des aspects désormais classiques de la physique du plasma, d'autre part des domaines encore peu explorés susceptibles d'avoir des prolongements intéressants.

La première partie comprend les phénomènes linéaires et non-linéaires, et les problèmes radiatifs. On y trouvera:

une description globale des ondes linéaires et des critères d'instabilité qui sont rigoureux et d'utilisation facile. Cet exposé est complété par une étude des principaux phénomènes non-linéaires. La théorie quasi-linéaire et l'approximation des phases aléatoires (Random Phase Approximation) y sont presentées de manière simple et logique.

une vue synthétique des problèmes radiatifs dans les plasmas. Les divers aspects du rayonnement des plasmas y sont développés sous une forme claire et pratique.

La deuxième partie, de nature prospective, couvre trois sujets principaux:

Le théorème fluctuation-dissipation et son application à l'étude des plasmas denses. L'intérêt de cette méthode est de fournir immédiatement une équation approchée pour la fonction de corrélation, indépendante de tout développement en puissance du paramètre plasma.

L'utilisation d'un modèle en physique des plasmas: celui des plasmas fortement magnétisés où l'approximation du centre guide permet une description simple, très complète du plasma et des phénomènes microscopiques (coefficients de diffusion, équation cinétique), moyennant des hypothèses très raisonnables.

Enfin, nous avons tenu à présenter un aspect en pleine évolution: celui de la physico-chimie des plasmas. Les processus dominants y sont clairement mis en évidence, ils sont ensuite étudiés de la manière la plus pratique possible.

Nous espérons que cet ouvrage rendra service à tous ceux, tant expérimentateurs que théoriciens, qui s'intéressent à la Physique des Plasmas en leur apportant, soit des idées nouvelles, soit une documentation précise.

Pour cette session et ce volume l'école a reçu des subventions du Ministère de l'Education Nationale, de la Division des Affaires Scientifiques de l'OTAN, au titre de son programme d'Instituts d'Etudes Avancées, et du Commissariat à l'Energie Atomique, gérée par le Service de Physique Théorique.

Jean Peyraud

Cécile DeWitt

## LIST OF PARTICIPANTS

| | |
|---|---|
| ADAM, Jean-Claude | Centre de Physique Théorique Ecole Polytechnique, Paris |
| BORNATICI, Marino | c/o ESRIN-FRASCATI, Centre d'Etudes de Limeil, Villeneuve-Saint-Georges |
| DELPECH, Jean-François | Institut d'Electronique, Orsay |
| DE MICHELIS, Claudio | PFC/S.P.P., Fontenay-aux-Roses |
| DEMOKAN, Ordal | Department of Physics, Middle East Technical University, Ankara |
| DEUTSCH, Claude | Laboratoire de Physique des Plasmas, Orsay |
| ESCANDE, D. | Laboratoire PMI, Ecole Polytechnique, Paris |
| FABRE, E. | Laboratoire PMI - Ecole Polytechnique, Paris |
| GOEDE, Adelbert | FOM-Instituutvoor Atoom-en Molecuulfysica, Amsterdam |
| GOERTZ, Christoph | Physics Department, Rhodes University, Grahamstown |
| GRAVIER, René | DPh-PFC/S.P.P., Fontenay-aux-Roses |
| GRESSILLON, D. | Laboratoire de Physique des milieux ionisés, Ecole Polytechnique, Paris |
| JOYCE, Glenn | Physics Research Center, University of Iowa, Iowa City |
| KATSONIS, K. | Laboratoire de Physique des Plasmas, Université Paris-Sud, Orsay |
| LAUNOIS, Daniel | DPh.PFC - Service du Confinement des Plasmas, Fontenay-aux-Roses |
| LAVAUD, Michel | CRPHT-CNRS, Orleans |
| LEMAIRE, J. | Institut d'Aéronomie Spatiale de Belgique, Bruxelles |
| LERCH, Lothar | Institut de Physique Théorique, Université d'Innsbruck, Innsbruck |

| | |
|---|---|
| MAHESH-PRASAD, Srivasta | DEPTT, Physique and Astrophysique University of Delhi, Delhi |
| MASSELOT, Christian | Centre d'Etudes de Limeil, Villeneuve-Saint-Georges |
| MATTHIEUSSENT, Gilles | Laboratoire de Physique des Plasmas, Faculté des Sciences, Orsay |
| MATTIOLI, Mario | DPh-PFC/S.P.P. Fontenay-aux-Roses |
| MIHAILESCU, Ion N. | Section de radiations et de plasmas, Institut de Physique Atomique, Bucarest |
| MISGUICH, Jacques | SIGI-CEN/FAR, Fontenay-aux-Roses |
| MISRA, Krishna | Physics Section, Institute of Technology, Banaras Hindu University, Varanasi |
| MONTE, Jacqueline | Laboratoire de Physique des Plasmas, Grenoble |
| MONTES, Carlos | Equipe de Mécanique Statistique, Observatoire de Nice, Nice |
| NGUYEN THE HUNG | Centre de Recherches Mathématiques, Université de Montréal, Montreal |
| OLIVAIN, Jacques | DPh-PFC/S.P.P., Fontenay-aux-Roses |
| PELLETIER, Guy | Département de la Fusion - C.E.N.G., Grenoble |
| PELLETIER, Jacques | Laboratoire de Physique des Plasmas, Grenoble |
| PERONA, Giovanni E. | Ist di Elettronica e Telecomunicazioni, Politecnico C. so Duca Degli Abruzzi 24, Torino |
| PESME, Denis | Centre de Physique Théorique, Ecole Polytechnique, Paris |

| | |
|---|---|
| ROBERSON, Charles | Yale University, Mason Laboratory, New Haven |
| SCHMITT, J. | Laboratoire PMI, Ecole Polytechnique, Paris |
| SEN, Abhijit | Physical Research Laboratory, Navrangpura, Ahmedabad |
| SINGH, R.P. | G.R.I. - C.N.E.T., Issy-les-Moulineaux |
| SLUIJTER, Frans W. | Technische Hogeschool, Eindhoven |
| SOULET, Yves | Physique Quantique, Université Paul Sabatier, Toulouse |
| STENFLO, L. | Department of Plasma Physics, University of Umeå, Umeå |
| VERHEEST, Frank | Seminarie voor analytische mechanica, Rijksuniversiteit Gent, Jozef, Gent |
| VIRMONT, J. | Laboratoire PLI, Ecole Polytechnique, Paris |
| WEYNANTS, R. | Laboratorium voor Plasmafysica, Koninklijke Militaire School, Bruxelles |
| YU, Y. | Department of Plasma Physics, Umeå University, Umeå |
| ZADWORNY, F. | Laboratoire de Physique des Plasmas, Grenoble |

# COLLECTIVE EMISSION PROCESSES IN UNMAGNETIZED PLASMAS

G. BEKEFI

*Department of Physics and Research Laboratory of Electronics*
*Massachusetts Institute of Technology*
*Cambridge, Massachusetts*

CONTENTS

# PREFACE

The fundamental concepts of non-equilibrium plasma kinetic theory have been laid down more than a decade ago and the fruits of these studies are being exploited still, by experimenters and theorists alike. Thus, "Applications of Plasma Kinetic Theory with Special Emphasis on Fluctuations and Radiation" could well make an appropriate subtitle for these notes. The word-application-should be stressed because the basic precepts of the theory are neither questioned nor examined on these pages. Indeed, we use the simplest and fairly intuitive approach usually referred to as the "method of dressed test particles" to get at the desired results, without even attempting to make the connection with the well known formal expansion techniques of kinetic theory. It is gratifying that the dressed test particle method leads to practical results which have hitherto been found always correct and sufficiently accurate. Those who wish to understand the foundations of plasma kinetic theory are referred to the recent text of Montgomery [1].

Our discussion is limited to highly ionized, almost collisionless (Vlasov) plasmas. The inclusion of collisions with atoms or even the presence of numerous large angle Coulomb collisions leads to difficulties many of which are neither well understood nor well studied. No externally applied magnetic field is allowed to act on the plasma. The presence of a magnetic field results in a whole host of additional modes of oscillation, time and length scales which spoil much of the intrinsic simplicity of the subject. Moreover, the new complexities are such that they cannot be given a fair treatment in a short lectures series like the present.

# 1. ELECTRICAL PROPERTIES OF A PLASMA

In this chapter we give an account of some well-known electrical properties of a plasma induced by the passage of a disturbance propagating through it. The discussion is restricted to unmagnetized media in which particle collisions, though possibly significant, are not of primary concern. The response of the plasma to a small disturbance is fully accounted for by a single quantity - the medium's dielectric coefficient. It is this quantity with which we shall be concerned most heavily.

## 1.1 *Equations of the Electromagnetic Field*

The electric and magnetic fields $E(r,t)$ and $H(r,t)$ of a plasma are generated by the motions of charged particles comprising the ionized medium. The current density $J(r,t)$ and the charge density $\rho(r,t)$ are the source terms used in introducing the plasma into Maxwell's equations,

$$\nabla \times E = -\mu_0 \frac{\partial H}{\partial t} \tag{1.1}$$

$$\nabla \times H = \varepsilon_0 \frac{\partial E}{\partial t} + J \tag{1.2}$$

$$\nabla \cdot E = \frac{\rho}{\varepsilon_0} \tag{1.3}$$

$$\nabla \cdot B = 0 \tag{1.4}$$

with $\varepsilon_0$ and $\mu_0$ as the permittivity and permeability of free space, respectively. Implicit in Eqs. (1.2) and (1.3) is the equation of continuity

$$\frac{\partial \rho}{\partial t} + \nabla \cdot J = 0. \tag{1.5}$$

At any point $r$ and at any time $t$, $J$ and $\rho$ may be usefully considered as sums of two components

$$J(r,t) = J_m(r,t) + J_s(r,t) \tag{1.6}$$

$$\rho(r,t) = \rho_m(r,t) + \rho_s(r,t), \tag{1.7}$$

the macroscopic current and charge densities $J_m$, $\rho_m$ representing the local average drift and charge accumulation *induced* in response to $E(r,t)$; and the remaining stochastic quantities $J_s$ and $\rho_s$ created by the random motions of the particles. Our primary involvement is with stochastic processes; nonetheless, because stochastic phenomena are strongly influenced by the average, induced behavior of the plasma, we begin by a summary of the latter. In so doing we assume for the time being that $J_s = \rho_s = 0$. One must then picture the electromagnetic fields which are interacting with the plasma as being generated by some external agency.

When the fields perturbing the gas are sufficiently weak, the macroscopic current $J_m$ and the electric field $E$ are linearly related through a function which, itself, is independent of the strength of the radiation field. This relationship can be written [2] in a very general form resembling Ohm's law

$$J_m(r,t) = \iint \sigma(r-r', t-t') \cdot E(r',t')\, d^3r'\, dt' \tag{1.8}$$

with $\sigma$ as a conductivity operator. Its tensor character comes about because of the possible anisotropy of the plasma medium. The above equation embodies the ideas of temporal and spatial dispersion. The presence of time dispersion in the medium is signalled by the fact that the current (and thus also the charge) at a given time $t$ depends on the values of the electric field during all previous times $t'$. Similarly, the presence of spatial dispersion is allowed for in that $J_m$ at the point $r$ depends not only on the value of $E$ at that point, but also on all other neighboring points $r'$. This so-called "nonlocal" phenomenon arises whenever the charged particles have motions of their own (due to thermal effects, for example) which are in addition to the motions they acquire in the imposed electric field $E$. The fact that $\sigma$ does not depend on the absolute position $r$ but only on the relative position $R \equiv r - r'$ comes from assuming that the medium is uniform and homogeneous.

On introducing in Eq. (1.8) the space and time Fourier transforms defined as

$$A(r,t) = \iint A(\mathbf{k},\omega)\, e^{j\omega t - j\mathbf{k}\cdot r} \frac{d^3\mathbf{k}}{(2\pi)^3} \frac{d\omega}{(2\pi)} \tag{1.9}$$

$$A(\mathbf{k},\omega) = \iint A(r,t)\, e^{-j\omega t + j\mathbf{k}\cdot r}\, d^3r\, dt \tag{1.10}$$

one readily obtains the more familiar form of Ohm's law written in the frequency and wave vector domain:

$$J(\mathbf{k},\omega) = \sigma(\mathbf{k},\omega)\cdot E(\mathbf{k},\omega) \tag{1.11}$$

The conductivity $\sigma(\mathbf{k},\omega)$ is connected to the operator $\sigma(\boldsymbol{r} - \boldsymbol{r}', t - t')$ of Eq. (1.8) through

$$\sigma(\mathbf{k},\omega) = \iint \sigma(R,\tau)\, e^{-j\omega\tau + j\mathbf{k}\cdot R}\, d^3R\, d\tau \qquad (1.12)$$

with $R \equiv \boldsymbol{r} - \boldsymbol{r}'$ and $\tau \equiv t - t'$.

In many situations of interest, the conductivity $\sigma(\mathbf{k},\omega)$ proves to be nearly purely imaginary (signifying but weak damping of the fluctuations) and for that reason it is frequently more convenient to work with a related quantity, namely the dielectric coefficient $K$ defined as

$$K(\mathbf{k},\omega) = 1 + \frac{\sigma(\mathbf{k},\omega)}{j\omega\varepsilon_0} \qquad (1.13)$$

where 1 is a unity matrix. With this coefficient, Maxwell's Eqs. (1.1) to (1.5) written for the Fourier components $E(\mathbf{k},\omega)$, $H(\mathbf{k},\omega)$ then become

$$\mathbf{k} \times E = \omega\mu_0 H \qquad (1.14)$$

$$\mathbf{k} \times H = -\,\omega\varepsilon_0 K\cdot E \qquad (1.15)$$

$$\mathbf{k} \cdot E = j\rho/\varepsilon_0 \text{ or } \mathbf{k} \cdot K \cdot E = 0 \qquad (1.16)$$

$$\mathbf{k} \cdot B = 0 \qquad (1.17)$$

$$\omega\rho = \mathbf{k} \cdot J \qquad (1.18)$$

Equations (1.16) are two forms of Poisson's equations, in which the second form is derived by use of Eqs. (1.11) and (1.18). In these equations $J$ and $\rho$ denote the macroscopic current and density discussed earlier, but we omit the subscript $m$ so as not to burden the notation. The wave equation for one monochromatic, plane wave component $\exp[j\omega t - j\mathbf{k}\cdot\boldsymbol{r}]$ is obtained by eliminating $H$ from Eqs. (1.14) and (1.15) with the result that

$$\mathbf{k} \times (\mathbf{k} \times E) + \left(\frac{\omega}{c}\right)^2 K \cdot E = 0 \qquad (1.19)$$

When the plasma is unmagnetized, Eq. (1.19) permits the existence of two entirely independent families of modes of propagation, each family being characterized by a different set of phase velocities $\omega/\mathrm{k}$ and possibly also different dielectric coefficients, $K$. Table 1.1 summarizes the properties of the two families of modes, derived from a scrutiny of Eqs. (1.14) to (1.19). The one family typifies the traditional transverse electromagnetic wave with the vectors **k**, $E$ and $H$ mutually orthogonal; the wave requires the full set of Maxwell's equations for its description. No density oscillations accompany this disturbance. The other family is comprised of longitudinal density oscillations with $\mathbf{k} \times E = 0$.

TABLE 1.1. COMPARISON OF THE TRANSVERSE ELECTROMAGNETIC AND LONGITUDINAL MODES OF OSCILLATION IN AN UNMAGNETIZED PLASMA

| Property | Transverse Electromagnetic Waves | Longitudinal Plasma Waves |
|---|---|---|
| Vector Directions | $\mathbf{k} \cdot E = 0$ 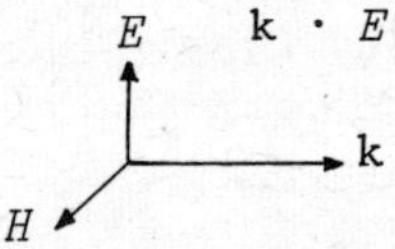  | $\mathbf{k} \times E = 0$ 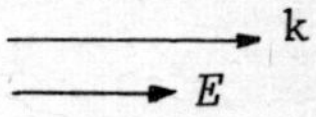  |
| Dispersion Relation | $k^2 c^2 = \omega^2 K_T(\mathbf{k},\omega)$ | $K_L(\mathbf{k},\omega) = 0$ |
| Electric field $E(\mathbf{k},\omega)$ | finite | finite |
| Magnetic field $H(\mathbf{k},\omega)$ | finite | zero |
| Current density $J(\mathbf{k},\omega)$ | finite | finite |
| Charge density $\rho(\mathbf{k},\omega)$ | zero | finite |

The modes do not have an *rf* magnetic field associated with them, and they obey Poisson's equation (1.16) alone; they are quasi-electrostatic and it therefore comes as no surprise that they are representative of the collective Coulomb interactions between the charged particles. Therefore, the longitudinal oscillations normally prevail over the transverse electromagnetic ones, both in their larger energy content and in their importance as a source of a variety of radiation processes. It is for this reason, that we shall be concerned most heavily with longitudinal fluctuations and the associated dielectric coefficient $K_L(\mathbf{k},\omega)$.

To make a comparison of the energies in the two types of modes, consider a "black" box of temperature $T$ filled with plasma. In thermal equilibrium, the energy density in the longitudinal oscillations* is approximately

* This energy is comprised of two parts: the energy density in the electric field and the oscillatory kinetic energy of the particles.

$$U_L \simeq \kappa T \frac{d^3\mathbf{k}_L}{(2\pi)^3}\left[\frac{\partial\omega K_L}{\partial\omega}\right]$$

$$= \frac{\kappa T}{2\pi^2}\,\mathbf{k}_L^2\,\frac{\partial \mathbf{k}_L}{\partial\omega}\,d\omega\left[\frac{\partial\omega K_L}{\partial\omega}\right],$$

and in the transverse electromagnetic oscillations it is

$$U_T \simeq \kappa T \frac{d^3\mathbf{k}_T}{(2\pi)^3}\left[\frac{\partial\omega K_T}{\partial\omega}\right]$$

$$= \frac{\kappa T}{2\pi^2}\,\mathbf{k}_T^2\,\frac{\delta \mathbf{k}_T}{\delta\omega}\,d\omega\left[\frac{\partial\omega K_T}{\partial\omega}\right].$$

We now note that $\delta\omega/\delta\mathrm{k} \equiv v_g$ is the group velocity, $\omega/\mathrm{k} \equiv v_p$ is the phase velocity and $Uv_g \equiv F$ is the radiation flux. In terms of these qualities, the ratio of the wave fluxes can be written in the form,

$$\frac{F_L}{F_T} = \left(\frac{V_{pT}}{V_{pL}}\right)^2 \frac{\partial\omega K_L/\partial\omega}{\partial\omega K_T/\partial\omega}\ .$$

The transverse electromagnetic wave is a fast wave and in an unmagnetized plasma it propagates with a velocity $c$ or greater. On the other hand, longitudinal waves are generally (but not always) slow, with velocities of the order of the thermal speeds of the particles, $v_0$. Thus, $F_L/F_T \approx (c/v_0)^2$ and one is usually justified in neglecting the transverse component of the oscillatory field [In the above equations, $\kappa T$ is the energy per mode, $d^3\mathrm{k}/(2\pi)^3$ in the density of modes in the interval $d\mathrm{k}$, and $\delta\omega K/\delta\omega$ is a term necessary to account correctly for the energy in a dispersive medium [3]].

## 1.2 *The Longitudinal Dielectric Coefficient* $K_L(\mathbf{k},\omega)$.

Hitherto the electrical conductivity and dielectric coefficient have been treated merely as phenomenological parameters which connect the plasma response ($J$) to the electrical disturbance ($E$). To go beyond this step, one must examine the detailed motions of the charges under the influence of the electromagnetic forces. To begin with, let us restrict ourselves to plasma electrons, assuming that the ions are immobile. This is justified whenever the typical frequencies of the exciting fields are so high that ions cannot readily follow the excursions.

When the forces are sufficiently weak, we can make the linearized Boltzmann-Vlasov equation

$$\frac{\partial f_1}{\partial t} + v \cdot \frac{\partial f_1}{\partial r} + \frac{q}{m} E \cdot \frac{\partial f}{\partial v} = \left(\frac{\delta F}{\delta t}\right)_{coll.} \tag{1.20}$$

as the basis of the computations. Here $f(v)$ is the time stationary, equilibrium (but *not* necessarily Maxwellian) distribution of electron velocities existing in the absence of the perturbing *rf* electric field $E$; $f_1(r,v,t)$ represents a small perturbation such that at any instant of time, the total distribution is $F = f + f_1$; $q$ is the charge and $m$ the mass of the particle species in question. Since the *rf* magnetic field in the longitudinal oscillation is zero (see Table 1.1), and there are no externally applied steady magnetic fields, no contribution from these forces need be considered in the above equation. The right-hand side is a phenomenological way of representing particle collisions. For now we assume that the plasma is collisionless and we set $(\delta F/\delta t)_{coll.}$ equal to zero.

To solve Eq. (1.20) for a single Fourier component of the perturbation, we write

$$f_1(r,v,t) = f_1(\mathbf{k},v,\omega)\, e^{j\omega t - j\mathbf{k}\cdot r} \tag{1.21}$$

and substituting in Eq. (1.20), obtain

$$f_1 = -\left(\frac{q}{m}\right) \frac{E \cdot \delta f/\delta v}{j(\omega - \mathbf{k}\cdot v)} \tag{1.22}$$

The fluctuating charge density $\rho(\mathbf{k},\omega)$ follows upon integration over all particle velocities in three-dimensional velocity space, with the result that

$$\rho = q\int f_1\, d^3v \tag{1.23}$$

$$= -\left(\frac{Nq^2}{m}\right)\int \frac{E \cdot \delta f/\delta v}{j(\omega - \mathbf{k}\cdot v)}\, d^3v$$

In writing this equation the distribution $f$ is assumed to be normalized to unity in such a way that

$$\int f\, d^3v = 1 \tag{1.24}$$

This has the consequence that the charge density $N$ appears explicitly in Eq. (1.23) as shown. Inserting it in the equation of continuity (1.18) and with the aid of Eq. (1.11), we arrive at the sought-after expression for the dielectric coefficient

$$K_L(\mathbf{k},\omega) = 1 + \frac{\omega_p^2}{k^2} \int \frac{\mathbf{k}\cdot\delta f(v)/\delta v}{\omega - \mathbf{k}\cdot v}\, d^3v \tag{1.25}$$

where $\omega_p^2 \equiv Nq^2/m\varepsilon_0$ is the square of the plasma frequency. The integral must be correctly defined at $\omega = \mathbf{k}\cdot v$ where a singularity occurs; following Landau's prescription [4] of integrating around this pole one obtains

$$K_L(\mathbf{k},\omega) = 1 + \frac{\omega_p^2}{k^2} P \int \frac{\mathbf{k}\cdot\delta f(v)/\delta v}{\omega - \mathbf{k}\cdot v}\, d^3v + j\pi \frac{\omega_p^2}{k^2} \int \delta(\omega - \mathbf{k}\cdot v)\, \mathbf{k}\cdot \frac{\delta f(v)}{\delta v}\, d^3v \tag{1.26}$$

where $P$ denotes the principal value of the integral. The residue imparts to the dielectric coefficient an imaginary part signifying the presence of damping. The energy loss from the *rf* fluctuations occurs through a resonance mechanism between the *rf* oscillations and those particles whose velocity equals the phase velocity of the wave. This fact is brought out clearly through the $\delta$-function appearing in the second term of Eq. (1.26). Because the *rf* charge density $\rho$ and the current density $J$ are linear sums of contributions from the different charged species, the generalization of Eq. (1.25) [or Eq. (1.26)] to a plasma composed of $i$ different charged species is simply,

$$K_L(\mathbf{k},\omega) = 1 + \sum_i \frac{\omega_{pi}^2}{k^2} \int \frac{\mathbf{k}\cdot\delta f_i(v)/\delta v}{\omega - \mathbf{k}\cdot v} \tag{1.27}$$

Up to this point no restriction was made on the actual form of the distribution function $f(v)$. Now, let us consider one species of particles (electrons) and assume that they have a Maxwellian velocity distribution

$$f(v) = \left(\frac{m}{2\pi\kappa T}\right)^{3/2} e^{-mv^2/2\kappa T} \tag{1.28}$$

at the temperature $T$. Inserting this distribution in Eq. (1.26) one finds (after somewhat lengthy manipulation) that the dielectric coefficient can be written as

$$K_L(\mathbf{k},\omega) \equiv 1 - \frac{\omega_p^2}{\omega^2}\,\phi(x) \tag{1.29}$$

where the quantity $\Phi(x)$ contains the error function, and is given by

$$\phi(x) = -2x^2\left(1 - 2x\, e^{-x^2} \int_0^x e^{t^2}\, dt - j\sqrt{\pi}\, x e^{-x^2}\right) \tag{1.30}$$

with x defined as follows:

$$x = \frac{\omega}{\sqrt{2}\ kv_0};\quad v_0 = \sqrt{\frac{\kappa T}{m}}.$$

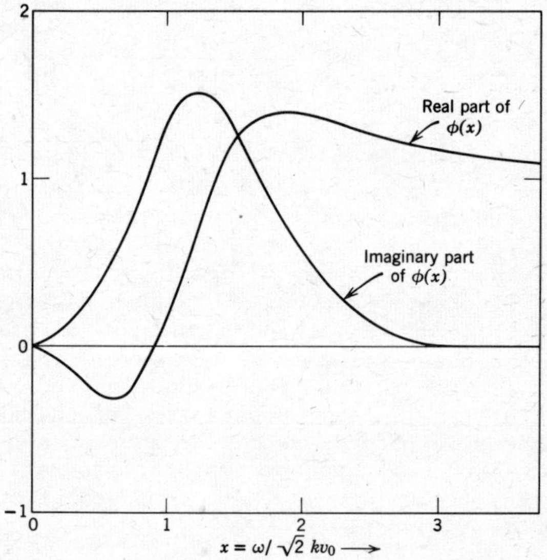


Fig. 1.1 The function $\phi$ in the expression for the longitudinal dielectric coefficient $K_L(k,\omega) = 1 - (\omega_p^2/\omega^2)\phi$, assuming a Maxwellian distribution of electron velocities and infinitely massive ions.

Figure 1.1 shows a plot of the real and imaginary parts of $\Phi(x)$. Note that the imaginary part (damping) is zero both at zero frequency and at very high frequencies. The reason for this is clear from the form of Eq. (1.26) which in effect states that the imaginary part of $K_L$

$$Im\ K_L(\mathbf{k},\omega) \propto \left[\frac{\delta f}{\delta v}\right]_{v = \frac{\omega}{k}} \tag{1.31}$$

is proportional to the derivative of $f(v)$ evaluated at $v = \omega/k$. Since for a Maxwellian distribution $\delta f/\delta v$ is zero at $v = 0$ and is also very small at large $v$, $Im\ K_L$ vanishes in these two limits. Indeed, in the two cases $\Phi(x)$ given above takes on a simpler form, with the following results:

$$K_L(\mathbf{k},\omega) \simeq 1 - \frac{\omega_p^2}{\omega^2}\left\{1 + \frac{3k^2v_0^2}{\omega^2} + j\sqrt{\frac{\pi}{2}}\left(\frac{\omega}{kv_0}\right)^3 e^{-\omega^2/2k^2v_0^2}\right\} \tag{1.32}$$

when $(\omega/\sqrt{2}kv_0 >> 1$, and

$$K_L(\mathbf{k},\omega) \simeq 1 + \frac{\omega_p^2}{k^2 v_0^2} - j\sqrt{\frac{\pi}{2}}\frac{\omega_p^2}{\omega^2}\left(\frac{\omega}{kv_0}\right)^3 \tag{1.33}$$

when $(\omega/\sqrt{2}kv_0) << 1$. Taking these equations to their limits, we see that at infinite wavelength (k = 0), Eq. (1.32) becomes

$$K_L(0,\omega) = 1 - \frac{\omega_p^2}{\omega^2} \tag{1.34}$$

showing that now spatial dispersion is nonexistent. This is not surprising since "nonlocal" phenomena occur on a scale-length in which the particle experiences displacements of the order of the wavelength. Result (1.34) where $K_L$ is independent of k (and $v_0$) is known as the "cold plasma approximation". When we go to the opposite extreme of zero frequency, we obtain from Eq. (1.33) the static value of the dielectric coefficient:

$$K_L(k,0) = 1 + \left(\frac{\omega_p}{kv_0}\right)^2 \tag{1.35}$$

The complete trend of $K_L$ for all values of x is found on combining the results of Fig. 1.1 with Eq. (1.29). The results of such computations are illustrated in Fig. 1.2.

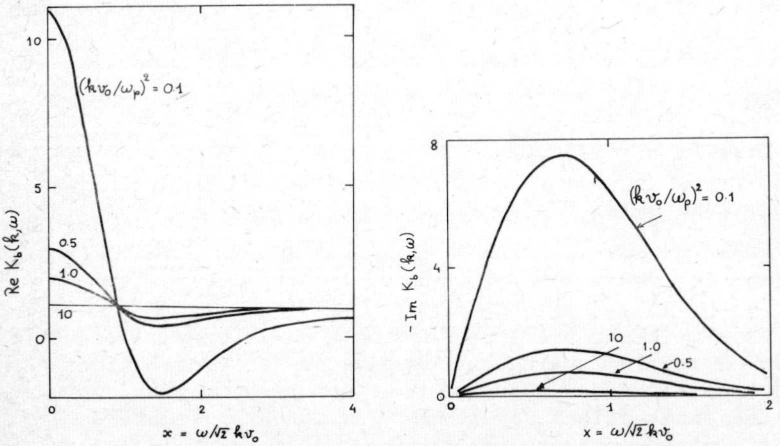


Fig. 1.2 The real part and the imaginary part of the longitudinal dielectric coefficient as a function of frequency [from reference 5].

It is often useful to express the value of the dielectric coefficient given by Eq. (1.30) in terms of a certain function $Z(\zeta)$ called the plasma dispersion function that has been tabulated by Fadeeva and Terent'ev [6] and Fried and Conte [7]. This function is closely related to the error function appearing in Eq. (1.30) but it is more general in that the argument $\zeta$ can be complex. It, and its derivative with respect to the argument $(Z'(\zeta))$, are defined as follows:

$$\begin{aligned} \sqrt{\pi}\, Z(\zeta) &= \int_{-\infty}^{\infty} \frac{e^{-t^2}}{t-\zeta}\, dt && \text{for } Im\,\zeta < 0 \\ &= P\int_{-\infty}^{\infty} \frac{e^{-t^2}}{t-\zeta}\, dt - j\pi\, e^{-\zeta^2} && \text{for } Im\,\zeta = 0 \qquad (1.36) \\ &= \int_{-\infty}^{\infty} \frac{e^{-t^2}}{t-\zeta}\, dt - j2\pi\, e^{-\zeta^2} && \text{for } Im\,\zeta > 0 \end{aligned}$$

and

$$Z'(\zeta) = -\,2[1 + \xi Z(\xi)] \qquad (1.37)$$

In terms of these functions, the longitudinal dielectric coefficient becomes

$$K_L(\mathrm{k},\omega) = 1 - \frac{\omega_p^2}{\omega^2}\, \mathrm{x}^2 Z'(\mathrm{x}) \qquad (1.38)$$

where as before x is given by $\mathrm{x} = \omega/\sqrt{2}\mathrm{k}v_0$. If $i$ species of charged particles participate in the motion then by Eq. (1.27)

$$K_L(\mathrm{k},\omega) = 1 - \sum_i \frac{\omega_{pi}^2}{\omega^2}\, \mathrm{x}_i^2\, Z'(\mathrm{x}_i) \qquad (1.39)$$

with $\mathrm{x}_i = \omega/\sqrt{2}\mathrm{k}v_{0i}$.

## 1.3 *The Influence of Collisions*

We saw in the previous section that in certain frequency ranges, collisionless Landau damping of the oscillations can be very small indeed. But we must not rush to the conclusion that the fluctuations now remain unattenuated; short-range collisions of electrons with neutral atom or collision with ions are to some extent ever present, and it is they that take over when Landau damping is too small. The proper way to allow for collisions is through the full Boltzmann collision integral (or its equivalent for Coulomb encounters). Since this leads to virtually insurmountable computational problems, various approximations are frequently adopted. The simplest is the relaxation model

$$\begin{aligned}\left(\frac{\delta F}{\delta t}\right)_{\text{coll.}} &= -\nu(v)\,[F - f] \\ &= -\nu(v)\, f_1\end{aligned} \tag{1.40}$$

with $F = f + f_1$ and $\nu(v)$ as a phenomenological collision frequency which may or may not be a function of the particles' speed. A collision term of this type forces the distribution function to relax upon each collision to the average distribution. As a result Eq. (1.40) does not conserve particles at every position and every instant of time, but it conserves particles only in an averaged sense. A probably somewhat better model is a variant of that due to Bhatnagar, Gross and Krook [8]. Here the collision term is artificially taylored so as to give precise particle conservation. The form*

$$\left(\frac{\delta F}{\delta t}\right)_{\text{coll.}} = -\nu(v)\,[F - f\,\frac{\int \nu(v) F\, d^3v}{\int \nu(v) f\, d^3v}] \tag{1.41}$$

has just this property, as can be easily checked by integrating each term over all velocities. When $\nu$ is constant and independent of speed, this collision term takes on a more manageable form

$$\left(\frac{\delta F}{\delta t}\right)_{\text{coll}} = -\nu\,[f_1 - f\,\frac{\int f_1 d^3v}{\int f d^3v}] \tag{1.42}$$

which we now apply in computing the dielectric coefficient $K_L(\mathbf{k},\omega)$.

We insert Eq. (1.42) on the right-hand side of (1.20) and following the procedure outlined by Eqs. (1.21) to (1.24), we obtain the following expression for the fluctuating charge density $\rho(\mathbf{k},\omega)$

$$\rho[1 - \nu\int \frac{f(v)\, d^3v}{j(\omega - \mathbf{k}\cdot v) + \nu}] = -\left(\frac{Nq^2}{m}\right)\int \frac{E\cdot \delta f(v)/\delta v}{j(\omega - \mathbf{k}\cdot v) + \nu}\, d^3v \tag{1.43}$$

and the associated dielectric coefficient:

$$K_L(\mathbf{k},\omega) = 1 + \frac{\omega_p^2}{\mathrm{k}^2}\,[\int \frac{\mathbf{k}\cdot \delta f(v)/\delta v}{(\omega - \mathbf{k}\cdot v) - j\nu}\, d^3v][1 + j\nu\int \frac{f(v)\, d^3v}{(\omega - \mathbf{k}\cdot v) - j\nu}]^{-1} \tag{1.44}$$

---

* For other more physical models see for example H. Derfler, *Phys. Fluids* *10*, 1636 (1967).

By analogy with Eq. (1.29), we again write that

$$K_L(\mathbf{k},\omega) = 1 - \frac{\omega_p^2}{\omega^2}\ \Theta\,(x,jy) \tag{1.45}$$

where now the new quantity $\Theta$ is a function of two variables,

$$x = \frac{\omega}{\sqrt{2}\ kv_0}\ ,\ y = \frac{\nu}{\sqrt{2}\ kv_0} \tag{1.46}$$

When expressed in terms of the $Z$-function defined in the previous section, $\Theta$ has the form,

$$\Theta = -\ 2x^2[1 + \frac{xZ(\xi)}{1 + jyZ(\xi)}] \tag{1.47}$$

with $\zeta = x + jy$. Figure 1.3 compares the behavior of $\Theta$ in the presence and absence of collisions. Observe that the case $y = 1$ we have chosen represents quite heavy collisional damping; since $v_0/\nu$ is approximately the particle's mean free path (MFP), it follows that $y = [\sqrt{2}\ k(\text{MFP})]^{-1}$ and for $y = 1$, the wavelength of the oscillation is about eight times longer than the mean-free path. In spite of this, the overall features of $\Theta$ have not been greatly altered by the presence of collisions. In passing we note that had we chosen the relaxation model (1.40) to represent collisional effects, Eq. (1.47) would have to be replaced by another function, namely

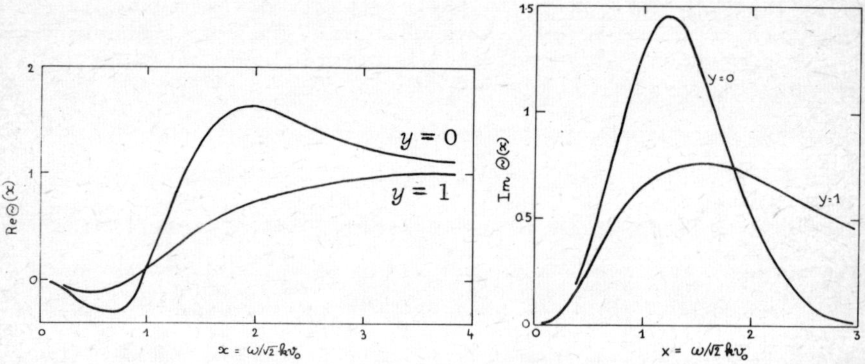


Fig. 1.3 The real part and the imaginary part of the function $\Theta\ (x,jy)$ in the expression $K_L = 1 - (\omega_p^2/\omega^2)\ \Theta$, showing the effect of collisions. When $y \equiv \nu/\sqrt{2}\ kv_0$ is zero, these result are the same as those shown in Fig. 1.1.

$$- 2x^2 \; [1 + xZ(\xi)] \tag{1.48}$$

## 1.4 *The Dispersion Relations*

The quantities ω and k appearing in the expressions for the dielectric coefficient $K_L$ are real numbers; indeed, there is no reason to expect them to be otherwise. Furthermore, ω and k bear no special relationship to one another, and their values can be picked arbitrarily. Matters are different when we impose the requirement that the oscillation be a natural, freely propagating mode of the system. In section 1.1 we showed that this is synonymous with

$$K_L(\mathbf{k},\omega) = 0, \tag{1.49}$$

a condition which imposes a definite relationship on ω and k; usually the condition cannot be met when ω and k are both real, and one or the other quantity must be permitted to be complex. This just says that the wave does not remain of constant amplitude, but that it decays (or grows) in time and, (or), in space.

Consider a collisionless plasma whose dielectric coefficient is given by Eq. (1.38). Then the corresponding dispersion relation is given by the transcendental equation

$$1 - \frac{\omega_p{}^2}{2k^2 v_0^2} Z'(\omega/\sqrt{2}\, k v_0) = 0 \tag{1.50}$$

which ordinarily requires a computer generated program for its full solution. Figure 1.4 shows the results of such computations, assuming that k is real and ω is complex. However, in the case of long wavelengths, $(\omega/\sqrt{2}\, k v_0) \gg 1$, the approximation (1.32) suffices and we then find that

$$\begin{aligned} \mathrm{Re}\omega &\simeq [\omega_p^2 + 3k^2 v_0^2]^{\frac{1}{2}} \\ \mathrm{Im}\omega &\simeq \mathrm{Re}\omega \left\{ \sqrt{\frac{\pi}{8}} \left(\frac{\omega_p}{k v_0}\right)^3 \exp -\left[\frac{\omega_p^2}{2k^2 v_0^2} + \frac{3}{2}\right] \right\} \end{aligned} \tag{1.51}$$

which is a result due to Landau.

In Fig. 1.4 we see that damping of the wave grows dramatically as the wavelength diminishes. Taking the value $\mathrm{Im}\omega/Re\;\omega \leq 0.1$ as "acceptable" we see that freely propagating waves are possible only in the range

$$1 < \frac{\omega}{\omega_p} \lesssim 1.4$$
$$\sqrt{2}\,\frac{kv_0}{\omega_p} \lesssim 0.6 \qquad (1.52)$$

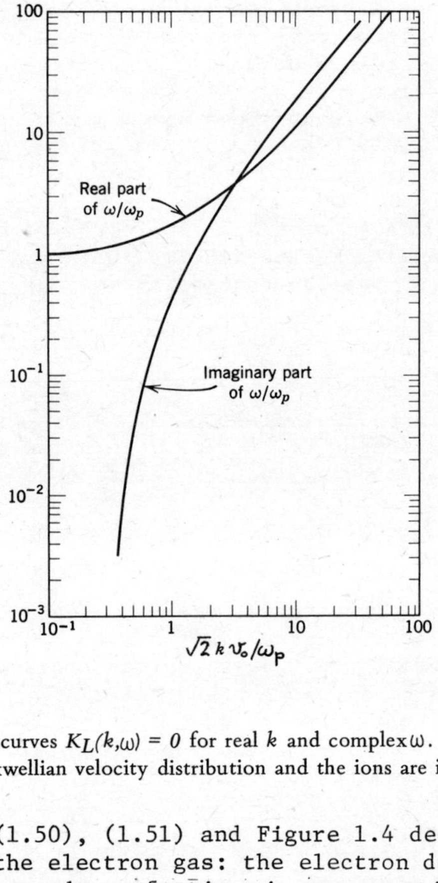


Fig. 1.4 Dispersion curves $K_L(k,\omega) = 0$ for real $k$ and complex $\omega$. The electrons have a Maxwellian velocity distribution and the ions are infinitely massive.

Equations (1.50), (1.51) and Figure 1.4 describe the disturbance in the electron gas: the electron density undergoes compressions and rarefactions in an essentially immobile ion background. At sufficiently low frequencies this model breaks down and ion motions must be included by writing in place of Eq. (1.38),

$$K_L(k,\omega) = 1 - \frac{\omega_{pe}^2}{2k^2 v_{oe}^2} Z'(\omega/\sqrt{2}\ kv_{oe}) - \frac{\omega_{pi}^2}{2k^2 v_{oi}^2} Z'(\omega/\sqrt{2}\ kv_{oi}) \quad (1.53)$$

where $\omega_{pe} = \sqrt{Ne^2/m\varepsilon_0}$ is the electron plasma frequency, $\omega_{pi} = \sqrt{Ne^2/M\varepsilon_0}$ is the ion plasma frequency with $M$ as the ion mass; and $v_{oe} = \sqrt{\kappa T_e/m}$ and $v_{oi} = \sqrt{\kappa T_i/M}$ are the electron and ion thermal speeds. Examination of Eq. (1.53) with $K_L = 0$ discloses that in addition to the high frequency electron plasma mode described above, the new dispersion relation permits of a second, independent wave which exists at low frequency. It is known as the ion-acoustic wave; here the electron and ion displacements are virtually identical, both species moving in phase with the applied electric field. When the ions are cold and the electrons hot, such that $v_{oe} \gg \omega/k \gg v_{oi}$, suitable expansions of the Z-functions can be made. With these, one finds that the real part of the dispersion characteristic is given by

$$\text{Re}\,\omega \simeq k\left[\frac{\kappa T_e}{M}\{1 + (kv_{oe}/\omega_{pe})^2\}^{-1} + 3\,\frac{\kappa T_i}{M}\right]^{\frac{1}{2}}. \qquad (1.54)$$

If one further assumes that $kv_{oe}/\omega_{pe} \ll 1$, which is a good approximation in many situations,

$$\text{Re}\,\omega \simeq k\sqrt{\frac{\kappa T_e + 3\kappa T_i}{M}} \qquad (1.55)$$

from which we see that the wave is dispersionless - it propagates with constant velocity independent of frequency. When the damping is weak and $kv_{oe}/\omega_{pe} \ll 1$, one also finds that

$$\text{Im}\,\omega \approx k\sqrt{\frac{\kappa T_e}{M}}\sqrt{\frac{\pi}{8}}\left\{\sqrt{\frac{m}{M}} + \left(\frac{T_e}{T_i}\right)^{3/2} e^{-T_e/2T_i}\right\} \qquad (1.56)$$

Exact computer solutions of Eq. (1.53) must be carried out whenever the electron and ion temperatures are comparable to one another. The results of such computations are shown in Fig. 1.5, where the reciprocal of Im $\omega$ is plotted as a function of $T_e/T_i$. For large values of this ratio the damping is weak and is primarily due to Landau damping on the electrons. As $T_i$ approaches $T_e$ the damping becomes very heavy and the contribution from ion Landau damping begins to dominate. A comparison of the graphed results with Eq. (1.56) shows that this equation is a rather poor approximation to the truth. On the other hand, Eq. (1.55) for the real part of $\omega$ agrees to within 3 percent with the exact computations for all values of $T_e/T_i$ from one to 40. In the limit of zero ion temperature, and at frequencies where the electron Landau damping is not too heavy, expression (1.53) for the dielectric coefficient can be approximated by,

$$K_L(k,\omega) \approx 1 - \frac{\omega_{pi}^2}{\omega^2} + \frac{\omega_{pe}^2}{k^2 v_{oe}^2}\left[1 - j\sqrt{\frac{\pi}{2}}\,\frac{\omega}{kv_{oe}}\right] \qquad (1.57)$$

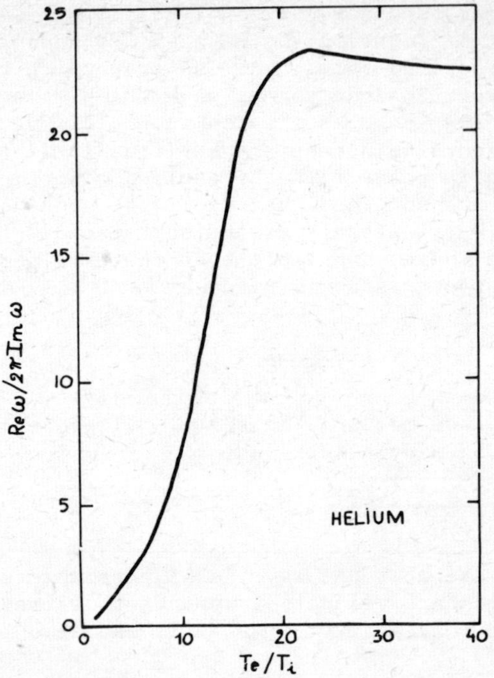


Fig. 1.5 Landau damping of an ion acoustic wave in a collisionless helium plasma as a function of the ratio of electron to ion temperatures. The quantity $\mathrm{Re}\,\omega/2\pi\, Im\,\omega$ also equals $\delta/\lambda$ where $\lambda$ is the wavelength and $\delta$ the damping length defined as: wave amplitude $\propto exp[-z/\delta]$. The curve shown is virtually independent of the absolute value of $T_i$, and is applicable in the wave number range, $kv_{oe}/\ _{pe} \ll 1$. [from reference 10].

## 2. TEST CHARGE IN A PLASMA

The motion of a single point charge passing through a highly ionized plasma can be understood only in terms of its simultaneous interaction with many of the neighboring charges. A study of these collective interactions and of their consequences, is the subject matter of this chapter.

### 2.1 *The Field Around a Charged Particle*

We wish to find the electric field $E(r,t)$ in the neighborhood of a "test charge" $q_t$ moving through an infinite homogeneous plasma. Since quasi-electrostatic fields are of present interest, Poisson's equation

$$\nabla \cdot E = \frac{\rho_m}{\varepsilon_0} + \frac{\rho_t}{\varepsilon_0} \tag{2.1}$$

is the only electrodynamic relation that needs to be considered. Here $q_m$ is the macroscopic charge density representing the presence of the plasma (see section 1.1) and $\rho_t$ is the density of the point test charge

$$\rho_t(\boldsymbol{r},t) = q_t\ \delta(\boldsymbol{r} - \boldsymbol{r}_t(t)) \tag{2.2}$$

with $\boldsymbol{r}_t(t)$ as its position at time $t$. We now restrict the motion of the particle to rectilinear translation at constant velocity*, with the result that

$$\rho_t(\boldsymbol{r},t) = q_t\ \delta(\boldsymbol{r} - \boldsymbol{r}_0 - vt) \tag{2.3}$$

where $v$ is the speed and $\boldsymbol{r}_0$ its initial position.

It is desired to eliminate $\rho_m$ from Eq. (2.1) as the "spokesman" for the background plasma, in favor of the dielectric coefficient, whose properties we now well understand. That such a replacement is possible has been proved already in the derivation of Eqs. 1.16. Note, however, that the dielectric coefficient is usefully defined only in the frequency, wave-vector domain into which Eq. (2.1) must now be cast. Introducing the longitudinal dielectric coefficient $K_L(\mathbf{k},\omega)$, we then find that

$$\mathbf{k} \cdot E(\mathbf{k},\omega)\ K_L(\mathbf{k},\omega) = j\rho_t(\mathbf{k},\omega)/\varepsilon_0 \tag{2.4}$$

Here $\rho_t(\mathbf{k},\omega)$ is the space-time Fourier component of $\rho_t(\boldsymbol{r},\mathrm{t})$ and it has the value

$$\rho_t(\mathbf{k},\omega) = 2\pi\ q_t\ \delta(\omega - \mathbf{k} \cdot v)\ e^{j\mathbf{k}\cdot\boldsymbol{r}_0} \tag{2.5}$$

Substitution of Eq. (2.5) in Eq. (2.4) yields the following expression for the electric field of the test charge:

$$E(\mathbf{k},\omega) = 2\pi j\ q_t\ \mathbf{k}\ \frac{\delta(\omega - \mathbf{k}\cdot v)}{|\mathbf{k}|^2\varepsilon_0\ K_L(\mathbf{k},\omega)}\ e^{j\mathbf{k}\cdot\boldsymbol{r}_0} \tag{2.6}$$

---

*A straight line path implies that the test charge suffers few large angle deflections; a constant velocity implies that the rate of loss of energy experienced by the test charge must be sufficiently small.

To know the field in real physical space, the above result must be inverted by means of a Fourier-k transform [see Eqs. (1.9) and (1.10)] with the result that

$$E(\boldsymbol{r},\boldsymbol{r}_0,\omega) = \frac{jq_t}{(2\pi)^2}\int \frac{\mathbf{k}\, d^3\mathbf{k}}{|\mathbf{k}|^2\varepsilon_0\, K_L(\mathbf{k},\omega)}\,\delta(\omega - \mathbf{k}\cdot v)e^{-j\mathbf{k}\cdot(\boldsymbol{r} - \boldsymbol{r}_0)} \tag{2.7}$$

Furthermore, transforming into real time and performing the delta-function integration over frequency as is indicated by the equation, the electric field at any time $t$ and position $\boldsymbol{r}$ is obtained:

$$E(\boldsymbol{r},\boldsymbol{r}_0,t) = \frac{jq_t}{(2\pi)^3}\int \frac{\mathbf{k}\, d^3\mathbf{k}}{|\mathbf{k}|^2\varepsilon_0\, K_L(\mathbf{k},\mathbf{k}\cdot v)}\,e^{-j\mathbf{k}\cdot(\boldsymbol{r} - \boldsymbol{r}_0 - vt)} \tag{2.8}$$

Using Poisson's equation the associated electrical potential $\phi$ follows directly

$$\phi(\boldsymbol{r},\boldsymbol{r}_0,t) = \frac{q_t}{(2\pi)^3}\int \frac{d^3\mathbf{k}}{|\mathbf{k}|^2\varepsilon_0\, K_L(\mathbf{k},\mathbf{k}\cdot v)}\,e^{-j\mathbf{k}\cdot(\boldsymbol{r} - \boldsymbol{r}_0 - vt)}\,. \tag{2.9}$$

The important point about the foregoing equations lies in the fact that, the presence of the plasma in which the charge is embedded is fully accounted for by a single quantity - the dielectric coefficient. Thus, what at first sight appears to be a thankless task, namely the computation of the interaction of $q_i$ with all other charges, turns out to be a straightforward problem. The electric field $E$ of Eq. (2.8) in effect replaces the charges' Coulomb field.

## 2.2 *Debye Shielding*

We can now derive the static field around a stationary test charge. We set $v = 0$ in Eq. (2.9) and for a start assume (incorrectly) that the plasma ions play no part. In this case the static dielectric coefficient is given by Eq. (1.35), and on substituting this in Eq. (2.9), we obtain

$$\phi(R) = \frac{q_t\, v_0^2}{(2\pi)^3\varepsilon_0}\int \frac{d^3\mathbf{k}}{\omega_p^2 + \mathbf{k}^2 v_0^2}\,e^{-j\mathbf{k}\cdot R}\quad ;\ R = \boldsymbol{r} - \boldsymbol{r}_0 \tag{2.10}$$

On carrying out this integration, we arrive at the familiar screened potential,

$$\phi(R) = \frac{q_t}{4\pi\varepsilon_0}\,\frac{e^{-R/L_D}}{R} \tag{2.11}$$

where $L_D$ is the screening distance (for electrons) given by

$$L_D = \sqrt{\frac{\kappa T_e \varepsilon_0}{Ne^2}}$$

$$= \frac{v_{0e}}{\omega_{pe}} \qquad (2.12)$$

$$= 7.43 \times 10^3 \sqrt{T_e/N} \text{ meter}$$

In the last equation $T_e$ is given in $eV$ and the electron density in meter$^{-3}$; $\omega_{pe}$ is the electron plasma frequency.

If we allow for the presence of one or more species ($i$) of ions in addition to the electrons the static dielectric coefficient becomes,

$$K_L(\mathrm{k},0) = \frac{\mathrm{k}^2 + L_D^{-2} + \Sigma_i L_{Di}^{-2}}{\mathrm{k}^2} \; ; \; L_{Di} = \sqrt{\frac{\kappa T_i \varepsilon_0}{N_i q_i^2}} \qquad (2.13)$$

with the result that the Debye distance $L_D$ appearing in the exponent of the expression for $\phi(R)$ must be now replaced by the total effective Debye length $L_{DT}$ defined as,

$$\left(\frac{1}{L_{DT}}\right)^2 = \left(\frac{1}{L_D}\right)^2 + \sum_i \left(\frac{1}{L_{Di}}\right)^2 \qquad (2.14)$$

When we have a neutral plasma of electron density $N$ with only one species of ions of charge $Ze$, whose temperature $T_i$ does not necessarily equal the electron temperature $T$, it follows that

$$\left(\frac{1}{L_{DT}}\right)^2 = \frac{Ne^2}{\varepsilon_0}\left[\frac{1}{\kappa T_e} + \frac{Z}{\kappa T_i}\right] \qquad (2.15)$$

from which we see that the colder of the two species of particles dictates the magnitude of the shielding distance.

We point out that under the static conditions ($v = 0$) considered here, it is imperative that the ions be properly included in determining the shielding distance. However, in the study of oscillatory phenomena and other nonstationary conditions the ions often do not appear in the expression for the shielding distance. The reason is that on a time scale for which the heavy ions cannot respond, they do not contribute to shielding.

When the test charge moves ($v \neq 0$) Eq. (2.9) can no

longer be evaluated with ease. Moreover, the shielding is rather different now and the characteristic exponential behavior indicated by Eq. (2.11) no longer occurs for all distances $R$. At small $R$ one finds pretty much exponential shielding, but at very large $R$ ($R \to \infty$) the potential varies inversely as the third power of the distance. This inverse power law suggests that a moving test charge with its shielding cloud constitute an electric quadrupole. [For more details, see Montgomery, Joyce and Sugihara, *Plasma Phys.* 10, 681 (1968); and Cooper, *Phys. Fluids* 12, 2707 (1969)].

## 2.3 *Energy Loss of Test Charge*

The loss of energy suffered by the particle stems from the fact that it polarizes the plasma medium and so does work on it. Since this test charge is not located at the center of the induced polarization, it suffers a dragging force from it. Of the total energy loss one can separate out approximately that part which is due to the *resonant* interaction between the test charge and the plasma particles; the loss rate by this mechanism must therefore correspond to the rate of energy gained by the collective plasma oscillations, that is, to the spontaneous emission of plasma waves. The remainder of the energy loss can be better pictured as being due to small angle Coulomb collisions of the test electron with the individual particles of the medium. In this section we address ourselves to the problem of calculating these energy losses, usually called polarization losses.

We now derive an expression for the polarization energy loss of a test charge of charge $q_t$ and mass $m_t$ moving with velocity $v$ through an isotropic plasma described by the longitudinal dielectric coefficient $K_L(\mathbf{k},\omega)$. The induced polarization field is obtained from Eq. (2.8) by subtracting from the total field that part which remains when the test charge moves through vacuum:

$$E_{\text{ind}}(\boldsymbol{r},\boldsymbol{r}_0,t) = \frac{jq_t}{(2\pi)^3}\int \frac{\mathbf{k}\, d^3\mathbf{k}}{|\mathbf{k}|^2}\, e^{-j\mathbf{k}\cdot(\boldsymbol{r}-\boldsymbol{r}_0-\boldsymbol{v}t)} \times \left[\frac{1}{\varepsilon_0\, K_L(\mathbf{k},\mathbf{k}\cdot\boldsymbol{v})} - \frac{1}{\varepsilon_0}\right] \tag{2.16}$$

The rate at which the test charge loses its energy to the plasma is obtained from

$$\left[\frac{dW}{dt}\right]_{\text{pol}} = q_t \left[\boldsymbol{v}\cdot E_{\text{ind}}\right]_{\boldsymbol{r}=\boldsymbol{r}_0+\boldsymbol{v}t} \tag{2.17}$$

where we note that $E_{ind}$ is to be evaluated at the instantaneous position of the charge, $r(t) = r_0 + vt$. Substituting Eq. (2.16) in (2.17) and using the "reality" condition for the electric field, namely that $E_{ind}(\mathbf{k},\omega) = E^*_{ind}(-\mathbf{k}, -\omega)$, we obtain,

$$\left[\frac{dW}{dt}\right]_{pol} = - \frac{q_t^2}{(2\pi)^3} \int \frac{d^3k \ \mathbf{k}\cdot \boldsymbol{v}}{\varepsilon_0 |k|^2} \operatorname{Im}\left[\frac{1}{K_L(\mathbf{k},\mathbf{k}\cdot\boldsymbol{v})}\right] \tag{2.18}$$

which is the sought-for result. Observe that the polarization energy loss is independent of the mass of the test charge and thus differs from zero even for a particle of infinite mass; observe also that if $q_t$ represents a tiny charge clump consisting of $N$ particles moving together, then the energy lost from the clump is proportional to $N^2$; on the other hand, when one has $N$ particles moving independently of one another as may be the case in a *very tenuous* beam, then the total energy lost is proportional to $N$. Given the dielectric coefficient of the medium, we can in principle compute "exactly" the energy loss of the particle. This normally requires numerical computation. However, the interaction with collective modes can be roughly separated out and an approximate result derived. For now we let the test charge $q_t$ interact with the plasma electrons only, and assume that the ions have an infinite mass.

To evaluate the above integral we write $d^3k = 2\pi k_\perp dk_\perp dk_\parallel$ where $k_\parallel$ and $k_\perp$ are the components of $\mathbf{k}$ along and perpendicular to the velocity $v$. Noting also that $k_\parallel v = \omega$, we obtain,

$$\left|\frac{dW}{dt}\right|_{pol} = - \frac{q_t^2 v}{4\pi^2\varepsilon_0} \int_0^\infty \int_{-\infty}^\infty \frac{k_\perp dk_\perp \omega d\omega}{\omega^2 + k^2 v^2} \operatorname{Im}\left[\frac{1}{K_L(\mathbf{k},\omega)}\right] \tag{2.19}$$

Let us now examine the quantity,

$$\operatorname{Im}\left|\frac{1}{K_L}\right| = - \frac{K_i}{K_r^2 + K_i^2} \tag{2.20}$$

where $K_r$ and $K_i$ are the real and imaginary parts of $K_L$. Now, there is a frequency $\omega = \omega_0 (\approx \omega_p)$ at which the denominator of Eq. (2.20) exhibits a minimum. For weak damping this occurs when $K_r(k,\omega_0) \approx 0$. Thus, here the level of the electric-field excitation is greatly enhanced. This is not surprising because this requirement is the approximate dispersion relation for a longitudinal wave and thus represent a "resonant" condition for the wave's existence as a natural mode of oscillation of the plasma medium. At frequencies $\omega \approx \omega_0$, in the immediate neighborhood of the resonant spike, we can expand the dielectric coefficient in a Taylor series about $\omega_0$ and, since $K_L(k,\omega_0) = 0$, write,

$$K_L(k,\omega) \simeq \left[\frac{\partial \mathrm{Re} K_L(k,\omega)}{\partial\omega}\right]_{\omega=\omega_0} (\omega-\omega_0) + j\,\mathrm{Im} K_L(k,\omega_0) \tag{2.21}$$

In the limit of weak damping ($\mathrm{Im}\,K_L(k,\omega_0) \to 0$) it then follows that,

$$\mathrm{Im}\left[\frac{1}{K_L}\right] \simeq \pi \left[\frac{\partial \mathrm{Re} K_L}{\partial\omega}\right]^{-1}_{\omega=\omega_0} \delta(\omega-\omega_0)$$

and (2.22)

$$\frac{1}{|K_L|^2} \simeq -\frac{\pi}{\mathrm{Im}\,K_L}\left[\frac{\partial \mathrm{Re} K_L}{\partial\omega}\right]^{-1}_{\omega=\omega_0} \delta(\omega-\omega_0)$$

For resonant interactions with an essentially lossless plasma we now replace $\mathrm{Im}(1/K_L)$ by a delta function as prescribed by Eq. (2.22). On carrying out the integrations we obtain,

$$\left[\frac{dW}{dt}\right]_{\mathrm{pol}} \simeq -\frac{q_t^{\,2}\omega_p^{\,2}}{8\pi\varepsilon_0 v}\ \ln\left[1 + \frac{k^2_{\perp\,\max} v^2}{\omega_p^{\,2}}\right] \tag{2.23}$$

a result due to Pines and Bohm [11]. The parameter $k_{\perp\max}$ comes from cutting off the integration over $k_\perp$, since the integral diverges logarithmically as $k_\perp \to \infty$. Collective modes are not excited unless $kL_D \lesssim 1$ and we set $k^2_{\perp\,\max} = 1/L_D^{\,2}$, with the result that,

$$\left[\frac{dW}{dt}\right]_{\mathrm{pol}} \simeq -\frac{q_t^{\,2}\omega_p^{\,2}}{8\pi\varepsilon_0 v}\ \ln\left[1 + \frac{v^2}{v_0^{\,2}}\right] \tag{2.24}$$

where $v_0 = \sqrt{\kappa T/m}$ is the thermal speed of the plasma electrons.

Polarization loss is not the only mechanism for energy exchange between a test charge and the plasma medium. There is an energy gain mechanism in which the test charge picks up energy from the fluctuating background electric fields. That some such counteracting mechanism must exist is clear, since a test charge with energy equal to the average thermal energy of the plasma is in equilibrium and neither loses nor gains energy. Whereas polarization loss is independent of the mass of the test particle, the rate of energy gain is inversely proportional to the mass. Thus stochastic energy gain plays an essential role in the motion of a light particle (electron) through the plasma, but it can often be safely neglected for ions. We shall not attempt to derive here the expression for this energy gain mechanism, but merely quote the result. In the case when the test charge interacts with Maxwellian plasma particles of temperature $T$, one finds that [5]

$$\left[\frac{dW}{dt}\right]_{\text{stoch}} = \frac{q_t^{\,2}\,\kappa T}{2\pi^2 \varepsilon_0 m_t v} \int dk\; k\; \mathrm{Im}\left[\frac{1}{K_L(\mathbf{k},\mathbf{k}\cdot v)}\right] \tag{2.25}$$

Thus the total energy exchange between the test charge $q_t$, $m_t$ and a single species of plasma particles, is the sum of terms (2.18) and (2.25).* If the interaction is with several species (say plasma electrons and ions), then the energy exchange rates must be properly summed over these species, and the dielectric coefficient $K_L$ must reflect their presence [see Eq. (1.27)]. We note that the integrals in both expressions diverge logarithmically for large values of $k$, and they must be appropriately cut off at some $k = k_{\text{max}}$. This divergence is typical of calculations made from the Vlasov equation which does not take proper account of short-range collisions. The value of $k_{\text{max}}$ normally chosen is the reciprocal of the classical distance of closest approach of the test particle and an average plasma particle (alternately, $k_{\text{max}}$ is taken as the reciprocal of the de Broglie wavelength, when quantum effects become important).

---

*We point out that if the Balescu-Lenard equation [1] is made the starting point of the computations, the complete energy exchange [polarization plus stochastic] emerges naturally [12].

An interesting limiting solution of Eqs. (2.18) and (2.25) can be obtained from the following observation [13]: when $kL_D \gg 1$, the magnitude of the dielectric coefficient $|K_L(\omega,\mathbf{k})|$ approaches unity (see Fig. 1.2). If then $|K_L(\mathbf{k},\omega)|$ of Eqs. (2.18) and (2.25) is replaced by 1 and the lower limit of integration is replaced by $1/L_D$, the equations become

$$\left[\frac{dW}{dt}\right]_{\substack{\text{pol+}\\ \text{stoch}}} = -\,4\pi\left(\frac{q_t q}{4\pi\varepsilon_0}\right)^2 \frac{N}{m}\sqrt{\frac{m}{2\kappa T}}\;\ell n\Lambda$$

$$\left[\Psi\left(\sqrt{\frac{mv^2}{2\kappa T}}\right) - \frac{2}{\sqrt{\pi}}\frac{m}{m_t}\,e^{-mv^2/2\kappa T}\right] \tag{2.26}$$

where $m$, $q$, $N$ and $T$ are the mass, charge, density, and temperature of the plasma particles in question, and the $\Psi$ is a function

$$\Psi(x) = \frac{\mathrm{erf}(x)}{x} - \frac{2}{\sqrt{\pi}}\,e^{-x^2} \tag{2.27}$$

with $erf(x)$ as the error integral. The function is shown plotted in Fig. 2.1. The quantity $\ell n\Lambda$ involves the ratio of the upper and lower cutoff parameters*

$$\ell n\Lambda = \ell n\left(\frac{k_{\max}}{k_{\min}}\right) = \ell n\left(L_D\left[\frac{Ze^2}{4\pi\varepsilon_0\kappa T_e}\right]^{-1}\right) \tag{2.28}$$

The noteworthy point about Eq. (2.26) is the fact that it is precisely the result [15] one obtains for the energy exchange rate from a simple binary calculation for the Coulomb collisions. This is not too surprising since setting $|K_L| = 1$ is equivalent to neglecting collective effects. Observe also that the expression (2.26) includes both energy loss and gain mechanisms — namely it contains the collisional equivalents of the polarization loss and the stochastic gain discussed earlier. The latter is represented by the term prefixed with $m/m_t$. When the test charge interacts with more than one plasma species, the total energy loss is made up of a sum of terms exactly like those shown on the right-hand side of Eq. (2.26).

---

*The definition of $\Lambda$ given by Eq. (2.28) is one of several used in the literature. They differ from one another by numerical constants of order unity, which has usually little effect, once the logarithm is taken.

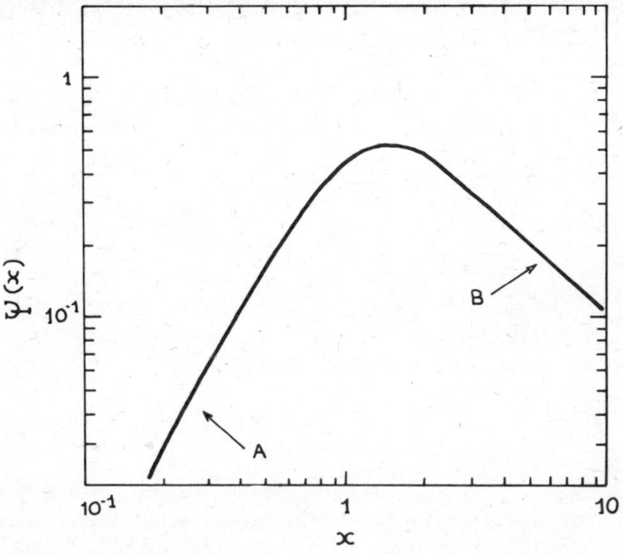


Fig. 2.1 The energy loss rate function $\Psi(x)$ in the expression (2.26). In the low velocity range $A[x \ll 1]$, $\Psi = 4x^2/3\sqrt{\pi}$; in the high velocity range $B$, $= 1/x$. Recall that Eq. (2.26), and thus also $(x)$, take account only of the collisional part of the total energy exchange; energy loss due to collective generation of waves is here excluded.

*Case of the test electron.* Consider $q_t$ to be a test electron interacting with a thermal background of plasma electrons; the ions are assumed to be infinitely heavy and any interaction with them is disregarded for now. Substituting Eqs. (1.29) and (1.30) in Eqs. (2.18) and (2.25), and performing the numerical integration,* leads to the results shown in Fig. 2.2. The lower dashed curve represents the stochastic energy gained by the electron, and the upper dashed curve the polarization loss. The solid line is the sum total of the two effects. We see that polarization loss predominates at high electron speeds ($v \gg v_{oe}$) and stochastic acceleration gain predominates at low speeds. When the energy of the test electron equals the mean thermal energy of the plasma electrons, the test electron is in thermal equilibrium with the particles of the medium and thus on the average it neither gains nor loses energy.

It is now instructive to establish contact between this more complete collective model for the energy exchange and results obtained from the simple binary model described by

---

* For example, in Eq. (2.28) we write $k$ in spherical coordinates such that $d^3k \quad 2\pi k^2 dk \sin\theta d\theta$; then integration over $k$ can be done analytically. The integration over $\theta$ requires numerical procedures.

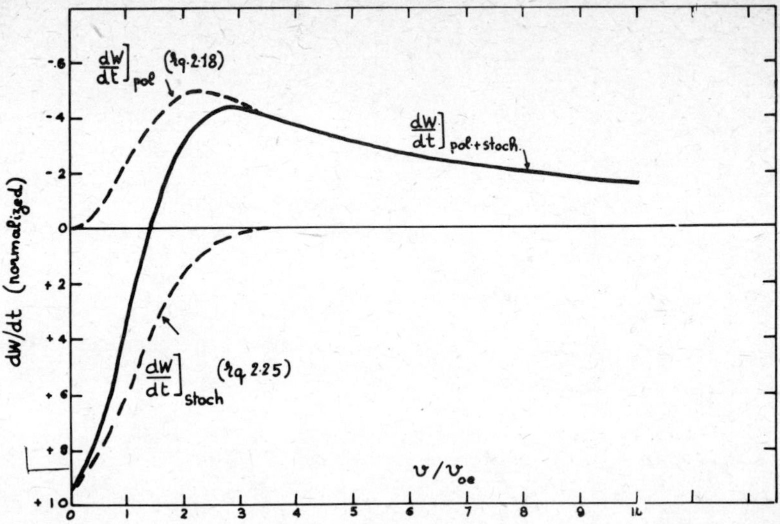


Fig. 2.2 The energy exchange rate of a test electron of speed $v$, exchanging energy with plasma electrons only. The assumed plasma parameters are $N = 10^{16}\,cm^{-3}, T_e = 4 \times 10^4\ {}^0K$. The quantity plotted is the true energy exchange rate $dW/dt$ divided by the following quantity:

$$(4\pi/\sqrt{2\pi})(q_t q/4\pi\varepsilon_0)^2 (N/m)(1/v_{oe}) \ln\Lambda$$

By means of this normalization, the curve shown is almost independent of the plasma parameters $N$ and $T_e$ chosen, and varies by less than 10 percent as $N$ and $T$ are changed [from reference 13].

Eq. (2.26). This comparison [13] is made in Fig. 2.3. The differences between the solid and dashed curves represent the energy exchange resulting from collective effects. We observe that when the test electron has low speed, $v \lesssim v_{oe}$, the binary collision model works very well. This is as expected, because collective excitations become significant only for the higher test particle velocities, as is clearly indicated by Eq. (2.24). The reasons for this can be understood in terms of the following physical picture. The energy loss of the test electron appears in the form of longitudinal plasma oscillations. These oscillations are excited by a Cerenkov-like mechanism, requiring the well known synchronism condition $k \cdot v = \omega$, or $\omega/k = v \cos\theta$. Since $\omega$ is of order of $\omega_p$, the plasma frequency, and $k$ cannot exceed $L_D^{-1}$ to assure weak Landau damping, we must have that $v/v_{oe} \gtrsim 1$. The magnitude of the collective wave mechanism is seen in Fig. 2.3. For example, when $v/v_{oe}$ equals 10, about one third of the total energy loss goes into waves. Comparison of this value with that obtained from the approximate equation (2.24) shows that the latter predicts too high a value for the wave part of the energy loss. But this is to be expected, because in the computational scheme adopted in obtaining Eq. (2.24),

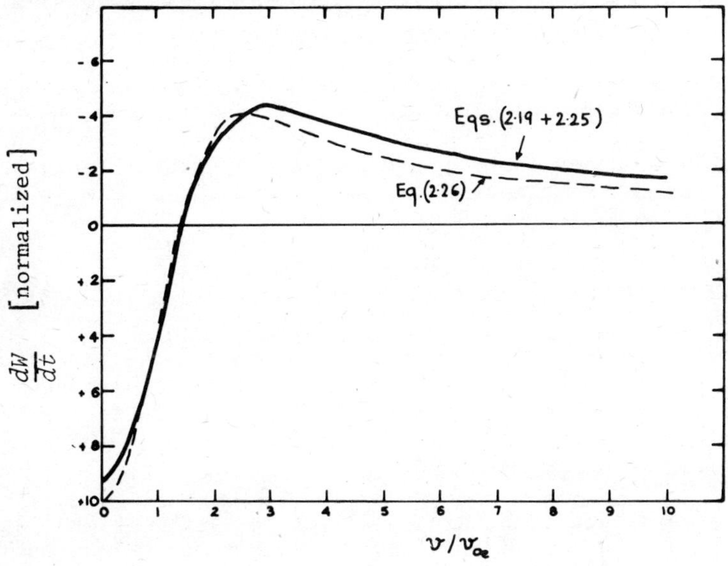


Fig. 2.3 Comparison of the normalized energy exchange rate [see caption to Fig. 2.2] with that due to binary collisions only. The solid curve is the total rate [collisional plus collective] and the dashed curve accounts for the collisional part only. [from reference 13].

some collisional contributions are undoubtedly included; moreover, the approximations were somewhat severe.

As the velocity of the test charge (electron or ion) diminishes relative to the thermal velocity of the plasma electrons, interactions with the plasma ions becomes progressively more important, and our numerical results eventually become invalid. An examination of Eq. (2.26) [with the ion terms included], shows that there is a characteristic velocity [15]

$$u = \left(3\sqrt{\frac{\pi}{2}}\, Z^2\right)^{1/3} v_{oe} \left(\frac{m}{M}\right)^{1/3} \tag{2.29}$$

at which the energy lost by the test charge is shared equally amongst the plasma electrons and ions. Thus, the test particle energy goes predominantly into the plasma electrons when $v > u$, and into the ions when $v < u$. In expression (2.29), $m$ is the electron and $M$ the ion mass, $Ze$ is the ion charge and $v_{oe} = \sqrt{\kappa T_e/m}$ is the electron thermal speed. Thus we see that for a hydrogen plasma, with $m/M = 1/1836$, $u \simeq 0.1\, v_{oe}$ and the computations shown in Figs. 2.2 and 2.3 are certainly good approximations when the speed of the test electron exceeds 0.1 times the electron thermal speed.

*Case of the test ion.* In many realistic situations, the velocity $v$ of the test ion exceeds the velocity $u$ defined by Eq. (2.29). Under these conditions the ion loses energy predominantly to the plasma electrons, and Eq. (2.26) simplifies to

$$\frac{dW}{dt} = -\, 4\pi \left(\frac{q_t q}{4\pi\varepsilon_0}\right)^2 \frac{N}{m} \sqrt{\frac{m}{2\kappa T_e}}\; \ell n\Lambda\; \Psi\left(\frac{mv^2}{2\kappa T_e}\right) \tag{2.30}$$

where we neglect a small term of order $m/m_t$. The variation of $dW/dt$ with test-ion energy is now described completely by the function $\Psi(x)$ shown plotted in Fig. 2.1. It is seen that the rate of loss is greatest when $x \equiv \sqrt{mv/2\kappa T_e}$ is approximately 1.5.

We recollect that in the derivation of Eq. (2.26) [and thus in the derivation of (2.30)], certain approximations and cutoff procedures in the integrals were made. Their effect is two-fold: first, collective interactions are removed from the resulting equations, and secondly the cutoff at close distances eliminates all energy losses caused by large angle collisions. One may well ask what the ensuing errors are when Eq. (2.30) is used in place of a so-called "fully convergent theory" which does not have these limitations. Although further discussion of such theories [16, 17] lies outside the scope of this review, their conclusions are worth noting. It is convenient to write the fully correct theory just as is shown in Eq. (2.30), but to replace the $\ell n\Lambda$ term by a new term defined as

$$\begin{aligned}(\ell n\Lambda)_{\text{eff}} &= \ell n\Lambda + \Delta_1(x) + \Delta_2(x) \\ &= \ell n\Lambda + \Delta(x)\end{aligned} \tag{2.31}$$

Here $\Delta_1(x)$ is the correction caused by including collective effects, and $\Delta_2(x)$ is the correction caused by close, large angle collisions made by the test ion. The sum total of the two corrections is given by the parameter $\Delta(x)$. Figure 2.4 illustrates the magnitudes of the effects as a function of the speed of the test ion. They are to be compared with the classical $\ell n\Lambda$ which is typically between 8 and 20. Thus we may conclude that the error made by using the approximate form (2.30) is usually not too severe.

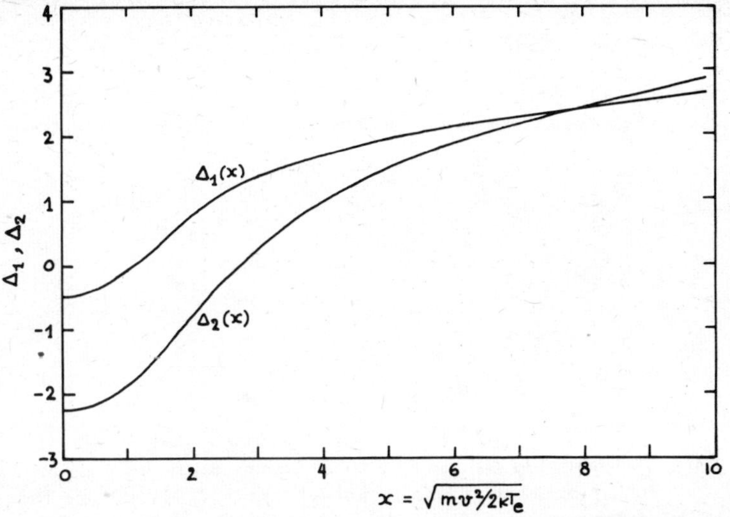


Fig. 2.4 Corrections to the energy loss rate formula (2.30) for the case of a test ion interacting with the plasma electrons. The correction is made as explained by Eq. (2.31); $\Delta_1$ accounts for collective effects and $\Delta_2$ for the close, large angle Coulomb encounters. [After reference 16]. $v$ **is the velocity of the test ion.**

Despite the large body of theoretical work done on the very fundamental problem of the energy exchange between a particle and a plasma, there is virtually no *definitive* experiment to check the theoretical models. One of the major reasons for this is the relatively small amount of energy exchange involved, which can easily be masked by several other experimental effects: scattering from plasma inhomogeneities, loss (or gain) of energy in the plasma sheaths which the test particle must traverse, excitation of atomic states in the ion, scattering from a background of neutral atoms, etc. Of the few attempts [14, 18] at experimental verification, the results of Klavan et al [19] are probably the best. Here the background medium is a highly ionized, thermal Cesium plasma ($T \simeq 2100°K$) in the form of a 7.5 cm long column, with a plasma density not exceeding $4 \times 10^{11} cm^{-3}$. Injected into the column is a monoenergetic beam of Cesium ions with energies between 35 and 150 eV. The beam is very tenuous ($\sim 10^{-10}$ Amp) such that the separation between neighboring beam ions is many Debye lengths. This is important, because under these conditions the test ions interact individually with the plasma, so that the well-known beam-plasma phenomena, in which interactions of waves on the beam with waves on the plasma, may be ignored.

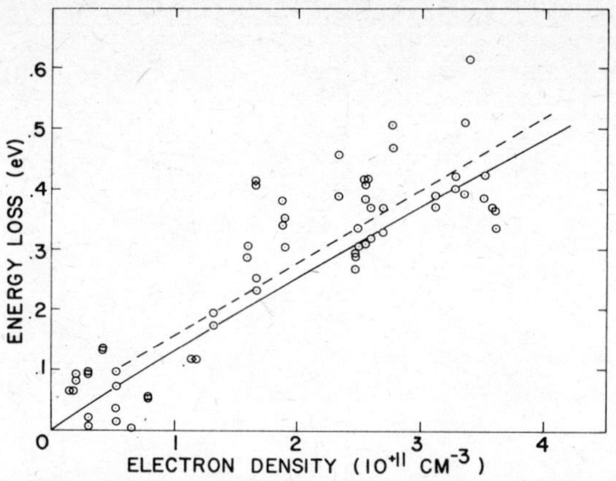


Fig. 2.5 The energy loss suffered by a 50 eV Cesium ion traversing a 7.5 cm long column of Cesium plasma of various densities; $T_e = 2100\ ^0K$. The circles are from experiment, the solid line is from theory and the dashed line is the best straight line passing through the experimental data [from reference 19].

Figure 2.5 shows the results of the experiments and their comparison with theory. Note the small energy loss, amounting to something like 1/10 of one percent of the initial energy lost per centimeter path length traversed by the test ions. Under conditions of this experiment, the energy loss is predominantly to the plasma electrons.

Despite the relatively small loss of energy, the phenomenon cannot be discounted in the design of forthcoming thermonuclear fusion reactors, and the question of plasma heating by energetic particles has recently received much renewed attention [12,20]. In toroidal devices like Tokamaks, ion heating is a very difficult question to resolve. Heating by mere Ohmic dissipation of the circulating current is probably insufficient; the additional heating effect of fusion generated $\alpha$-particles (born at 3.5 MeV in a $D$-$T$ reaction) can help to alleviate these difficulties. In addition, heating by the injection of energetic neutrals with energies $\sim$ 100 keV is being contemplated in several laboratories.

Consider then a typical Tokamak discharge with a density of $5 \times 10^{13}$ cm$^{-3}$, and electron temperature $T_e$ = 1 keV and an ion temperature $T_i$ = 0.5 keV. Energetic neutral particles are injected into the toroidally confined plasma, and protons are generated by charge exchange inside the medium. As the proton's initial energy decays with time, electrons are heated predominantly in the early stages [i.e. $v >> u$ of Eq. (2.29)], but later, near thermalization where $v < u$, ion heating becomes significant. This is a fortunate circumstance in view of the fact that one wishes to deposit as much energy as possible in the plasma ions.

Given an expression for $dW/dt$ [say, Eq. (2.26)], the proton energy $W$ at a time $t$ is determined from

$$t(W) = \int_{W_0}^{W} dW \left(\frac{\partial W}{\partial t}\right)^{-1} \tag{2.32}$$

where $W_0$ is the initial proton energy at the time $t = 0$. The time for the proton to thermalize, that is for it to become one of the average plasma particles, is infinite. Therefore one defines a more useful time $t_{th}$, as that required for the test particle to decay to *twice* the thermal energy of the plasma particles. The total energy exchange is then given by the relation

$$\Delta W = \int_{0}^{t_{th}} \left(\frac{dW}{dt}\right) dt \tag{2.33}$$

which can be broken up into energy lost to the electrons and energy lost to the ions. Numerical computations are shown in Fig. 2.6, which exhibits the relative fractions of injection energy going into the electrons and ions respectively. The break-even point is seen to occur when the initial proton energy $W_0$ is about 44 keV. As $W_0$ increases, the fraction going into ions decreases, as expected; however, the absolute value of energy going into the ion (also shown plotted) keeps rising until it saturates at a value $W_0 \simeq 70$ keV. This then appears to be the optimum injection energy. It means, for example, that an addition of 1 percent of 70 keV protons to the Tokamak plasma would increase the ion energy from 0.5 keV to about 1 keV. Note, however, that this density of injected particles constitutes a very hefty beam, namely, one with a current density of some 30 amp per square centimeter.

Figure 2.7 shows the time required for the injected proton to thermalize, that is, drop in energy from $W_0$ to $2\kappa T_e = 2$ keV. When $W_0 = 70$ keV, the time is about 20 milliseconds which is conveniently short, being but a fraction of the plasma lifetime.

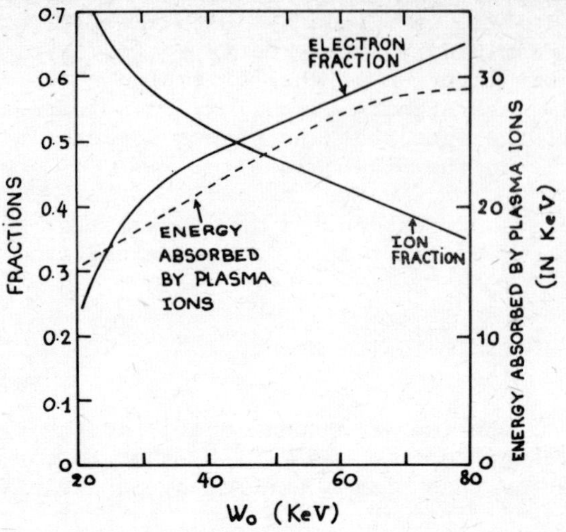


Fig. 2.6 The two solid curves give the relative amounts of injected ion energy deposited into plasma electrons and ions, as a function of the injection energy *W*. The dashed curve shows the absolute value of the energy deposited into the ions. The initial Tokomak-like plasma is characterized by $N = 5 \times 10^{13}$ $cm^{-3}$, $T_e = 1$ *keV*, $Ti = 0.5$ *keV*. [From reference 12].

Several effects have been neglected in the above model. For one, the plasma particles of a Tokamak are in a strong magnetic field; this modifies the assumed straight test particle trajectories and also the plasma dielectric coefficient. However, both of these effects on $dW/dt$ are believed to be fairly small. Secondly, for the large beam currents needed for heating, the single test-charge model may be inapplicable and beam-plasma interactions (and possible accompanying instabilities) may have to be taken into account. One suspects that our single test-charge model is valid as long as the separation $s$ between the beam particles greatly exceeds the Debye length $L_D$ of the plasma particles, namely

$$s > L_D \tag{2.34}$$

If then $N_b$ is the density of beam particles, $s \simeq N_b^{1/3}$ and Eq. (2.34) can be cast into the form

$$\frac{N_b}{N} < \frac{4\pi}{3N_D} \tag{2.35}$$

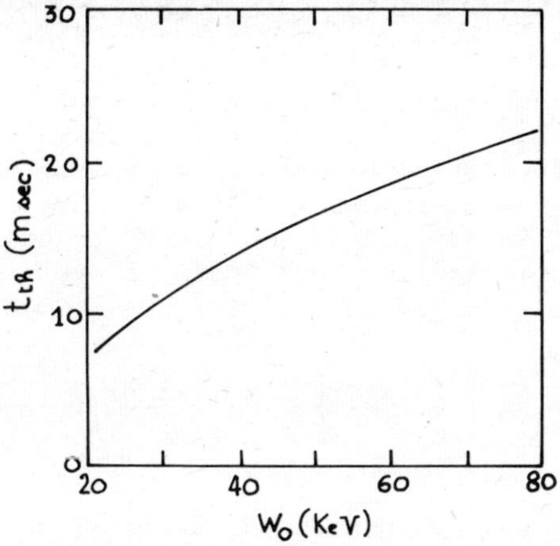


Fig. 2.7 Thermalization time as a function of injection energy $W_0$ for the same conditions as those given in the caption to Fig. 2.6.

where $N$ is the density of plasma particles, and $N_D$ is the number of these particles in a Debye sphere. In Tokamak plasmas, for example, $N_D \gtrsim 10^6$ and therefore $N_b < 10^{-6} N$

*Influence of an rf field on the energy loss.* We recall that a large fraction of the total polarization loss suffered by a test charge comes from binary Coulomb collisions of the test charge with the plasma particles. It seems unlikely that this part of the loss can be changed appreciably except by a modification of the distribution function of the plasma particles. Barring this, one may attempt to influence the collective (radiation) part of the polarization loss. It has been suggested [21,22] that the presence of a large amplitude *rf* field may indeed increase the loss through an enhanced emission of longitudinal plasma oscillations (Langmuir waves). One such mechanism can be the nonlinear coupling of the plasma oscillations (at a frequency $\sim \omega_p$) with the high frequency field (at $\omega_H$), causing the generation of waves at frequencies $n\omega_H \pm \omega_p$ ($n = 1,2,3 \ldots$). Such an *rf* field may well be part of the turbulent spectrum of a plasma, or it may be imposed externally.

Consider then a test charge $q_t$ subjected to a uniform sinusoidal field

$$E_H \sin \omega_H t \tag{2.36}$$

of frequency $\omega_H$. This field has a two-fold effect: (a) it changes [21] the trajectory of the test charge from rectilinear motion to one given by [cf. Eq. (2.1)]

$$r(t) = r_0 + vt - \left(\frac{q_t}{m_t}\right) \frac{E_H}{\omega_H^2} \sin \omega_H t \tag{2.37}$$

and (b) is changes [22] the dielectric coefficient $K_L(\mathbf{k},\omega)$ of the plasma medium. If the test charge is massive (an ion) and the frequency $\omega_H$ high, effect (a) may be disregarded; if one further assumes that the ions of the plasma are immobile so that interactions with electrons only need be considered, the polarization loss averaged over a period of the high frequency *rf* field is [22] [cf. Eq. (2.18)]

$$\left|\frac{dW}{dt}\right|_{\text{pol}} = -\frac{q_t^2}{(2\pi)^3} \int \frac{d^3k\,\mathbf{k}\cdot v}{\varepsilon_0 k^2} \sum_{n=-\infty}^{\infty} J_n^2\left(\frac{\mathbf{k}\cdot v_H}{\omega_H}\right) \operatorname{Im}\left[\frac{1}{K_L'(\mathbf{k},\mathbf{k}\cdot v)}\right] \tag{2.38}$$

Here $J_n$ is the Bessel function of order $n$, whose argument contains the velocity

$$v_H = \frac{q}{m} \frac{E_H}{\omega_H} \tag{2.39}$$

induced in the plasma electrons by the *rf* field. The quantity $K_L'$ is the modified dielectric coefficient given by [cf. Eq. (1.25)]

$$K_L' = 1 + \frac{\omega_{pe}^2}{k^2} \int \frac{\mathbf{k}\cdot\partial f(\mathbf{W})/\partial \mathbf{W}}{kv + n\omega_H - \mathbf{k}\cdot\mathbf{W}}\, d^3W \tag{2.40}$$

Observe that when the perturbing *rf* field is set equal to zero, Eqs. (2.38) and (2.40) correctly revert back to their original values (2.18) and (1.25), respectively.

To determine (approximately) the collective part of the polarization loss, one now proceeds exactly as indicated below Eq. (2.18). In the special case where the test particle speed and the applied electric field are parallel ($v \parallel v_H$) one then finds that

$$\left|\frac{dW}{dt}\right|_{\text{pol}} = -\frac{q_t^2\omega_p^2}{8\pi\varepsilon_0 v} \sum_{n=-\infty}^{n=\infty} \left(1 + n\frac{\omega_H}{\omega_p}\right) \times J_n^2\left(\frac{v_H}{v}\left\{n + \frac{\omega_p}{\omega_H}\right\}\right) \ln\left[1 + \frac{v^2}{v_0^2}\left(1 + n\frac{\omega_H}{\omega_p}\right)^{-2}\right] \tag{2.41}$$

where $v_0 = \sqrt{\kappa T_e/m}$ is the thermal speed of the plasma electrons. This result is to be compared with Eq. (2.24)

$$\left|\frac{dW_0}{dt}\right|_{pol} = -\frac{q_t^2\omega_p^2}{8\pi\varepsilon_0 v} \ln\left[1 + \frac{v^2}{v_0^2}\right] \tag{2.42}$$

for the polarization loss in the absence of the *rf* electric field. The ratio of Eqs. (2.41) to (2.42)

$$\left|\frac{dW\ dt}{dW_0/dt}\right|_{pol} = J_0^2\left(\frac{v_H}{v}\frac{\omega_p}{\omega_H}\right) + \sum_{n\neq 0}\left(1 + n\frac{\omega_H}{\omega_p}\right) J_n^2\left(\frac{v_H}{v}\left\{n + \frac{\omega_p}{\omega_H}\right\}\right)$$

$$\times \frac{\ln\left[1 + \frac{v^2}{v_0^2}\left(1 + n\frac{\omega_H}{\omega_p}\right)^{-2}\right]}{\ln\left[1 + \frac{v^2}{v_0^2}\right]} \tag{2.43}$$

then gives the change in the energy loss caused by the presence of the field. For example, in the case of a relatively weak high frequency field for which $\omega_H >> \omega_p$ and $v_H < v$, and expansion of the Bessel functions in a power series yields to lowest order in $(v_H/v)$

$$\left[\frac{dW/dt}{dW_0/dt}\right]_{pol} \simeq 1 + \frac{3}{2}\left(\frac{v_H}{v}\right)^2 \frac{\ln\left[\frac{v}{v_0}\frac{\omega_p}{\omega_H}\right]}{\ln\left[\frac{v}{v_0}\right]} \tag{2.44}$$

showing that there is a small enhancement proportional to the square of the applied *rf* field.

Consider now the case of strong high frequency fields for which $\omega_H/\omega_p >> v_H/v >> 1$. The resulting asymptotic expansion of the Bessel functions must then be summed; for the resonance condition

$$\frac{v_H}{v} = \pi\left|\frac{1}{2} + s\right|;\ s \gtrsim 1 \tag{2.45}$$

we find that

$$\left|\frac{dW/dt}{dW_0/dt}\right|_{pol} \simeq 1 + \frac{2\left(\frac{v}{v_0}\frac{\omega_p}{\omega_H}\right)}{\ln\left(\frac{v}{v_0}\right)} \tag{2.46}$$

Here, the enhancement can be considerable. This is due to the fact that when the resonance condition (2.45) is obeyed the waves radiated at frequencies $n\omega_H \pm \omega_p$ act in phase.

Let us note that appreciable enhancement requires that the induced particle velocity be of the order of or greater than the electron thermal velocity, that is

$$|v_H| = \frac{e}{m} \frac{|E_H|}{\omega_H} \gtrsim v_0 \tag{2.47}$$

Such a requirement may not be difficult to fulfill using a powerful laser as the source of the high frequency field. For example, a $CO_2$ laser operating at a wavelength of 10.6μ can provide an electric field $E_H \approx 10^8$ V/cm, giving a value of $v_H \simeq 10^9$ cm/sec. This is to be compared with $v_0 \simeq 6 \times 10^7$ cm/sec which is the thermal velocity of a 1 eV electron.

## 3. PLASMA FLUCTUATIONS

Our understanding of the motion of a single test charge in a plasma provides a good background for a discussion of plasma fluctuations in general. Equations (2.6) to (2.9) describe quantitatively the shielding of the test charge by the plasma charges, whose combined effects is expressed by the dielectric coefficient $K_L(\mathbf{k},\omega)$. The test charge "dressed" in its dielectric cloud produces an electric field which effectively replaces its "collective" Coulomb field. Our present task is to calculate the total fluctuating field due to all such particles.

### 3.1 *Time and Space Averages*

The charges, currents and fields of a plasma all undergo more or less random fluctuations about a mean; their values are usually zero when averaged over time and space. Thus a meaningful quantity is not the average charge, current or field, but the square of these quantities. Let $A(\mathbf{r},t)$ be the characteristic parameter and $A(\mathbf{k},\omega)$ its Fourier transform, as defined by Eqs. (1.9) and (1.10). Using an extension of Parseval's theorem applicable to both time and space, the average of $A^2(r,t)$ may be written in the form

$$\langle A^2(\mathbf{r},t)\rangle = \iint A^2(\mathbf{k},\omega)d\omega \frac{d^3k}{(2\pi)^3} \tag{3.1}$$

where $A^2(\mathbf{k},\omega)$, the "spectral power density" associated with $A$, is defined as

$$A^2(\mathbf{k},\omega) = \frac{1}{2\pi} \lim_{\substack{V\to\infty \\ T\to\infty}} \frac{1}{TV} \left| A(\mathbf{k},\omega)\cdot A^*(\mathbf{k},\omega) \right| \tag{3.2}$$

Here $T$ is the time and $V$ the volume intervals involved in the averaging processes. If, for example, $A(\boldsymbol{r},t)$ stands for the electric field, then $\langle A^2(\boldsymbol{r},t)\rangle$ is the space-time average of the electrical energy density, and $A^2(\mathbf{k},\omega)$ is its spectrum in wave-vector and frequency space.

An alternate way of deriving the average of $A^2(\boldsymbol{r},t)$ is to take the ensemble average of the product of the Fourier transform and of its complex conjugate, with the result that,

$$\begin{aligned}\langle A^2(\boldsymbol{r},t)\rangle &= \langle A(\boldsymbol{r},t)\cdot A^*(\boldsymbol{r},t)\rangle \\ &= \frac{1}{(2\pi)^8} \iiiint \langle A(\mathbf{k},\omega)\cdot A^*(\mathbf{k}',\omega')\rangle \\ &\qquad \exp(j(\omega-\omega')t - j(\mathbf{k}-\mathbf{k}')\cdot\boldsymbol{r})\,d\omega\, d\omega'\, d^3k\, d^3k'\end{aligned} \tag{3.3}$$

In a time-stationary and spatially homogeneous system the two methods of averaging are equivalent. Therefore we can equate the right-hand sides of Eqs. (3.1) and (3.3) and obtain a relation between the spectral power density $A^2(\mathbf{k},\omega)$ and the spectral correlation function $A(\mathbf{k},\omega)\cdot A^*(\mathbf{k}',\omega')$:

$$\langle A(\mathbf{k},\omega)\cdot A^*(\mathbf{k}',\omega')\rangle = (2\pi)^5\, A^2(\mathbf{k},\omega)\, \delta(\mathbf{k}-\mathbf{k}')\, \delta(\omega-\omega') \tag{3.4}$$

Thus knowledge of the correlation function for the quantity in question tells us all we need to know about the system; its derivation for different types of systems is the problem to which we now address ourselves.

## 3.2 *Electron Fluctuations in a Plasma with Immobile Ions*

In this section we restrict ourselves to electron fluctuations at such high frequency that the ions can be considered as immobile, and they simply provide a neutralizing background for the electrons. The correlation function of the fluctuating electric field is obtained by taking the dot product $E\cdot E^*$ of Eq. (2.6) and ensemble-averaging over all electrons:

$$\langle E(\mathbf{k},\omega)\cdot E^*(k',\omega')\rangle = \frac{\mathbf{k}\cdot\mathbf{k}'}{\varepsilon_0^2|k|^2|k'|^2K_L(\mathbf{k},\omega)K_L^*(\mathbf{k}',\omega')}\langle\sum_{ij}\rho_i(\mathbf{k},\omega)\rho_j^*(\mathbf{k}',\omega')\rangle \quad (3.5)$$

where

$$\langle\sum_{ij}\rho_i(\mathbf{k},\omega)\rho_j^*(\mathbf{k}',\omega')\rangle = (2\pi)^2\sum_{ij} e_i e_j\ \delta(\omega-\mathbf{k}\cdot v_i)\delta(\omega'-\mathbf{k}'\cdot v_j)\langle\exp(j(\mathbf{k}\cdot r_{0i}-\mathbf{k}'\cdot r_{0j})\rangle \quad (3.6)$$

To proceed with the problem, one now invokes a familiar theorem [23,24] due to Hubbard and Rostoker which states that to "good approximation" the electrons shielded by their Debye clouds are uncorrelated unless they are in contact. With this as basis, only terms with $i = j$ are retained in the above sums, with the result that the double summation reduces to a single summation. Also using the identity

$$\langle\exp(j(\mathbf{k}-\mathbf{k}')\cdot r_{0i})\rangle = (2\pi)^3\ \delta(\mathbf{k}-\mathbf{k}')V^{-1} \quad (3.7)$$

where $V$ is the plasma volume, and

$$\delta(\omega-\mathbf{k}\cdot v_i)\delta(\omega'-\mathbf{k}'\cdot v_i) = \delta(\omega-\mathbf{k}\cdot v_i)\delta(\omega-\omega')\delta(\mathbf{k}-\mathbf{k}') \quad (3.8)$$

it follows that

$$\langle\sum_{ij}\rho_i(\mathbf{k},\omega)\rho_j^*(\mathbf{k}',\omega')\rangle = (2\pi)^5 e^2\delta(\mathbf{k}-\mathbf{k}')\delta(\omega-\omega')\langle\sum_i \delta(\omega-\mathbf{k}\cdot v_i)\rangle V^{-1} \quad (3.9)$$

The ensemble average over velocities $v_i$ can now be expressed in terms of the velocity distribution function $f(v)$ with the result that,

$$\langle\sum_{ij}\rho_i(\mathbf{k},\omega)\rho_j^*(\mathbf{k}',\omega')\rangle = (2\pi)^5 Ne^2\delta(\mathbf{k}-\mathbf{k}')\delta(\omega-\omega')\int f(v)\delta(\omega-\mathbf{k}\cdot v)d^3v \quad (3.10)$$

where $N$ is the electron density and $f(v)$ the distribution normalized so that

$$\int f(v)d^3v = 1 \quad (3.11)$$

To complete the calculation, we insert Eq. (3.10) in Eq.

(3.5) and find,

$$\langle E(\mathbf{k},\omega)\cdot E^*(\mathbf{k}',\omega')\rangle$$

$$= \frac{(2\pi)^5 Ne^2}{|k|^2\varepsilon_0^2|K_L(\mathbf{k},\omega)|^2}\,\delta(\mathbf{k}-\mathbf{k}')\delta(\omega-\omega')\int f(v)\delta(\omega-\mathbf{k}\cdot v)d^3v \tag{3.12}$$

A more compact form of the same result is obtained in terms of the spectral power density [see Eqs. (3.2) and (3.3)], namely

$$E^2(\mathbf{k},\omega) = \frac{Ne^2}{|k|^2\varepsilon_0^2|K_L(\mathbf{k},\omega)|^2}\int f(v)\delta(\omega-\mathbf{k}\cdot v)d^3v \tag{3.13}$$

We note the synchronism condition which states that only those particles whose velocity equals the "phase velocity" of the fluctuation contribute to the spectrum. We also see that so far the result is quite general, and holds for any distribution of electron velocities, as long as this distribution does not cause any instabilities. In the presence of instabilities the spectrum of $E^2(\mathbf{k},\omega)$ is no longer space or time-stationary since a nonlinear saturation mechanism is *not* built into the above derivation. The dielectric coefficient $K_L$ is given by Eq. (1.26) for the distribution $f(v)$ in question.

From a knowledge of the spectral density of electric field fluctuations, one can readily generate the corresponding quantities for the density and current fluctuations. Using Eqs. (1.16) and (1.18), it follows that,

$$\begin{aligned}\rho^2(\mathbf{k},\omega) &= |k|^2\varepsilon_0^2E^2(\mathbf{k},\omega)\\ J^2(\mathbf{k},\omega) &= \omega^2\varepsilon_0^2E^2(\mathbf{k},\omega)\end{aligned} \tag{3.14}$$

where $E^2(\mathbf{k},\omega)$ is given by Eq. (3.13). However, a word of warning: the spectra of $E^2(\mathbf{k},\omega)$ and $J^2(\mathbf{k},\omega)$ represent only those portions of the full expressions which are associated with longitudinal wave excitations; parts of $E^2$ and $J^2$ which come from transverse electromagnetic oscillations, and which are associated with the transverse dielectric coefficient are not included here.

It is instructive to put Eq. (3.13) into a somewhat different form by replacing $1/K_L^2$ with

$$\frac{1}{K_L^2} = -\frac{\mathrm{Im}(1/K_L)}{\mathrm{Im}(K_L)} \tag{3.15}$$

and substituting for Im $K_L$ from Eq. (1.26):

$$E^2(\mathbf{k},\omega) = -\frac{m}{\pi\varepsilon_0}\,\mathrm{Im}\left(\frac{1}{K_L}\right)\frac{\int\delta(\omega-\mathbf{k}\cdot\boldsymbol{v})\,f(\boldsymbol{v})\,d^3v}{\int\delta(\omega-\mathbf{k}\cdot\boldsymbol{v})\,\mathbf{k}\cdot\frac{\partial f(\boldsymbol{v})}{\partial\boldsymbol{v}}\,d^3v} \tag{3.16}$$

In thermal equilibrium $f(\boldsymbol{v})$ is Maxwellian, and Eq. (3.16) becomes,

$$E^2(\mathbf{k},\omega) = \frac{\kappa T}{\pi\omega\varepsilon_0}\,\mathrm{Im}\left[\frac{1}{K_L(\mathbf{k},\omega)}\right] \tag{3.17}$$

with $T$ as the electron temperature. This equation is seen to be a form of the fluctuation-dissipation theorem [25] (or Nyquist's theorem) which relates the power spectrum of the fluctuating field to the mediums' temperature and "resistance". We note that this result refers to the classical regime $\hbar\omega \ll \kappa T$; in the quantum-mechanical regime, Eq. (3.17) still applies with $\kappa T/\omega$ replaced by $\hbar/[\exp(\hbar\omega/\kappa T)-1]$.

*Thermal fluctuations.* As an example, let us examine the spectrum of charge density fluctuations in the electron plasma. From Eqs. (3.14) and (3.17) we obtain

$$\rho^2(\mathbf{k},\omega) = e^2N^2(\mathbf{k},\omega) = \frac{k^2\varepsilon_0}{\pi}\,\frac{\kappa T}{\omega}\,\mathrm{Im}\left[\frac{1}{K_L(\mathbf{k},\omega)}\right] \tag{3.18}$$

This is shown plotted in Fig. 3.1 for various values of the parameter $kL_D$. In the figure $N^2(\mathbf{k},\omega)$ is normalized to the mean electron density $N$. We see that when $kL_D$ is large compared with unity, low frequency fluctuations dominate. Indeed, under these conditions the spectrum approaches a Gaussian

$$\rho^2(\mathbf{k},\omega) = e^2N^2(\mathbf{k},\omega) = \frac{Ne^2}{\sqrt{2\pi}\,kv_0}\exp(-\omega^2/2k^2v_0^2) \tag{3.19}$$

which is typical of Doppler-broadening of individual, uncorrelated electrons. In other words, collective interactions do not occur here. On the other hand, for small value of $kL_D$ the fluctuations are concentrated at high frequencies centered around the electron plasma frequency, or more

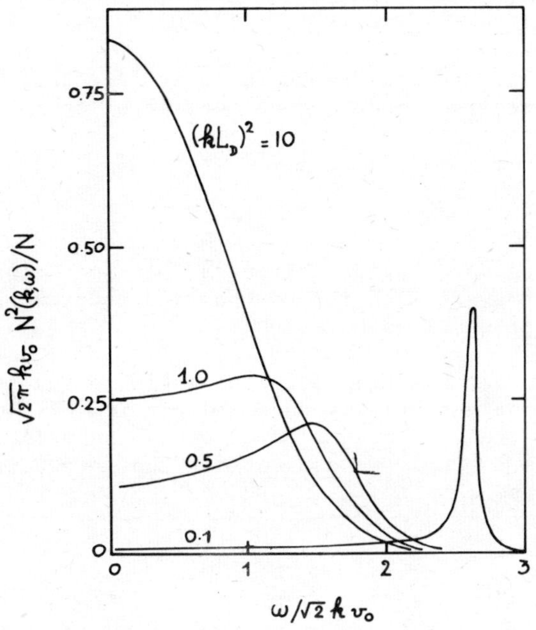


Fig. 3.1 Spectra of electron density fluctuations in a thermal plasma with immobile ions. $L_D = v_0/\omega_p$ is the electron Debye length. The peaks in the spectra occur approximately at frequencies $\omega = \omega_0$ and wavenumbers $k = k_0$ given by the dispersion relation $\omega_0^2 = \omega_p^2\,[1 + 3(k_0 L_D)^2]$. The quantity $\sqrt{2\pi} k v_0 N^2(k,\omega)/N$ plotted along the ordinate is dimensionless.

accurately, around the frequency given by dispersion relation (1.51). This is not surprising, because when $kL_D$ is small, Landau damping is weak and large amplitude oscillations readily build up at frequencies corresponding to a natural modes of oscillation of the medium.

The energy density of the fluctuations is given by [see Eqs. (3.1) and (3.17)]

$$\frac{1}{2}\varepsilon_0 \langle E^2(\boldsymbol{r},t)\rangle = \frac{\kappa T}{2\pi} \iint \frac{d\omega}{\omega} \frac{d^3k}{(2\pi)^3} \, \mathrm{Im}\left[\frac{1}{K_L(k,\omega)}\right] \tag{3.20}$$

When the integration over frequency $\omega$ is carried out, one finds that

$$\frac{1}{2}\varepsilon_0 \; \langle E^2(\mathbf{r}) \rangle = \frac{\kappa T}{2} \int \frac{d^3k}{(2\pi)^3} \; \frac{1}{1 + (kL_D)^2} \tag{3.21}$$

Since $d^3k/(2\pi)^3$ represents the density of modes, it follows that the energy per mode can be considered as

$$\frac{\kappa T}{2} \; \frac{1}{1 + (kL_D)^2} \tag{3.22}$$

Thus, at long wavelengths the energy per mode approaches $\kappa T/2$, in agreement with the equipartion theorem, but is much less at short wavelengths.

One can now use Eq. (3.21) to estimate the total energy density in the essentially undamped portion of the fluctuating spectrum. Since undamped waves can exist only for wave numbers $k \lesssim L_D^{-1}$, we can cut off the integration at $k_{\text{max}} = L_D^{-1}$ and obtain

$$\frac{1}{2}\varepsilon_0 \; \langle E^2 \rangle = \frac{\kappa T}{4\pi^2 L_D^3} \int_0^1 \frac{x^2 dx}{1 + x^2} = \frac{0.215}{4\pi^2 L_D^3} \; \kappa T \tag{3.23}$$

It is instructive to compare the wave energy density with the kinetic energy density contained in the electrons themselves. The result of taking the ratio of these two quantities is

$$z \equiv \frac{(1/2)\varepsilon_0 \langle E^2 \rangle}{(3/2) N \kappa T} \simeq \frac{0.015}{N_D} \tag{3.24}$$

The parameter

$$N_D = (4/3)\pi N L_D^3 \tag{3.25}$$

$$= 1.72 \times 10^{12} \frac{[T(eV)]^{3/2}}{[N(m^{-3})]^{1/2}}$$

is the number of particles in the Debye sphere; since $N_D$ is a fairly large quantity for most gaseous plasmas, it follows that the wave energy of thermal fluctuations is normally small compared with the kinetic energy of the particles.

To compute the magnitude of density fluctuations, we insert Eq. (3.17) in Eq. (3.14) and integrating over frequencies, we obtain that

$$\langle \rho^2(r,t) \rangle = \kappa T \varepsilon_0 \int \frac{d^3k}{(2\pi)^3} \frac{k^2}{1 + (kL_D)^2}$$

Restricting the integration once again to the "collective" regime, $0 \leq k \lesssim L_D^{-1}$, it follows that

$$\frac{\langle N^2(r\ t) \rangle}{N^2} = \frac{2}{3\pi N_D} \int_0^1 \frac{x^4 dx}{1+x^2} = \frac{2}{3\pi N_D} \times 0.119$$

Thus the fractional root-mean-square density fluctuation is

$$\sqrt{\frac{\langle N^2(r,t) \rangle}{N^2}} \approx \frac{0.16}{\sqrt{N_D}}$$

The accompanying Table lists some of the pertinent characteristics of three rather different plasmas: the first is a typical discharge encountered in gaseous electronics, the second is a plasma of thermonuclear fusion interest, and the third is a plasma in a highly condensed state as may sometimes be obtained in exploding-wire experiments or in the production of plasmas by means of high-power lasers. In this last case we note the high value of the fluctuating electric field and the small value of $N_D$; indeed, at such small values of $N_D$ the entire statistical model presented on these pages will require revision. However, little is known either in theory or experiment concerning this state of matter.

TABLE 3.1 Characteristic Properties of Three Plasmas; $z$ is the Ratio of the Electrical Energy in Collective Fluctuations to the Particle Kinetic Energy [Eq.(3.24)] and $E$ is the Corresponding Field Strength.

| Medium | $N$ $cm^{-3}$ | $T$ eV | $L_D$ cm | $N_D$ | $z$ | $\sqrt{\langle E^2 \rangle}$ V/cm |
|---|---|---|---|---|---|---|
| Gas discharge | $10^{12}$ | 1 | $7\times10^{-4}$ | $1.7\times10^3$ | $9\times10^{-6}$ | 7 |
| Fusion plasma | $10^{14}$ | $10^3$ | $2\times10^{-3}$ | $5.3\times10^6$ | $3\times10^{-9}$ | 39 |
| Laser-produced plasma | $10^{18}$ | 1 | $7\times10^{-7}$ | 1.7 | $9\times10^{-3}$ | $2\times10^5$ |

The frequency spectrum of the field fluctuations is obtained by carrying out the integration in Eq. (3.20) over all wavenumbers $k$. This task requires numerical evaluation of the integral. The integral diverges logarithmically at large $k$ and for that reason it is cut off at the customary $k = k_{max}$ discussed in Chapter 2. Figure 3.2 illustrates the type of spectra derived. The bumps near the plasma frequency

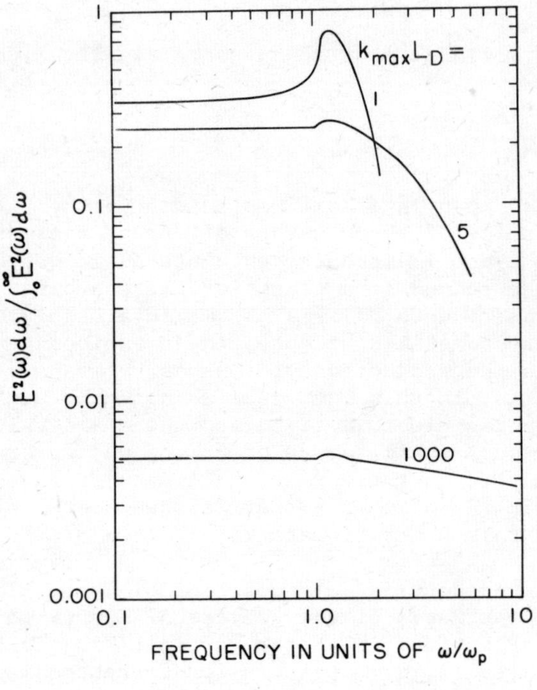


Fig. 3.2 Spectra of electric field fluctuations in a thermal plasma with immobile ions. The quantity $E^2(\omega)d\omega$ is obtained by integrating the "spectral power density" $E^2(k,\omega)$ over all wavenumbers $k$. The quantity $k_{max} = [e^2/4\pi\varepsilon\kappa T]^{-1}$

represent collective plasma oscillations superimposed on a continuum generated by small angle electron-ion collisions. Each graph refers to a different value of the parameter $k_{max}L_D$ where $k_{max}$ is approximately the reciprocal of the classical impact parameter for a 90° collision, $1/k_{max} \approx (e^2/4\pi\varepsilon_0\kappa T)$. Note that the product $k_{max}L_D$ is roughly equal to the number of particles in the Debye sphere $N_D$, and we thus see from the figure that as $N_D$ decreases the "collective" bump near $\omega_p$ becomes more pronounced in accord with the

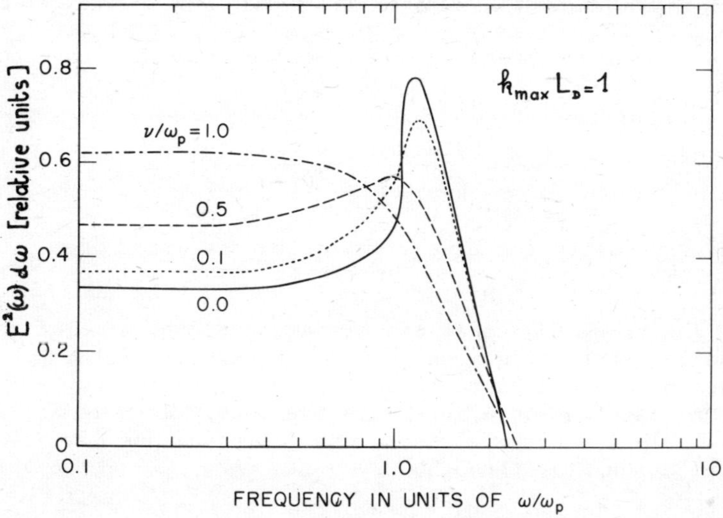


Fig. 3.3 Spectra of electric field fluctuations, showing the effects of short range (electron-atom) collisions.

predictions of Eq. (3.24). We must note however that in the case $k_{max}L_D = 1$ (which corresponds to $N_D \sim 1$) the theory used is really no longer strictly valid. In some experiments the plasma is relatively weakly ionized and short-range electron-atom collisions play an important role. The calculations of Fig. 3.2 do not allow for such collisions, since $K_L(k,\omega)$ used in expression (3.20) is that obtained from the collisionless Vlasov equation. However, using the dielectric coefficient as given by Eqs. (1.45) to (1.47), inserting it in (3.20) and performing the integration over $k$, leads to the results shown in Fig. 3.3. It is observed that when the collision frequency $\nu$ approaches the plasma frequency, collective oscillations are virtually destroyed.

*Nonthermal fluctuations.* A non-Maxwellian distribution of electron velocities can have quite a dramatic effect on the fluctuation spectrum. Equation (3.13), or its equivalent, Eq. (3.16) are the starting point of our discussion. Indeed, the equations can be thought of as more general forms of the fluctuation-dissipation theorem applicable to such nonequilibrium systems. By anology with Eq. (3.17) we may write Eq. (3.16) as

$$E^2(\mathbf{k},\omega) = \frac{\kappa T_{\text{eff}}}{\pi\omega\varepsilon_0} \quad \text{Im}\left[\frac{1}{K_L(\text{k},\omega)}\right] \tag{3.26}$$

where $T_{\text{eff}}$ now represents an effective "fluctuation" temperature

$$\frac{\kappa T_{\text{eff}}}{m} = -\frac{\int\delta(\omega - \mathbf{k}\cdot\boldsymbol{v})f(\boldsymbol{v})d^3v}{\int\delta(\omega - \mathbf{k}\cdot\boldsymbol{v})\frac{\mathbf{k}}{\omega} \cdot \frac{\partial f(\boldsymbol{v})}{\partial\boldsymbol{v}}\, d^3v} \tag{3.27}$$

which reduces to the true kinetic temperature $T$ when $f(\boldsymbol{v})$ is Maxwellian.

The foregoing equations are ameneable to simplification by noting that two components of the particle velocity transverse to **k** do not participate in any manipulations over **k**, and one can therefore integrate them out. If $u$ is the particle velocity parallel to **k**, one defines a one dimensional distribution $F(u)$ through

$$F(u) = \int f(u,v_\perp)\, d^2v_\perp \tag{3.28}$$

with $v_\perp$ as the component of $\boldsymbol{v}$ orthogonal to **k**. In terms of $F(u)$ then, Eq. 3.27 becomes

$$\frac{\kappa T_{\text{eff}}}{m} = -\left(\frac{\omega}{\text{k}}\right) \frac{[F(u)]_{u\,=\,\omega/\text{k}}}{\left[\frac{\partial F(u)}{\partial u}\right]_{u\,=\,\omega/\text{k}}} \tag{3.29}$$

It is also readily shown from Eq. (1.26) that

$$\text{Im}\, K_L(\text{k},\omega) = \frac{\pi\omega_p^2}{\text{k}^2}\left[\frac{\partial F(u)}{\partial u}\right]_{u\,=\,\omega/\text{k}} \tag{3.30}$$

Since $T_{\text{eff}}$ is inversely proportional to the derivative of $F(u)$ with respect to $u$ [as is shown by Eq. (3.29)], the level and spectrum of the fluctuations are sensitively dependent on the shape of the distribution function in the neighborhood of $u = \omega/\text{k}$. Of course, the same thing can be seen happening when one examines the equivalent relation (3.13). In the neighborhood of **k** and $\omega$ where $Re\ K_L(\text{k},\omega) \to 0$, the fluctuation intensity becomes inversely proportional to Im $K_L$. Thus, in the vicinity of weakly damped normal modes of the system, the fluctuations can exhibit large enhancements. As an example, let us compute what happens in the vicinity of the plasma frequency where, as we know, longitudinal plasma waves are likely to be emitted in abundant quantity. To take a concrete case, let the distribution function be isotropic and of the form

$$f(v) = \text{constant} \times \exp(-bv^p) \qquad (3.31)$$

where $p$ is some positive number. Application of Eq. (3.29) to this situation gives a fluctuation temperature in terms of $\Gamma$ functions with argument $(\omega/kv_0)$. To obtain the frequency spectrum of $T_{eff}$, the wave number $k$ is eliminated from the expression for $T_{eff}$ by use of the approximate dispersion relation (3.51), namely

$$\omega^2 \simeq \omega_p^2 + 3k^2v_0^2 \qquad (3.32)$$

If now a kinetic temperature is defined by means of $\kappa T_K = (1/3)m \langle v^2 \rangle$ where $\langle v^2 \rangle$ is the mean-square electron velocity, the resulting variation of $T_{eff}/T_K$ with frequency is as shown in Fig. 3.4. Note that a resonance behavior is observed in $T_{eff}$ near the plasma frequency, with enhancement for $p < 2$

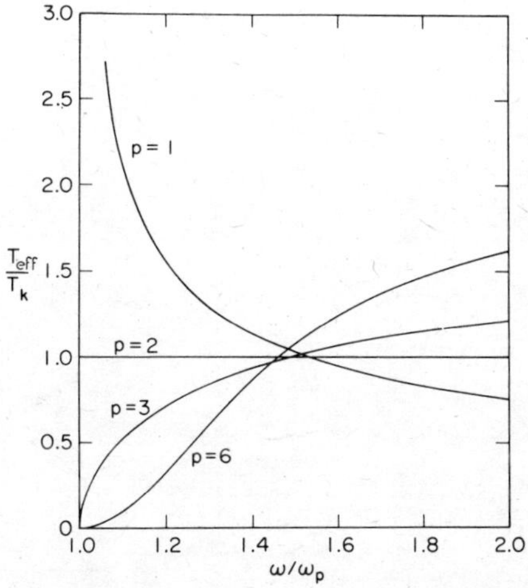


Fir. 3.4 The effective fluctuation temperature $T_{eff}$ as a function of frequency, near the electron plasma frequency $\omega_p$. $T_k$ is the kinetic temperature. The parameter $p$ specifies the form of the electron velocity distribution function $f(v) \propto \exp[-bv^p]$. [From reference 26].

and supression for $p > 2$. But $T_{eff}$ is undefined for $\omega < \omega_p$ since no propagating modes exist in this region; also, the assumption of weak damping is violated when $\omega$ is much in excess of $\omega_p$, and our results are inapplicable for $\omega \gtrsim 2\omega_p$.

It is not too surprising that clever tayloring of the velocity distribution can yield (in theory at least) a tremendous enhancement of the fluctuations. The prescription is to have in the region $v = \omega/k >> v_0$ a very flat velocity distribution with many particles. An example [27,28] of this is a two-temperature plasma which is composed of a cold dense background of electrons with temperature $T_C$ and density $N_C$; superposed on this is a group of hot electrons of temperature $T_H$ and density $N_H$. Both components have Maxwellian distributions, with $T_H >> T_C$ and $N_H << N_C$. The frequency spectrum of the energy density computed by numerical integration of Eq. (3.26) over all $k$ (up to $k = k_{max}$) is illustrated in Fig. 3.5, for the situation $T_H/T_C = 10^2$, $N_H/N_C = 10^{-8}$, and $k_{max}L_D = 10^7/\sqrt{2}$. The large enhancement of the plasma spike is best appreciated by comparing this result with the computations of thermal fluctuations shown in Fig. 3.2. As the peak increases in height, it also narrows in frequency; nevertheless, under non-thermal conditions the total area under the peak can be many times larger than it is in thermal equilibrium. Note that only a minute number of hot electrons is required to obtain this enhancement. However, a word of warning is in order: the calculations are for an entirely collisionless plasma, and one suspects that collisional

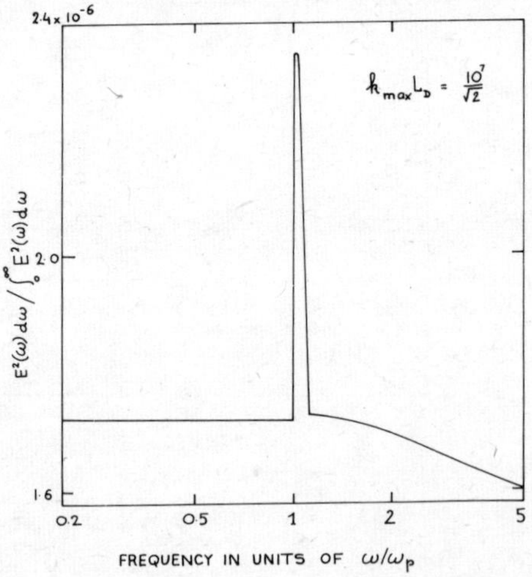


Fig. 3.5 Spectrum of electric field fluctuations of a nonthermal two-temperature plasme in the vicinity of the electron plasma frequency $\omega_p$. [After reference 28].

effects will set in long before such levels of enhancement are ever achieved in practice, even if the assumed velocity distribution can be maintained.

### 3.3 *Fluctuations with Ion Motions Included*

In the absence of ion motion, the fluctuation in the electron density $N_e^2(\mathbf{k},\omega)$ follows directly from Eqs. 3.13 and 3.14:

$$N_e^2(\mathbf{k},\omega) = \frac{N}{|K_L(\mathbf{k},\omega)|^2}\int f_e(v)\delta(\omega-\mathbf{k}\cdot v)d^3v \tag{3.33}$$

where $f_e(v)$ denotes the distribution of electron velocities. When ion dynamics is permitted in the computation, the electron fluctuation is found to depend both on $f_e(v)$ and the ion velocity distribution $f_i(v)$. This is not unexpected because the electrons and ions are correlated through their Coulomb interactions. The results of the calculations [29], done in a manner similar to those given in section 3.2, are the following:

$$N_e^2(\mathbf{k},\omega) = N\left|\frac{1+P_i}{1+P_e+P_i}\right|^2\int f_e(v)\delta(\omega-\mathbf{k}\cdot v)d^3v$$

$$+ N\left|\frac{P_e}{1+P_e+P_i}\right|^2\int f_i(v)\delta(\omega-\mathbf{k}\cdot v)d^3v \tag{3.34}$$

Here the dielectric coefficient for the two charged species is defined as [cf. Eq. (1.27)]

$$K_L(\mathbf{k},\omega) = 1 + P_e + P_i \tag{3.35}$$

with $P_e$ and $P_i$ as the electron and ion "susceptibilities" given by

$$P_e = \frac{\omega_{pe}^2}{k^2}\int d^3v\,\frac{\mathbf{k}\cdot\partial f_e(v)/\partial v}{\omega-\mathbf{k}\cdot v}$$

$$P_i = \frac{\omega_{pi}^2}{k^2}\int d^3v\,\frac{\mathbf{k}\cdot\partial f_i(v)/\partial v}{\omega-\mathbf{k}\cdot v} \tag{3.36}$$

The corresponding expression for the ion density fluctuations is identical to Eq. 3.34 with the electron and ion terms interchanged:

$$N_i^2(\mathbf{k},\omega) = N\left|\frac{1+P_e}{1+P_e+P_i}\right|^2 \int f_i(v)\delta(\omega-\mathbf{k}\cdot v)d^3v \tag{3.37}$$

$$N\left|\frac{P_i}{1+P_e+P_i}\right|^2 \int f_e(v)\delta(\omega-\mathbf{k}\cdot v)d^3v$$

The spectrum of electric field fluctuations is

$$E^2(\mathbf{k},\omega) = \sum_{e,i} \frac{Ne^2}{|k|^2\varepsilon_0^2|K_L(\mathrm{k},\omega)|^2} \int f_{e,i}(v)\delta(\omega-\mathbf{k}\cdot v)d^3v \tag{3.38}$$

with $K_L(\mathbf{k},\omega)$ given by Eqs. (3.35) and (3.36). This result can be derived by simple inspection of Eq. (3.13).

In the special case of a thermal plasma with equal electron and ion temperatures $T$,

$$N_e^2(\mathbf{k},\omega) = -\frac{|k|^2\varepsilon_0\kappa T}{\pi e^2\omega}\,\mathrm{Im}\left[\frac{P_e(1+P_i)}{1+P_e+P_i}\right] \tag{3.39}$$

a result that follows readily from Eq. (3.34) when Maxwellian distributions are substituted for $f_e$ and $f_i$ and when use is made of Eq. (3.30). Neglect of ion motions (that is $P_i \to 0$) yields the spectrum of density fluctuations in an "electron" plasma. In this case the term Im [ ] appearing in the above equation reduces to the familiar factor $-\mathrm{Im}[1/K_L(k,\omega)]$.

With equal electron and ion temperatures, the electron and ion fluctuation spectra integrated over frequency are equal and have values given by [5]

$$\langle N_e^2(\boldsymbol{r},t)\rangle = \langle N_i^2(\boldsymbol{r},t\rangle = N\int \frac{d^3k}{(2\pi)^3}\,\frac{1+(kL_D)^2}{2+(kL_D)^2}$$

The "cross-correlation" between electron and ion fluctuations is

$$\langle N_eN_i\rangle = N\int \frac{d^3k}{(2\pi)^3}\,\frac{1}{2+(kL_D)^2}$$

From these results we see that the density fluctuation per wave mode has two limiting cases: for large values of $kL_D$

$$\langle N_e^2\rangle = \langle N_i^2\rangle = N;\ \langle N_eN_i\rangle \to 0$$

showing that there is no correlation between the electron and ion motions, in accord with expectations. On the other hand, when $kL_D \ll 1$ then the fluctuation per wave mode is given by

$$<N_e^2> = <N_i^2> = \frac{1}{2}N; \quad <N_e N_i> = \frac{1}{2}N$$

exhibiting strong electron-ion correlations. These characteristics are to be compared with those found in section 3.2 where the ions were assumed to be infinitely heavy:

$$<N_e^2(\mathbf{r},t)> = N \int \frac{d^3k}{(2\pi)^3} \frac{(kL_D)^2}{1+(kL_D)^2}; \quad <N_i^2(\mathbf{r},t)> = 0$$

Of course, ion fluctuations also contribute to the total electric field. From Eq. (3.14) we have that

$$E^2(\mathbf{k},\omega) = \frac{e^2}{k^2\varepsilon_0^2}[N_e^2(\mathbf{k},\omega) + N_i^2(\mathbf{k},\omega) - 2N_e(\mathbf{k},\omega)N_i(\mathbf{k},\omega)]$$

From the above results we then find in place of Eq. (3.24) that now

$$\frac{\frac{1}{2}\varepsilon_0<E^2>}{\frac{3}{2}N\kappa T} \simeq \frac{0.018}{N_D}$$

showing that the ions contribute but little to the collective fluctuations of the electric field in a thermal plasma with $T_e = T_i$.

The spectrum obtained from a numerical evaluation of Eq. (3.39) is illustrated in Fig. 3.6, for different values of the parameter $kL_D$ ($L_D$ is the electron Debye length). The frequency variable is specified by the quantity $x = \omega/\sqrt{2}kv_0$ where $v_0 = \sqrt{\kappa T_e/m}$ is the electron thermal speed. The plateau in the range $0.1 \lesssim x \lesssim 1$, together with the spike at $x \gtrsim \sqrt{1.5}$, represents the electron feature of the spectrum. It is virtually identical to the plots of Fig. 3.1, and is hardly affected by the ion motions. On the other hand, the low frequency portion of Fig. 3.6 is generally (but not always) dominated by the ions. At small values of $kL_D$, the low frequency part is essentially a Gaussian profile representative of a Doppler broadened line having a width characteristic of *ion* thermal speeds. However, as $kL_D$ is increased, this ion line shape broadens out until it takes on

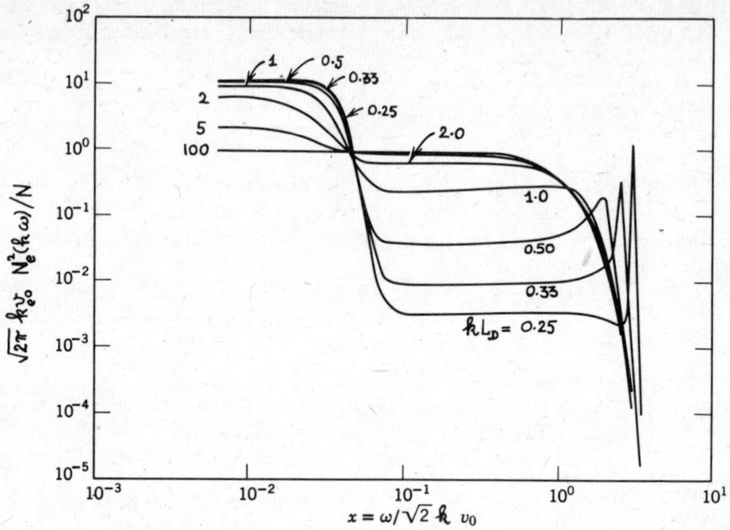


Fig. 3.6 Spectra of electron density fluctuations allowing also for ion motions. The calculations refer to a thermal plasma of $H$ ions with equal electron and ion temperatures. [After reference 30].

Gaussian shape of width typical of *electron* thermal speeds. Matters are different when the electron and ion temperatures are unequal. When $T_e \lesssim T_i$, there is basically no modification of the spectrum from that shown in Fig. 3.6. But, when $T_e >> T_i$, weakly damped ion-acoustic waves can exist (see section 1.4), and as a result they are readily excited spontaneously by the random motions of the charges. Thus a new resonance appears in the spectrum, in addition to the electron plasma resonance seen in Fig. 3.6. This resonance occurs approximately at a frequency and wave-number given by Eq. (1.54) and is shown plotted in Fig. 3.7. Again, just like the electron plasma resonance, the ion acoustic resonance can be greatly enhanced by proper choice of the distribution functions $f_e(v)$ and $f_i(v)$. One such example is when the plasma is on the verge of being unstable as a result of electrons drifting past the ions [31,32].

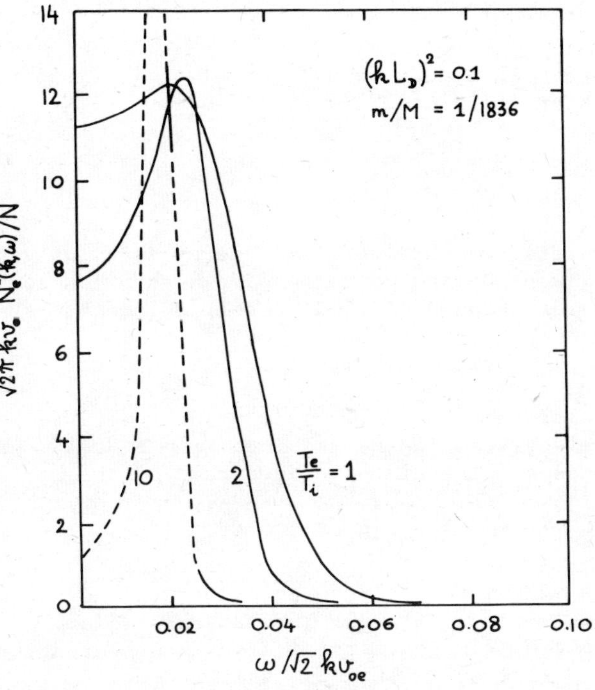


Fig. 3.7 Spectra of electron density fluctuations in a thermal plasma of *H* ions with unequal electron and ion temperatures. Only the low frequency parts of the spectra are shown. The high frequency "electron feature" is the same as that shown in Fig. 3.1 [After reference 5].

## 3.4 *Kirchhoff's Law*

Kirchhoff's law provides a relationship between the direct (emitting) and inverse (absorbing) processes going on in the medium. The direct and inverse processes are linearly related through the local value of the average energy density of the electromagnetic field. Customarily, Kirchoff's law is defined for transverse waves, but clearly the same law must also apply for the longitudinal fields of the ionized gas. The usefulness of the law lies in the fact that if any two of the quantities, — emission, absorption, energy density — are known, the third can be determined.

Consider then longitudinal fluctuations alone and let

$$U(\mathbf{k},\omega)\ \frac{d^3k}{(2\pi)^3}\ d\omega = \frac{1}{2}\ \varepsilon_0\ E^2(\mathbf{k},\omega)\ \frac{d^3k}{(2\pi)^3}\ d\omega \tag{3.40}$$

be the spectral energy density of electric field fluctuations in the interval $[d^3k/(2\pi)^3]d\omega$. Also let

$$j(\mathbf{k},\omega)\ \frac{d^3k}{(2\pi)^3}\ d\omega \tag{3.41}$$

be the rate at which electrical energy is being generated, and let $\alpha(\mathbf{k},\omega)$ be the corresponding absorption coefficient (in units of seconds $^{-1}$). Kirchhoff's law then states that

$$j(\mathbf{k},\omega) = U(\mathbf{k},\omega)\ \alpha(\mathbf{k},\omega) \tag{3.42}$$

An expression for $\alpha(\mathbf{k},\omega)$ is readily derived by computing the electrical energy absorbed from a "test" wave launched in the plasma. The time-averaged rate of absorption is

$$\langle P_{abs}\rangle = \langle J\cdot E\rangle = \frac{1}{4}\left[E\cdot J^* + E\cdot J^*\right] \tag{3.43}$$

where the second form of the equation will now be used when a sinusoidal signal travels through the medium. Since $J$, $E$, and the dielectric coefficient $K_L$ are connected through Eqs. (1.11) and (1.13), it follows that

$$\begin{aligned}\langle P_{abs}\rangle &= \frac{1}{4}\ j\omega\varepsilon_0\ E^2(\mathbf{k},\omega)\ [K_L - K_L{}^*] \\ &= -\ \frac{1}{2}\ \omega\varepsilon_0\ E^2(\mathbf{k},\omega)\ \mathrm{Im}\ K_L(\mathbf{k},\omega)\end{aligned} \tag{3.44}$$

Since the electrical energy density equals $(1/4)\varepsilon_0\ E^2$, we have that

$$\alpha(\mathbf{k},\omega) = -\ 2\omega\ \mathrm{Im}\ K_L(\mathbf{k},\omega) \tag{3.45}$$

[Note that $(1/4)\varepsilon_0\ E^2$ is the energy density in the longitudinal electric field only. Energy residing in the oscillatory kinetic energy of the particles is not being considered here. Had we included the latter, the total energy density would then read [3] $(1/4)\varepsilon_0\ E^2\ \partial\omega K_L/\partial\omega$]. Substituting Eq. (3.45) in Eq. (3.42), we have that

$$j(\mathbf{k},\omega) = -\ 2\omega\ U(\mathbf{k},\omega)\ \mathrm{Im}\ K_L(\mathbf{k},\omega) \tag{3.46}$$

from which $j$ can be determined if $U$ is known, and vice versa.

Let us compute $U$ from a knowledge of $j$. The energy loss

experienced by a single test charge [Eq. (2.18)] can be written in the form

$$\frac{dW}{dt} = -\frac{q_t^2}{(2\pi)^3}\iint \frac{d^3k\ \omega d\omega}{\varepsilon_0\ k^2}\,\delta(\omega - \mathbf{k}\cdot v)\ \mathrm{Im}\left[\frac{1}{K_L(\mathbf{k},\omega)}\right]. \tag{3.47}$$

Assuming that $q_t$ refers to electrons, and summing over all plasma electrons having the velocity distribution $f(v)$ yields the coefficient $j(\mathbf{k},\omega)$:

$$j(\mathbf{k},\omega) = \frac{Ne^2\omega}{\varepsilon_0\ k^2}\ \mathrm{Im}\ \frac{1}{K_L(\mathbf{k},\omega)}\int f(v)\ \delta(\omega - \mathbf{k}\cdot v)\ d^3v \tag{3.48}$$

In making this summation, the "test" charges are assumed to be uncorrelated which conforms to the "dressed" test particle approach described earlier in this chapter. When one now inserts Eq. (3.48) in Eq. (3.46) one finds that

$$U(\mathbf{k},\omega) \equiv \frac{1}{2}\,\varepsilon_0\ E^2(\mathbf{k},\omega)$$

$$= \frac{Ne^2}{2\varepsilon_0\ k^2}\ \frac{1}{|K_L(\mathbf{k},\omega|^2}\int f(v)\ \delta(\omega - \mathbf{k}\cdot v)\ d^3v \tag{3.49}$$

which is identical with Eq. (3.13), derived earlier in a very different way.

At frequencies ($\omega \sim \omega_{pe}$) where collective interactions are important, $j$ and $\alpha$ have simple physical meanings: the emission is that of longitudinal Langmuir oscillations, which is being balanced by Landau damping. This occurs at frequencies $\omega_0$ and wave vectors $\mathbf{k}_0$ for which Re $K_L(\mathbf{k}_0,\omega_0) \to 0$.

## 4. ELECTROMAGNETIC INTERACTIONS

One of the principal ways of observing the properties of a plasma is to study the magnitude and the spectrum of its longitudinal fluctuations, of the kind studied in the previous chapter. Since only transverse electromagnetic waves can reach an externally situated observer, a transformation mechanism from one wave type to the other must be invoked. The nonlinear terms in the equations governing the response of the plasma to a given perturbation provide the necessary coupling between waves*; this leads to frequency conversion

---

*Sharp plasma boundaries or plasma inhomogeneities may lead to a *linear* transformation of wave types. However, this subject lies outside the scope of the present article.

and energy exchange between them. In probing the plasma, one or more waves may be imposed externally, or they may all be part of the overall fluctuation spectrum. Examples of the various situations are discussed in this chapter.

## 4.1 *Scattering of Waves from Density Fluctuations*

Efficient nonlinear interaction between two waves is expected to occur when they beat together in such a way that their sum or difference frequencies and wave vectors just match the frequency and wave vector of the third wave; that is,

$$\begin{aligned} \omega_1 + \omega_2 &= \omega_3 \\ \mathbf{k}_1 + \mathbf{k}_2 &= \mathbf{k}_3 \end{aligned} \tag{4.1}$$

The interaction is illustrated schematically in Fig. 4.1 which shows waves 1 and 2 scattering of one another and combining to form the third wave. By reversing the arrows on

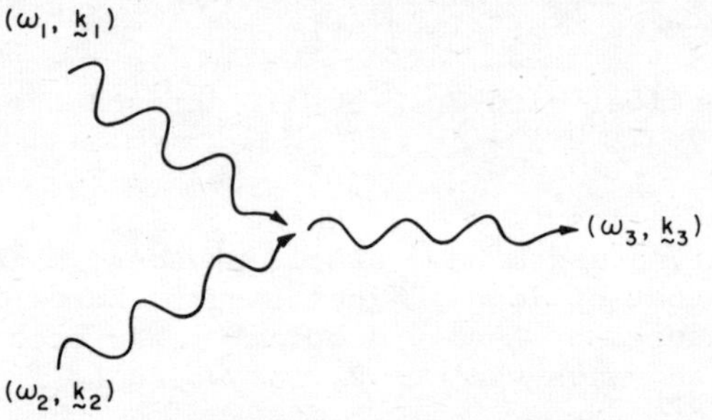


Fig. 4.1 Schematic drawing of the nonlinear coupling of three waves.

the diagram, we obtain the situation wherein wave 3 breaks up and decays into two other waves. These interactions may involve only transverse electromagnetic or only longitudinal plasma waves or, in the general case, a mixture thereof. By interpreting $\omega$ as the energy and $\mathbf{k}$ as the momentum of a single quantum, the resonance conditions (4.1) are seen to be nothing more than conservation requirements on energy and momentum for the system in question.

Let us then consider the interaction of an incident, sinusoidal transverse or longitudinal wave

$$E_i \, e^{j\,\omega_i t \,-\, j\,\mathbf{k}_i \cdot \mathbf{r}} \tag{4.2}$$

with a density fluctuation $\delta N(\mathbf{r},t)$ of the type described in Chapter 3. The electric field $E_i$ of the incident wave is assumed to be large compared with the fluctuating electric field but small enough so as not to perturb the character of the intrinsic fluctuations. Also, the fluctuation in electron density only will be considered, with $\delta N(\mathbf{r},t)$ as the deviation of the electron density from its equilibrium value $N$. As a result of the interaction of $E_i$ with $\delta N$, a scattered (transverse and/or longitudinal) wave with a Fourier component

$$E_s\, e^{j\,\omega_s t\, -\, j\mathbf{k}_s\cdot\mathbf{r}} \tag{4.3}$$

can arise. The computation of its intensity and spectral distribution is the subject matter of this section. In general, one must proceed via the Boltzmann-Vlasov equation and retain in it the appropriate nonlinear terms which are responsible for transforming part of the incident wave into the scattered wave. But, if the phase velocities $\omega_i/k_i$ and $\omega_s/k_s$ of both waves are large compared with the electron thermal speed $v_0 = \sqrt{KT_e/m}$, Landau damping is negligible and the coupling problem can be solved more easily [33] by means of the moment equations. In this framework then, the incident, the scattered wave and the background fluctuations drive a current in the plasma whose density is

$$\begin{aligned} J_{\text{total}}(\mathbf{r},t) &= qN(\mathbf{r},t)\; v(\mathbf{r},t) \\ &= q[N + N_i(\mathbf{r},t) + N_s(\mathbf{r},t) + \delta N(\mathbf{r},t)] \qquad (4.4) \\ &\quad \times [v_i(\mathbf{r},t) + v_s(\mathbf{r},t) + \delta v(\mathbf{r},t)] \end{aligned}$$

The second form of the equation shows the components of $N$ and $v$ explicitly: $N_i$ and $v_i$ are the density and velocity induced by the incident wave, $N_s$ and $v_s$ the values induced by the scattered wave; and $\delta N$ and $\delta v$ are the amplitudes of the intrinsic density and velocity fluctuations, respectively.

Of the total current $J_{\text{total}}$, only a few terms appearing on the right hand side of Eq. (4.4) contribute to the current $J_s$ associated with the wave scattered from the fluctuations, and thus of interest here. The current $[N + N_i]\ [v_i]$ associated with the incident wave alone, and the current $[N + \delta N]\ [\delta v]$ caused by the fluctuations alone are extraneous to the problem and can be subtracted out. Also neglecting quantities which are of second and higher order in the fluctuation parameters (note that $N >> N_i$, $N_s$ and $\delta N$) one obtains

$$J_s(\mathbf{r},t) = q[v_i(\mathbf{r},t)\delta N(\mathbf{r},t) + N_i(\mathbf{r},t)\delta v(\mathbf{r},t) + Nv_s(\mathbf{r},t)] \tag{4.5}$$

Multiplication of both sides of Eq. (4.5) by $\exp[-j\omega_s t + j\mathbf{k}_s\cdot\mathbf{r}]$ and integration over $d^3r\, dt$ readily yields the Fourier component of the scattered wave [see Eqs. (1.9) and (1.10)]:

$$\begin{aligned} J_s(\mathbf{k}_s,\omega_s) &= q[v_i(\mathbf{k}_i,\omega_i)\delta N(\mathbf{k}_s-\mathbf{k}_i;\ \omega_s-\omega_i)] \\ &+ q[N_i(\mathbf{k}_i,\omega_i)\delta v(\mathbf{k}_s-\mathbf{k}_i;\ \omega_s-\omega_i)] \\ &+ q[Nv_s(\mathbf{k}_s;\ \omega_s)] \end{aligned} \tag{4.6}$$

Observe that the Fourier components of the density and velocity fluctuations $\delta N$, $\delta v$ are to be evaluated at

$$\begin{aligned} \omega &= \omega_s - \omega_i \\ \mathbf{k} &= \mathbf{k}_s - \mathbf{k}_i \end{aligned} \tag{4.7}$$

in conformity with the selection rules (4.1). Thus, by picking different combinations of $(\mathbf{k}_i,\omega_i)$ and $(\mathbf{k}_s,\omega_s)$ of the incident and scattered waves, one can in principle study selectively any portion of the fluctuation spectrum $N(\mathbf{k},\omega)$. Using relations (4.7) one can now rewrite Eq. (4.6) in the following abbreviated form which we shall use henceforth:

$$\begin{aligned} J_s(\mathbf{k}_s,\omega_s) &= q[v_i(\mathbf{k}_i,\omega_i)\delta N(\mathbf{k},\omega)] \\ &+ q[N_i(\mathbf{k}_i,\omega_i)\delta v(\mathbf{k},\omega)] \\ &+ q[Nv_s(\mathbf{k}_s,\omega_s)] \end{aligned} \tag{4.8}$$

It is now required to express the right hand side of Eq. (4.8) solely in terms of the incident electric field $E_i$ and the spectrum of the density fluctuations $\delta N(\mathbf{k},\omega)$. To this purpose one eliminates $\delta v(\mathbf{k},\omega)$ in favour of $\delta N(\mathbf{k},\omega)$ through use of the continuity equations (1.5); the result for longitudinal fluctuations is

$$\delta v(\mathbf{k},\omega) = \frac{k}{|k|^2}\frac{\omega}{N}\,\delta N(\mathbf{k},\omega) \tag{4.9}$$

The induced velocities $v_i$ and $v_s$ are derived from the momentum conservation equations

$$\frac{\partial v}{\partial t} + (v\cdot\nabla)v = \frac{q}{m}\left|E + v \times B\right| \tag{4.10}$$

one for $v_i$ and one for $v_s$, which when Fourier analyzed yield to lowest order,

$$v_i(\mathbf{k}_i,\omega_i) = -\frac{jq}{m\omega_i} E_i \tag{4.11}$$

and [33]

$$v_s(\mathbf{k}_s,\omega_s) = -\frac{jq}{m\omega_s}\left| E_s + \delta v(\mathbf{k},\omega) \times B_i \right|$$

$$+ \delta v(\mathbf{k},\omega)\frac{\mathbf{k}\cdot v_i(\mathbf{k}_i,\omega_i)}{\omega_s}$$

$$+ v_i(\mathbf{k}_i,\omega_i)\frac{\mathbf{k}_i\cdot\delta v(\mathbf{k},\omega)}{\omega_s} \tag{4.12}$$

Inserting Eqs. (4.9), (4.11) and (4.12) in Eq. (4.8) then gives the sought-after value of the current density $J_s$ associated with the fluctuations $\delta N$:

$$J_s(\mathbf{k}_s,\omega_s) = -\frac{jq^2}{m\omega_i}\,\delta N(\mathbf{k},\omega)\left[E_i + \frac{\mathbf{k}}{|k|^2}\frac{\omega}{\omega_i}\left(\mathbf{k}_i\cdot E_i\right) + \frac{k_s}{|k|^2}\frac{\omega}{\omega_s}\left(\mathbf{k}\cdot E_i\right)\right] \tag{4.13}$$

$$\equiv -\frac{jq^2}{m\omega_i}\,\delta N(\mathbf{k},\omega)Q$$

In deriving this result, the term $(jqE_s/m\omega_s)$ of Eq. (4.12) has been omitted because it is not connected with the fluctuation spectrum. We point out that the first term of Eq. (4.13) describes the current induced by the interaction of the density fluctuation $\delta N$ with the velocity $v_i$[see Eq. (4.8)], the second term describes the current induced by the interaction of the velocity fluctuation with $N_i$ and the third term is the current generated by the presence of the scattered wave. The vector $Q$ appearing in the second form of the equation (4.13) denotes the function in the square brackets; it in a sense measures the amplitude and polarization of the incident wave.

The current $J_s$ of the scattered wave can be decomposed

$$J_s = J_s(\parallel) + J_s(\perp)$$

$$= \mathbf{k}_s \frac{(\mathbf{k}_s \cdot J_s)}{k_s^2} - \frac{\mathbf{k}_s \times (\mathbf{k}_s \times J_s)}{k_s^2} \tag{4.14}$$

according to whether it is oriented along or perpendicular to the propagation constant of the scattered wave. In the former case the scattered wave is longitudinal ($\mathbf{k}_s \times E_s = 0$), and in the latter case it is transverse electromagnetic ($\mathbf{k}_s \cdot E_s = 0$) (see Table 1.1]. Likewise, the incident electric vector $E_i$ may be that of a purely longitudinal wave ($\mathbf{k}_i \times E_i = 0$), or a purely transverse electromagnetic wave ($\mathbf{k}_i \cdot E_i = 0$). Thus, we

TABLE 4.1. Values of the Vector $Q$ in the Equation

$$J_s(\mathbf{k}_s, \omega_s) = -\frac{jq^2}{m\omega_i}\, \delta N(\mathbf{k}, \omega) Q$$

for Different Combinations of Incident and Scattered Waves

| | INCIDENT WAVE: T, $\mathbf{k}_i \cdot E_i = 0$ | INCIDENT WAVE: L, $\mathbf{k}_i \times E_i = 0$ |
|---|---|---|
| SCATTERED WAVE: T, $\mathbf{k}_s \cdot E_s = 0$ | $Q = -\dfrac{\mathbf{k}_s \times (\mathbf{k}_s \times E_i)}{k_s^2}$ | $Q = -\dfrac{\mathbf{k}_s \times (\mathbf{k}_s \times E_i)}{k_s^2} \times \left[1 - \left(\dfrac{k_i}{k}\right)^2 \dfrac{\omega}{\omega_i}\right]$ |
| SCATTERED WAVE: L, $\mathbf{k}_s \times E_s = 0$ | $Q = \mathbf{k}_s \left(\dfrac{\mathbf{k}_s \cdot E_i}{k_s^2}\right) \times \left[1 + \left(\dfrac{k_s}{k}\right)^2 \dfrac{\omega}{\omega_s}\right]$ | Full expression as given by Eq. (4.13) |

T = Transverse
L = Longitudinal

see that encompassed within Eq. (4.13) are four entirely separate physical processes: the transformation of an incident transverse into another scattered transverse wave (T → T); the transformation of a transverse into a longitudinal wave (T → L); the formation of a transverse scattered from a longitudinal incident wave (L → T); and finally the generation of a scattered longitudinal wave from an incident longitudinal wave (L → L). Table 4.1 gives the values of $J_s(\mathbf{k}_s,\omega_s)$ for each of the four processes. To compute the power in the scattered wave, products like $[J_s(\mathbf{k}_s,\omega_s) \cdot E_s(\mathbf{k}_s,\omega_s)]$ must be formed and thus the relationship between $J_s$ and $E_s$ is needed. To this end, the wave equation

$$\mathbf{k}_s \times (\mathbf{k}_s \times E_s) + \left(\frac{\omega}{c}\right)^2 K(\mathbf{k}_s,\omega_s)E_s = j\omega_s\mu_0 J_s \tag{4.15}$$

is used, yielding the result

$$E_s(\mathbf{k}_s,\omega_s) = \frac{j}{\omega_s\ \varepsilon_0\ K_{\text{eff}}(\mathbf{k}_s,\omega_s)}\ J_s(\mathbf{k}_s,\omega_s). \tag{4.16}$$

Here an effective dielectric coefficient $K_{\text{eff}}$ is employed to distinguish between longitudinal and transverse excitation, namely

$$K_{\text{eff}}(\mathbf{k}_s,\omega_s) = \begin{cases} K_L(\mathbf{k}_s,\omega_s) & \text{longitudinal} \\ K_T - \left(\dfrac{\mathbf{k}_s c}{\omega_s}\right)^2 & \text{transverse} \end{cases} \tag{4.17}$$

Note that $K_L$ is the same longitudinal dielectric coefficient as that discussed at length in Chapter 1, while $K_T$ is the coefficient for transverse waves and is to good approximation given by

$$K_T(\mathbf{k}_s,\omega_s) = 1 - \frac{\omega_p^2}{\omega_s^2} \tag{4.18}$$

The scattered power is obtained by integrating $J_s(r,t) \cdot E_s(r,t)$ over spatial coordinates and by averaging over a period T of the scattered wave [cf Eqs. (3.1) to (3.5)]:

$$\left\langle\frac{dW}{dt}\right\rangle = \frac{1}{T}\iint J_s(r,t)\cdot E_s(r,t)\ d^3r\ dt \tag{4.19}$$

Now, $J_s(r,t)$ and $E_s(r,t)$ are replaced by their Fourier transforms as prescribed by Eq. (1.10); then, using Eqs. (4.13) and (4.16) and carrying our the indicated integrations over the delta functions one finds that

$$\left\langle\frac{dW}{dt}\right\rangle = -V\left(\frac{q^2}{m\omega_i}\right)^2 \iint \frac{d^3k_s}{(2\pi)^3}\frac{d\omega_s}{\omega_s}\frac{N_e{}^2(\mathbf{k},\omega)\,|Q|^2}{\varepsilon_0}\ \text{Im}\left(\frac{1}{K_{\text{eff}}(\mathbf{k}_s,\omega_s)}\right) \tag{4.20}$$

Here $V$ is the volume of the scattering region and $Q$ is the vector listed in Table 4.1 for the specific interaction being considered. The spectral density $N_e^2(\mathbf{k},\omega)$ is precisely the function discussed in Chapter 3. The subscript $e$ denotes that we are considering scattering from electron fluctuations only, but we must recall that because of the Coulomb force between the electrons and the ions, $N_e^2(\mathbf{k},\omega)$ also carries information about the ion motions. Direct scattering from ions is usually negligibly small, in the ratio of the square of the electron to the ion masses [see Eq. (4.20)]. Equation (4.20) can now be used as the basis for computing the rate of scattering associated with the different processes discussed earlier, and two examples are given below.

*Scattering of a transverse electromagnetic wave.* Table 4.1 shows that the scattering of an incident transverse wave and the subsequent generation of a second transverse wave is characterized by a value of $Q$ given by

$$Q = \frac{\mathbf{k}_s \times (\mathbf{k}_s \times E_i)}{k_s^2} \tag{4.21}$$

The scattered wave is assumed to be virtually undamped and thus the only contribution to the integral of Eq. (4.20) comes from the pole of the integrand. If $\mathbf{k}_{s0}$ and $\omega_{s0}$ are the wave vector and frequency of the freely propagating scattered wave then by the same arguments that lead to Eqs. (2.22) we have that

$$\begin{aligned} \mathrm{Im}\left[\frac{1}{K_{\mathrm{eff}}}\right] &= \pi\left[\frac{\partial \mathrm{Re}K_{\mathrm{eff}}}{\partial k_s}\right]^{-1} \delta(\mathbf{k}_s - \mathbf{k}_{s0}) \\ &= -\frac{\pi\omega_{s0}^2}{2k_{s0}c^2}\,\delta(\mathbf{k}_s - \mathbf{k}_{s0}) \end{aligned} \tag{4.22}$$

The spectrum of the scattered radiation is now obtained by integrating over the modulus of $k_s$ but not over angles or frequencies. Noting that $d^3k_s = k_s^2\,dk_s\,d\Omega$, where $d\Omega$ is an element of solid angle of the scattered radiation and by the use of Eqs. (4.21) and (4.22) it readily follows that

$$\frac{dW(\omega_s)}{Vdtd\Omega d\omega_s} = \varepsilon_0 c r_0^2 \left(\frac{\omega_s}{\omega_i}\right)^2 \sqrt{K_T(\omega_s)}\, N_e^2(\mathbf{k},\omega) \left|\frac{\mathbf{k}_s \times (\mathbf{k}_s \times E_i)}{k_s^2}\right|^2 \tag{4.23}$$

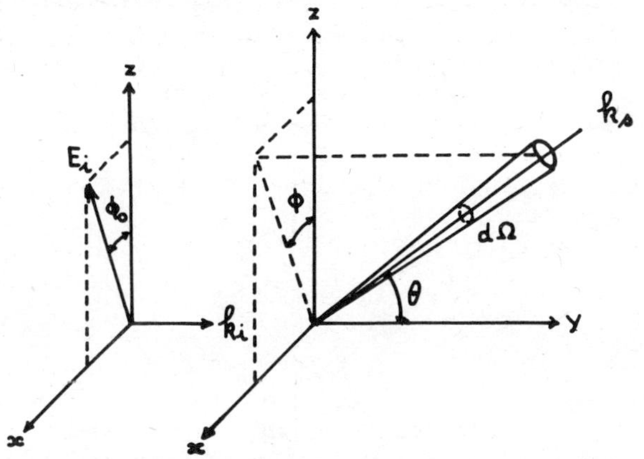


Fig. 4.2 Coordinate system for calculating the scattering by a plasma. The subscript *i* refers to the linearly polarized incident wave and the subscript *s* to the scattered wave.

This basic equation for all scattering studies* gives the power scattered per unit volume of medium, per unit frequency interval $d\omega_s$ per unit solid angle $d\Omega$, in terms of the spectrum $N_e^2(\mathbf{k},\omega)$ of the electron density fluctuations. The quantity

$$r_0^2 = \left(\frac{e^2}{4\pi\varepsilon_0 mc^2}\right)^2 \tag{4.24}$$

$$= 7.94 \times 10^{-30} \text{ meter}^2$$

represents the square of the classical electron radius. The term

$$\sqrt{K_T(\omega_s)} = \sqrt{1 - \frac{\omega_p{}^2}{\omega_s{}^2}} \tag{4.25}$$

---

*The equation is restricted to plasmas that are not too hot, such that $v_0/c \ll 1$ where $v_0$ is the electron thermal velocity. For the case of mildly relativistic plasmas, see the work of J. Sheffield, *Plasma Physics, 14*, 783 (1972).

takes account of the fact that the scattered wave travels through a dispersive medium; and the term $(\omega_s/\omega_i)^2$ allows for the possibility that the incident and scattered waves may have very different frequencies. In practice however, $K_T$ is close to unity as is the ratio $(\omega_s/\omega_i)^2$. The reason is that experiments are normally conducted under conditions where $\omega_i \gg \omega_p, \omega$. The term $\mathbf{k}_s \times (\mathbf{k}_s \times E_i)$ specifies the direction of the electric field of the scattered wave, and the quantity

$$\left|\frac{\mathbf{k}_s \times (\mathbf{k}_s \times E_i)}{k_s^2}\right|^2 = [1 - \sin^2\theta\cos^2(\phi_0 - \phi)]|E_i|^2 \tag{4.26}$$

illustrates the radiation pattern of the scattered wave. The angles $\theta$, $\phi_0$ and $\phi$ are defined in Fig. 4.2; the radiation pattern is characteristic of dipole radiation with maximum emission at $\theta = 0$ and $\pi$ and zero emission at $\theta = \pi/2$ and $3\pi/2$ for $[\phi_0 = \phi]$. Note that $|E_i|^2$ is the value of the electric field squared inside the dispersive plasma and is related to the time-averaged Poynting flux of the incident wave through

$$<|S_i|> = \frac{1}{2}\,\varepsilon_0 c\,\sqrt{K_T(\omega_i)}\;|E_i|^2 \tag{4.27}$$

In view of the fact that the spectrum of scattered radiation is proportional to the function (see Figs. 3.1, 3.6, 3.7)

$$N_e^2(\mathbf{k},\omega) = N_e^2(\mathbf{k}_s - \mathbf{k}_i;\ \omega_s - \omega_i) \tag{4.28}$$

described in detail in Chapter 3, little needs to be added here. Let us just note that when $\omega_i \gg \omega_p$, then $|\omega_s/k_s| \simeq |\omega_i/k_i| \simeq c$ and

$$\begin{aligned} |k| &= 2k_i \sin(\theta/2) \\ &= 2(\omega_i/c)\sin(\theta/2) \end{aligned} \tag{4.29}$$

where $\theta$ is the scattering angle illustrated in Fig. 4.2. Collective features in the longitudinal fluctuation spectrum can occur only for $|\mathbf{k}L_D| \lesssim 1$ with $L_D$ and the electron Debye length (see Figs. 3.1, 3.6 and 3.7). Thus it follows from Eq. (4.29) that in order for these features to appear in the scattered spectrum, it is necessary that

$$2L_D(\omega_i/c)\sin(\theta/2) \lesssim 1 \tag{4.30}$$

When this inequality is not obeyed, the scattered spectrum is characteristic of single particle scattering from uncorrelated electrons. Experimental studies [34] of both types of spectra using microwave or laser radiation are today almost common practice in a large number of laboratories. Scattering of radio waves from the ionosphere is also being carried out [35]. Much information can be derived from such studies. When the electrons and ions have Maxwellian distributions of velocities, an analysis of the scattered spectrum can give the electron and ion temperatures as well as the plasma density. When the medium is mildly nonthermal (but stable) the velocity distribution can be estimated. In particular, when the electron feature of the scattered spectrum is viewed the effective fluctuation temperature of Eq. (3.29)

$$\frac{\kappa T_{\text{eff}}}{m} = -\left(\frac{\omega}{k}\right)\frac{[F(u)]_{u\,=\,\omega/k}}{\left|\dfrac{\partial F(u)}{\partial u}\right|_{u\,=\,\omega/k}}$$

can be deduced for a series of values of $\omega = \omega_s - \omega_i$ and $\mathbf{k} = \mathbf{k}_s - \mathbf{k}_i$. From such measurement, the distribution function $F(u)$ of the plasma electrons can be inferred.

On the other hand, when the medium is turbulent as a result of plasma instabilities, the longitudinal fluctuation spectrum $N_e^2(\mathbf{k},\omega)$ can no longer be related with ease to the particle dynamics as we were able to do in chapter 3, and as a consequence the amount of information to be gleaned from such scattering experiments is substantially less. Nontheless, the importance of such measurements must not be underestimated. For example, we refer to recent observations [36] on collisionless shocks in z-pinches and θ-pinches which often exhibit "power" spectra of the form

$$N^2(\mathbf{k}) \equiv \int N_e^2(\mathbf{k},\omega)d\omega \propto \frac{1}{k^3}\ln\left(\frac{1}{kL_D}\right)$$

That such a law has been predicted by Kadomtsev [37] from a nonlinear theory of the current-driven ion-acoustic instability is gratifying.

*Conversion of longitudinal into transverse waves.* The second example to which we address ourselves is the conversion of an incident longitudinal wave with electric vector $E_i$ into a transverse electromagnetic wave, as a consequence of scattering from plasma fluctuations. In this case the vector $Q$ of Table 4.1 has the form

$$Q = -\frac{\mathbf{k}_s \times (\mathbf{k}_s \times E_i)}{\mathbf{k}_s^2}\left[1 - \left(\frac{k_i}{k}\right)^2 \frac{\omega}{\omega_i}\right] \tag{4.31}$$

Then, proceeding exactly as we did in the previous subsection, use of Eq. (4.20) gives for the scattered power the result

$$\frac{dW(\omega_s)}{Vdtd\Omega d\omega_s} = \varepsilon_0 c\, r_0^2 \left(\frac{\omega_s}{\omega_i}\right)^2 \sqrt{K_T(\omega_s)}\left[1 - \left(\frac{k_i}{k}\right)^2 \frac{\omega}{\omega_i}\right]^2 \times N_e^2(\mathbf{k},\omega)\left|\frac{\mathbf{k}_s \times (\mathbf{k}_s \times E_i)}{k_s^2}\right| \tag{4.32}$$

which is seen to be rather similar to Eq. (4.23). Observe, however, that the term responsible for the character of the radiation pattern, namely

$$\left|\frac{\mathbf{k}_s \times (\mathbf{k}_s \times E_i)}{k_s^2}\right|^2 = \sin^2\theta\, |E_i|^2 \tag{4.33}$$

has a different angular dependence, as a result of the fact that now $(\mathbf{k}_i \times E_i) = 0$. Thus, the radiation in the forward and backward directions is zero and is maximum for $\theta = \pi/2$, $3\pi/2$.

Implicit in the above calculation is the assumption that the incident longitudinal wave has negligible damping. For this to be so, Eqs. (1.52) tell us that the frequency of the wave is limited to the range

$$\omega_p \lesssim \omega_i \lesssim 1.4\omega_p \tag{4.34}$$

and thus, as a consequence of the selection rules (4.7), the frequency of the scattered transverse electromagnetic wave is confined to lie in the range

$$(\omega + \omega_p) < \omega_s \lesssim (\omega + 1.4\omega_p) \tag{4.35}$$

where $\omega$ is the frequency of a fluctuation. We now recall that $N_e^2(\mathbf{k},\omega)$ exhibits peaked spectra when $\omega$ lies close to the frequency of a natural mode of propagation of a longitudinal wave; in an unmagnetized plasma this occurs when $\omega$ corresponds to the frequency of an ion acoustic wave (which is almost zero frequency on our scale), or when $\omega$ corresponds to the frequency of an electron-plasma wave, in which case

$$\omega_p \lesssim \omega \lesssim 1.4\omega_p \tag{4.36}$$

Therefore, the scattering process will cause emission of transverse radiation preferentially in two bands: one centered in the range $\omega_p \lesssim \omega_s \lesssim 1.4\omega_p$, and the other centered in the range $2\omega_p \lesssim \omega_s \lesssim 2.8\omega_p$. Such bands are expected to be most pronounced when the fluctuations are nonthermal, and in particular when the plasma is on the verge of being unstable for the frequencies in question. Thus, the physical picture that emerges is the following: because of nonlinear coupling of waves, the high frequency longitudinal plasma wave (Langmuir oscillation) scatters off an ion-accoustic fluctuation and is converted into a transverse wave at a frequency $\sim\omega_p$; or it scatters off another Langmuir fluctuation and is converted into a transverse wave at $\sim 2\omega_p$. Type II solar bursts often exhibit such bands in their emission spectra and the observed structure is sometimes explained [27, 38] in the manner just described; however, the observations are not fully understood and the entire matter is far from being settled at this time.

## 4.2 *Bremsstrahlung*

Bremsstrahlung refers to a process in which transverse electromagnetic waves are emitted as a result of the acceleration of the free plasma electrons in the quasi static Coulomb fields of the neighboring charges. The traditional point of view is one of two-body collisions; by that we mean that the electron is deflected in the field of a nearby charge and radiates. Because of the long range of the Coulomb force, this binary collision model is oversimplified, and the simultaneous interaction of an electron with many charges needs to be examined.

Consider than an ensemble of charges moving randomly under the action of the stochastic forces acting on them. If all motions are nonrelativistic, the power radiated is given by the formula of Larmor

$$\frac{dW(\mathbf{r},t)}{dt} = \frac{\sqrt{K_T}}{6\pi\varepsilon_0 c^3}\left|\ddot{d}(\mathbf{r},t)\right|^2 \tag{4.37}$$

with $\ddot{d}$ as the second time derivative of the electric dipole moment $d$ of the entire system of $N$ charges

$$\ddot{d} = \sum_{p=1}^{N} q_p\, \dot{v}_p \tag{4.38}$$

having accelerations $\dot{v}_p$. The quantity $K_T(\mathbf{r},t)$ is a dielectric coefficient operator which takes account of the fact that the charges radiate in a dispersive plasma medium rather than in vacuum. After Fourier inversion, $K_T$ will transform into the usual dielectric coefficient

$$K_T \simeq \sqrt{1 - \frac{\omega_p^2}{\omega^2}}$$

The acceleration $\dot{v}_p$ is caused by the electric fields of neighboring charges and so Eq. (4.38) may be written as

$$\begin{aligned} \ddot{d} &= \sum_p \sum_{p'} q_p E_{pp'} \left(\frac{q_p}{m_p}\right) \\ &= \frac{1}{2} \sum_p \sum_{p'} q_p E_{pp'} \left|\frac{q_p}{m_p} - \frac{q_{p'}}{m_{p'}}\right| \end{aligned} \tag{4.40}$$

where $E_{pp'}$ is the electric field at particle $p$ produced by particle $p'$. The second form of the equation follows from Newton's third law, $q_p E_{pp'} = - q_{p'} E_{p'p}$. Observe that a system composed of particles all of which have the same charge to mass ratio does not radiate by dipole emission. This can be made clearer by casting Eq. (4.38) into the form

$$\begin{aligned} \ddot{d} &= \frac{d^2}{dt^2} \sum_p \left(\frac{q_p}{m_p}\right) m_p \boldsymbol{r}_p(t) \\ &= \frac{q_p}{m_p} \frac{d^2}{dt^2} \sum_p m_p \boldsymbol{r}_p(t) \end{aligned} \tag{4.41}$$

But $\Sigma m_p \boldsymbol{r}_p(t) = R(t) \Sigma m_p$ where $R$ is the radius vector of the center of mass. If the latter is at rest or in uniform motion relative to the observer, $\ddot{d} = 0$. Thus, for example, interactions amongst electrons alone do not emit, but only encounters of electrons with ions*.

Consider then a plasma composed of electrons [charge $-e$, mass $m$] and one species of ions [charge $Ze$, mass $M$]. It then follows from Eq. (4.40) that

$$\begin{aligned} \ddot{d} &= \sum_e \sum_i Ze\, \boldsymbol{E}_{ie} \left|\frac{Ze}{M} + \frac{e}{m}\right| \\ &\simeq \frac{Ze^2}{m} \sum_e \sum_i \boldsymbol{E}_{ie} \end{aligned} \tag{4.42}$$

*Higher order quadrupole radiation from electron-electron impacts becomes significant when the particle energy is of the order of or greater than $mc^2[\simeq 0.5 MeV]$, but this regime lies outside the scope of this article.

where the double sum is over electrons and ions and $E_{ie}$ is the electric field at the $i$ th ion due to the $e$ th electron. If now all ion motions are neglected one can readily find the total electric field at an ion and then sum over all ions. If this field is denoted by $E(\mathbf{r}_i, t)$, Eq. (4.42) can be reduced to a single sum of the form,

$$\ddot{d} = \frac{Ze^2}{m} \sum_i E(\mathbf{r}_i,\ t) \tag{4.43}$$

Substitution of this result in formula (4.37) and ensemble averaging yields the time averaged power in bremsstrahlung:

$$\left\langle \frac{dW}{dt} \right\rangle = \frac{Z^2 e^4}{6\pi\varepsilon_0 m^2 c^3} \left\langle \sqrt{K_T(\mathbf{r},\ t)}\ E(\mathbf{r}_i,\ t) \cdot E^*(\mathbf{r}_{i'},\ t) \right\rangle \tag{4.44}$$

If the sites of the stationary ions are taken to be distributed at random, (that is, there is no spatial correlation in the positions of the ions), the averaging procedure outlined above leads to the result [39]

$$\left\langle \frac{dW(\mathbf{r},\ t)}{dt} \right\rangle = \frac{Z^2 e^4 NV}{6\pi\varepsilon_0 m^2 c^3} \iint d\omega\ \frac{d^3k}{(2\pi)^3}\ \sqrt{K_T(\omega)}\ E^2(\mathbf{k},\omega) \tag{4.45}$$

where $V$ is the plasma volume, $N$ the ion density and $E^2(\mathbf{k},\omega)$ is the spectral power density of longitudinal electric field fluctuations. The last named quantity is precisely the function defined by Eq. (3.13). To obtain the bremsstrahlung spectrum, the integration over frequency is not carried out, giving

$$\frac{dW(\omega)}{Vdtd\omega} = \frac{Z^2 e^4 N}{6\pi\varepsilon_0 m^2 c^3} \sqrt{K_T(\omega)} \left[ \int \frac{d^3k}{(2\pi)^3} E^2(\mathbf{k},\omega) \right] \tag{4.46}$$

Observe that knowledge of the frequency spectrum

$$E^2(\omega) \equiv \int \frac{d^3k}{(2\pi)^3} E^2(\mathbf{k},\omega) \tag{4.47}$$

of longitudinal fluctuations discussed in chapter 3 provides direct knowledge of the bremsstrahlung spectrum of the transverse electromagnetic radiation. Thus, for example, when the electrons have a Maxwellian distribution of velocities, $E^2(\omega)$ is as shown plotted in Fig. 3.2; when thermal equilibrium does not occur spectra like that shown, in Fig. 3.5 are appropriate.

The dielectric coefficient $K_T$ appearing in Eq. (4.46) "weights" the longitudinal spectrum $E^2(\omega)$ in a rather important way: while it has little influence at high frequencies where $\omega \gg \omega_p (K_T \to 1)$, it cuts-off the spectrum of bremsstrahlung at frequencies $\omega \leq \omega_p$ since here $K_T$ is purely imaginary. This is just as expected in view of the fact that a transverse electromagnetic wave cannot propagate at frequencies below the electron plasma frequency in an unmagnetized essentially collisionless plasma.

The close relationship between $E^2(\omega)$ and the bremsstrahlung spectrum is a manifestation of the fact that the last named is but a result of converting longitudinal to transverse oscillations. More specifically, the "bump" in the bremsstrahlung spectrum at a frequency just above $\omega_p$ can be viewed as the scattering of a Langmuir fluctuation from an ion acoustic wave and conversion to a transverse electromagnetic wave, via the same mechanism as was already discussed in section 4.1. Of course by taking the ions to be immobile and uncorrelated, the predictions of Eq. (4.46) cannot be entirely correct because the equation as it stands takes proper account only of the high frequency electron feature of the fluctuations (be these thermal or nonthermal). Allowance for ion motions has been recently attempted [40] and spectra obtained for the case of Maxwellian electrons and ions. One finds that as the ratio of electron to ion temperature increases the "bump" in the bremsstrahlung spectrum likewise increases. This is to be expected on physical grounds because only at large values of $Te/Ti$ is Landau damping of ion acoustic waves small as was shown in Chapter 1, and therefore the fluctuation spectrum is strongly peaked as is shown in Fig. 3.7. By the same token, one may well expect a large enhancement [41] of the bremsstrahlung "bump" in the presence of nonthermal ion sound waves, particularly when these are on the verge of becoming unstable. However, a detailed study of the bremsstrahlung spectrum in the presence of nonthermal ion sound waves is at present not available.

*Kirchhoff's law and the radiation temperature.* In section 3.4 we discussed Kirchhoff's law for thermal and nonthermal plasmas as it applies to longitudinal fluctuations. There, the direct (or forward) process of Cerenkov-like emission was balanced by the inverse process caused by Landau damping. In the present case of bremsstrahlung we can again write by analogy with Eq. (3.42) that

$$j(\omega) = U(\omega)\,\alpha(\omega) \tag{4.48}$$

where now however, $j(\omega)$ is the bremsstrahlung emission coefficient, $U$ is the energy density of transverse electromagnetic waves; and $\alpha(\omega)$ is the absorption of transverse waves due to small angle Coulomb collisions and conversion of transverse waves to longitudinal waves. Using Eqs. (4.46) and (3.13) we obtain for the emission coefficient the result that

$$j(\omega) \equiv \frac{1}{V}\left\langle\frac{dW(\omega)}{dt\,d\omega}\right\rangle = \frac{[Ze\omega_p^2]^2}{6\pi\varepsilon_0 c^3}\sqrt{K_T(\omega)}\iint\frac{d^3k}{(2\pi)^3}\,d^3v f(v)\,\frac{\delta(\omega-\mathbf{k}\cdot v)}{k^2|K_L(\mathbf{k},\omega)|^2} \tag{4.49}$$

with $f(v)$ as the distribution of electron velocities. The corresponding absorption coefficient [42] (which we do not derive here) is given by

$$\alpha(\omega) = -\frac{\pi[Ze\omega_p^2]^2}{3m\varepsilon_0\omega^2}\iint\frac{d^3k}{(2\pi)^3}\,d^3v\,\frac{\mathbf{k}\cdot\partial f(v)}{\omega\;\partial v}\,\frac{\delta(\omega-\mathbf{k}\cdot v)}{k^2|K_L(\mathbf{k},\omega)|^2} \tag{4.50}$$

which applies to any isotropic, but not necessarily Maxwellian distribution of velocities, $f(v)$.

The energy density follows on substituting Eqs. (4.49) and (4.50) in Eq. (4.48). The result can be written as

$$U(\omega) = \frac{\omega^2\sqrt{K_T(\omega)}}{2\pi^2 c^3}\left[\kappa T_{\text{rad}}\right] \tag{4.51}$$

where the phenomenological radiation temperature $T_{\text{rad}}$ is defined as

$$\frac{\kappa T_{\text{rad}}}{m} = -\frac{\displaystyle\iint d^3k\,d^3v\,f(v)\,\frac{\delta(\omega-\mathbf{k}\cdot v)}{k^2|K_L(\mathbf{k},\omega)|^2}}{\displaystyle\iint d^3k\,d^3v\,\frac{\mathbf{k}\cdot}{\omega}\,\frac{\delta f(v)}{\delta v}\,\frac{\delta(\omega-\mathbf{k}\cdot v)}{k^2|K_L(\mathbf{k},\omega)|^2}} \tag{4.52}$$

Observe that when $f(v)$ is Maxwellian, $T_{\text{rad}}$ reduces to the true kinetic temperature $T$ of the electrons. Under these conditions, $U(\omega)$ of Eq. (4.51) becomes the Rayleigh-Jeans formula for the energy density of black-body radiation in a medium of refractive index given by $\sqrt{K_T(\omega)} = \sqrt{1 - \omega_p^2/\omega^2}$. Thus, Eq. (4.51) can be viewed as a generalization of the classical black-body formula to nonthermal situations involving fully ionized plasmas. In passing we point out that in weekly ionized plasmas where collisions dominate and collective effects are negligible, Eq. (4.51) still holds but now $T_{\text{rad}}$ is given by a different formula, namely [43]

$$\frac{\kappa T_{\text{rad}}}{m} = -\frac{\displaystyle\int_0^\infty \frac{\nu(v)}{\nu^2(v)+\omega^2}\, v^4 f(v)\,dv}{\displaystyle\int_0^\infty \frac{\nu(v)}{\nu^2(v)+\omega^2}\, v^3\, \frac{\partial f}{\partial v}\, dv} \tag{4.53}$$

Here $\nu(v)$ is the collision frequency for momentum transfer between electrons and atoms; it is an empirically known quantity which may or may not be a function of the electron speed $v$.

The radiation temperature $T_{\text{rad}}$ plays a similar role in emission measurements as the fluctuation temperature $T_{\text{eff}}$ plays in scattering experiments [see section 4.1]. $T_{\text{rad}}$, like $T_{\text{eff}}$, is a directly measurable quantity [26] and knowledge of its frequency dependence can be used in estimating the distribution of electron velocities. For example, using a radiometer tuned to a frequency of 3000 MHz, Skarsgard et al. [44] observe an enhanced radiation temperature at $\omega$ near $\omega_p$, from the positive column of a pulsed helium discharge. Enhancements by factors of 2 to 4 in excess of the kinetic temperature are measured. The results are interpreted in the following way: the ionization of helium metastables in the decaying afterglow plasmas generates hot nonthermal electrons in the tail of an otherwise Maxwellian velocity distribution. Equation (4.52) is then assumed to be applicable to the experimental situation, and is used to relate the distribution function to the measured radiation temperature. It is only fair to say that these studies are of a preliminary nature and need to be confirmed by much more exhaustive experiments. In particular one may well question the applicability of the equation that was derived for a fully ionized collisionless Vlasov plasma to an experiment where collisions are certainly not negligible. On the other hand, very convincing experiments have been reported for weakly ionized collision dominated plasmas [26, 45]; here Eq. (4.53) [or its equivalent] has been used with success in deducing the distribution function.

## 4.3 *Optical Transitions Induced by Plasma Fluctuations*

Hitherto, our discussion dealt with interaction of fields with the free charges of the plasma. Here, we turn to the various effects caused when the fluctuating fields act on the bound electrons of atoms (or ions). These fields are substantial and when they act on an assembly of atoms, the resultant Stark splitting of energy levels leads to a modification of their emission or absorption spectra. The different manifestations of these phenomena is the subject matter of the following pages. We begin by what may be termed classical plasma spectroscopy — a condition where particle collisions prevail and plasma turbulence is absent.

*Collisional phenomena.* When the stochastic electric fields $E_c$ are due primarily to collisions between charges, one is often (but not always) justified in adopting the following model. The motion of the sluggish ions is neglected and their field distribution around the emitting atom calculated as if they were at rest. Thus, the ions are responsible for the virtually static Stark splitting of energy levels. The fast electrons on the other hand are treated as classical projectiles which, as they pass the atom with various impact parameters, produce a perturbation in atomic potential. This causes a line broadening which, when weighted over the distribution $W(E_c)$ of ion field strengths yields the final line shape. If then $I(\omega)$ is the line intensity at the frequency $\omega$, one finds that [46]

$$I(\omega) = \frac{1}{\pi} \operatorname{Re} \int dE_c \, W(E_c) \, d_{\alpha\beta} \times \langle\alpha\beta\left| \left\{ j\omega - \frac{j[H_a(E_c) - H_b(E_c)]}{\hbar} - \phi_{ab} \right\} \right|\alpha'\beta'\rangle \, d^*_{\alpha'\beta} \tag{4.54}$$

Here $d_{\alpha\beta}$ is the dipole moment operator, $H_a(E_c)$ and $H_b(E_c)$ are the Hamiltonians perturbed by the electric field $E_c$ of the two atomic states $a$ and $b$; and $\alpha$ and $\beta$ are substates of levels $a$ and $b$, respectively. The quantity $\phi_{ab}$ is an operator which describes the electron impact broadening.

The mean value of the ion field strength $\langle E_c \rangle$ depends on the ion density $N$. With $N^{-1/3}$ as the interparticle distance an estimate of $\langle E_c \rangle$ follows from Coulomb's law:

$$<E_c> \approx \frac{e}{4\pi\varepsilon_0} <\frac{1}{r^2}> = \frac{eN^{2/3}}{4\pi\varepsilon_0} \qquad (4.55)$$

This comes close to the true value (given by Holtsmark [47]), namely

$$<E_c> = \frac{2.61\ eN^{2/3}}{4\pi\varepsilon_0} \qquad (4.56)$$

The distribution $W(E_c)$ itself is shown plotted in Fig. 4.3. Its applicability, however, is restricted to relatively low plasma densities since in its derivation the collective interaction between neighboring charged perturbers was neglected. If such particle correlations are taken into account [46], one obtains distributions like those shown by the remaining two curves of Fig. 4.3. It is not surprising that the effect of the correlations is specified by $N_D$, the number of particles in the Debye sphere. Note that even when $N_D$ is quite large, departures from the Holtsmark profile are quite appreciable. The case $N_D = 1.95$ shown in the figure must be close to the limit of validity of the theory.

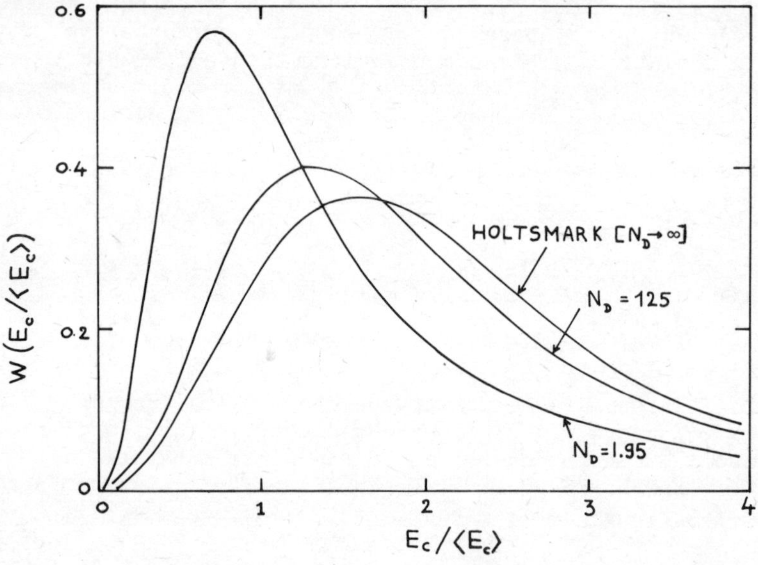


Fig. 4.3 Probability distribution of the quasistatic ion field around a netural atom, with and without charged particle correlations. [After reference 46].

At a sufficiently high plasma density $N$, the line broadening given by Eq. (4.54) dominates all other effects (like Doppler broadening); for this to be so, typical densities are $N > 10^{15}$ cm$^{-3}$ for hydrogen atoms, and $N > 10^{16}$ cm$^{-3}$ for helium atoms. In these density regimes the fitting of a measured line profile to that predicted by Eq. (4.54) furnishes a most valuable method of accurately determining the plasma density since the line width $\Delta$ is a direct measure of $N$: In hydrogenic atoms that exhibit a linear Stark effect, $N \alpha \Delta^{3/2}$ and in other atoms $N \propto \Delta$ where the constants of proportionality are weak functions of temperature and in hydrogen also of density. Figure 4.4 illustrates the extent to which the experimental and theoretical profiles can be fitted with $N$

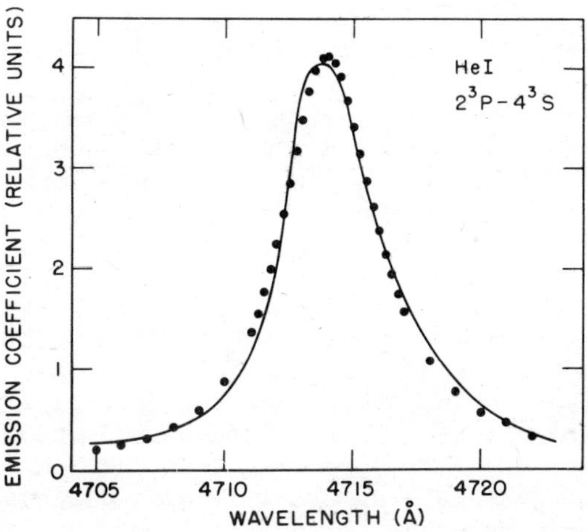


Fig. 4.4 The isolated *HeI* $2^3P$ - $4^3S$ spectral line. The points are Abel inverted experimental data as obtained from a laser-produced helium plasma. The solid curve is from theory for the case $N = 3.8 \times 10^{16}$ $cm^{-3}$, $T_e = 34.000^0 K$. [From reference 48].

as the parameter. These measurements were made on a plasma produced by a pulsed $CO_2$ laser in an arrangement [48] shown schematically in Fig. 4.5. The laser radiation is focused into a gas cell containing helium at a pressure of ~ 0.75 atm. The light from the helium plasma generated by the laser pulses is focused onto the slits of a scanning spectrometer. The output signal from the photomultiplier is fed into a boxcar integrator and then to graphic display equipment. The gate of the boxcar can be set to any desired time delay relative to the time the laser is fired. The results of

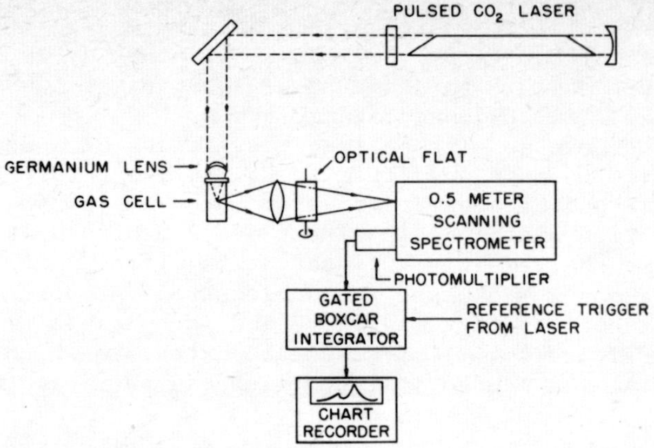


Fig. 4.5 Experimental arrangement in spectroscopic studies of laser produced plasmas [From reference 48].

Fig. 4.4 refer to a time of 5 μ sec in the plasma afterflow. The optical flat shown in Fig. 4.5 can be rotated so that different portions of the plasma fireball can be focused on the spectrometer slits. Such scans across the more or less cylindrical plasma followed by an Abel inversion of the spectral line yields the "true" intensity profile at any given radial position $r$ across the plasma. The measurement in Fig. 4.4 corresponds to a radial distance $r \simeq 0.11$ mm from the plasma axis, where $N$ is found to be $3.8 \times 10^{16}$ $cm^{-3}$ (and the electron temperature $T_e$ = 34.000° K).

The energy levels of the line just discussed are far removed from some other perturbing levels — in this sense the spectral line is "isolated". Of much more recent interest is the Stark profile of an allowed line interacting with a nearby forbidden transition. As is well known the forbiddenness of an optical transition is broken by a sufficiently strong external perturbation of the atom, provided for example by the electric fields of the neighboring charges. Not only is the combined line shape of the "allowed-forbidden" combination of basic theoretical interest, but its thorough understanding may provide a powerful diagnostic method of determining plasma densities. Here, not only the width of the Stark broadened line, but the intensity ratio and relative shift of the two components may well be used in a sensitive determination of plasma density. Figure 4.6 illustrates the theoretical predictions for the allowed $2^1P - 3^1D$ helium line at 6678 Å and the nearby forbidden $2^1P - 3^1P$ component at 6632 Å.

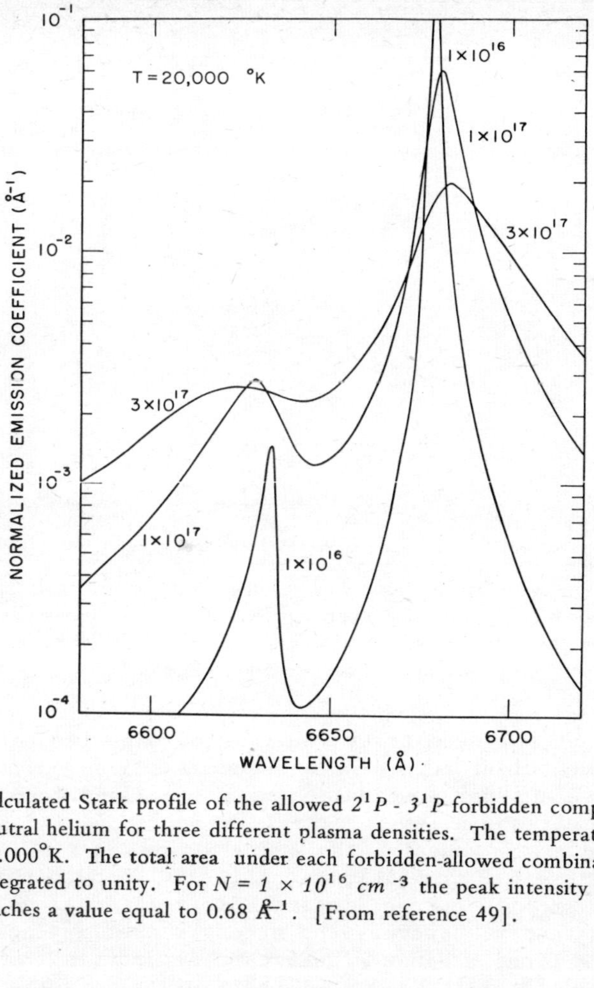


Fig. 4.6 Calculated Stark profile of the allowed $2^1P$ - $3^1P$ forbidden component of neutral helium for three different plasma densities. The temperature is 20.000°K. The total area under each forbidden-allowed combination is integrated to unity. For $N = 1 \times 10^{16}$ $cm^{-3}$ the peak intensity at ~ 6678Å reaches a value equal to 0.68 $Å^{-1}$. [From reference 49].

We see that as the plasma density $N$ increases the lines not only shift but the ratio of their peak intensities changes greatly. Figure 4.7 shows a comparison between theory and measurements [49] made on the same laser produced plasma as that described above. We see that the agreement is good at and near the allowed line, but substantial disagreements occur at the forbidden component. The reasons for the differences are not fully understood at this time;* however,

* For recent new experiments and theory see: R. J. Hawryluk, G. Bekefi and E. V. George, *Phys. Rev. A10*, 265 (1974); also C. Deutsch, C. M. Sassi and G. Couland, *Ann. Phys. 83*, 555 (1974).

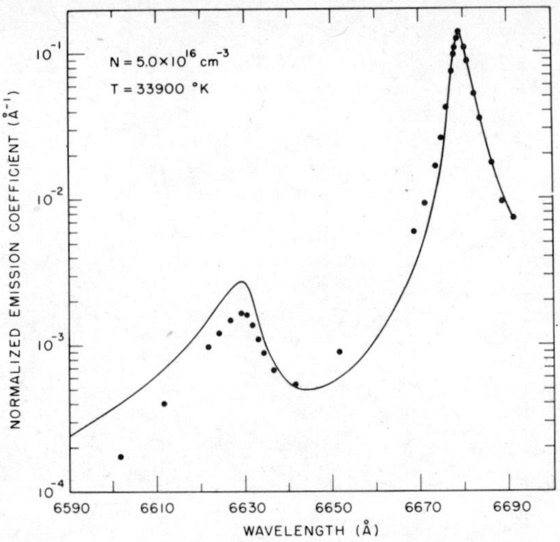


Fig. 4.7 The allowed and forbidden line profiles of the same spectral lines as those shown in Fig. 4.6. The points represent Abel inverted experimental results; the line is from theory so chosen as to give the best fit with experiment along the allowed line. [From reference 49].

they are not believed to be due to some experimental errors or instrumental effects. As a second example [50], we show in Fig. 4.8 another combination of allowed and forbidden lines, namely the allowed $2^3P - 4^3D$ line at 4471 Å and its two forbidden components, the $2^3P - 4^3F$ transition at 4470 Å and the $2^3P - 4^3P$ transition at 4517 Å. We note that the intensity profile of the distant 4517 Å line agrees quite well with theory; however, the strongly overlapping 4470 Å and 4471 Å lines exhibit significant departures from computations, albeit their overall half-width is correct. It is clear that our understanding of this subject is just beginning and that further theoretical and experimental work is needed. The need for more understanding of this area of classical plasma spectroscopy becomes even more urgent in view of its relevance to turbulent plasmas discussed in the following subsection.

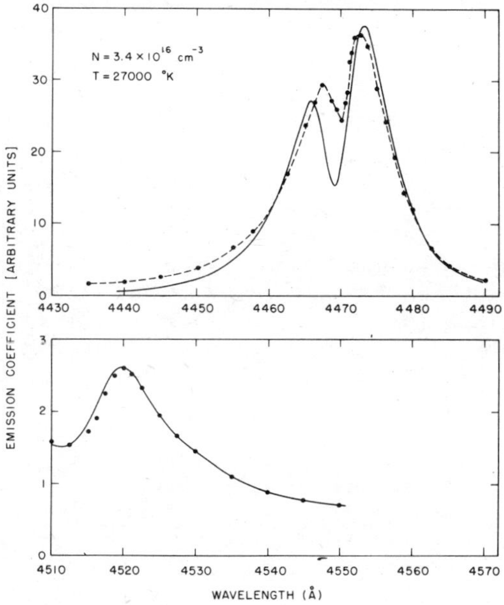


Fig. 4.8 Profiles of the allowed $2^3P$ - $4^3D$ and the forbidden $2^3P$ - $4^3F$ transitions (shown above) and the forbidden $2^3P$ - $4^3P$ transition (shown below). The points are Abel inverted experimental data and the dashed line merely connects the pertinent experimental values. The solid line is from theory for the case $N = 3.4 \times 10^{16}$ $cm^{-3}$, $T = 27.000°K$, which are plasma parameters that have been determined independently. The theory is normalized to the experiments at one point only, the peak of the allowed line. [After reference 50].

*Low-frequency turbulent phenomena.* We have seen on earlier pages that collective plasma oscillations are quite a common occurrence. The accompanying electric fields can be substantial especially when the oscillations are unstable or on the verge of becoming unstable. When these fields become comparable with the Holtsmark field of Eq. (4.56), modifications of line shapes of spectral lines can be expected. For example when the plasma density $N = 10^{15}$ $cm^{-3}$, the Holtsmark field $\langle E_C \rangle = 3.7 \times 10^3$ $V/cm$; turbulent fields $E(\omega)$ of this magnitude and higher have been observed.

The electric field $E(\omega)$ causes a Stark splitting of energy levels $\Delta U$; in hydrogenic atoms or ions $\Delta U$ is proportional to $E$ and we may write that $\Delta U = aE(\omega)$ where $a$ is a constant of proportionality. The type of effect which $E(\omega)$ produces depends on the ratio

$$\times = \frac{\hbar\omega}{aE(\omega)} \tag{4.57}$$

When $\times << 1$, which is the regime being considered in this subsection, the quasi static approximation is expected to be generally good. Thus, very low frequency turbulence acting on an emitting hydrogenic atom or ion will manifest itself mainly in that it will cause an additional line broadening of a spectral line, over and above its "classical" broadening. Such a broadening has been recently observed in plasmas produced in a θ-pinch [51], in a plasma gun [52] and in a toroidal device [53]. The additional broadening can be used to infer the field strength of the turbulence. The experiments are as yet fairly crude and no detailed theoretical understanding of the modified line *shape* is available.*

In addition to experiments being done on hydrogenic atoms and ions, measurements have been reported on the effects of low frequency turbulence in helium atoms, with emphasis on forbidden transitions. In accord with expectations, one finds [54] that sufficiently strong turbulence breaks the forbiddenness. A measurement made on a conical $z$-pinch plasma having a density $N \simeq 6 \times 10^{12}$ cm$^{-3}$ and electron temperature $T_e \simeq 35$ eV, is illustrated in Fig. 4.9. The spectral lines in question are the allowed $2^3P - 4^3D$ line at 4471 Å and the $2^3P - 4^3F$ transition at 4470 Å, - the same combination as that discussed in Fig. 4.8. Observe the strong forbidden line and the large broadening of the allowed line. Both effects are almost entirely due to turbulence, because at a density of $6 \times 10^{12}$ cm$^{-3}$ the forbidden line would be negligibly small and the Stark broadening of the allowed line would be of no importance. The electric field causing this effect is approximately $7kV$/cm.

No theory of the effect of low frequency turbulence on the line shape of forbidden and allowed lines has yet been made.† In lieu of this one may pose the following question: pretending that the line shape of Fig. 4.9 is due to classical particle-produced fields as was the case in the previous subsection, what *hypothetical* plasma density $N$ and temperature $T_e$ would be needed to obtain a best fit with observations? The answer is $N$(hypothetical) = $10^{15}$ cm$^{-3}$ and $T_e$ (hypothetical) = 40.000 °K, and the fit is shown as the solid line of Fig. 4.9. It is curious that the fit is so good. This

---

* For preliminary calculations see G. V. Sholin, *Soviet Physics-Doklady 15*, 1122 (1971).

† For recent attempts see G. Bekefi and C. Deutsch, *Comments on Plasma Physics and Controlled Fusion* (to be published).

suggests one of two things: either the turbulent field has a distribution $W(E)$ similar to that shown in Fig. 4.3, or the spectral line shape is very insensitive to the precise form of $W(E)$.

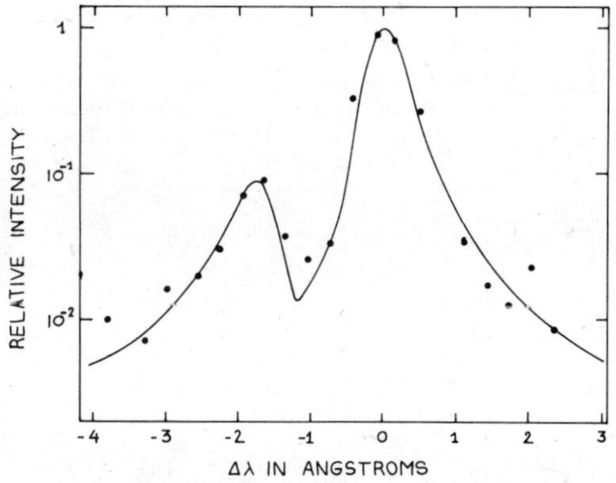


Fig. 4.9 The *HeI* $2^3P$ - $4^3D$ allowed transition and the $2^3P$ - $4^3F$ forbidden transition in a turbulent z-pinch plasma having a density $N \simeq 6 \times 10^{12} cm^{-3}$ and a temperature $T_e \simeq 35eV$. The solid points are from experiment. The line is from classical stochastic Stark theory corresponding to $N = 10^{15} cm^{-3}$ and $T_e = 40.000°K$. [From reference 54].

*High-frequency turbulent phenomena.* When the quantity × of Eq. (4.57) is not negligibly small, time-dependent Stark effects come into play and changes of the optical spectra now become functions of both the amplitude and frequency $\omega$ of the turbulent field. Therefore, it is not surprising that such optical studies are able to provide us with a diagnostic method through which the strength and spectral content of the collective plasma fields may in principle be deduced.

Of the many different time-dependent Stark effects presumably possible, the scheme proposed by Baranger and Mozer [55] appears most attractive. The authors examine the effects induced in the vicinity of an optically forbidden transition with an allowed line situated nearby. For simplicity, they assume an atomic system comprised of only three levels, $i,j,k$ (see Fig. 4.10) such that transition $j \to k$ is allowed by the usual dipole section rule, but transition $i \to k$ is forbidden (i.e. the change in the angular quantum number $\Delta\ell = 0$ or 2; note, however, that both levels $i$ and $k$ must be connected to $j$ via allowed dipole transitions).

Now, an atom situated in state $i$, and under the influence

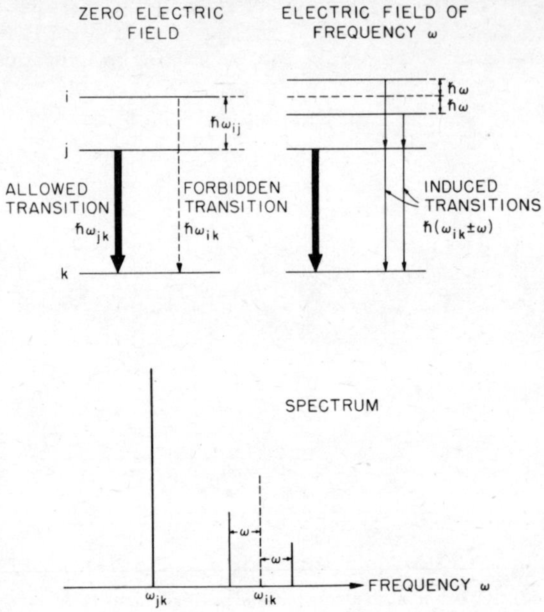


Fig. 4.10 Energy level diagram for a three-level atom (above), and the resulting spectrum (below), illustrating the physics of the time-dependent Stark effect.

of an electric field $E(\omega)$ of frequency $\omega$, undergoes a transition via the virtual state $j$ to the lower state $k$. In the process it emits a photon of frequency $\omega_{ik} + \omega$ or $\omega_{ik} - \omega$. Thus, two so-called "satellites" appear in the spectrum, flanking the (non-existent) forbidden line and separated from it by a frequency $\pm\omega$ (see the lower diagram of Fig. 4.10). If, for example, the electric field in question is associated with a Langmuir oscillation, we see that the process described is just another example of the coupling of longitudinal waves (plasmons) and transverse electromagnetic waves (photons). (The inverse-absorption process in which the atom in the lower state $k$ is subjected simultaneously to a plasmon and photon is discussed by Reinheimer [56]). The coupling is a nonlinear process and thus the transition probability for the event is expected to be proportional to the square (and higher powers) of the electric field $E$. Second-order, time-dependent perturbation calculations [55] yield the sought-for satellite intensity [see Appendix]. The result is most conveniently expressed as the ratio $S$ of the integrated satellite intensity to that of the allowed line - a quantity readily amenable to direct experimental measurement:

$$S_{\pm}(\omega) = \langle E^2(\omega)\rangle[4\pi\epsilon_0\hbar^2/6m^2e^2][\omega_{ij}\pm\omega]^{-2}R_{\ell\ell'} \qquad (4.58)$$

Here $\langle E^2(\omega)\rangle$ is the time average of the electric field squared, and $\omega_{ij}$ is the frequency separation between the forbidden and allowed lines (see Fig. 4.10). The dimensionless parameter $R_{\ell\ell'}$ is proportional to the square of an integral over the coordinate operator of the atomic electron and its value is fairly readily determined in terms of the angular and magnetic quantum numbers of the allowed and forbidden levels ($\ell$, $m$; $\ell'$, $m'$). The minus sign refers to the "near" satellite, namely the one which lies closest to the allowed line, and the plus sign is associated with the other, "far" satellite. We observe that as a result of the frequency-dependent denominator, the near satellite is the more intense one, but since the theory is restricted to small perturbations, we require that $S_{\pm} \ll 1$; therefore the resonance case $(\omega_{ij} - \omega) \to 0$ must be excluded.*

In addition to the appearance of satellites, the electric field also causes a quadratic Stark shift of the allowed line. For a three-level atom its magnitude is [57]

$$\Delta\omega \approx - 2\omega_{ij}[S_+S_-]^{1/2} \qquad (4.59)$$

This shift can be used as an additional and rather simple way of estimating the rms electric field strength [58]. But it can also lead to some confusion in the determination of satellite positions if these are measured relative to the (now shifted) allowed line.

When the electric field has a preferential direction, the satellites are polarized [random polarization is assumed in the derivation of Eq. 4.58] but the effect is weak. This and several other predictions of second-order Stark theory discussed above have been very convincingly tested by Cooper and Ringler [58] in model experiments in which a weakly ionized gas was subjected to a known microwave electric field. The presence of a magnetic field adds additional effects [59] [see the Appendix]. Now the satellite profiles exhibit characteristic Zeeman patterns which can be used with profit in determining the dominant polarization of the electric field.

The smallness of the satellites is one of their drawbacks and thus their use as a tool in measuring $E(\omega)$ is restricted to a rather special class of plasmas. As an example, consider that $E$ is generated by Langmuir waves at $\omega \approx \omega_p$. In this case Eq. 4.58 can be cast into the form

* But see D. Prosnitz and E. V. George, *Phys. Rev. Letters 32*, 1282 (1974).

$$S_{\pm} = 0.25\, z[T/V_H][\omega_p/(\omega_{ij} \pm \omega_p)]^2 R_{\ell\ell'} \qquad (4.60)$$

where $T$ is the electron temperature in eV and $V_H = 13.5$ eV is the ionization energy of hydrogen. The quantity $z$ is the ratio of the electrical energy residing in the oscillations, to the particle kinetic energy,

$$z = \left|\frac{1}{2}\,\varepsilon_0 \langle E_p^2\rangle\right| \left|\frac{3}{2}\, N\kappa T\right|^{-1} \qquad (4.61)$$

The $R_{\ell\ell'}$ parameter of atomic lines has values between ~50 and ~500. With $T$ equal to a few eV and with $[\omega_p/(\omega_{ij}\pm\omega)]^2$ of order unity, the satellite strength (relative to the allowed line) is of order $z$. Because of various experimental constraints, values of $S_{\pm}$ in excess of 0.01 are desirable. Thus, $z$ of Eq. (4.61) should not be much less than ~ 1%. As an illustration, we consider the case where $E_p$ is driven by thermal fluctuations of the electrons. In this circumstance [see Eq. (3.24)] $z(\text{thermal}) \approx 0.015/N_D$ where $N_D$ is the number of particles in the Debye sphere. To achieve our 1% level for $z$ therefore requires a thermal plasma with $N_D$ of order unity, which means high density ($N \gtrsim 10^{18}$ cm$^{-3}$) and low temperature ($T \approx$ 1eV) [see Table 3.1]. Plasmas with these somewhat pathological conditions are found in experiments with exploding wires [60] and in some plasmas produced by illuminating a gas or solid target with high-power lasers. Experiments in this regime of $N_D \lesssim 1$ are of considerable interest, because here the very foundations of conventional kinetic theory as presented in chapters 2 and 3 are in question, and better theories are as yet unavailable.

Figure 4.11 illustrates the spectrum achieved by exploding a lithium wire in air. Although this spectrum is time-integrated, the observed emission comes predominantly from a short time interval about 1 μsec after termination of the explosion. At this time the plasma density is about $10^{18}$ cm$^{-3}$ and the electron temperature about 1 eV. In the figure, A denotes the self-reversed allowed line and F the forbidden component. The broadening of the allowed line and the appearance of the forbidden line come largely from classical, collision-produced electric fields as described at the beginning of this section. The spectral line $S$ is the distant plasma satellite; the near satellite is presumably buried within the broad wings of the allowed line. The integrated intensity of the satellite corresponds closely to what one would expect from thermal plasma fluctuations. Observe that in addition to Eq. (4.58), which compares the satellite intensity to the allowed-line intensity, we can

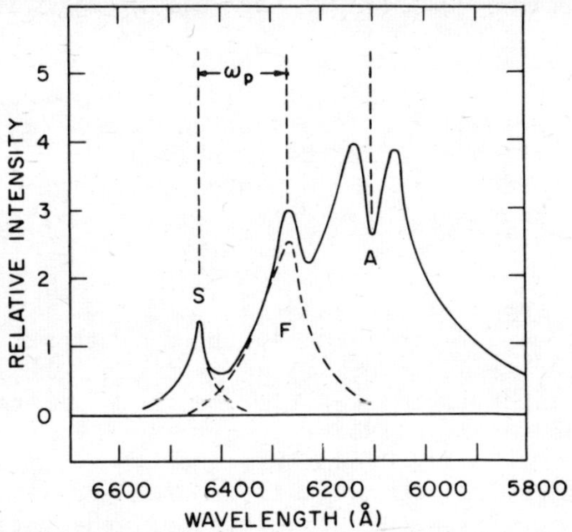


Fig. 4.11 Spectrum from an exploded lithium wire, after subtraction of the continuum. *A* is the allowed *2P - 3D* line of *Li I*. *F* is the *2P - 3P* forbidden line and *S* the plasma staellite. The dashed curves represent an attempt to unfold the partially overlapping lines. [After reference 60].

make an even more convenient comparison with the forbidden line. For a three-level atom, the intensity of the forbidden line is obtained from Eq. (4.58) by setting $\omega = 0$, replacing $\langle E^2(\omega)\rangle$ by the Holtsmark field $\langle E_c^2\rangle$ [or its modification shown in Fig. 4.3] and multiplying by two:

$$S(\text{forbidden}) = 2\langle E_c^2\rangle[4\pi\varepsilon_0\hbar^2/6m^2e^2\omega_{ij}^2]R_{\ell\ell}, \qquad (4.62)$$

Thus, the ratio of satellite intensity to forbidden line intensity takes on the form

$$\frac{I(\text{satellite})}{I(\text{forbidden})} = \frac{1}{2}\,\frac{\langle E_p^2\rangle}{\langle E_c^2\rangle}\left[\frac{\omega_{ij}}{\omega_{ij}\pm\omega_p}\right]^2 \qquad (4.63)$$

which can be used to advantage in making rough comparisons of experiment and theory.

The majority of plasmas have $N_D \gtrsim 10^2$ and it is clear that the satellite method of studying $E$ is mainly confined to systems which exhibit at least one strong nonthermal mode: the oscillatory electrical energy should exceed the thermal level by two to three orders in magnitude. Note that we do not ask for the unlikely situation $z \sim 1$, but for the more modest requirement $z$(thermal) << $z$ << 1. Experiments in which satellites were presumably detected fall into the aforementioned category. They include θ-pinches [61,62,63], the $z$-pinch [64], the plasma gun [65], the electron-beam excited plasma [59], and other turbulent systems [66,67,68, 69]. In the majority of these experiments the collective fluctuations occured in the neighborhood of the electron plasma frequency, and the electric fields were typically in the range $10^3$-$10^5$ V/cm.

To date, none of the existing experiments exhibit a sufficiently clean satellite line profile to warrant a detailed interpretation of its shape.* Yet the profile is a potential asset to the experimenter. An extension [28, 57] of the Baranger and Mozer theory to allow for non-sinusoidal $E(\omega)$ fields shows that when the allowed and forbidden lines are not excessively broadened to cause appreciable merging with the satellites, the satellite profile reproduces faithfully the energy spectrum of the plasma fluctuations. Thus, for example, when the fluctuations are of thermal origin the spectrum of a satellite would be as indicated in Fig. 3.2; under nonthermal conditions, it may be like that shown in Fig. 3.5.

Satellite intensities can be expected to obey Eq. (4.58) only as long as $S_\pm$ remains sufficiently small. Some existing experiments indicate a need for better numbers. Griem's [57] computations to fourth order yield the following results for the corrected intensities $S_-$ and $S_+$, expressed in terms of the $S_\pm$ values of Eq. (4.58):

$$\begin{aligned} S_- &= S_- - 2S_-^2 \{1 + (\omega_{ij}/\omega) + 4[\omega_{ij}/(\omega_{ij} + \omega)]^2\} \\ S_+ &= S_+ + 2S_+^2 \{-1 + (\omega_{ij}/\omega) - 4[\omega_{ij}/(\omega_{ij} - \omega)]^2\} \end{aligned} \qquad (4.64)$$

An examination of these equations shows that the higher-order terms reduce the near satellite quite noticeably without appreciably changing the far satellite. The absence of the near satellite from some experiments has, in fact, been blamed [61, 62] on just this reduction.

Other fourth- and higher-order effects of considerable potential interest concern the generation of satellites at harmonic frequencies. For example, a satellite at 2ω can be interpreted [70] as a three-quantum process involving two plasmons and one photon. Here the atomic selection rule

* However, see D. Prosnitz and E. V. George, *Phys. Rev. Letters* *32*, 1282 (1974).

positions the satellites symmetrically on either side of the *allowed* line and separated from it by $\pm 2\omega$. The satellite strengths are given by [57,70]

$$S(2\omega) = (1/2)S_+S_-[\omega_{ij}\pm\omega)/\omega]^2$$

where the minus sign refers to the satellite stationed on the same side of the allowed line as the forbidden component. A recent analysis free of perturbation theory is described in the Appendix.

At the present time the experiments lag well behind our theoretical understanding of the time-dependent Stark effect. The observations are still quite crude; often only one satellite appears, and at other times the collective oscillations are too low in frequency to be resolved; they merge into a single spectrum centered around the forbidden component. When either of these situations occurs, considerable care must be exercised to establish unequivocally that what is being observed is in fact a satellite: "classical" particle-produced fields can cause an emergence of the forbidden line. Doppler shifts because of mass motion of the plasma, plasma inhomogeneities along the direction of sight, and impurity lines are some of the phenomena that may be mistakenly ascribed to satellites. Most experiments have been made on a few spectral lines of neutral helium; and while satellites have been observed in other nonhydrogenic atoms [60,70], (and they may also exist in nonhydrogenic ions) a systematic search has yet to be made.

In hydrogenic atoms (or ions) the large linear Stark broadening makes observations of the Baranger and Mozer type

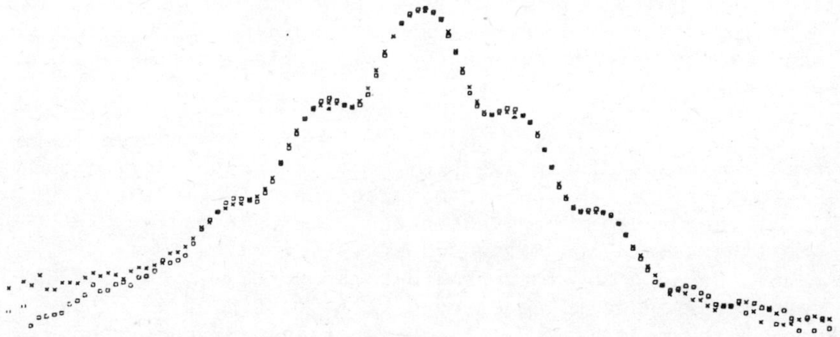

Fig. 4.12 Frequency spectrum of the hydrogen $H$ line. The modulation of the line is caused by subjecting the radiating atoms to a sinusoidal microwave field having a frequency of 35 GHz and an amplitude of ~$1kV/cm$. [From reference 73].

of satellites unfavourable — the width of the allowed line (expressed in frequency units) is comparable with the frequency of the *rf* electric fields $E(\omega)$. Under these conditions the theoretical model described above is no longer applicable, and one must turn to analyses like those given by Blockinzew [71] and Townes [72]. These calculations predict a fairly complicated modulation of the line profile with peaks and valleys separated approximately by the frequency $\omega$ of the *rf* field. Figure 4.12 shows the spectrum of the hydrogen $H_\alpha$ line. The observed modulation of the line is produced by a sinusoidal *rf* field of about 35 GHz applied externally and acting on the radiating atoms of a fairly tenuous discharge in hydrogen. The depth of the modulation is a measure of the strength of the *rf* field, but its relationship to $|E(\omega)|$ can be quite complex particularly if $E(\omega)$ is not monochomatic. Thus the spectrum of a turbulent field may be very difficult to unfold from the measured line shape [74,75].

## 5. SUMMARY

The calculations presented on the foregoing pages were made for an infinite, homogeneous, electrically neutral, fully ionized collisionless plasma composed of classical point charges. In practice these idealizations can of course be satisfied only partially, a fact which puts certain limits on the range of useable plasma parameters. To exhibit this range, at least in part, we refer to Fig. 5.1 which represents a map of plasma temperature $T$ versus density $N$. The central region of this map containing the compass needle deleneates the domain wherein our computations are quite well satisfied. However, the precise positions of the boundaries shown should not be considered sancrosanct nor are they as sharply defined as the drawing suggests.

From the North our domain is bounded by an area where realitivistic effects are prevalent, corresponding to the region in which the particle (electron) kinetic energy $U_k$ is of the order of or greater than the rest energy $m_0c^2$. The line $A$ is drawn at the conservative value, $U_k = 0.1\ m_0c^2$. We note however, that our theoretical model can be readily extended to accomodate relativistic effects (by use of the relativistic Vlasov equation) and that therefore the crossing of line $A$ offers no really serious challenge to the theorist.

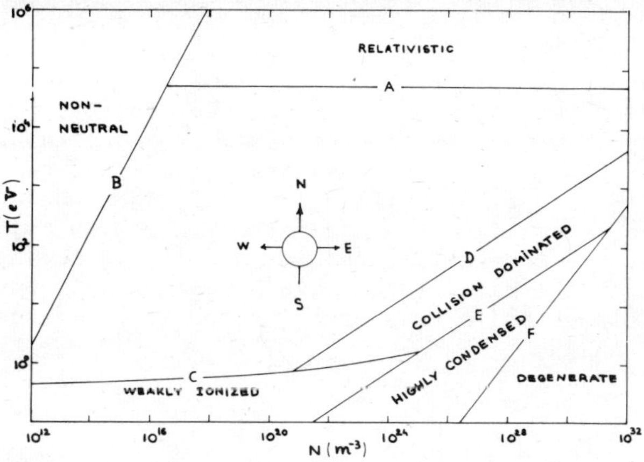


Fig. 5.1 A map of plasma temperature versus density. The central region containing the compass needle approximately delineates the domain of applicability of these notes.

From the West and North-West our domain is constrained by the fact that an electrically neutral plasma, and one able to exhibit its varied collective effects must extend spatially over many Debye lengths

$$L_D = 7.43 \times 10^3 \frac{\sqrt{T(eV)}}{\sqrt{N(m^{-3})}} \text{ meters} \tag{5.1}$$

In laboratory produced plasmas for example, electrically non-neutral sheaths of thickness $5L_D$-$10L_D$ form at the container walls. Curve $B$ of the figure is drawn for the assumed value $L_D = 1$ cm. Beyond this boundary then, vessels whose size is smaller than say $50L_D$ (50 cm) can still contain an ionized gas in which however, electrical non-neutrality is the rule rather than the exception.

In the South our domain borders on an area where weakly ionized plasmas are the norm. Here collisions of the charged particles with atoms or molecules are more frequent than collisions of the charged particles amongst themselves. Consider for example a hydrogen plasma with $\nu_{ea}$ as the collision frequency for momentum transfer between electrons and atoms and $\nu_{ei}$ as the collision frequency of electrons with ions. One then finds that the ratio of the respective collision frequencies for this gas is approximately

$$\frac{\nu_{ei}}{\nu_{ea}} \approx 23\left(\frac{N}{N_a}\right)\frac{\ln\Lambda}{T^{3/2}} \tag{5.2}$$

where $N_a$ is the density of atoms, $N$ the density of electrons, $T$ the temperature of the electrons in eV and $\Lambda$ the quantity defined by Eq. (2.28). Setting $\ln\Lambda = 10$, $T = 1$ eV we see that $(\nu_{ei}/\nu_{ea}) \approx 2.3 \times 10^2(N/N_a)$. Thus, when the degree of ionization is as little as one percent, Coulomb collisions have already overtaken electron-atom collisions. A second relation for the ratio $(N/N_a)$ can be obtained if one assumes that the plasma is in Saha equilibrium (most plasmas are not!), so that

$$\frac{N^2}{N_a} = 6.04 \times 10^{27}\,\frac{g_i}{g_a}\,T^{3/2}\,\exp(-e\phi_i/\kappa T)\ m^{-3} \tag{5.3}$$

where $g_i$ and $g_a$ are the statistical weights of ion and atom, $T$ is the temperature in eV and $\phi_i$ is the ionization potential. On combining this equation with Eq. (5.2) yields

$$N \simeq 1.4 \times 10^{29}\,\frac{g_i}{g_a}\,\ln\Lambda\,\frac{\nu_{ea}}{\nu_{ei}}\,\exp(-e\phi_i/\kappa T)\ m^{-3} \tag{5.4}$$

Curve $C$ of Fig. 5.1 is drawn by setting $(\nu_{ea}/\nu_{ei}) = 0.1$ and plotting $N$ as a function of $T$ from Eq. (5.4). This curve then serves in separating (at least in some cases) the region of weak ionization from the region of strong ionization. We point out that the kinetic theory of weakly ionized gases is a relatively neglected field of research despite the fact that a great many plasmas are in such a state. It is only recently [78] that the theoretical work of Rostoker and others for highly ionized media is beginning to be extended to the weakly ionized regime, and that some experimental work is also in progress [79].

Although the plasma may be highly ionized, frequent Coulomb collisions can make the Vlasov description inappropriate. For this not to happen the electron-ion collision frequency

$$\nu_{ei} \simeq 8.7 \times 10^{-2}\,\omega_{pe}\left(\frac{\ln 3N_D}{N_D}\right) \tag{5.5}$$

with $N_D$ as the number of particles in the Debye sphere

$$N_D = 1.72 \times 10^{12} \frac{[T(\text{eV})]^{3/2}}{[N(m^{-3})]^{1/2}} \tag{5.6}$$

must be much less than the frequency $\omega$ of the oscillations in question. [Sometimes one must also judge the importance of electron-electron collisions and if ions participate in the fluctuations, also ion-ion collisions]. In Figure 5.1 we see that the East and South-East are surrounded by collision dominated territory. Curve $D$ is plotted by requiring that $\nu_{ei}$ be equal to 1/100 of the electron plasma frequency $\omega_p$.

As one penetrates more deeply into the South-East corner of the map, one discovers almost virgin territory. Curve $E$ corresponds to the case where there is only one particle in the Debye sphere ($N_D = 1$); this is equivalent to saying that here the Debye length $L_D$ equals the interparticle distance $(4\pi N/3)^{-1/3}$. Plasmas in stellar interiors and plasmas produced by intense laser illumination of solid targets lie in this parameter range. Unfortunately little as yet is known about this important domain either in theory or in experiment (but see [80,81]).

When one explores the South-East corner even more deeply one comes across degenerate land — the curve $F$ is drawn at a value where the kinetic energy of an electron ($3\kappa T/2$) equals the Fermi energy

$$U_F = \frac{\hbar}{2m} (3\pi^2 N)^{2/3}. \tag{5.7}$$

## 6. APPENDIX

## SPECTROSCOPIC DETECTION OF WAVES: A Theoretical Description

*by* C. Deutsch

### I. *Introduction*

Usually the stochastic Stark effect arising from the thermal fluctuations is used to determine the electron number density, and also to a lesser extent the electron temperature [46]. The purpose of this note is to provide evidence for the detection of waves through their harmonic Stark effect in bound systems (ions or atoms) immersed in an ionized medium. In order to simplify the presentation, we shall take advantage of the fact that waves, in contradistinction to thermal fluctuations, are mostly observed in nonthermal situations, where atoms are able to coexist with high electron temperatures. As a result, we shall restrict ourselves to neutral emitters. Also, we shall put the emphasis on the time-dependent Stark effect and the corresponding informations provided by it through the modulations of the line shape of partially degenerate neutral lines as first emphasized by Baranger and Mozer in their pionneering paper [55]. A detailed account of the stochastic Stark effect of partially degenerate neutral lines may be found in the papers listed in Ref. 76. Also, we shall sketch out the possibilities afforded by a full (and not only perturbative) treatment of the time-dependent Stark effect in the presence of a steady magnetic field, along the lines of the work of Hicks et al [77]. The very interest in the methods we are presenting is the same as that one afforded by ordinary Stark effect. It provides a very elegant means of diagnostic that does not perturb the emitting plasma. Moreover, we shall see that atoms prove to be very sensitive detectors of the frequencies and amplitudes of electric fluctuations.

## II. *Perturbation Approach*

In order to get a straightforward and physically vivid approach to the nonthermal Stark effect, we shall follow the Beranger-Mozer perturbation approach recently extended to the fourth order of the atomic evolution operator expansion [57, 62].

So, let us consider a three-levels atom* (Fig. 4.10) with two closely spaced upper levels (1 and 2) and one lower level (0) in the perturbing electric field

$$E = \sum_k E_k \cos(\Omega t + \delta_k) \; ..... \tag{1}$$

where $E_k$ and $\delta_k$ are arbitrary amplitudes and phases respectively. Moreover, we make the initial assumption that the electric fields are uniform over the spatial extension of an excited atom.

---

*The nomenclature of section 4.3 and Fig. 4.10 is here replaced by the following notation: levels $i$, $j$, $k$ are here called 2, 1, 0, respectively; $\omega_{ij}$ is now called $\Delta$ and the frequency $\omega$ of the $rf$ field is here written as $\Omega$. The allowed transition ($j \to k$) is transition $1 \to 0$; the forbidden transition ($i \to k$) is now $2 \to 0$.

Further, we assume that only the transition 0-1 and 1-2 are allowed according to electric dipole selection rules ($\Delta\ell = \pm 1$, $\ell$ being the angular momentum quantum number). Then, the semiclassical interaction Hamiltonian between the atom and the plasma oscillation is

$$V(t) = - e\,\boldsymbol{r}\cdot\boldsymbol{E} = - e \sum_{\alpha = x,y,z;k} x_{\alpha k} \cos(\Omega t + \delta_k) \tag{2}$$

when $\boldsymbol{r}$ denotes the position operator of the optical electron. The spectrum in the vicinity of the 0-1 transition follows from

$$\mathcal{I}(\omega) = \mathrm{Re} \int_0^\infty dt\, e^{i\omega t} \langle U_{11}(t) \rangle_{Av} \tag{3}$$

$\omega$ being the measured frequency relative to the frequency for the unperturbed 0-1 transition. The average is over all $k$ modes, and the relevant matrix elements of the evolution operator $U$ (in the interaction representation) obey the time-dependent Schrödinger equation

$$\begin{aligned} \dot{U}_{11} &= \frac{1}{i\hbar} V_{12}(t)\, e^{i\Delta t}\, U_{21} \\ \dot{U}_{21} &= \frac{1}{i\hbar} V_{21}(t)\, e^{-i\Delta t}\, U_{11} \end{aligned} \tag{4}$$

subject to the initial conditions $U_{11}(0) = 1$, $U_{21}(0) = 0$, because the atom may be assumed to be either in state 1 or in state 2, initially.

The system (4) is now solved by iteration which yields

$$U_{11} = 1 + \left(\frac{1}{i\hbar}\right)^2 \int_0^t dt_1\, \tilde{V}_{12}(t_1) \int_0^{t_1} dt_2\, \tilde{V}^*_{21}(t_2) +$$

$$\left(\frac{1}{i\hbar}\right)^4 \int_0^t dt_1\, \tilde{V}_{21}(t_1) \int_0^{t_1} dt_2\, \tilde{V}^*_{21}(t_2) \int_0^{t_2} dt_3\, \tilde{V}_{12}(t_3)$$

$$\int_0^{t_3} dt_4\, \tilde{V}^*_{21}(t_4) + \ldots \tag{5}$$

when $\tilde{V}(t) = \exp(i\Delta t)\, V(t)$, and * denotes the complex conjugate. On substitution of Eq. (2) all integrations can be performed leading to

$$U_{11} = 1 + \left(\frac{e}{2\hbar}\right)^2 \sum_k (\boldsymbol{r}_{12}\cdot\boldsymbol{E}_k)(\boldsymbol{r}_{21}\cdot\boldsymbol{E}_k)\left\{\frac{\exp[i(\Delta+\Omega)t]}{(\Delta+\Omega)^2} + \frac{\exp[i(\Delta-\Omega)t]}{(\Delta-\Omega)^2}\right\}$$

$$+ \left(\frac{e}{2\hbar}\right)^4 \sum_{k,l} \Bigg[(\boldsymbol{r}_{12}\cdot\boldsymbol{E}_k)(\boldsymbol{r}_{21}\cdot\boldsymbol{E}_k)(\boldsymbol{r}_{12}\cdot\boldsymbol{E}_l)(\boldsymbol{r}_{21}\cdot\boldsymbol{E}_l) +$$

$$(\boldsymbol{r}_{12}\cdot\boldsymbol{E}_k)(\boldsymbol{r}_{21}\cdot\boldsymbol{E}_l)(\boldsymbol{r}_{12}\cdot\boldsymbol{E}_k)(\boldsymbol{r}_{21}\cdot\boldsymbol{E}_l)\left\{\frac{\exp[i(\Delta+\Omega)t]}{(\Delta+\Omega)^4}(\Delta/\Omega + 1) -\right.$$

$$\left.\frac{\exp[i(\Delta-\Omega)t}{(\Delta-\Omega)^4}(\Delta/\Omega - 1)\right\}\Bigg] + \ldots \tag{6}$$

where the phases and amplitudes are assumed to be random, and terms corresponding to the usual line broadening are omitted. The leading term 1 corresponds to the allowed line, all profiles being $\delta$ functions in the present approximation. The factors associated with $\exp[i(\Delta\pm\Omega)t$ thus give according to Eq. (3) the relative intensities $S_+$ and $S_-$ of far and near satellites respectively. Then it only remains to sum over modes and to evaluate the product of $\boldsymbol{r}$-matrix elements. In the first (Baranger-Moser) correction term, we can write:

$$\sum_k (\boldsymbol{r}_{12}\cdot\boldsymbol{E}_k)(\boldsymbol{r}_{21}\cdot\boldsymbol{E}_k) = r_{12}r_{21} \sum_k E_k^2 \cos^2\theta_k$$

$$= \frac{1}{2} R_{12} \sum_k E_k^2 = \frac{2}{3} R_{12} \langle E^2 \rangle \tag{7}$$

introducing the magnitudes of the two vectors and the angle $\theta_k$ between them, over which we average assuming spatial isotropy. The product of radial matrix elements is in the Hydrogenic approximation

$$R_{12} = \frac{\ell_>}{2\ell_>+1}\ \frac{9}{4}\, n^2(n^2-\ell_>^2)a_0^2 \tag{8}$$

$\ell_> = \max(\ell_1,\ell_2)$, $n$ being the main quantum number. Finally, $\langle E^2\rangle$ is the average over time and sum over modes $k$ of the square of the electric field oscillating with frequency $\Omega$. The fourth-order correction term is seen to involve products of $\cos^2\theta_k$ and $\sin^2\theta_k$ which are independent when the number of contributing modes is large, i.e. when the $k = \ell$ contribution is small; similarly evaluating the remaining sums and introducing $R_{12}$ in the 4th order term, we obtain Eqs. (4.64) of the main text. In the process, Fourier transforms are taken of Eq. (6) and $R_{12}$ is replaced by the quantity $R_{\ell\ell'} = R_{12}/a_0^2$.

## III. *Complete Treatment*

The above derivation is quite straightforward but it does not allow one to consider an arbitrary $\Omega$ because when $\Omega = \Delta$ expressions (6) and (4.64) blow up. Moreover, the perturbation approach is not able to account for an eventual shift of the satellites. If one wants to remove these shortcomings, one has to consider the atomic line shape arising from the time-dependent Stark effect in a more fundamental way. Such an approach is now available with the very recent and comprehensive work of Hicks et al. [77], who generalize a previous Autler and Townes' three-levels treatment [72].

The first and essential step in this method consists in the replacement of the time-dependent Schrödinger equation by an infinite set of algebraic equations. Therefore, let us consider

$$i\dot{\psi} = H(\boldsymbol{r},t)\psi = (H_0 + H_1 + H_2)\psi \tag{9}$$

with $H_0$ the unperturbed atomic Hamiltonian such that

$$H_0 U_j = \omega_j U_j, \; j = 1,2,3,\ldots .N \tag{10}$$

Now introduce a static magnetic field with $H_1$ such that

$$(H_0+H_1)U_j = \omega_j' U_j, \; \omega_j' = \omega_j + m_j\omega_L \tag{11}$$

where $\omega_L = 1/2\ |eB/m_e c|$ is the Larmor frequency. $B$ allows for more accurate specifications of the satellites Stark patterns, and will give additional support for the neglect of the thermal Stark effect. The oscillating electric part can now be written as

$$H_2(\boldsymbol{r},t) = H_2^+(\boldsymbol{r})\exp(i\omega t) + H_2^-(\boldsymbol{r})\exp(-i\omega t) \tag{12}$$

We now look for a general solution of the form

$$\psi(\boldsymbol{r},t) = \sum_{j=1}^{N} T_j(t)U_j(\boldsymbol{r}) \tag{13}$$

with $j$, running over the multiplicity spectrum, and

$$T_j(t) = \exp(-i\lambda t) \sum_{s=-\infty}^{\infty} C_{js} \exp(-is\omega t) \tag{14}$$

where $\lambda$ and $C_{js}$ are respectively real-valued and complex-valued scalars. Now, we are allowed to evaluate $S(\omega_\gamma)d\omega_\gamma d\Omega$, the number of photons emitted into the solid angle $d\Omega$ per second, with polarization $e_\gamma$ and frequency range $(\omega_\gamma, \omega_\gamma + d\omega_\gamma)$, in terms of the emission spectrum $S(\omega_\gamma)$

$$S_i^k(\omega_\gamma) = \frac{e^2\omega_\gamma}{2\pi\hbar c^3} \sum_{s=-\infty}^{\infty} \delta(\omega_k' + \omega_\gamma - \lambda_i - s\omega)$$

$$\times \sum_{j,j'=1}^{N} \omega_{j'k}' \omega_{j'k}' \xi_j^k \xi_{j'}^{k*} C_{js}^i C_{j's}^{i*} \tag{15}$$

for the transition $i \to k$, with

$$\xi_j^k \equiv \int d^3r\, U_k^* \,\hat{e}_j^* \cdot \mathbf{r} U_j, \quad \omega_{jk}' \equiv \omega_k' - \omega_j';$$

$\hat{e}_j$ is the polarization vector of the emitted photon, and $U_k$ (resp. $U_j$) the spatial part of the corresponding wave function.

Until now, no perturbation assumptions have been introduced, so that Eq. (15) may be considered as an exact result so far as the Stark spectrum of the Schrödinger Eq. (9) is taken as exact. Nevertheless, if one desires to obtain an accurate estimate without too much computational effort, it appears useful to introduce some smallness assumptions for the matrix elements $\beta_{j'i}^{\pm} = \langle j|H_2^{\pm}|j'\rangle$. So if the $i \to k$ transition is visualized as a two-photon process, with the first one $\pm\hbar\Omega$ exchanged with the oscillating field, Eq. (15) may be approximated by

$$S_i^k(\omega_\gamma) = \frac{e^2\omega_\gamma}{2\pi\hbar c^3} \delta(\omega_\gamma - \omega_{i'k}' - \omega_i^S \mp \omega) \sum_j \frac{\omega_{jk}'^2}{(\omega_{ij}' \pm \omega)^2} |\beta_{ji}^{\mp}|^2 |\xi_j^k|^2 \tag{16}$$

with

$$\omega_i^S = \sum_{j'} \left[\frac{|\beta_{j'i}^-|^2}{\omega_{ij'}' + \omega} + \frac{|\beta_{j'i}^+|^2}{(\omega_{ij'}' - \omega)}\right] \tag{17}$$

denoting a frequency-dependent Stark constant for the upper site $i$.

The indices $j$ and $j'$ run over all intermediate states allowed by the dipole electric selection rule ($\Delta\ell = \pm1$). The lower state $k$ is considered to be so tightly bound that one could neglect its Stark perturbations. Finally, as in the foregoing section, we are mainly interested in the ratio $S_+(\theta,\zeta)$ of the number of photons emitted per second into the solid angle $d\Omega$ through the forbidden transition $i \to k$ ($\Delta\ell = \pm2$) to the number of photons emitted in the allowed transition $j \to k$ ($\Delta\ell = \pm1$) written as

$$S(\theta,\zeta) = \frac{e^2 E_{rms}}{240\hbar^2} \frac{\ell}{2\ell - 1} \left|R_{n\ell}^{n'\ell-1}\right|^2 \frac{\int d\omega_\gamma\{[1]_{\pi F} + [2]_{\sigma F}\}}{\int d\omega_\gamma\{[1]_{\pi A} + [2]_{\sigma A}\}} \tag{18}$$

with $\theta$ = angular direction of observation relative to $B$
$\zeta$ = angle between the electric fluctuation and $B$
$\ell$ = angular quantum number of the upper state $i$.

The brackets indexed with $\pi$ and $\sigma$ respectively refer to a photon polarization parallel and perpendicular to $B$. They are given by

$$[1]_{\pi F} = \left|\frac{\cos^2\zeta}{(\Delta\pm\omega)^2}(\ ) + \frac{\sin^2\zeta}{(\Delta\pm\omega-\omega_L)^2}(\ ) + \frac{\sin^2\zeta}{(\Delta\pm\omega+\omega_L)^2}(\ )\right|_{\pi F} \tag{19}$$

$$[2]_{\sigma F} = \left|\frac{\cos^2\zeta}{(\Delta\pm\omega)^2}(\ ) + \frac{\sin\ \zeta}{(\Delta\pm\omega-\omega_L)^2}(\ ) + \frac{\sin^2\zeta}{(\Delta\pm\omega+\omega_L)^2}(\ )\right|_{\sigma F} \tag{20}$$

where ( ) are different functions of $\theta$ and the quantity $\varepsilon_{mij,mjk} \equiv \omega_{ik}' \omega_{jk}' \delta(\omega_\gamma - \omega_{ik}' - \omega_i^S + \omega)$. Also, the parameter $R_{n\ell}^{n'\ell-1}$ is defined as

$$R_{n\ell}^{n'\ell'} \equiv \int dr\ r^3 R_{n'\ell'}(r) R_{n\ell}(r)$$

and

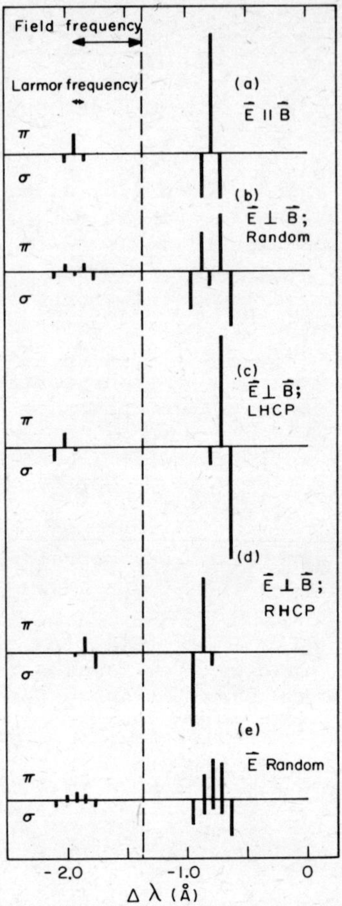


FIG. 1. Normal Zeeman patterns of the satellites of the forbidden transition $4\,^1F \rightarrow 2\,^1P$ of He I, calculated from perturbation theory and showing both $\sigma$ and $\pi$ polarization for various configurations of the electric field, all for direction of observation perpendicular to $\vec{B}$, field frequency of 2.35 cm$^{-1}$, and magnetic field strength of 7 kG. $\Delta\lambda = 0$ corresponds to the allowed line $4\,^1D \rightarrow 2\,^1P$; the vertical dashed line denotes the position of the forbidden $4\,^1F \rightarrow 2\,^1P$ transition. Shown are (a) $\vec{E}$ parallel to $\vec{B}$, (b) $\vec{E}$ perpendicular to $\vec{B}$ and random in azimuth, (c) $\vec{E}$ perpendicular to $\vec{B}$ and left-hand circularly polarized, (d) $\vec{E}$ perpendicular to $\vec{B}$ and right-hand circularly polarized, and (e) $\vec{E}$ random in direction.

$$[1]_{\pi A} + [2]_{\sigma A} = [\eta_1 \cos^2\theta + \eta_{-1}\sin^2\theta + 2\eta_0 \sin^2\theta]_{\pi A} + [\eta_1 + \eta_{-1}]_{\sigma A}$$

$$\eta_{mjk} = \omega'^{3}_{jk}\ \delta(\omega_\gamma - \omega_{jk} - \omega^S_{jk}) \qquad (21)$$

IV. *Stark Patterns of the Satellite* ($B \neq 0$)

As an example of the usefulness of Eq. (18) we reproduce in Fig. 1 the calculation of Hicks et al. [77] for the ratio of the intensities of components of the Zeeman pattern of the satellites of the forbidden line ($4^1F \rightarrow 2^1P$) to the intensity of the central $\pi$ component ($m_{jk} = 0$) of the 4922 Å *HeI* allowed line ($4^1D \rightarrow 2^1P$) for various configurations of electric and magnetic fields with, $\theta = \pi/2$, $\omega_{ij} = 5.63\ \text{cm}^{-1}$ ($4^1D \rightarrow 4^1F$ separation of *HeI*), $\omega = 2.35\ \text{cm}^{-1}$ (electric field frequency = 70.5 GHz), $\omega_L = 0.33\ \text{cm}^{-1}$ ($|B| = 7$ kG), and $E_{rms} = 1$ kV/cm.

The various spectra given in Fig. 1 clearly exhibit the dependence of the satellite line shape on the direction of the oscillating electric field taken with respect to *B*.

If *E* is entirely random in directions (see Fig. 1(e)) we must replace $\cos^2\zeta$ and $\sin^2\zeta$ in Eqs. (18) and (19) by their average values

$$\langle \sin^2\zeta \rangle_{AV} = (4\pi)^{-1} \int \sin^2\zeta\ d\Omega = \frac{2}{3}$$

$$\langle \cos^2\zeta \rangle_{AV} = (4\pi)^{-1} \int \cos^2\zeta\ d\Omega = \frac{1}{3}$$

With these substitutions Eq. (18) reduces in the limit $B \rightarrow 0$ to the expression for random fields ($S_\pm$) given by Baranger and Mozer (Eq. 4.60).

If $\omega \rightarrow 0$, the Zeeman patterns of the two satellites merge, the resulting pattern consisting of three $\pi$ components and five $\sigma$ components in the Zeeman pattern of the forbidden line produced by the quasistatic fields of the ions in a plasma.

## 7. REFERENCES

1. MONTGOMERY, D.C., *Theory of Unmagnetized Plasma*, Gordon and Breach, 1971.

2. LANDAU, L.D. and LIFSHITZ, E.M., *Electrodynamics of Continuous Media*, Pergamon Press and Addison-Wesley, 1960.

3. ALLIS, W.P., BUCHSBAUM, S.J. and BERS, A., *Waves in Anisotropic Plasmas*, M.I.T. Press, 1963.

4. LANDAU, L.D., *J. Physics (U.S.S.R.) 10*, 25 (1946).

5. SITENKO, A.G., *Electromagnetic Fluctuations in Plasma,* Academic Press, 1967.

6. FADDEYEVA, V. and TERENTEV, N.M., *Tables of Values of the Probability Integral in Complex Arguments,* G I T T L, Moscow, 1954.

7. FRIED, B.D. and CONTE, S.D., *The Plasma Dispersion Function*, Academic Press, 1961.

8. BHATNAGAR, P.L., GROSS, E.P. and KROOK, M., *Phys. Rev. 94*, 511 (1954).

9. LIEBERMAN, M.A. and BERS, A., Quarterly Progress Report No. 77, p. 141 (1965), Research Laboratory of Electronics, M.I.T.

10. MIX Jr., L.P., M.S. Thesis, Dept. of Physics (1969), M.I.T.

11. PINES, D. and BOHM, D., *Phys. Rev., 85*, 338 (1952).

12. SIGMAR, D.J. and JOYCE, G., *Nuclear Fusion, 11*, 447 (1971).

13. JOYCE, G., SIGMAR, D.J. and HINES, K.C., Oak Ridge National Laboratory Unpublished Report, 1967.

14. HALVERSON, W., Report No. EUR-CEA-FC-386 (1966) and Report No. EUR-CEA-FC-472 (1968), Association-Euratom-CEA, Fontenay-aux-Roses.

15. BUTLER, S.T. and BUCKINGHAM, J.J., *Phys. Rev., 126*, 1 (1962).

16. MAY, R.M., *Phys. Letters*, *29*, 577 (1969); also MAY, R.M. and CRAMER, N.F., *Phys. Letters*, *30A*, 10 (1969); also MAY, R.M., *Australian J. Phys.*, *22*, 687 (1969).

17. AONO, O., *J. Phys. Soc. Japan*, *17*, 853 (1962); KIHARA, T. and AONO, O., *J. Phys. Soc. Japan*, *18*, 837 (1963); ITIKAWA, Y. and AONO, O., *Phys. Fluids*, *9*, 1259 (1966); AONO, O., *Phys. Fluids*, *11*, 341 (1968); *11*, 832 (1968).

18. ORMROD, J.H., Report AECL-2669 (January 1967), Chalk River, Canada; also SMITH, M.R. and JOHNSON, W.B., Plasma Research Report No. A-38 (June 1965), Case Institute of Technology.

19. KLEVAN, I.L. and COX, D.M., BROWN Jr., H.H. and BEDERSON, B., *Phys. Rev. Letters*, *28*, 1254 (1972).

20. ROSE, D.J., *Nuclear Fusion*, *9*, 183 (1969); also papers by ROSE, D.J., PETRAVIC, L.K., PETRAVIC, M. and WATSON, C.; OLIPHANT, T., in the Conference on *Nuclear Fusion Reactors*, Culham, England, 1969.

21. TAVDGIRIDZE, T.L. and TSINTSADZE, N.L., *Soviet Physics (J.E.T.P.)*, *31*, 524 (1970).

22. ALIEV, Yu.M., GORBUNOV, L.M. and RAMAZASHVILI, R.R., *Soviet Physics (J.E.T.P.)*, *34*, 785 (1972).

23. ROSTOKER, N., *Nuclear Fusion*, *1*, 101 (1961).

24. HUBBARD, J., *Proc. Roy. Soc.*, *A 260*, 114 (1961).

25. LANDAU, L.D. and LIFSHITZ, E.M., *Statistical Physics*, Addison-Wesley, 1958.

26. WRIGHT, B.L. and BEKEFI, G., *Phys. Fluids*, *14*, 1764 (1971); also *14*, 1773 (1971).

27. TIDMAN, D.A. and DUPREE, T.H., *Phys. Fluids*, *8*, 1860 (1965).

28. CHAPPELL, W.R., COOPER, J. and SMITH, E.W., J.Q.S.R.T. *Journal Quant. Spectrosc. Rad. Transfer*, *10*, 1195 (1970).

29. BEKEFI, G., *Radiation Processes in Plasmas*, J. Wiley & Sons, 1966.

30. GERRY, E.T. and PATRICK, R.M., *Phys. Fluids*, *8*, 208 (1965).

31. ICHIMARU, S., PINES, D. and ROSTOKER, N., *Phys. Rev. Letters, 8*, 231 (1962).

32. ARUNASALAM, V. and BROWN, S.C., *Phys. Rev., 140*, A471 (1965).

33. AKHIEZER, A.I., AKHIEZER, I.A., POLOVIN, R.V., SITENKO, A.G. and STEPANOV, K.N., *Collective Oscillations in a Plasma*, Pergamon Press, 1967.

34. KUNZE, H.S., in *Plasma Diagnostics*, Lochte-Holtgreven (Ed.), North-Holland Press, 1968.

35. EVANS, J.V. and LOEWENTHAL, M., *Planetary Space Science, 12*, 915 (1964); PERKINS, F.W., SALPETER, E.E. and YNGVESSON, K.O., *Phys. Rev. Letters, 14*, 579 (1965); also *Phys. Rev., 139*, A55 (1965).

36. DAUGHNEY, C.C., HOLMES, L.S. and PAUL, J.W.M., *Phys. Rev. Letters, 25*, 497 (1970); also KEILHACKER, M. and STEUER, K.H., *Phys. Rev. Letters, 26*, 694 (1971); also AMAGISHI, Y., *Phys. Rev. Letters, 29*, 405 (1972).

37. KADOMTSEV, B., *Plasma Turbulence*, Academic Press, 1965.

38. TIDMAN, D.A., *Planetary Space Science, 13*, 781 (1965).

39. MERCIER, R.P., *Proc. Phys. Soc., 83*, 819 (1964).

40. HIRSHFIELD, J.L., RATNER A.S. and D.L. De ANGELIS, *5th Europ. Conf. Controlled Fusion and Plasma Physics*, Grenoble 1972, page 172.

41. DAWSON, J. and OBERMAN, C., *Phys. Fluids, 6*, 394 (1963).

42. DAWSON, J. and OBERMAN, C., *Phys. Fluids, 5*, 517 (1962); also BIRMINGHAM, T., DAWSON, J. and OBERMAN, C., *Phys. Fluids, 8*, 297 (1965).

43. BEKEFI, G., HIRSHFIELD, J.L. and BROWN, S.C., *Phys. Fluids, 4*, 173 (1961).

44. SKARSGARD, M.H., STOLL, I. and WARNOCK, A.C.C., *Phys. Rev. Letters, 23*, 1023 (1969).

45. HEYMANN, P., *Beitr. Plasma Physik, 8*, 233 (1968); NOON, J.H., BLACZAK, P.R. and HOLT, E.H., *J. Appl. Phys., 39*, 9 (1968); GREGORY, B.C., BERGEVIN, B. and DROUGHET, M.G., *Tenth Int. Conf. Phenomena in Ionized Gases*, Oxford, 1971. Paper 5.1.

46. GRIEM, H.R., *Plasma Spectroscopy*, McGraw Hill, 1964.

47. HOLTSMARK, J., *Ann. Physik, 58*, 577 (1919).

48. GEORGE, E.V., BEKEFI, G. and YA'AKOBI, B., *Phys. Fluids, 14*, 2708 (1971).

49. YA'AKOBI, B., GEORGE, E.V., BEKEFI, G. and HAWRYLUK, R.J., *J. Physics* (Atomic Physics, Part B) 1972.

50. BEKEFI, G., GEORGE, E.V. and YA'AKOBI, B., Quarterly Progress Report, M.I.T., Research Laboratory of Electronics No. 103, p. 86, October 1971.

51. GRIEM, H.R. and KUNZE, H.J., *Phys. Rev. Letters, 23*, 1279 (1969).

52. BEN-YOSEF, N. and RUBIN, A.G., *Phys. Letters, 33A*, 222 (1970).

53. LEVINE, M.A. and GALLAGHER, C.C., *Phys. Letters, 32A*, 14 (1970).

54. MATT, D.R. and SCOTT, F.R., *Phys. Fluids, 15*, 1047 (1972); see also KEILHACKER, M., KORNHERR, M., NIEDERMEYER, H., STEUER, K.H. and CHODURA, R., Fourth Conf. Plasma Physics and Controlled Nuclear Fusion Research, Madison, Wisconsin, June 1971, Paper CN-28/J-10

55. BARANGER, M. and MOZER, B., *Phys. Rev., 123*, 25 (1961).

56. REINHEIMER, J., *J. Quant. Spectr. and Radiative Transfer, 4*, 671 (1964).

57. GRIEM, H.R., *Notes on Broadening of Spectral Lines by Charges Particles*, unpublished.

58. COOPER, III, W.S. and RINGLER, H., *Phys. Rev., 179*, 226 (1969).

59. COOPER, III, W.S. and HESS, R.A., *Phys. Rev. Letters, 25*, 433 (1970).

60. YA'AKOBI, B. and BEKEFI, G., *Phys. Letters, 30A*, 539 (1969).

61. KUNZE, H.J. and GRIEM, H.R., *Phys. Rev. Letters, 21*, 1048 (1968).

62. KUNZE, H.J., GRIEM, H.R., De SILVA, A.W., GOLDENBAUM, G.C. and SPALDING, I.J., *Phys. Fluids, 12*, 2669 (1969).

63. KEILHACKER, M., KORNHERR, M., NIEDERMEYER, H., STEUER, K.H. and CHODURA, R., Fourth Conf. Plasma Physics and Controller Nuclear Fusion Research, Madison, Wisconsin, June 1971; Paper CN-28/J-10

64. ZELENIN, G.V., KUTSYN, A.A., MAZNICHENKO, M.E., PAVLICHENKO, O.S. and SUPRUNENKO, V.A., *J.E.T.P. (Soviet Physics), 31*, 1009 (1970).

65. BEN-YOSEF, N. and RUBIN, A.G., *J. Quant. Spect. Radiative Transfer, 11*, 1 (1971).

66. SCOTT, F.R., NEIDIGH, R.V., McNALLY, J.R. and COOPER, III, W.S., *J. Appl. Phys., 41*, 5327 (1970).

67. ALEXEFF, I., ESTABROOK, K., HIROSE, A., JONES, W.D., NEIDIGH, R.V., OLSEN, J.N., SCOTT, F.R., STIRLING, W.L., WIDNER, M.M. and WING, W.R., *Phys. Rev. Letters, 25*, 848 (1970).

68. BEREZIN, A.B., DUBOVOY, L.B. and LYUBLIN, B.V., Fourth Conf. Plasma Physics and Controlled Nuclear Fusion Research, Madison, Wisconsin, June 1971, Paper CN-28/E-8

69. BARAVIAN, G., BENATTAR, R., BRETAGNE, J., GODART, J.L. and SULTAN, G., *Phys. Letters, 30A*, 198 (1969).

70. BURGESS, D.D., *J. Phys.*, B (Atomic and Molecular Physics) *4*, L7 (1971).

71. BLOCKINZEW, D.I., *Phys. Z. S owjetunion, 4*, 501 (1933).

72. AUTLER, S.H. and TOWNES, C.H., *Phys. Rev., 100*, 703 (1955); also, TOWNES, C.H. and SHAWLOW, A.L., *Microwave Spectroscopy*, McGraw-Hill, New York, 1955, Chap. 10.

73. COOPER, III, W.S., This figure has been kindly lent to G.B. from Dr. Cooper's extensive collection of data.

74. GALLAGHER, C.C. and LEVINE, M.A., *Phys. Rev. Letters, 27*, 1693 (1971).

75. COHN, A., BAKSHI, P. and KALMAN, G., *Phys. Rev. Letters, 29*, 324 (1972).

76. For a thorough discussion of partially overlapping neutral lines, see: GRIEM, H.R., *Astrophys. J.*, *154*, 111 (1968); DEUTSCH, C., HERMAN, C. and DRAWIN, H.W., *Phys. Rev.*, *178*, 261 (1969) and *186*, 204 (1969); BARNARD, A.J., COOPER, J. and SHAMEY, L.J., *Astron. Astrophys.*, *1*, 28 (1969); COULAUD, G., DEUTSCH, C. and SASSI, M., *Phys. Rev. A*, September 1972.

77. HICKS, W.W., HESS, R.A. and COOPER, III, W.S., *Phys. Rev.*, *5*, 490 (1972).

78. WILLIAMS, R.H. and CHAPPELL, W.R., *Phys. Fluids*, *14*, 591 (1971); *15*, 1059 (1972).

79. GEISSLER, K.H., GREENWALD, R.A. and CALVERT, W., *Phys. Fluids*, *15*, 96 (1972).

80. COHEN, E.G.D. and MURPHY, T.J., *Phys. Fluids*, *12*, 1404 (1969); *13*, 216 (1970); DeWITT, H.E., *Phys. Rev.*, *140A*, 466 (1965); BRUSH, S.G., SAHLIN, H.L.S. and TELLER, E., *J. Chem. Phys.*, *45*, 2102 (1966).

81. KATO, M., *Phys. Fluids*, *15*, 460 (1972).

# LINEAR WAVES AND INSTABILITIES*

ABRAHAM BERS

*Department of Electrical Engineering and Research Laboratory of Electronics, Massachusetts Institute of Technology, Cambridge, Massachusetts, U.S.A.*

*This work was supported in part by the National Science Foundation (Grant GK-28282X1), and in part by the Atomic Energy Commission (Contract AT(11-1)3070).

CONTENTS

# 0. INTRODUCTION

The <u>macroscopic</u> dynamics of highly-ionized plasmas is a complex many-body problem involving the interaction of electromagnetic fields and charged particles in a continuum. The large variety of phenomena that can be exhibited by a plasma are not usefully described by any one single model for the dynamics of the particles, but rather by a variety of models, each one suited for bringing out the important physics of a particular class of plasma phenomena. On the other hand, electromagnetic fields are described in all cases by Maxwell's equations. Hence, in seeking a unified description of macroscopic plasma phenomena it is natural to choose an electrodynamic description, i.e., the set of Maxwell's equations in which the dynamics of these particles is represented by current densities and charge densities that are functions of the electric and magnetic fields. This description of a plasma has its obvious drawbacks in that it emphasizes the dynamics of the electromagnetic fields rather than the dynamics of the charged particles which are acted upon by, and produce, these fields. Nevertheless, most of the high-frequency and short wavelength linear, as well as weakly non-linear, <u>coherent</u> wave phenomena are best described in this unified manner, independent of plasma model. Turbulence phenomena and details of wave-particle interactions will not be treated in this chapter.

Section I outlines the electrodynamic equations for small-amplitude waves and their dispersion relation in a homogeneous plasma. For such waves, energy and momentum, and their flow and transformation, are described in Section II, which brings out the concept of an active medium, i.e., one with potential for instability. Perturbation theory of waves is treated in Section III, where it is applied to linear coupling of waves, and the resulting instabilities from such interactions between active and passive waves. Linear stability analysis in time and space is described in Section IV, where the time-asymptotic, time-space Green's function for an arbitrary dispersion relation is developed. Section V applies the perturbation theory of waves (Section III) to nonlinear coupling, with particular emphasis on pump-driven (parametric) interactions of waves. Details of the time-space evolution of instabilities due to coupling (either linear, Section III, or nonlinear, Section V) are given in Appendix 1. Finally, since most of the chapter is developed for homogeneous plasmas, Appendix 2 summarizes extensions of the formalisms to weakly inhomogeneous plasmas.

The contents of this chapter differ from the lectures presented to students at the 1972 Les Houches School. In the lectures the material was presented by sketching the general formalism and emphasizing the detailed application of the analytical techniques to specific examples and plasma models. In contrast, here we shall take the opposite approach and emphasize the formal structure for studying wave propagation, coupling, and instabilities, independent of plasma model or specific example one may choose. Neither will be claimed to have been done in a complete or satisfactory manner. Specific, "important" examples of these aspects of plasma physics are too numerous and varied in physical content to be dealt with in a short write-up such as this, and specialized reviews in journals and books have appeared in abundance (see references at end of this chapter).

# I. WAVES IN PLASMAS

## A. *Linear Response Functions for a Homogeneous Plasma*

In an electrodynamic description of plasmas, it is convenient to distinguish two types of response functions. The first type of response function gives the electric current density as a function of (in general an operator on) the electromagnetic fields. This response function depends upon the model which one chooses as appropriate for describing the plasma particle dynamics. We shall call this the internal response function. This response function and Maxwell's equations give a self-consistent electrodynamic description of the plasma. This overall description of the plasma can also be characterized by a response function which describes the natural and driven responses due to external electromagnetic fields, or external current and charge densities. These ideas are schematically illustrated in Fig. IA-1.

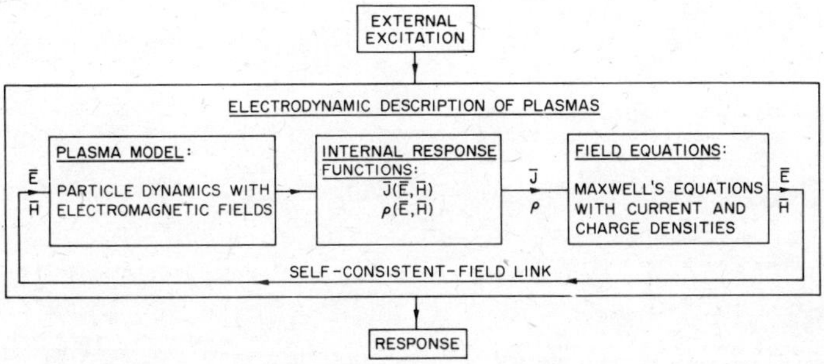


Fig. I A-1

For a linearized description of a homogeneous plasma the internal response function must be of the general form of a convolution integral in space and time. Thus, for example, the current density is related to the electric field by

$$\bar{J}(\bar{r},t) = \int d^3r' \int dt' \,\bar{\bar{\sigma}}(\bar{r}',t')\cdot\bar{E}(\bar{r}-\bar{r}',t-t') \qquad \text{(IA-1)}$$

where $\overline{\overline{\sigma}}(\bar{r},t)$ is the conductivity tensor (internal response function) which expresses the dependence of the current upon the time-space "history" of the electric field. The integrations in Eq. IA-1 must be performed with due regard to causality and relativity. The latter makes the spatial and temporal integrations interdependent. Another useful internal response function is the dielectric tensor, which relates the electric displacement vector to the electric field,

$$\bar{D}(\bar{r},t) = \int d^3r' \int dt' \; \overline{\overline{\epsilon}}(\bar{r}',t') \cdot \bar{E}(\bar{r}-\bar{r}',t-t') \qquad \text{(IA-2)}$$

The response functions in Eqs. IA-1 and IA-2 are related since for a plasma we define $\bar{D}$ so that

$$\frac{\partial \bar{D}}{\partial t} = \epsilon_o \frac{\partial \bar{E}}{\partial t} + \bar{J} \qquad \text{(IA-3)}$$

taking the appropriate Fourier transforms in space and time of Eqs. IA-1 and IA-2 we have

$$\bar{J}(\bar{k},\omega) = \overline{\overline{\sigma}}(\bar{k},\omega) \cdot \bar{E}(\bar{k},\omega) \qquad \text{(IA-4)}$$

$$\bar{D}(\bar{k},\omega) = \overline{\overline{\epsilon}}(\bar{k},\omega) \cdot \bar{E}(\bar{k},\omega) \qquad \text{(IA-5)}$$

and using Eq. IA-3 we find

$$\overline{\overline{\epsilon}}(\bar{k},\omega) = \epsilon_o \left( \overline{\overline{\delta}} + i \frac{\overline{\overline{\sigma}}(\bar{k},\omega)}{\omega \epsilon_o} \right) \equiv \epsilon_o \overline{\overline{K}}(\bar{k},\omega) \qquad \text{(IA-6)}$$

where $\overline{\overline{\delta}}$ is the identity tensor and $\overline{\overline{K}}$ is the normalized dielectric tensor.

Since Maxwell's equations together with Eq. IA-1 are also linear and homogeneous, we can deduce that the natural responses of a plasma must satisfy an integral equation of the form

$$\int d^3r' \int dt' \; \overline{\overline{D}}(\bar{r}',t') \cdot \bar{E}(\bar{r}-\bar{r}',t-t') = 0 \qquad \text{(IA-7)}$$

where $\overline{\overline{D}}(\bar{r},t)$ is the dispersion tensor. The space-time Fourier transform of Eq. IA-7 gives

$$\overline{\overline{D}}(\bar{k},\omega) \cdot \bar{E}(\bar{k},\omega) = 0 \qquad \text{(IA-8)}$$

which for nontrivial solutions requires

$$\det[\overline{\overline{D}}(\bar{k},\omega)] = D(\bar{k},\omega) = 0 \qquad \text{(IA-9)}$$

Equation IA-9 is the dispersion relation, giving $\omega = \omega(\bar{k})$ which describes the dependence of the frequency upon wave vector. Thus the natural responses of the plasma can be written as a superposition of complex waveforms exp $i(\bar{k}\cdot\bar{r}-\omega t)$ with $\omega$ and $\bar{k}$ related by the dispersion relation Eq. IA-9. This superposition becomes clearly defined when we consider the external excitation of a plasma. The response of the plasma to an external source is described by

$$\bar{\bar{D}}(\bar{k},\omega)\cdot\bar{E}(\bar{k},\omega) = \bar{S}_{ext}(\bar{k},\omega) \qquad \text{(IA-10)}$$

where $\bar{S}_{ext}(\bar{k},\omega)$ is the space-time Fourier transform of the source vector. Solving for the electric field gives

$$\bar{E}(\bar{r},t) = \int\frac{d\omega}{2\pi}\, e^{-i\omega t}\int\frac{d^3k}{(2\pi)^3}\, e^{i\bar{k}\cdot\bar{r}}\, \frac{\bar{\bar{D}}_{adj}\cdot\bar{S}_{ext}}{D(\bar{k},\omega)} \qquad \text{(IA-11)}$$

Hence, it is convenient to introduce a Green's function for the plasma

$$G(\bar{r},t) = \int\frac{d\omega}{2\pi}\, e^{-i\omega t}\int\frac{d^3k}{(2\pi)^3}\, e^{i\bar{k}\cdot\bar{r}}\, \frac{1}{D(\bar{k},\omega)} \qquad \text{(IA-12)}$$

which can be said to describe the natural responses of the plasma, and gives meaning to the dispersion function $D(\bar{k},\omega)$ of Eq. IA-9. Equations IA-11 and IA-12 require that the integration contours be specified so that causality is satisfied. We shall postpone this until chapter IV where Eq. IA-12 forms the basis of our stability analysis.

The dispersion tensor of Eq. IA-8, and the dispersion function of Eq. IA-9, can be expressed in terms of the dielectric tensor of Eq. IA-5. Using Maxwell's equations

$$\nabla \times \bar{E} = -\frac{\partial\bar{B}}{\partial t} \qquad \text{(IA-13)}$$

$$\nabla \times \bar{H} = \frac{\partial\bar{D}}{\partial t} \qquad \text{(IA-14)}$$

where $\bar{B} = \mu_o\bar{H}$ and $\bar{D}$ is given by Eqs. IA-1 and IA-3, we find their space-time Fourier transform

$$\bar{k} \times \bar{E} = \omega\mu_o\bar{H} \qquad \text{(IA-15)}$$

$$\bar{k} \times \bar{H} = -\omega\bar{\bar{\epsilon}}(\bar{k},\omega)\cdot\bar{E} \qquad \text{(IA-16)}$$

Eliminating $\bar{H}$ we obtain

$$\bar{k} \times (\bar{k} \times \bar{E}) + \frac{\omega^2}{c^2} \bar{\bar{K}}(\bar{k},\omega) \cdot \bar{E} = 0 \qquad \text{(IA-17)}$$

or in the form of Eq. IA-8

$$\left[\frac{c^2}{\omega^2} (\bar{k}\,\bar{k} - k^2\,\bar{\bar{\delta}}) + \bar{\bar{K}}(\bar{k},\omega)\right] \cdot \bar{E} = 0 \qquad \text{(IA-18)}$$

For later convenience we have thus chosen our dispersion tensor to be

$$\bar{\bar{D}}(\bar{k},\omega) = \frac{c^2}{\omega^2} (\bar{k}\,\bar{k} - k^2\,\bar{\bar{\delta}}) + \bar{\bar{K}}(\bar{k},\omega) \qquad \text{(IA-19)}$$

and the dispersion relation as

$$D(\bar{k},\omega) = \det\left[ \frac{c^2}{\omega^2} (\bar{k}\,\bar{k} - k^2\,\bar{\bar{\delta}}) + \bar{\bar{K}}(\bar{k},\omega)\right] \qquad \text{(IA-20)}$$

So far we have discussed an electrodynamic description of a plasma based on the full set of Maxwell's equations. This is frequently unnecessarily complicated. Various approximate descriptions are often physically more illuminating. We shall make use of the quasi-electrostatic (QES) approximation and the magneto-hydrodynamic (MHD) approximation. In the former the magnetic field is considered small and the electric field is self-consistently related to the plasma charge density through Poisson's equation. In the MHD approximation, which is useful only for low frequencies and long wavelengths, the plasma current density and the magnetic field are linked through Ampere's equation, and the electric field is assumed small. In this case the formulation of a conductivity tensor, Eq. IA-4, although still valid, emphasizes the electric field which is small. We shall therefore find it more convenient to use a different formulation.

In the *QES approximation* we assume

$$\nabla \times \bar{E} \approx 0 \qquad \text{(IA-21)}$$

and hence

$$\bar{E} \approx -\nabla\Phi \qquad \text{(IA-22)}$$

Using Gauss' law

$$\nabla \cdot \bar{D} = 0 \qquad \text{(IA-23)}$$

together with the Fourier transformed fields and Eq. (IA-5), we obtain

$$\bar{k} \cdot \bar{\bar{\epsilon}}(\bar{k},\omega) \cdot \bar{k}\,\Phi = 0 \qquad \text{(IA-24)}$$

It is convenient to introduce the longitudinal dielectric function

$$\epsilon_\ell(\bar{k},\omega) \equiv \bar{k} \cdot \bar{\bar{\epsilon}}(\bar{k},\omega) \cdot \bar{k}/k^2 \qquad \text{(IA-25)}$$

or its normalized form $K_\ell(\bar{k},\omega)=\epsilon_\ell(\bar{k},\omega)/\epsilon_o$ Equation (IA-24), for nontrivial solutions to $\Phi$, requires

$$K_\ell(\bar{k},\omega) = 0 \qquad \text{(IA-26)}$$

which is the QES dispersion relation. The QES approximation assumes that the electric field is entirely along the direction of the wavevector. In general, as Eqs. (IA-15 and 16) show, the electric field has components both along ($\bar{E}_\ell$) and across ($\bar{E}_t$) the wavevector; this is shown schematically in Fig. IA-2.

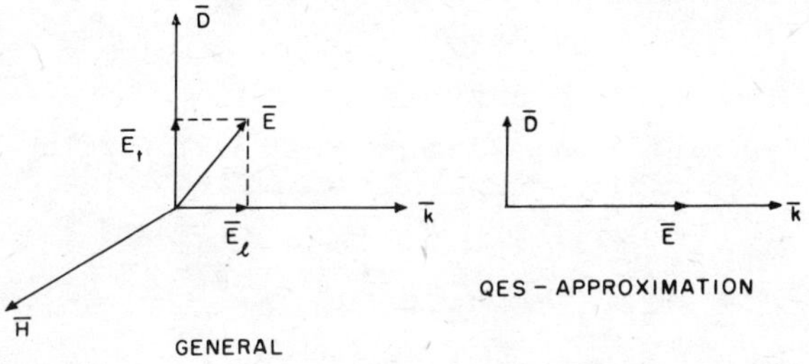


Fig. I A-2

Using the general dispersion relation Eq. (IA-18) we find that these components of $\bar{E}$ are related by

$$\left[\frac{c^2}{\omega^2} k^2 \bar{\bar{\delta}} - \bar{\bar{K}}(\bar{k},\omega)\right] \cdot \bar{E}_t = \bar{\bar{K}} \cdot \bar{E}_\ell \qquad \text{(IA-27)}$$

Hence, if

$$k^2 >> \frac{\omega^2}{c^2} \left|K_{ij}(\bar{k},\omega)\right| \qquad \text{(IA-28)}$$

the component of Eq. (IA-27) along $\bar{k}$ gives the QES dispersion relation, Eq. (IA-26). Equation (IA-28) provides a condition for the validity of the QES approximation: it is valid for sufficiently slow waves provided they occur in $(\omega,\bar{k})$ regions where the dielectric tensor elements are not too large, i.e. not too near a resonance. Under these conditions, Eq. (IA-27) also provides the correction to the QES electric field, i.e. a transverse electric field

$$\bar{E}_t \cong \frac{\omega^2}{c^2k^2}\,\bar{\bar{K}}\cdot\bar{E}_\ell \qquad \text{(IA-29)}$$

and through Faraday's law, Eq. (IA-15) and Eq. (IA-29) a magnetic field.

## II. CONSERVATION EQUATIONS

Associated with wave propagation is also a transport of energy and momentum. We shall study these within the framework of the small-amplitude equations for a plasma. This is particularly useful for understanding instabilities in plasmas that result from perturbation or coupling of waves.

### A. *Energy and Power*

In electrodynamics energy and power are related by Poynting theorem, which follows from Maxwell's equations,

$$\nabla\cdot(\bar{E}\times\bar{H}) + \frac{\partial}{\partial t}\left(\frac{1}{2}\mu_o H^2 + \frac{1}{2}\epsilon_o E^2\right) = -\bar{E}\cdot\bar{J} \qquad \text{(IIA-1)}$$

In general the right-hand side of this equation must be interpreted to express not only the dissipation or generation of electrical energy, but contains also the energy and energy flow associated with the medium. For the general, linear electrodynamic description of a plasma, as given in Sec. IA, it is not possible to derive a conservation equation for the instantaneous and local energy density. This is so because in general the linear current is not related to the electric field in a local and instantaneous manner (see Eq. IA-1). We therefore seek a conservation relation for the time-averaged energy and its flow. In order for these to be well-defined we must restrict ourselves to waves that are only weakly damped or growing, i.e. $\omega$ and $\bar{k}$ to have imaginary parts that are small compared to their real parts ($|\omega_i|\ll|\omega_r|$ and $|\bar{k}_i|\ll|\bar{k}_r|$). Hence, from the dispersion tensor, Eqs. (IA-18) and (IA-19), we shall need that

$$|\epsilon^a_{\alpha\beta}(\bar{k}_r,\omega_r)| \ll |\epsilon^h_{\alpha\beta}(\bar{k}_r,\omega_r)| \qquad \text{(IIA-2)}$$

and, by Eq. (IA-6),

$$|\sigma^h_{\alpha\beta}(\bar{k}_r,\omega_r)| << |\sigma^a_{\alpha\beta}(\bar{k}_r,\omega_r)| \qquad \text{(IIA-3)}$$

where we have let $\bar{\bar{\epsilon}}=\bar{\bar{\epsilon}}^h+i\bar{\bar{\epsilon}}^a$ and $\bar{\bar{\sigma}}=\bar{\bar{\sigma}}^h+i\bar{\bar{\sigma}}^a$ From Eqs. (IIA-1) and (IA-4) we see that this implies that the time-averaged electrical power dissipated or generated by the medium

$$p = \frac{1}{2}\,\bar{E}^* \cdot \bar{\bar{\sigma}}^h\,(\bar{k}_r,\omega_r) \cdot \bar{E} \qquad \text{(IIA-4)}$$

be small.

In order to obtain a Poynting-type equation for the total time-averaged energy we need a dynamic set of equations for the complex field amplitudes. We can obtain this by allowing the wave field amplitudes to be slowly varying functions of space and time:

$$\bar{E}(\bar{r},t) = \mathrm{Re}\,[\bar{\mathscr{E}}(\bar{r},t)\,e^{i(\bar{k}\cdot\bar{r}-\omega t)}] \qquad \text{(IIA-5)}$$

$$\bar{H}(\bar{r},t) = \mathrm{Re}\,[\bar{\mathscr{H}}(\bar{r},t)\,e^{i(\bar{k}\cdot\bar{r}-\omega t}] \qquad \text{(IIA-6)}$$

where

$$\left|\frac{\partial\bar{\mathscr{E}}}{\partial t}\right| << |\omega\bar{\mathscr{E}}| \qquad \text{(IIA-7)}$$

$$\left|\frac{\partial\mathscr{E}_\alpha}{\partial\bar{r}}\right| << |\bar{k}\mathscr{E}_\alpha| \qquad \text{(IIA-8)}$$

and similarly for the variation of $\mathscr{H}$. The current density is related to the electric field by Eq. (IA-1). Using Eq. (IIA-5) and expanding $\bar{\mathscr{E}}$ up to first order in its variation we obtain

$$\bar{J}(\bar{r},t) = \mathrm{Re}\left\{\left[\bar{\bar{\sigma}}(\bar{k},\omega)\cdot\bar{\mathscr{E}} + \frac{\partial\bar{\bar{\sigma}}}{\partial\omega}\cdot\frac{\partial\bar{\mathscr{E}}}{\partial t} - i\,\frac{\partial\sigma_{\bar{\alpha}\beta}}{\partial\bar{k}}\cdot\frac{\partial\mathscr{E}_\beta}{\partial\bar{r}}\right]e^{i(\bar{k}\cdot\bar{r}-\omega t)}\right\} \qquad \text{(IIA-9)}$$

where in the last term the repeated indices are to be summed, the bar over the index indicates the vector components of the expression, and the dot product is of the vectors $\partial/\partial\bar{k}$ and $\partial/\partial\bar{r}$. Substituting Eqs. (IIA-5), (IIA-6), and (IIA-9) into Maxwell's equations, and separating the orders in accord with

Eqs. (IIA-3), (IIA-7) and (IIA-8) we obtain

$$\bar{k} \times \bar{\mathscr{E}} - \omega\mu_o \bar{\mathscr{H}} = 0 \tag{IIA-10}$$

$$\bar{k} \times \bar{\mathscr{H}} + \omega\bar{\bar{\epsilon}}^h(\bar{k},\omega) \cdot \bar{\mathscr{E}} = 0 \tag{IIA-11}$$

and

$$\nabla \times \bar{\mathscr{E}} + \mu_o \frac{\partial \bar{\mathscr{H}}}{\partial t} = 0 \tag{IIA-12}$$

$$\nabla \times \bar{\mathscr{H}} - \frac{\partial \omega \bar{\bar{\epsilon}}^h}{\partial \omega} \cdot \frac{\partial \bar{\mathscr{E}}}{\partial t} + \omega \frac{\partial \epsilon^h_{\alpha\beta}}{\partial \bar{k}} \cdot \frac{\partial \mathscr{E}_\beta}{\partial \bar{r}} - \bar{\bar{\sigma}}^h \cdot \bar{\mathscr{E}} = \bar{\mathscr{J}}_{ex} \tag{IIA-13}$$

where $\bar{\mathscr{J}}_{ex}$ is a small perturbing external current density, and the small damping or growth effects due to $\bar{\bar{\sigma}}^h$ have been associated with the slow variations of the fields. Equations (IIA-10) and (IIA-11) lead to the dispersion relation (Eq. (IA-19)) for the modes with $\omega$ and $\bar{k}$ real. The slowly varying amplitudes of these modes are determined by Eqs. (IIA-12) and (IIA-13). Dot-multiplying Eq. (IIA-12) by $-\bar{\mathscr{E}}$ and Eq. (IIA-13) by $\bar{\mathscr{H}}^*$, and adding these together with their complex conjugates we find

$$\frac{\partial w}{\partial t} + \nabla \cdot \bar{s} + p = p_{ex} \tag{IIA-14}$$

where

$$w = \frac{1}{4}\mu_o |\bar{\mathscr{H}}|^2 + \frac{1}{4}\bar{\mathscr{E}}^* \cdot \frac{\partial \omega \bar{\bar{\epsilon}}^h}{\partial \omega} \cdot \bar{\mathscr{E}}. \tag{IIA-15}$$

$$\bar{s} = \text{Re}\, \frac{1}{2}\bar{\mathscr{E}} \times \bar{\mathscr{H}}^* - \frac{\omega}{4} \frac{\partial \epsilon^h_{\alpha\beta}}{\partial \bar{k}} \mathscr{E}^*_\alpha \mathscr{E}_\beta \tag{IIA-16}$$

$$p = \frac{1}{2}\bar{\mathscr{E}}^* \cdot \bar{\bar{\sigma}}^h \cdot \bar{\mathscr{E}} = \frac{\omega}{2} \bar{\mathscr{E}}^* \cdot \bar{\bar{\epsilon}}^a \cdot \bar{\mathscr{E}} \tag{IIA-17}$$

(IIA-18)

are respectively the time-averaged energy density, power flow density, density of power dissipated, and density of power supplied by the external current, to be evaluated for $\omega$ and $\bar{k}$ real. Using Eq. (IIA-10) we can eliminate the $\mathcal{H}$-field, and using the dispersion tensor introduced in Eqs. (IA-8) and (IA-19), we can express the time-averaged energy density as

$$w = \frac{1}{4\omega} \bar{\mathcal{E}}^* \cdot \frac{\partial \omega^2 \bar{\bar{\epsilon}}^h}{\partial \omega} \cdot \bar{\mathcal{E}}$$

$$= \frac{\omega \epsilon_o}{4} \bar{\mathcal{E}}^* \cdot \frac{\partial \bar{\bar{D}}^h}{\partial \omega} \cdot \bar{\mathcal{E}} \qquad \text{(IIA-19)}$$

We note that w may be either positive or negative, depending upon the positive definite or negative definite character of the tensor $(\partial/\partial\omega)(\omega^2\bar{\bar{\epsilon}}^h)$, or equivalently of $(\partial/\partial\omega)(\bar{\bar{D}}^h)$. If we think of Eq. (IIA-14) as describing the excitation of the wave by an external source, the difference between a positive and negative energy wave is that the former requires that the external source supply power to the medium while the latter requires that the external source extract power from the medium. Similarly we find the group velocity

$$\bar{v}_g = \frac{\partial \omega}{\partial \bar{k}} = \frac{-\dfrac{\partial D}{\partial \bar{k}}}{\dfrac{\partial D}{\partial \omega}} = \frac{\bar{s}}{w} \qquad \text{(IIA-20)}$$

being equal to the energy velocity and thus relating $\bar{s}$ and w. A wave with positive group velocity will have its power flow vector in the direction of the group velocity if its energy is positive, and in the opposite direction of the group velocity if its energy is negative.

Equation (IIA-14) with $p_{ex}$=0 may be specialized to the case where the slow variation is due to the dissipation or growth mechanisms of the medium as expressed by the small $\bar{\bar{\sigma}}^h$ or $\bar{\bar{\epsilon}}^a$. We thus find the small damping (or growth) rate of the mode ($\bar{k}$ real, $\omega=\omega_r - i\gamma$) to be given by

$$\gamma = \frac{p}{2w} = \frac{\omega^2 \bar{\mathcal{E}}^* \cdot \bar{\bar{\epsilon}}^a \cdot \bar{\mathcal{E}}}{\bar{\mathcal{E}}^* \cdot \dfrac{\partial \omega^2 \bar{\bar{\epsilon}}^h}{\partial \omega} \cdot \bar{\mathcal{E}}}$$

$$= \frac{\bar{\mathcal{E}}^* \cdot \bar{\bar{D}}^a \cdot \bar{\mathcal{E}}}{\bar{\mathcal{E}}^* \cdot \dfrac{\partial \bar{\bar{D}}^h}{\partial \omega} \cdot \bar{\mathcal{E}}} \qquad \text{(IIA-21)}$$

For a negative energy wave that is stable ($\gamma>0$) the power dissipation will be found to be negative, while if it is unstable ($\gamma<0$) the power dissipation will be positive. The converse will hold for a positive energy wave. Hence the sign of (internal) power dissipated alone cannot determine the (marginal) stability of a wave.

Using our equations for $\overline{v}_g$ and $\gamma$, Eq. (IIA-14) becomes

$$\frac{\partial w}{\partial t} + \nabla \cdot (\overline{v}_g w) + 2\gamma w = p_{ex} \tag{IIA-22}$$

If we regard this equation as expressing the perturbation of the wave by an external medium, and assume the slow variation is only in time ($\overline{k}$ real, $\omega=\omega_r+i\omega_i$ ), we find

$$\omega_i = -\gamma + \frac{p_{ex}}{2w} \tag{IIA-23}$$

Hence, if the external medium presents positive dissipation ($p_{ex}<0$) to a stable ($\gamma>0$) negative energy wave, it will tend to destabilize it. A positive energy mode that is weakly growing ($\gamma<0$) may be stabilized by perturbation with an external medium of positive dissipation ($p_{ext}<0$). On the other hand a weakly growing negative energy wave can only be stabilized by negative dissipation in an external medium.

For quasi-electrostatic waves, having $\overline{E}$ approximately along $\underline{k}$, the energy relations can be expressed directly in terms of the longitudinal dielectric function, which is also the dispersion function (see Eqs. IA-21 through 26). With

$$K_\ell(\overline{k},\omega) = \text{Re } K_\ell + i \text{ Im } K_\ell \tag{IIA-24}$$

and $|\text{Im } K_\ell(\overline{k},\omega)| << |\text{Re } K_\ell(\overline{k},\omega)|$ , we find

$$w = \frac{1}{4} \epsilon_o |\mathscr{E}_\ell|^2 \omega \frac{\partial(\text{Re}K_\ell)}{\partial\omega} \tag{IIA-25}$$

Similarly, the wave damping rate is given by

$$\gamma = \frac{\text{Im } K_\ell}{\dfrac{\partial(\text{Re } K_\ell)}{\partial\omega}} \tag{IIA-26}$$

The power flow density must be evaluated with extra care since the electromagnetic power flow is not necessarily negligible, and may sometimes be the dominant power flow (e.g. in a cold plasma the second term of Eq. (IIA-16) is zero). Expressing the transverse electric field in terms of the longitudinal one (Eq. (IA-29)) we find

$$\mathrm{Re}\,(\tfrac{1}{2}\overline{E} \times \overline{H}^*) \cong -\tfrac{1}{4}\epsilon_o|\overline{E}_\ell|^2\,\frac{\omega}{k}(\overline{\overline{K}}^h + \overline{\overline{K}}^h_t)\cdot\frac{\overline{k}}{k} \qquad \text{(IIA-27)}$$

where the subscript t denotes the transpose of the tensor. Equation (IIA-27) shows that the electromagnetic power flow is in a direction perpendicular to $\overline{k}$. The total power flow, Eq. (IIA-16), is hence

$$\overline{s} = -\tfrac{1}{4}\epsilon_o|\mathscr{E}_\ell|^2\omega\left[\frac{\left(\overline{\overline{K}}^h + \overline{\overline{K}}^h_t\right)\cdot\overline{k}}{k^2} + \frac{k_ik_j}{k^2}\frac{\partial K^h_{ij}}{\partial\overline{k}}\right] \qquad \text{(IIA-28)}$$

*B. Momentum and Stress*

The conservation equation for electromagnetic momentum and stress in the presence of currents ($\overline{J}$) and charge ($\rho$) densities is usually obtained by expressing the electrical force density of the particles on the fields through Maxwell's equations. One thus obtains

$$\frac{\partial}{\partial t}(\overline{D}\times\overline{B}) + \nabla\cdot[\ \tfrac{1}{2}(\overline{D}\cdot\overline{E} + \overline{B}\cdot\overline{H})\overline{\overline{\delta}} - (\overline{D}\overline{E} + \overline{B}\overline{H}) \qquad \text{(IIB-1)}$$

$$= -\ (\rho\overline{E} + \overline{J}\times\overline{B})$$

where for fields in free space $\overline{D} = \varepsilon_o\overline{E}$, $\overline{B} = \mu_o\overline{H}$ , and the right-hand side is interpreted to give the dissipation or generation of electromagnetic momentum. In a dispersive medium, the right-hand side will, in addition, contain the momentum and momentum flow associated with the dynamics of the medium.

For the same reasons that were discussed in relation to energy (see Sec. II-A), we seek a conservation equation for the total time-averaged momentum in a medium with small dissipation. Allowing the fields to have slowly varying complex amplitudes, as in Eqs. (IIA-5 through 8), we now wish to separate the full set of Maxwell's equations into a set for the rapidly varying part, and another set for the slowly varying field amplitudes. For this, we need to express the charge density as a function of the electric field, and this can be done using Eq. (IIA-9). We then arrive at the following sets of Maxwell's equations, which are conveniently written in Minkowski-type field variables: $\overline{\mathcal{B}}=\mu_0\overline{\mathcal{H}}$ and $\overline{\mathcal{D}}^h=\overline{\overline{\epsilon}}^h(\overline{k},\omega)\cdot\overline{\mathcal{E}}$,

$$\overline{k} \times \overline{\mathcal{E}} - \omega\overline{\mathcal{B}} = 0 \tag{IIB-2}$$

$$\overline{k} \times \overline{\mathcal{H}} + \omega\overline{\mathcal{D}}^h = 0 \tag{IIB-3}$$

$$\overline{k} \cdot \overline{\mathcal{D}}^h = 0 \tag{IIB-4}$$

$$\overline{k} \cdot \overline{\mathcal{B}} = 0 \tag{IIB-5}$$

$$\nabla \times \overline{\mathcal{E}} + \frac{\partial \overline{\mathcal{B}}}{\partial t} = 0 \tag{IIB-6}$$

$$\nabla \times \overline{\mathcal{H}} - \frac{\partial}{\partial \omega}\frac{\partial \omega\overline{\mathcal{D}}^h}{\partial t} + \omega\frac{\partial}{\partial \overline{k}} \cdot \frac{\partial \overline{\mathcal{D}}^h_{\overline{\alpha}}}{\partial \overline{r}} - \omega\overline{\mathcal{D}}^a = \overline{\mathcal{J}}_{ex} \tag{IIB-7}$$

$$\nabla \cdot \overline{\mathcal{D}}^h - \overline{k} \cdot \frac{\partial}{\partial \omega}\frac{\partial \overline{\mathcal{D}}^h}{\partial t} + k_\alpha \frac{\partial}{\partial \overline{k}} \cdot \frac{\partial \overline{\mathcal{D}}^h_\alpha}{\partial \overline{r}} - \overline{k} \cdot \overline{\mathcal{D}}^a = \rho_{ex} \tag{IIB-8}$$

$$\nabla \cdot \overline{\mathcal{B}} = 0 \tag{IIB-9}$$

Equations (IIB-2), (IIB-3), (IIB-6) and (IIB-7) are identical to Eqs. (IIA-10) through (IIA-13), respectively, Eq. (IIB-8) is new and contains the charge density expressed to first order in variation and dissipation, and the other equations are obvious. Using these equations we now express the time averaged force density by the external sources on the fields as follows: $\overline{\mathcal{D}}^h \times (\text{IIB-6})^* + \overline{\mathcal{B}}^* \times (\text{IIB-7}) - \overline{\mathcal{E}}^*(\text{IIB-8}) - \overline{\mathcal{H}}(\text{IIB-9})^*$ plus their complex conjugates, and using Eq. (IIB-2) we find

$$\frac{\partial \overline{g}}{\partial t} + \nabla \cdot \overline{\overline{t}} + \overline{f} = \overline{f}_{ex} \tag{IIB-10}$$

where

$$\overline{g} = \text{Re}\left(\frac{1}{2}\overline{\mathcal{D}}^h \times \overline{\mathcal{B}}^*\right) + \overline{k}\frac{\partial}{\partial \omega}\left(\frac{1}{4}\overline{\mathcal{E}}^* \cdot \overline{\mathcal{D}}^h\right) \tag{IIB-11}$$

$$\overline{\overline{t}} = \overline{\overline{\delta}}\,\frac{1}{4}(\overline{\mathcal{E}}^* \cdot \overline{\mathcal{D}}^h + \overline{\mathcal{H}}^* \cdot \overline{\mathcal{B}}) - \text{Re}\,\frac{1}{2}(\overline{\mathcal{E}}^*\overline{\mathcal{D}}^h + \overline{\mathcal{H}}^*\overline{\mathcal{B}}) - \overline{k}\frac{\partial}{\partial \overline{k}}\left(\frac{1}{4}\overline{\mathcal{E}}^* \cdot \overline{\mathcal{D}}^h\right) \tag{IIB-12}$$

$$\bar{f} = \bar{k}\,\frac{1}{2}\,\bar{\mathscr{E}}^* \cdot \bar{\mathscr{D}}^a \qquad \text{(IIB-13)}$$

$$\bar{f}_{ex} = -\,\mathrm{Re}\,\frac{1}{2}(\rho_{ex}\bar{\mathscr{E}}^* + \bar{\mathscr{J}}_{ex} \times \bar{\mathscr{B}}^*) \qquad \text{(IIB-14)}$$

are respectively the time-averaged densities of momentum, stress, dissipative force, and external force. As in Sec. IIA, we can introduce the group velocity $\bar{v}_g$ and the small damping (or growth) rate $\gamma$ of the mode, and show that

$$\bar{\bar{t}} = \bar{g}\,\bar{v}_g \qquad \text{(IIB-15)}$$

and

$$\bar{f} = 2\gamma\bar{g} \qquad \text{(IIB-16)}$$

so that the conservation Equation (IIB-10) becomes

$$\frac{\partial\bar{g}}{\partial t} + \nabla \cdot (\bar{g}\,\bar{v}_g) + 2\gamma\bar{g} = \bar{f}_{ex} \qquad \text{(IIB-17)}$$

In addition, using the energy and power expressions found in Sec. IIA we can show that for a given mode

$$\bar{g} = \bar{k}\,\frac{W}{\omega} \qquad \text{(IIB-18)}$$

$$\bar{\bar{t}} = \frac{\bar{k}\,\bar{s}}{\omega} = \bar{k}\,\bar{v}_g\,\frac{W}{\omega} \qquad \text{(IIB-19)}$$

and

$$\bar{f} = \frac{\bar{k}\,p}{\omega} = \bar{k}2\gamma\,\frac{W}{\omega} \qquad \text{(IIB-20)}$$

Hence introducing the action density for a mode

$$n_{\bar{k}} = \frac{W}{\omega} \qquad \text{(IIB-21)}$$

Eq. (IIB-17) becomes

$$\frac{\partial}{\partial t}\,(\bar{k}n_{\bar{k}}) + \nabla \cdot (\bar{k}\,\bar{v}_g n_{\bar{k}}) + \bar{k}2\gamma n_{\bar{k}} = \bar{f}_{ex} \qquad \text{(IIB-22)}$$

It is frequently useful to think of an external source as applying a force to the medium. Such forces may either stabilize or destabilize certain modes. If we regard the interaction with the external force to give rise to the slow variation of the amplitudes in time only ($\bar{k}$ real, and $\partial/\partial t \to \omega_i$) we find from Eq. (IIB-22)

$$\omega_i = -\gamma + \frac{\bar{k} \cdot \bar{f}_{ex}}{2k^2 n_{\bar{k}}} \tag{IIB-23}$$

The effect of the external force is seen to depend upon both its direction with respect to $\bar{k}$, and the sign of the energy of the mode. Equation (IIB-23) can be used in evaluating the effectiveness of various schemes of feedback stabilization of plasma instabilities.

The equations for energy (Eq. (IIA-22)) and momentum (Eq. (IIB-17)) of modes which we have found form the basis of the quasiparticle description of a medium, and are fundamental in the study of both linear and nonlinear interactions of waves in a medium. For a (weakly) inhomogeneous and (slowly) time-varying medium we shall find that the conservation of action density is the fundamental conservation equation. (See Appendix 2.)

### *C. Coordinate Transformations*

In Sec. IIA we found that the small-amplitude energy of a wave may be negative. Since energy depends upon the frame of reference, it is frequently possible to uncover (or "generate") such waves by a transformation of coordinates. It is also of interest to inquire into the manner in which the quasiparticle picture is affected by transformation of coordinates.

From the invariance of the wave phase and the Lorentz transformations of coordinates and time we obtain the transformations of $\omega$ and $\bar{k}$ to a primed reference frame moving with uniform velocity $\bar{u}$ with respect to the unprimed frame:

$$\bar{k}' = \bar{k}_\perp + \gamma_u \left(\bar{k}_\parallel - \frac{\omega\bar{u}}{c^2}\right) \tag{IIC-1}$$

$$\omega' = \gamma_u(\omega - \bar{k} \cdot \bar{u}) \tag{IIC-2}$$

where

$$\gamma_u = \left(1 - \frac{u^2}{c^2}\right)^{-1/2} \tag{IIC-3}$$

and the subscripts $\parallel$ and $\perp$ refer to components respectively parallel and perpendicular to $\bar{u}$. From Eqs. (IIC-1) and (IIC-2) we note that $c^2\bar{k}$ and $\omega$ transform like a four-vector. We shall find it convenient to introduce the four-vector

$$\bar{k}^4 = [\bar{k}, \omega i/c] \qquad \text{(IIC-4)}$$

The equations for the slow-variation of energy and momentum can also be cast in four-notation. For this we assume the small-damping rate to be either absent or included in the time-derivative and write the four-vector gradient,

$$\nabla^4 = [\nabla, \frac{1}{ic}\frac{\partial}{\partial t}] \qquad \text{(IIC-5)}$$

the four-vector external force density

$$\bar{f}^4_{ex} = [\bar{f}_{ex}, \frac{i}{c}\, p_{ex}] \qquad \text{(IIC-6)}$$

and the small-amplitude energy-momentum tensor

$$\bar{\bar{T}}^4 = \begin{bmatrix} \bar{\bar{t}} & \frac{i}{c}\,\bar{s} \\ ic\bar{g} & -w \end{bmatrix} \qquad \text{(IIC-7)}$$

The energy-momentum Equations (IIA-14 and IIB-10) become in four-notation

$$\nabla^4 \cdot \bar{\bar{T}}^4 = \bar{f}^4_{ex} \qquad \text{(IIC-8)}$$

However, since these equations were derived from a linearization, expansion, and truncation of Maxwell's equations we are no longer guaranteed that $\bar{\bar{T}}^4$ is a four-tensor. We now prove that $\bar{\bar{T}}^4$ in Eq. (IIC-7) is indeed a four-tensor. Using the action density (Eq. (IIB-21)) we can write Eq. (IIC-7) as

$$\bar{\bar{T}}^4 = \begin{bmatrix} \bar{k}\ \bar{v}_g n_{\bar{k}} & \frac{i}{c}\ \bar{v}_g \omega n_{\bar{k}} \\ ic\bar{k}n_{\bar{k}} & -\omega n_{\bar{k}} \end{bmatrix} \qquad \text{(IIC-9 )}$$

Both $n_{\bar{k}}$ and $\bar{v}_g$ can, however, be expressed in terms of the hermitian part of the dispersion tensor $\bar{\bar{D}}^h$ (see Eq. (IIA-19)) as follows,

$$n_{\overline{k}} = \frac{\epsilon_o}{4}\frac{\partial D^h}{\partial\omega} \qquad \text{(IIC-10)}$$

$$\overline{v}_g = \frac{-\dfrac{\partial D^h}{\partial\overline{k}}}{\dfrac{\partial D^h}{\partial\omega}} \qquad \text{(IIC-11)}$$

where we have used the abbreviation

$$D^h = \overline{\mathscr{E}}^* \cdot \overline{\overline{D}}^h \cdot \overline{\mathscr{E}} \qquad \text{(IIC-12)}$$

Using Eqs. (IIC-10) and (IIC-11) in Eq. (IIC-9) we finally have

$$\overline{\overline{T}}^4 = -\frac{\epsilon_o}{4}\begin{bmatrix} \overline{k}\,\dfrac{\partial D^h}{\partial\overline{k}} & \dfrac{i}{c}\,\omega\,\dfrac{\partial D^h}{\partial\overline{k}} \\ \overline{k}\,\dfrac{\partial D^h}{\partial\omega i/c} & \dfrac{i}{c}\,\omega\,\dfrac{\partial D^h}{\partial\omega i/c} \end{bmatrix} \qquad \text{(IIC-13)}$$

which shows that $\overline{\overline{T}}^4$ is made up of the transpose product of two four-vectors $[\overline{k}, \omega i/c]$ and $[\partial D^h/\partial\overline{k}, \partial D^h/\partial\omega i/c]$. Hence $\overline{\overline{T}}^4$ is a four-tensor, and thus obeys the Lorentz transformation laws. Since we know the transformation laws for $\overline{k}$ and $\omega$ (Eqs. IIC-1 and 2), and $\overline{v}_g$ (according to Einstein's law of velocities), we need only find the transformation law for $n_{\overline{k}}$. Applying the Lorentz transformation for the 4-4 term of Eq. (IIC-9) we find

$$n'_{\overline{k}} = n_{\overline{k}}\,\gamma_u\left(1 - \frac{\overline{v}_g\cdot\overline{u}}{c^2}\right) \qquad \text{(IIC-14)}$$

Hence, the small-amplitude action density is in general not an invariant in a coordinate transformation. Using the equations for the transformation of $\overline{v}_g$ we can rewrite Eq. (IIC-14) to show that

$$n'_{\overline{k}}\left[1 - \left(\frac{\overline{v}'_g\cdot\overline{u}}{cu}\right)^2\right]^{1/2} = n_{\overline{k}}\left[1 - \left(\frac{\overline{v}_g\cdot\overline{u}}{cu}\right)^2\right]^{1/2} \qquad \text{(IIC-15)}$$

therefore is the invariant. The fact that the action density itself is not invariant shows that the quasiparticle picture is not complete when interactions are viewed from moving reference frames. The quasiparticle picture is, however, useful for studying wave interactions within a reference frame. We also note that for quasi-electrostatic modes($\overline{H} \approx 0$)and ($c \to \infty$) the action density *is* invariant

$$\left.\begin{aligned} [n'_{\overline{k}}]_{QES} &= [n_{\overline{k}}]_{QES} \\ \frac{w'}{\omega'}_{QES} &= \frac{w}{\omega}_{QES} \end{aligned}\right\} \qquad \text{(IIC-16)}$$

An important consequence of Eqs. (IIC-15) and (IIC-16) is that the sign of the energy density of a wave may change with a change of the reference frame. Thus, using Eq. (IIC-2) and Eq. (IIC-15) or (IIC-16),

$$\frac{w}{w'} \sim \frac{\omega}{\omega'} \sim \frac{\omega}{\omega - \overline{k} \cdot \overline{u}} \qquad \text{(IIC-17)}$$

Hence, positive energy waves in the primed frame will become negative energy waves in the unprimed frame provided their dispersion relation is such that $\omega - \overline{k} \cdot \overline{u} < 0$ i.e. provided that their phase velocity along $\overline{u}$ is less than u. So far we have applied our results to coordinate systems that are in translation relative to each other. For nonrelativistic motion we can equally well consider coordinate systems that rotate with respect to each other and establish (through torque invariance)

$$\frac{w}{w'} \sim \frac{\omega}{\omega'} \sim \frac{\omega}{\omega - m\,\omega_R} \qquad \text{(IIC-18)}$$

where $\omega_R$ is the angular velocity of the primed system and m is the integer phase number in the cyclic coordinate. Finally, it should be noted that negative energy waves cannot always be associated with a transformation of reference frames. We shall illustrate all of these possibilities in the next section.

### D. *Examples of Active Media*

We shall define a (plasma) medium to be active if it exhibits a small-amplitude time-averaged stored energy density or dissipates power density that makes it potentially unstable in interaction with a passive (plasma) medium. Active media by themselves are stable and easy to characterize. When they are combined with (i.e. coupled to) other media the overall stability or instability can frequently be predicted, and better understood, by studying separately the active and/or passive nature of the individual (i.e. uncoupled) physical parts. We shall distinguish and illustrate two types of active media: lossfree and dissipative.

In a lossfree or even weakly dissipative medium we have shown that the small-amplitude, time-averaged, energy density (Eq. IIA-19) is well defined. We have also shown that when this energy density is negative the perturbation of the medium by coupling to a weakly dissipative "external" medium leads to instability (Eq. IIA-23). We shall also find in the next section that the perturbation of negative energy by coupling to even lossfree "external" media can also lead to instability. Hence we shall define lossfree media exhibiting negative energy waves as active media. The appearance of such negative energy waves must of course be traced to the fact that they can only occur in media that are not in thermal equilibrium, and for which a source of free energy can be identified. The simplest example is provided by the model of a cold, collisionless beam of charged particles drifting with velocity $v_o$ that was provided by an energy source. The dispersion relation for electrostatic waves propagating along the direction of $v_o$ is

$$D(k,\omega) = 1 - \frac{\omega_b^2}{(\omega - \bar{k}\cdot\bar{v}_o)^2} = 0 \qquad \text{(IID-1)}$$

where $\omega_b$ is the plasma frequency of the beam. From this we find that two types of waves can propagate (see Fig. IID-1)

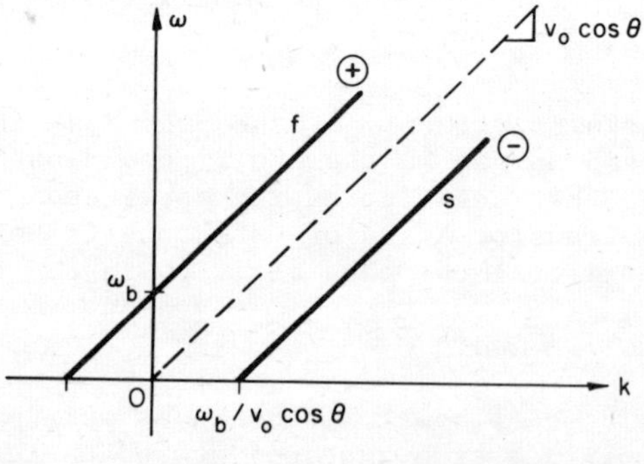


Fig. II D-1

$$\omega_f - \overline{k}_f \cdot \overline{v}_o = \omega_b \quad \text{(IID-2)}$$

and

$$\omega_s - \overline{k}_s \cdot \overline{v}_o = -\omega_b \quad \text{(IID-3)}$$

the first one with subscript f ("fast") having phase velocities greater than $v_o$, and the second one with subscript s ("slow") having phase velocities smaller than $v_o$. Using Eq. (IIA-19) (or IIA-25) we find

$$w_f = \frac{2\omega_f}{\omega_b} \frac{1}{4} \varepsilon_o |E|^2 \quad \text{(IID-4)}$$

and

$$w_s = -\frac{2\omega_s}{\omega_b} \frac{1}{4} \varepsilon_o |E|^2 \quad \text{(IID-5)}$$

Thus for positive frequencies the "slow" wave is a negative energy wave. (We indicate this by a circled sign along the s branch of the dispersion curve in Fig. IID-1). A physical picture of why the sign of the energy is different for the two waves can be obtained by examining the differences in the phases between the density and velocity perturbations of the two waves. Using the continuity equation and Eqs. (IID-2) and (IID-3) we can easily see that in the f-wave the density and velocity perturbations are in phase, while in the s-wave they are 180° out of phase. Hence we find that the excitation of the fast wave requires that energy be supplied to the beam, while the slow wave is excited when energy is removed from the beam. In the latter case the energy extracted must of course come from the drift energy of the beam, i.e. $v_o$ would have to decrease; this is of course a nonlinear effect, and is not described by the small-amplitude equations which assume $v_o$ constant. The amazing thing about the small-amplitude energy formalism is that it allows us to predict that such an energy transformation is potentially possible, and how it can be effected, i.e. by coupling to the s-wave, the negative energy wave.

In the previous section we have shown that the small-amplitude energy density, and in particular its sign, depend upon the frame of reference. The simplest illustration of this is also with the beam. Thus in a coordinate system moving with the beam velocity (Eqs. IIC-1 and 2) the beam waves become just oscillations in a cold stationary plasma for which the energy is clearly positive, as can also be shown directly with Eq. (IIA-19) (or IIA-25). The change in the energy between the two frames is given by Eq. (IIC-17), which also shows that the resultant "slow" wave, for which $\omega - kv_o < 0$, will be a negative energy wave.

Although the sign of the energy changes with uniformly moving reference frames, it is not always possible to find such a reference frame in which the energy of all waves is positive. This is best illustrated by a plasma having a beam-type velocity distribution function perpendicular to a constant applied magnetic field $\overline{B}_o$:

$$f_o(\overline{v}) = \frac{1}{2\pi v_{o\perp}}\,\delta(v_\perp - v_{o\perp})\,\delta(v_\parallel) \qquad \text{(IID-6)}$$

For quasi-electrostatic waves propagating at an arbitrary angle to $\overline{B}_o$ the dispersion relation is found to be

$$D(\overline{k},\omega) = 1 - \frac{\omega_p^2}{k^2}\sum_{n=-\infty}^{\infty}\left[\frac{k_\parallel^2 J_n^2}{(\omega - n\Omega)^2} + \frac{k_\perp^2 (J_{n-1}^2 - J_{n+1}^2)}{2\Omega(\omega - n\Omega)}\right]$$
$$= 0 \qquad \text{(IID-7)}$$

where $\omega_p$ is the plasma frequency, $\Omega$ is the cyclotron frequency and the argument of the Bessel functions is $(k_\perp v_{o\perp}/\Omega)$. The wave solutions of Eq. (IID-7) together with the sign and magnitude of their energies, as determined by Eq. (IIA-19) (or IIA-25), are shown in Fig. IID-2. Both the dispersion relation and the energies of these waves can be made to change with a change of reference frame. Thus in a reference frame moving with respect to the direction of the magnetic field at a nonrelativistic velocity $v_{o\parallel}$ we find, with the aid of Eqs. (IIC-1 and 2), and (IIC-17), the results shown in Fig. IID-3. This would pertain to a plasma with an unperturbed velocity distribution function

$$f_o(\overline{v}) = \frac{1}{2\pi v_{o\perp}}\,\delta(v_\perp - v_{o\perp})\,\delta(v_\parallel - v_{o\parallel}) \qquad \text{(IID-8)}$$

We note that it is not possible to find a reference frame for this plasma so that all its waves are positive energy. Physically, the reason for this is that the available free energy in a plasma with a distribution function of Eq. (IID-6) is partly random: although all particles have the same velocity magnitude $v_{o\perp}$, at any one point in space the direction of $v_{o\perp}$ is random in time.

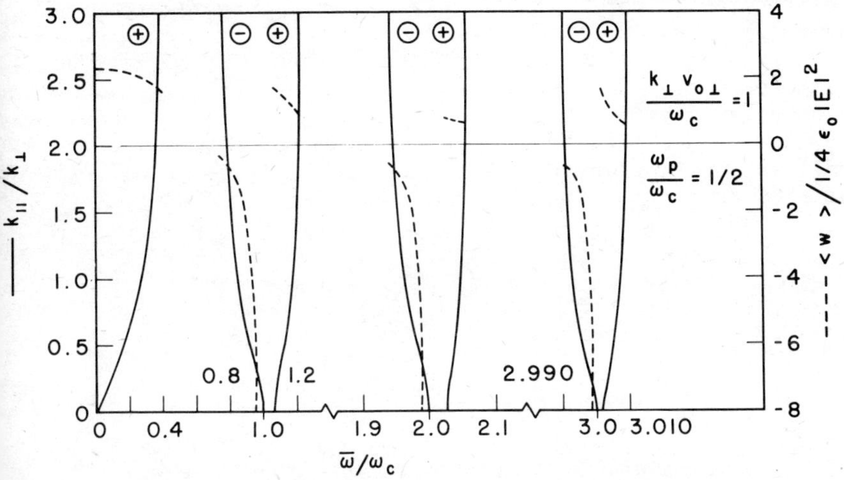


Fig. II D-2

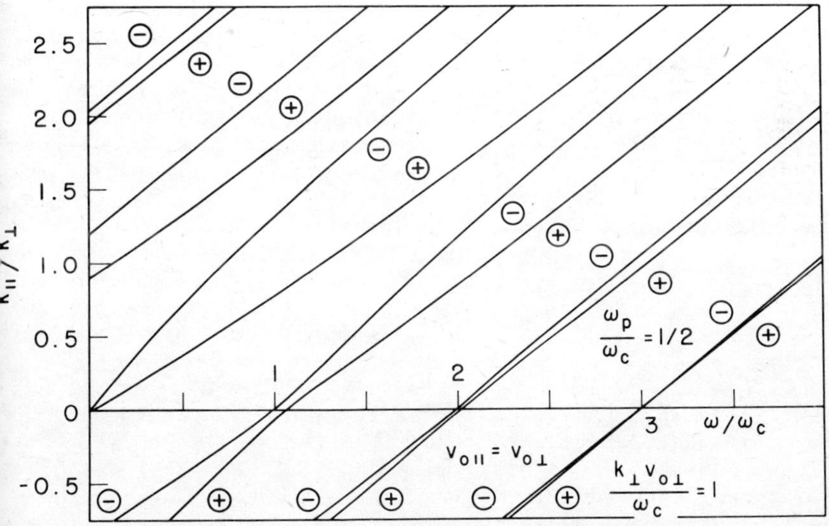


Fig. II D-3

In a dispersive medium with large dissipation the wave description is not well defined, and the concepts of time-averaged energy and power cannot be defined from an electrodynamic description (i.e. from $\bar{\bar{\sigma}}(\bar{k},\omega)$) but must be derived from the detailed dynamic equations of the medium. This is particularly true of media in which spatial dispersion is important. On the other hand, the density of power dissipated follows straightforwardly from the time-average of Eq. (IIA-1) We thus have

$$p = \frac{1}{2}\bar{E}^* \cdot \bar{\bar{\sigma}}^h(\bar{k},\omega) \cdot \bar{E}$$

$$= \frac{\omega}{2}\bar{E}^* \cdot \bar{\bar{\varepsilon}}^a(\bar{k},\omega) \cdot \bar{E} \qquad \text{(IID-9)}$$

which is identical in form to Eq. (IIA-17). However, for a dissipative medium in which the self-modes are heavily damped the time-average must be interpreted to be with respect to periodic fields that are imposed (or coupled to) externally. For such fields (with $\bar{k}$ and $\omega$ real) a dissipative medium will be defined as active when $p < 0$, i.e. when $\bar{\bar{\sigma}}^h(\bar{k}, \omega)$ is negative-definite. Conversely when $p > 0$, i.e. when $\bar{\bar{\sigma}}^h(\bar{k}, \omega)$ is positive-definite, the medium is passive. Thus if for real $\bar{k}$ and $\omega$ $\bar{\bar{\sigma}}^h(\bar{k}, \omega)$ is negative-definite, the coupling of an electric field, having such $\bar{k}$ and $\omega$, to this medium can induce a power flow from the medium to this field. A simple example of such a medium is a collision dominated drifting plasma described by the simplest fluid equations (continuity and momentum). For electrostatic fields with frequence $\omega$ small compared to the collision frequency $\nu$, and wavelengths $\lambda = 2\pi/k$ long compared to the particle mean-free paths $(v_T/\nu)$, the small-amplitude (longitudinal) conductivity is

$$\sigma_\ell(\bar{k},\omega) = \frac{\omega\sigma_o}{\omega - \bar{k}\cdot\bar{v}_o + ik^2 D} \qquad \text{(IID-10)}$$

where $\sigma_o = \omega_p^2\varepsilon_o/\nu$ is the dc conductivity, $D = v_T^2/\nu$ is the diffusion constant, and $\bar{v}_o$ is the drift velocity. It is easy to show that under the stated conditions the self-modes (i.e. the solutions of the dispersion relation $\sigma_\ell(\bar{k}, \omega) = \omega\varepsilon_o$) are heavily damped; this then is a dissipative medium. On the other hand, Eq. (IID-10) shows that for $\bar{k}$ and $\omega$ real Re $\sigma_\ell(\bar{k}, \omega) < 0$ if $\omega - kv_o \cos\theta < 0$, i.e. if the drift velocity exceeds the phase velocity parallel to it of an external electric field.

Under these conditions this dissipative medium is active and can give up power to such an external wave and make it grow in amplitude (i.e. unstable). The energy that is thus removed from the dissipative medium comes of course from the free energy source that gives rise to the assumed drift velocity.

Another example is a collisionless, maxwellian electron plasma having a drift velocity $v_o$ which is much smaller than its thermal velocity $v_e = (2\kappa T_e/m_e)^{1/2}$. The longitudinal dielectric function for phase velocities $(\omega/k)$ that are small compared to $v_e$ is

$$K_\ell(\bar{k},\omega) \cong 1 + \frac{2\omega_p^2}{k^2 v_e^2}\left(1 + i\sqrt{\pi}\,\frac{\omega - \bar{k}\cdot\bar{v}_o}{kv_e}\right) \quad \text{(IID-11)}$$

The imaginary part of $K_\ell$ is due to the Landau damping mechanism, which gives rise to dissipation. This dissipation becomes negative when again $\omega - \bar{k}\cdot\bar{v}_o < 0$, and waves satisfying this condition will be driven unstable when coupled to this plasma.

From the transformation equations in Sec. IIC we note that the power dissipated is also a function of the choice of reference frames, and in particular will transform like $p/p' \sim \omega/\omega'$ as in Eq. (IIC-17). This is also clear from the examples we have given.

## III. WEAK COUPLING OF WAVES

In the preceding two sections we have developed the linear wave description of a plasma. We shall now describe the interactions that are possible with these waves. This is useful in understanding and predicting instabilities in multicomponent plasma systems and in bounded systems, and in describing weakly nonlinear systems. Hence we shall develop a linear perturbation theory of coupling of waves. This same approach also forms the basis for studying the nonlinear interactions by perturbation theory. (See Section V.)

### A. *Energy and Momentum Formulations of Coupling*

In Chapter II we have shown that the conservation equations for energy and momentum also provided us with a means for studying the perturbation effects by external sources. From the conservation equations we can only deduce the perturbation of the squared magnitude of the field components.

We now seek a more detailed theory for finding the perturbation of the complex field amplitude and polarization.

We start by assuming a plasma to be perturbed by current and charge densities, whose detailed description we postpone for the moment. We consider the effects of these perturbations only on the weakly damped waves of the plasma, and assume that these acquire slowly varying complex vector amplitudes as given in Eqs. (IIA-5) through (IIA-8). Now, however, we make explicit these slowly varying complex vector amplitudes into a complex amplitude and polarization vector, and write

$$\bar{\mathscr{E}}(\bar{r},t) = a(\bar{r},t)\,[\bar{e}_o + \bar{e}(\bar{r},t)] \qquad \text{(IIIA-1)}$$

where $\bar{e}_o$ is the unperturbed polarization vector, and $a(\bar{r},t)$ and $\bar{e}(\bar{r},t)$ respectively the slowly varying complex amplitude and polarization due to the perturbation. These satisfy conditions similar to Eqs. (IIA-7 and 8). Similarly, for the slowly varying complex vector magnetic field we write

$$\bar{\mathscr{H}}(\bar{r},t) = a(\bar{r},t)\,[\bar{h}_o + \bar{h}(\bar{r},t)] \qquad \text{(IIIA-2)}$$

where the complex amplitude is the same as for $\bar{\mathscr{E}}$, and the slowly varying polarization vector $\bar{h}(\bar{r},t)$ also satisfies conditions similar to Eqs. (IIA-7 and 8). We now proceed as in Sec. IIA and derive the plasma current density using Eq. (IA- and (IIIA-1), and separate Maxwell's equations into a set for the rapid variations at $(\omega,\bar{k})$ of the mode

$$\bar{k} \times \bar{e}_o - \omega\mu_o\bar{h}_o = 0 \qquad \text{(IIIA-3)}$$

$$\bar{k} \times \bar{h}_o + \omega\bar{\bar{\epsilon}}^h(\bar{k},\omega) \cdot \bar{e}_o = 0 \qquad \text{(IIIA-4)}$$

and another set for the slow variations, which turns out to b

$$\nabla a \times \bar{e}_o + \mu_o\bar{h}_o \frac{\partial a}{\partial t} + i\bar{k} \times \bar{e}a - i\omega\mu_o\bar{h}a = 0 \qquad \text{(IIIA-5)}$$

$$\nabla a \times \bar{h}_o - \frac{\partial\omega\bar{\bar{\epsilon}}^h}{\partial\omega}\,\bar{e}_o\,\frac{\partial a}{\partial t} + \omega\,\frac{\partial\epsilon^h_{\alpha\beta}}{\partial\bar{k}}\,e_{o\beta} \cdot \frac{\partial a}{\partial\bar{r}} - \bar{\bar{\sigma}}^h \cdot \bar{e}_o a$$

$$+ i\bar{k} \times \bar{h}a + \omega\bar{\bar{\epsilon}}^h \cdot \bar{e}a = \bar{J}_p e^{i(\bar{k}_p-\bar{k})\cdot\bar{r}_e - i(\omega_p-\omega)t} \qquad \text{(IIIA-6)}$$

where the perturbing current density is assumed at a frequency $\omega_p$ and wavevector $\overline{k}_p$ close to respectively the mode frequency $\omega$ and wavevector $\overline{k}$, and $\overline{j}_p$ is its slowly varying complex vector amplitude. Equations (IIIA-3) and (IIIA-4) give the dispersion tensor (and dispersion relation) for the unperturbed mode

$$\overline{\overline{D}}^h \cdot \overline{e}_o = 0 \qquad \text{(IIIA-7)}$$

as was shown in Sec. IA. Equations (IIIA-5) and (IIIA-6) contain both the perturbed amplitudes and perturbed polarizations. These can be separated by noting that on the left-hand sides of these equations the last two terms can be combined into $\overline{\overline{D}}^h \cdot \overline{e}$. Hence, dot-multiplying the equations by the adjoint of the unperturbed solutions (respectively, Eq. (IIIA-5) by $\overline{h}_o^*$ and Eq. (IIIA-6) by $-\overline{e}_o^*$) and adding the resultant equations we obtain

$$w_o \frac{\partial a}{\partial t} + \overline{s}_o \cdot \nabla a + 2p_o a =$$

$$- \frac{1}{4}\overline{e}_o^* \cdot \overline{j}_p |a_o|^2 e^{i(\overline{k}_p - \overline{k})\cdot\overline{r}} e^{-i(\omega_p - \omega)t} \qquad \text{(IIIA-8)}$$

where we have introduced the unperturbed mode time-averaged densities of energy, power flow, and power dissipated as defined in Eqs. (IIA-15 through 17) with $\mathcal{E} \rightarrow \overline{e}_o a_o$ and $a_o$ the unperturbed mode amplitude. Equation (IIIA-8) is the differential equation for the perturbed complex amplitude $a(\overline{r},t)$ for a given current perturbation $\overline{j}_p$. Once this equation is solved for $a(\overline{r},t)$ we can substitute it into Eqs. (IIIA-5 and 6) and obtain a coupled set of equations for finding the perturbed polarization vectors $\overline{e}(\overline{r},t)$ and $\overline{h}(\overline{r},t)$. Here we shall only concern ourselves with the solution for the perturbed amplitude. Introducing the unperturbed group velocity ($\overline{v}_{go}$) and weak damping (or growth) rate ($\gamma_o$) of the mode, as found in Eqs. (IIA-20 and 21) we can rewrite Eq. (IIIA-8) in terms of only the unperturbed energy density and electric field of the mode, as follows

$$\left(\frac{\partial}{\partial t} + \overline{v}_{go} \cdot \nabla + \gamma_o\right) u(\overline{r},t) = \frac{-\frac{1}{4}\overline{E}_o^* \cdot \overline{J}_p}{w_o} \qquad \text{(IIIA-9)}$$

where

$$\overline{E}_o = a_o \overline{e}_o \exp i(\overline{k}\cdot\overline{r} - \omega t),\quad \overline{J}_p = \overline{j}_p(\overline{r},t) \exp i(\overline{k}_p \cdot \overline{r} - \omega_p t),$$

$$u(\overline{r},t) = \frac{a(\overline{r},t)}{a_o} \qquad \text{(IIIA-10)}$$

is the slowly-varying perturbed amplitude normalized to the

unperturbed mode amplitude. The right-hand side of the Eq. (IIIA-9) can be viewed as a complex interaction power density, per unit energy density of the unperturbed mode, that gives rise to the slow time-space dependence of the field amplitude.

In an analogous manner to the derivation of Eq. (IIIA-9), but using the full set of Maxwell's equations to include perturbations by both a current density $\overline{J}_p$ and related charge density $\rho_p$, we can derive the differential equation for the perturbed amplitude as related to a complex interaction force density:

$$\overline{g}_o \left(\frac{\partial}{\partial t} + \overline{v}_{go} \cdot \nabla + \gamma_o\right) u(\overline{r},t) = -\frac{1}{4}(\rho_p \overline{E}_o^* + \overline{J}_p \times \overline{B}_o^*) \quad \text{(IIIA-11)}$$

where $\overline{g}_o$ is the momentum density of the unperturbed mode as given in Eq. (IIB-11). Introducing the relation between energy and momentum of Eq. (IIB-18), and the action density of Eq. (IIB-21), we can also write Eq. (IIIA-11) as

$$\left(\frac{\partial}{\partial t} + \overline{v}_{go} \cdot \nabla + \gamma_o\right) u(\overline{r},t) = \frac{-\frac{1}{4}(\rho_p \overline{E}_o^* + \overline{J}_p \times \overline{B}_o^*)\cdot \overline{k}}{k^2 n_{\overline{k}o}} \quad \text{(IIIA-12)}$$

Equations (IIIA-9) and (IIIA-12) show that associated with an interaction power density there is always an interaction force density in the direction of $\overline{k}$, i.e.

$$\overline{E}_o^* \cdot \overline{J}_p = \omega(\rho_p \overline{E}_o^* + \overline{J}_p \times \overline{B}_o^*) \cdot \overline{k}/k^2 \quad \text{(IIIA-13)}$$

In other words, for weakly coupled systems (i.e. described by perturbation theory) the transfer of energy is always associated with a transfer of momentum. Equation (IIIA-13) can of course also be proved by using Maxwell's equations for the unperturbed fields together with the continuity equation for $\rho_p$ and $\overline{J}_p$. Equations (IIIA-9) and (IIIA-12) also provide us with two different means of thinking about, and evaluating, coupling coefficients. In the following, and for simplicity, we shall pursue our discussion on the basis of Eq. (IIIA-9).

Up to now we have not had to describe the detailed nature of the perturbing current $\overline{J}_p$. It is useful to distinguish the following possibilities. (a) The current $\overline{J}_p$ may be taken to represent a system whose modes are not well defined, as is, for example, a resistive medium (with only damped modes) or a reactive medium (with only evanescent modes). In these cases the coupling will be called nonresonant. (b) For a system whose modes are well-defined, $\overline{J}_p$ can be expressed in terms of these modes, and the perturbation of these modes has to be treated on an equal footing with the modes that we have so far considered as perturbed by $\overline{J}_p$. In such cases we shall find it convenient to distinguish between resonant coupling, when modes are in exact phase synchronism, and coupling off-resonance. (c) In many cases, particularly for interpenetrating media, the coupling off-resonance is mixed with resonant coupling, or the coupling in different parameter regimes changes from resonant to nonresonant, or vice-versa. (d) Finally, in media where dispersion dominates over nonlinearities, i.e. weakly nonlinear media, the finite-amplitude description of such media can be carried out by using perturbation theory on the small-amplitude modes with $\overline{J}_p$ representing the current density due to nonlinear effects. This then gives us the description of media by quasiparticle interactions. In the following we shall develop and illustrate each of these uses of the perturbation theory presented. In this chapter we shall only treat the linear coupling of waves and emphasize the instabilities that such coupling can produce. Coupling of waves due to nonlinear effects will be described in Section V.

### B. *Nonresonant Interaction*

Suppose that a plasma with well-defined modes is weakly coupled to a medium whose modes are not well-defined. The simplest example is that of coupling to a resistive-reactive medium with dispersion described by its conductivity function $\overline{\overline{\sigma}}_p(\overline{k}, \omega)$, or dielectric function $\overline{\overline{\epsilon}}_p(\overline{k}, \omega)$, which, for the $\overline{k}$ and $\omega$ values of the well defined plasma modes, have hermitian and antihermitian parts that are comparable to each other. To lowest order in the coupling we can assume that each mode of the plasma is independently perturbed by this medium, and use Eq. (IIIA-9) to find the evolution of the mode amplitude due to such perturbation. Suppose that the plasma and the perturbing medium are interpenetrating. Then the current $\overline{J}_p$ which flows in the perturbing medium is actually generated by the electric field of the plasma mode. As an example, assume the plasma mode is quasi-electrostatic. The electric field of this mode, $\overline{E}_o u(\overline{r},t)$ will induce an electric field $\overline{E}_p$ in the perturbing medium which is given, through Gauss' equation, by

$$E_{p\ell} = \frac{E_{o\ell}u(\overline{r},t)}{K_{p\ell}(\overline{k},\omega)} \qquad \text{(IIIB-1)}$$

where $K_{p\ell}(\overline{k},\omega)$ is the normalized longitudinal dielectric function as defined in Sec. IA. This electric field is just the mode electric field "dynamically shielded" by the perturbing medium. It now causes a current to flow in the perturbing medium which is just given by Eq. (IA-4), i.e.

$$\bar{J}_p = \frac{\bar{\bar{\sigma}}_p(\bar{k},\omega) \cdot \bar{k}}{kK_{p\ell}(\bar{k},\omega)} E_{o\ell}u(\bar{r},t) \qquad \text{(IIIB-2)}$$

Substituting Eq. (IIIB-2) into Eq. (IIIA-9) we find

$$(\frac{\partial}{\partial t} + \bar{v}_{go} \cdot \nabla + \gamma_o)u(\bar{r},t) = C_p(\bar{k},\omega)u(\bar{r},t) \qquad \text{(IIIB-3)}$$

where

$$\begin{aligned} C_p(\bar{k},\omega) &= -\frac{1}{4}\frac{|E_o|^2}{W_o}\frac{\sigma_{p\ell}(\bar{k},\omega)}{K_{p\ell}(\bar{k},\omega)} \\ &= -\frac{\sigma_{p\ell}(\bar{k},\omega)/\epsilon_{p\ell}(\bar{k},\omega)}{\omega \dfrac{\partial[\text{Re } K_\ell(\bar{k},\omega)]}{\partial\omega}} \end{aligned} \qquad \text{(IIIB-4)}$$

where we have used Eq. (IIA-25) to express the energy density of the unperturbed mode in terms of the longitudinal dielectric function of the plasma. For determining the stability of this interaction we look at the slow variation in time only, and simplify Eq. (IIIB-3) to

$$(\frac{\partial}{\partial t} + \gamma_o)\ u(t) = C_p\ \dot{u}(t) \qquad \text{(IIIB-5)}$$

The solution of Eq. (IIIB-5) is

$$u(t) = Ue^{-i\nu t} \qquad \text{(IIIB-6)}$$

where

$$\begin{aligned} \nu &= -i\gamma_o + iC_p \\ &= -\text{ Im } C_p + i\ [\text{Re } C_p - \gamma_o] \end{aligned} \qquad \text{(IIIB-7)}$$

Hence, if the unperturbed mode is stable, $\gamma_o > 0$, the perturbation will make it unstable if Re $C_p > \gamma_o$. A necessary condition for this to happen is that Re $C_p > 0$. From Eq. (IIIB-4), and expressing $K_{p\ell}$ in terms of $\sigma_{p\ell}$ by Eq. (IA-6), this necessary condition may be satisfied in either one of two circumstances: (1) If the unperturbed mode is a positive-energy mode, $w_o > 0$, then we need Re $\sigma_{p\ell} < 0$, i.e. the perturbing medium must present a negative real conductivity at the $(\omega,\bar{k})$ for the mode. (2) If the unperturbed mode is a negative-energy mode, $w_o < 0$, then the requirement is that Re $\sigma_{p\ell} > 0$, i.e. a positive real conductivity at $(\omega,\bar{k})$ may make it unstable. These conclusions are identical to the ones found from the conservation equation (see e.g. Eq. (IIA-23)). In addition, Eq. (IIIB-7) shows that there is also a shift in the real frequency of the mode. This we did not obtain from the conservation equation. Thus the mode-amplitude perturbation theory provides us with a modified dispersion relation for the mode under consideration.

Since our results are based on perturbation theory calculation they are valid only if $|C_p| << |\omega|$, consistent with our assumption of slow variation of the amplitudes. For a finite mode-energy, and the assumption of non-resonance $K_{p\ell}(\bar{k},\omega) \neq 0$, this can be satisfied for sufficiently small values of $|\sigma_{p\ell}(\bar{k},\omega)|$. It frequently happens, however, that the perturbing medium (usually also a plasma) is nonresonant only in certain $(\omega,\bar{k})$ regions of the unperturbed plasma wave spectrum, and in other $(\omega,\bar{k})$ regions it may be nearly resonant $(K_{p\ell} \approx 0)$. For such situations we can rewrite our perturbed dispersion relation by recognizing that $\nu = \omega' - \omega$, where $\omega'$ is the perturbed (complex) frequency and $\omega$ is the unperturbed (real) frequency, and that $K_{p\ell}$ should be allowed to vary in the vicinity of $\omega$. In this manner we extend our results and rewrite Eq. (IIIB-7) as

$$[\omega' - (\omega - i\gamma_o)]K_{p\ell}(\omega',\bar{k}) \cong - \frac{i}{4}\frac{\sigma_{p\ell}(\omega,\bar{k})}{w_o/|E_o|^2} \qquad \text{(IIIB-8)}$$

We shall find this very useful in understanding the possible instabilities in plasmas where various particle components have different dispersion characteristics.

A simple example that illustrates an instability driven by a nonresonant interaction is that of a cold, collisionless electron beam (Eq. IID-1) passing through a stationary collisionless electron plasma whose thermal velocity $v_e$ is much larger than the beam velocity $v_o$. If we ignore the ion-dynamics, for the phase velocities of the beam $\omega/k \sim v_o << v_e$ the electron plasma behaves as a passive dissipative medium

(Eq. IID-11 with $v_o = 0$). Hence from Eqs. (IIIB-4) through (IIIB-7) we find that the negative energy wave on the beam will be driven unstable, and the growth rate can be calculated from Eq. (IIIB-7).

A much more complicated example is the cold, collisionless beam-plasma interaction. For a beam density that is small compared to the plasma density we might consider the perturbation of the plasma oscillations by the beam. But since $K_b \sim 0$ in the region of strong interaction we find that Eqs. (IIIB-3 and 4) must be taken as in Eq. (IIIB-8) where $\sigma_{p\ell}$ and $K_{p\ell}$ become the beam conductivity and beam dielectric functions. The net result is an approximate overall dispersion relation

$$(\omega - \omega_p)(\omega - kv_o)^2 \approx \frac{\omega_b^2 \omega_p}{2} \qquad \text{(IIIB-9)}$$

which is correct for the resonance region $\omega \sim \omega_p \sim kv_o$. The solution of (IIIB-9) gives the well-known results

$$\frac{\omega_{im}}{\omega_p} \cong \frac{\sqrt{3}}{2^{4/3}} \left(\frac{n_b}{n_p}\right)^{1/3} \qquad \text{(IIIB-10)}$$

$$\frac{\omega_r}{\omega_p} \cong 1 - \frac{1}{2^{4/3}} \left(\frac{n_b}{n_p}\right)^{1/3} \qquad \text{(IIIB-11)}$$

where Eq. (IIIB-10) is the maximum growth rate. The complete solution of the exact dispersion relation is shown in Fig. IIIB-1, where $k_{rc} = (\omega_p/v_o)[1+(n_b/n_p)^{1/3}]^{3/2}$. We note that the unstable interaction also exists for frequencies below $\omega_p$ but not above $\omega_p$. In these regions we must regard this as the perturbation of the beam waves by the presence of the reactive medium which the plasma presents. Here, below $\omega_p$ we find that the positive and negative energy waves on the beam couple to each other to produce the instability, which for $\omega << \omega_p$ is given by

$$\omega_i \approx kv_o \left(\frac{n_b}{n_p}\right)^{1/2} \qquad \text{(IIIB-12)}$$

$$\omega_r \approx kv_o \qquad \text{(IIIB-13)}$$

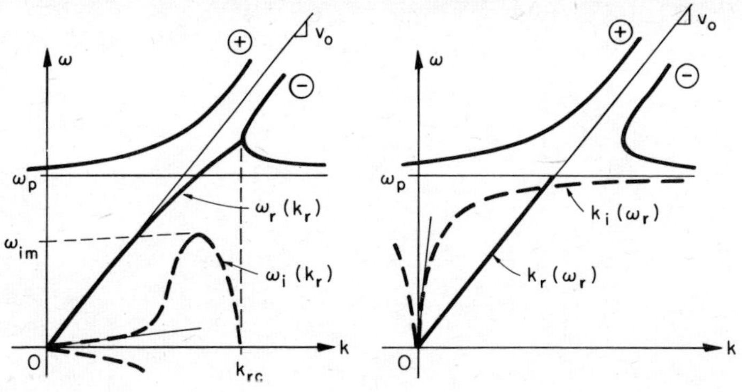


Fig. III B-1

Finally, we note from Fig. IIIB-1 that when the beam-plasma interaction is regarded in space, for a steady state real frequency, the growth rate is singular. Clearly any perturbation analysis would break down, and this is indeed signaled by Eq. (IIIB-3) since $v_g = 0$ for the cold plasma. This singular behavior can of course be removed by adding finite temperature effects to the plasma model, but we shall find that the true nature of this instability can only be obtained from a detailed stability analysis.

Reactive, nonresonant coupling can also occur in plasmas with anisotropic velocity distributions. Thus for example an ion beam having a beam-type velocity distribution function as in Eq. (IID-6) can be driven unstable by the presence of cold electrons that provide a reactive medium which couples the positive and negative energy modes of the ions.

## C. *Resonant Coupling*

The coupling of two systems, where each one has well defined modes in the absence of coupling, must be dealt with separately. In this case each system must be treated as perturbed by the other, so that in general all the modes (waves) of one system couple to all the modes of the other system. The simplest situation in this case would be the coupling of one wave from one system to one wave of the other system, so that we have a coupled set of equations of the form (IIIA-9) which now are:

$$(\frac{\partial}{\partial t} + v_1 \cdot \nabla + \omega_{i1})u_1 = K_{12}u_2 e^{-i\delta\omega t} e^{i\overline{\delta}_k \cdot \overline{r}} \quad \text{(IIIC-1)}$$

$$(\frac{\partial}{\partial t} + v_2 \cdot \nabla + \omega_{i2})u_2 = K_{21}u_1 e^{i\delta_\omega t} e^{-i\bar{\delta}_k \cdot \bar{r}} \quad \text{(IIIC-2)}$$

where

$$K_{12} = \frac{-\frac{1}{4}\bar{E}^*_{10} \cdot \bar{J}^{(1)}_{20}}{w_1} \triangleq \frac{p_{12}}{w_1} \quad \text{(IIIC-3)}$$

$$K_{21} = \frac{-\frac{1}{4}\bar{E}^*_{20} \cdot \bar{J}^{(2)}_{10}}{w_2} \triangleq \frac{p_{21}}{w_2} \quad \text{(IIIC-4)}$$

$$\delta_\omega = \omega_2 - \omega_1 \quad \text{(IIIC-5)}$$

$$\bar{\delta}_k = k_2 - k_1 \quad \text{(IIIC-6)}$$

where $\bar{J}^{(1)}_{20}$ is the current in 2 that interacts with 1, and $\bar{J}^{(2)}_{10}$ is the current in 1 that interacts with 2, and $v_1$ and $v_2$ are the group velocities of the unperturbed waves.

The interaction is clearly strongest when $\delta_\omega = 0$, $\bar{\delta}_k = 0$ which we call resonance, and for waves that are undamped, $\omega_{i1,2} = 0$. We shall first study such interactions. If in addition we restrict ourselves to interactions that are conservative, i.e. those for which the perturbed fields also satisfy the conservation of energy equation:

$$\sum_{i=1,2} (\frac{\partial}{\partial t} + v_i \cdot \nabla) w_i |u_i|^2 = 0 \quad \text{(IIIC-7)}$$

then, direct substitution of Eqs. (IIIC-1 and 2) into Eq. (IIIC-7) shows that

$$w_2 K_{21} = - K^*_{12} w_1 \quad \text{(IIIC-8)}$$

or

$$p_{21} = - p^*_{12}$$

and only one coupling coefficient is needed to specify the interaction. This allows us to classify all possible interactions without needing any particular model. We restrict ourselves to one spatial dimension. It is convenient to first separate the interactions so that there is only either coupling in time or space. We then find that the couplings are either stable or unstable depending upon whether the energies of the two modes are of the same sign or of opposite sign.

1. *Stable Couplings:* $w_1 w_2 > 0$. For coupling in time we find that the energy is coupled back and forth between the two waves as shown in Fig. IIIC-1a. The minimum time for a complete energy transfer is $\pi/2\nu$ where

$$\nu = \frac{|p_{12}|}{(w_1 w_2)^{1/2}} \qquad \text{(IIIC-9)}$$

For coupling in space we must distinguish two possible cases.

(a) Group velocities in the same direction: $v_1 v_2 > 0$, in which case the power coupling is also periodic as in Fig. IIIC-1a, and the minimum distance for total power transfer from one system to the other is $\pi/2\beta$ where

$$\beta = \frac{|p_{12}|}{(s_1 s_2)^{1/2}} = \frac{\nu}{(v_1 v_2)^{1/2}} \qquad \text{(IIIC-10)}$$

(b) Group velocities in the opposite direction: $v_1 v_2 < 0$ in which case the coupling is exponential in space as shown in Fig. IIIC-1b. Here we find the power transfer from 2 to 1 in a coupling distance L to be

$$\frac{s_1(L)}{|s_2(L)|} = \tanh^2 \alpha L \qquad \text{(IIIC-11)}$$

where

$$\alpha = \frac{|p_{12}|}{|s_1 s_2|^{1/2}} = \frac{\nu}{|v_1 v_2|^{1/2}} \qquad \text{(IIIC-12)}$$

Note the exponential behavior of $s_{1,2}$ does not imply an unstable growth in space.

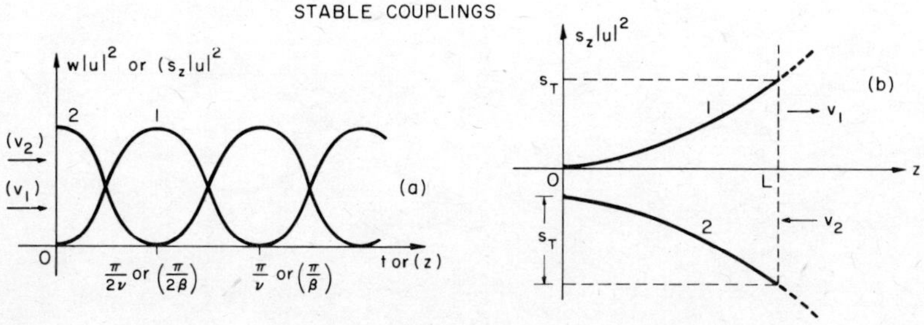


Fig. III C-1

2. *Unstable Couplings:* $w_1 w_2 < 0$. For coupling in time we find the energy in both waves increasing exponentially as shown in Fig. IIIC-2a, at a rate given by

$$\gamma = \frac{|p_{12}|}{|w_1 w_2|^{1/2}} \tag{IIIC-13}$$

Note that the total energy is conserved.

For coupling in space we must again distinguish between two cases.

(a) Group velocities in the same direction: $v_1 v_2 > 0$, for which the power also grows exponentially, as shown in Fig. IIIC-2a, at a rate

$$\alpha = \frac{|p_{12}|}{|s_1 s_2|^{1/2}} \tag{IIIC-14}$$

This is an example of spatial amplification.

(b) Group velocities in opposite direction: $v_1 v_2 < 0$. Now the power transfer is periodic as shown in Fig. IIIC-2b, but because of the opposing group velocities the power transfer over an interaction length L is

$$\frac{s_2(0)}{s_2(L)} = \frac{1}{\cos^2 \beta L} \tag{IIIC-15}$$

where

$$\beta = \frac{|p_{12}|}{(s_1 s_2)^{1/2}} \tag{IIIC-16}$$

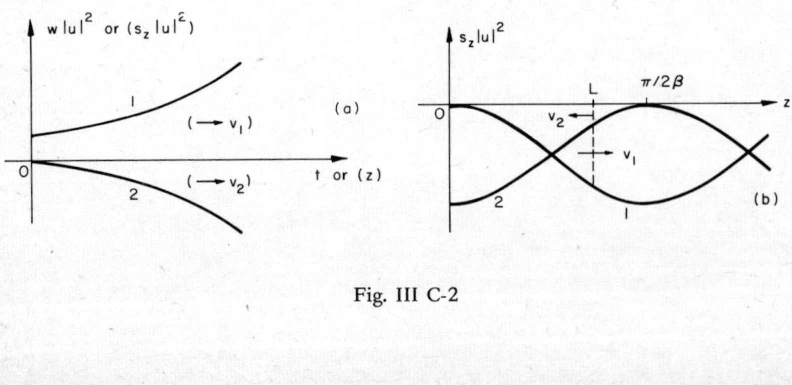


Fig. III C-2

For $L < \pi/2\beta$ we obtain an amplification of $s_2(0)$, but for $L = \pi/2\beta$ the gain becomes infinite indicating the possibility of start of oscillations from e.g. noise. Of course for $L > \pi/2\beta$ we always have the possibility for oscillations. The mechanism which makes this coupling unstable is the giving up of power by the negative-energy mode 1 to the positive-energy mode 2 which carries it in the direction opposite to 1 and therefore reexcites mode 1 where its amplitude was smaller; we have here a continuous feedback which for a critiical length $L = \pi/2\beta$, i.e. for sufficient feedback, will make the coupled system oscillate.

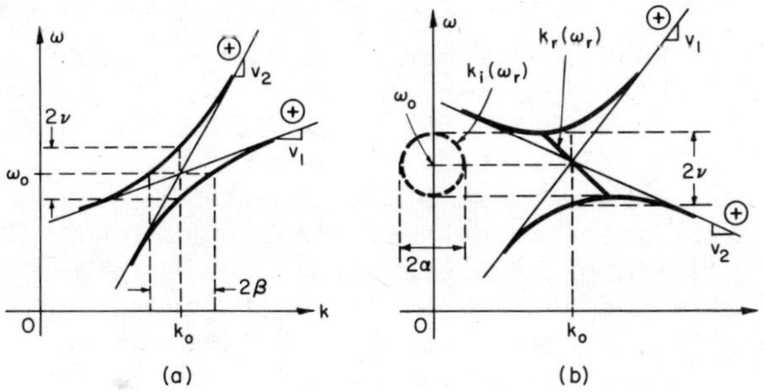


Fig. III C-3

It is instructive to also look at the coupling off-resonance, i.e. for $\delta_\omega \neq 0$ and $\delta_k \neq 0$. In these cases the coupled equation still leads to quadratic dispersion relations for the coupled system. The solutions of these dispersion relations are straightforward, and the results are illustrated in Fig. IIIC-3 for the stable couplings when $w_1 w_2 > 0$, and in Fig. IIIC-4 for the unstable couplings when $w_1 w_2 < 0$. As before, in each of these cases we must further distinguish between couplings for which $v_1 v_2 > 0$ and couplings for which $v_1 v_2 < 0$. The important thing to note is that solutions of even the simplest dispersion relations for either $\omega_i(k_r)$ or $k_i(\omega_r)$ alone can be misleading. Thus in the case of Fig. IIIC-3a the modes $k_i(\omega_r)$ do not imply that the coupling is unstable. Conversely the fact that in Fig. IIIC-4a for real $\omega$ all the k solutions are real does not guarantee that the coupling is stable.

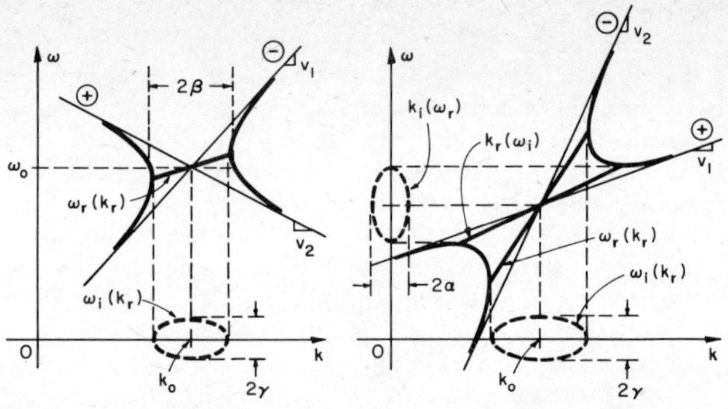


Fig. III C-4

In the weak coupling of modes the manner in which some couplings are stable or unstable could be resolved by the application of appropriate boundary conditions. In general this is difficult and we first study systems that are of infinite extent. In that case we must study the stability of the system in a manner that is independent of the boundaries at infinity. This implies that imposing either real $\omega$ or real k on the coupled system is not advisable. We must then study the evolution of localized excitations in both time and space. Returning to the coupled mode Eqs. (IIIC-1 and 2) with $\omega_{i1,2} = 0$, $\delta_\omega = 0$, $\bar{\delta}_k = 0$, and one spatial dimension, we find the Green's function response ($u_1 = g_1(z,t)$ and $u_2 = g_2(z,t)$) for the set of equations when system 2 is excited by a delta function at $z = 0$, $t = 0$. This is found most easily by Fourier Laplace transform techniques. We summarize the results one obtains for the unstable cases when $w_1 w_2 < 0$:

(a) $v_1 v_2 > 0$

$$g_1(z,t) = \frac{K_{12}}{v_2 - v_1} I_o[C(\tau_A^2 - \frac{z^2}{v_s^2})^{1/2}] \times [u(\tau_2) - u(\tau_1)]u(z) \qquad \text{(IIIC-16)}$$

(b) $v_1 v_2 < 0$

$$g_1(z,t) = \frac{K_{12}}{v_1 - v_2} I_o[C(\tau_A^2 - \frac{z^2}{v_s^2})^{1/2}] \times [u(\tau_1)u(z) + u(\tau_2)u(-z)] \qquad \text{(IIIC-17)}$$

where

$$c^2 = 4\,\frac{|v_1 v_2|}{(v_2 - v_1)^2}\,\frac{|p_{12}|^2}{|w_1 w_2|} \tag{IIIC-18}$$

$$\tau_i = t - \frac{z}{v_i} \tag{IIIC-19}$$

$$v_A^{-1} = \frac{1}{2}\,(v_1^{-1} + v_2^{-1}) \tag{IIIC-20}$$

$$v_s^{-1} = \frac{1}{2}\,(v_1^{-1} - v_2^{-1}) \tag{IIIC-21}$$

and the u's are the unit step functions. These are shown calculated in Fig. IIIC-5 which clearly indicates the difference between them. In case (a) the unstable pulse convects away from the origin, while in case (b) the instability is encompassing more and more of the system. This is the difference between a convective and absolute instability which we will study more generally in the next section.

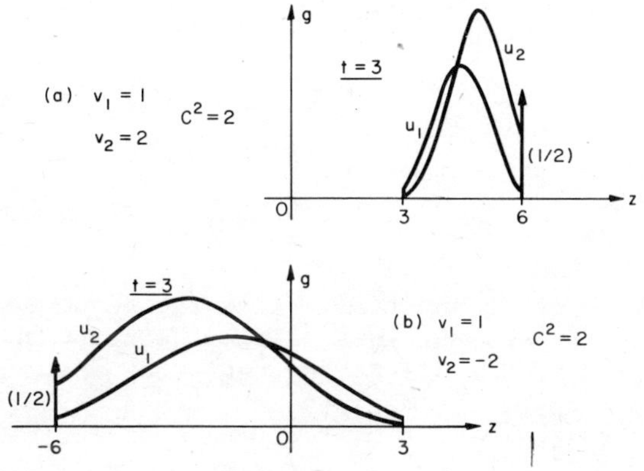


Fig. III C-5

## IV. STABILITY ANALYSIS

We now turn to the linear stability analysis of a plasma. As shown in Section I, for *any* plasma model the Green's function is related to the dispersion relation (Eq. IA-12). We shall relate the stability of a plasma to the time-asymptotic behavior of the Green's function, and thus to the complex $\omega$

and $\bar{k}$ plane characteristics of the dispersion relation.

For simplicity we consider the one-dimensional space case first. Equation (IA-12) becomes

$$G(z,t) = \int_L \frac{d\omega}{2\pi} e^{-i\omega t} \int_F \frac{dk}{2\pi} e^{ikz} \frac{1}{D(k,\omega)} \qquad \text{(IV-1)}$$

where $D(k,\omega)=0$ is the appropriate dispersion relation. The integrations in Eq. (IV-1) must be performed on contours placed in regions of absolute convergence in the complex $\omega$ and $k$ planes. These can be chosen as follows. Since the response in space may be bidirectional (i.e. for $z > 0$ and $z < 0$) the region of absolute convergence in the complex k-plane is a strip, as shown in Fig. IV-1a. The strip must include the

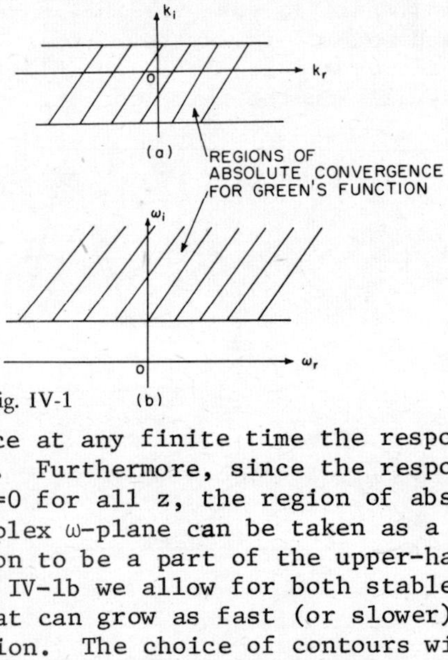


Fig. IV-1

real k-axis, since at any finite time the response must be limited in space. Furthermore, since the response must be causal, $G(z,t<0)=0$ for all z, the region of absolute convergence in the complex $\omega$-plane can be taken as a half-plane. Taking this region to be a part of the upper-half $\omega$-plane as shown in Fig. IV-1b we allow for both stable responses and unstable ones that can grow as fast (or slower) than the lower limit of the region. The choice of contours within these regions of absolute convergence are in general interdependent. ($D(k,\omega)$ is almost never the product of a function of k only times a function of $\omega$ only). Suppose we choose the F contour for Eq. (IV-1) on the real k-axis, as shown in Fig. IV-2a. Then, focusing our attention on the $\omega$-plane integration in Eq. (IV-1) we examine the function

$$\int_L \frac{d\omega}{2\pi} e^{-i\omega t} \frac{1}{D(k_r,\omega)} \equiv H(k_r,t) \qquad \text{(IV-2)}$$

The singularities of the integrand are given by

$$D(k_r,\omega) = 0 \tag{IV-3}$$

or equivalently by the branches

$$\omega(k_r) = \omega_r(k_r) + i\ \omega_i\ (k_r) \tag{IV-4}$$

which we show in Fig. IV-2b. This allows us to choose the L-contour so that causality will be satisfied: L must lie above the highest branch of $\omega(k_r)$, as is also shown in Fig. IV-2b. Note that for L so chosen, its map onto the complex k-plane through

$$D(k,\ \omega_L) = 0 \tag{IV-5}$$

i.e.

$$k(\omega_L) = k_r(\omega_{Lr},\ \omega_{Li}) + i\ k_i(\omega_{Lr},\omega_{Li}) \tag{IV-6}$$

cannot have any branches that cross the real k-axis. Thus, the k-plane integration in Eq. (IV-1)

$$\int_F \frac{dk}{2\pi}\ e^{ikz}\ \frac{1}{D(k,\omega_L)} \equiv I(z,\omega_L) \tag{IV-7}$$

exhibits the proper division between modes that are generated for z>0 and those generated for z<0.

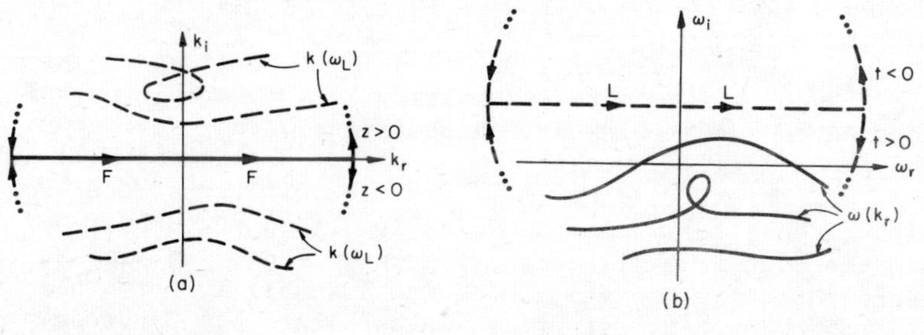


Fig. IV-2

At this stage the integrations in Eq. (IV-1) are completely specified, and in principle G(z,t) could be determined for all z and t. In practice, $D(k,\omega)$ is rarely simple enough for this to be carried out. For determining the stability or instability nature of the response we do not need to know this much detail. We shall only need to know the time-asymptotic response of G(z,t) for all z. Thus, we shall define a plasma (model) to be *absolutely unstable* if

$$\lim_{t\to\infty} G(z,t) \to \infty \tag{IV-8}$$

for all z. Conversely, if for all z (see Fig. IV-3)

$$\lim_{t\to\infty} G(z,t) = 0 \tag{IV-9}$$

a plasma (model) will be classified *not absolutely unstable*; under these circumstances the plasma may exhibit *either stable or convectively unstable* responses. Referring to Fig. IV-2b we see that the distinction between Eqs. (IV-8 and 9) hinges upon whether the region of absolute convergence cannot or can be extended to just below the real $\omega$-axis. Clearly, *if for all* $k_r$ *and all branches of Eqs. (IV-3 and 4)* $\omega_i(k_r)<0$ *the plasma (model) is stable.* In contrast, if for any one branch we find $\omega_i(k_r)>0$ for some $k_r$ values the plasma is said to be unstable; in this case one must further distinguish between absolute and convective instabilities.

The importance of distinguishing between absolute and convective instabilities is evident from Fig. IV-3. Since these instabilities are linear, their importance is primarily in the manner in which they take the perturbations into the nonlinear regime; this may be very much different in the case of a convective instability from the case of an absolute instability, since e.g. in the latter at every point in space the perturbations eventually grow to large amplitudes. As we shall see, an absolute instability is an unstable normal mode, and as such represents the build-up of an oscillator, while a convective instability may stably amplify signals and noise. Finally, since most systems of interest are of finite extent, the manner in which localized perturbations reach the boundaries is very important. Thus, e.g. in the presence of reflections a convective instability (or amplifier) can become absolute (i.e. an oscillator). Conversely, an absolute instability in a limited spatial extent may be stabilized, so that only convective growth be possible. Hence the direction of propagation of waves in an unstable medium is of great importance, and we shall find that the techniques usually used for determining this direction in a stable medium (i.e. the sign of the group velocity) may no longer be applicable.

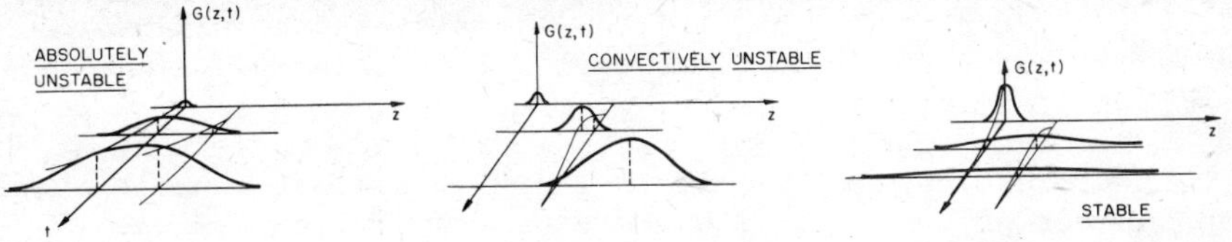


Fig. IV-3

## A. *Absolute Instabilities - Normal Modes*

In order to determine the existence of absolute instabilities we examine the singularities in the $\omega$-integrand of Eq. (IV-1), i.e. the singularities of

$$I(z,\omega) = \int_F \frac{dk}{2\pi} \frac{e^{ikz}}{D(k,\omega)} \qquad \text{(IV.10)}$$

in the $\omega$-plane between the L-contour and the real $\omega$-axis (Fig. IV-2). It is these singularities that could prevent us from extending the region of absolute convergence into the lower-half $\omega$-plane. One such obvious singularity is provided by the $\omega(k_r)$ branches that extend into the upper-half $\omega$-plane; these are lines of discontinuity in the function $I(z,\omega)$. To better see this, let us explicit the integral of Eq. (IV-10) for the simplest situation. Assume that: (a) $D^{-1}$ is single-valued and has only isolated singularities; (b) $\lim_{k\to\infty} D^{-1} \to 0$ at least as, or faster than, $k^{-1}$. Then, for $\omega$ within the region of absolute convergence, the integral in Eq. (IV-10) can be carried out by closure ($|k|\to\infty$) in the upper k-plane for z>0 and in the lower k-plane for z<0, and using residue evaluation. We thus obtain

$$I(z,\omega) = iu(z)\sum_u \frac{e^{ik_u(\omega)z}}{\left(\frac{\partial D}{\partial k}\right)_{k_u}} - iu(-z)\sum_\ell \frac{e^{ik_\ell(\omega)z}}{\left(\frac{\partial D}{\partial k}\right)_{k_\ell}} \qquad \text{(IVA-1)}$$

where u(z) is the unit-step function in z. In Eq. (IVA-1) the residues are written for first-order poles in the k-plane, and $\omega$ considered so chosen in the convergence region. For other $\omega$'s within the region of absolute convergence the k-plane singularities may of course be higher-order poles; their residues would be obtained by a limiting process of coalescing poles *within each* of the sums of Eq. (IVA-1). As is evident from Fig. IV-2, each term in Eq. (IVA-1) represents a spatially decaying function of z, either for z>0 or for z<0. If we attempt to define the region of absolute convergence in $\omega$ to be below the highest $\omega(k_r)$ branch we find of course that we would violate causality, and in addition discontinuously change the number of modes for z>0 and z<0. It may be possible to avoid this if we can first deform F to $\tilde{F}$ so that the changed branches, $\omega(k_r)$ to $\omega(k_{\tilde{F}})$, lie lower in the $\omega$-plane. This is shown in Figs. IVA-1a and b. This type of contour deformation would be inhibited if at some $\omega$ the $\tilde{F}$ contour becomes "*pinched*" by two (or more) $\tilde{L}$ contours, as shown in Fig. IVA-1c.

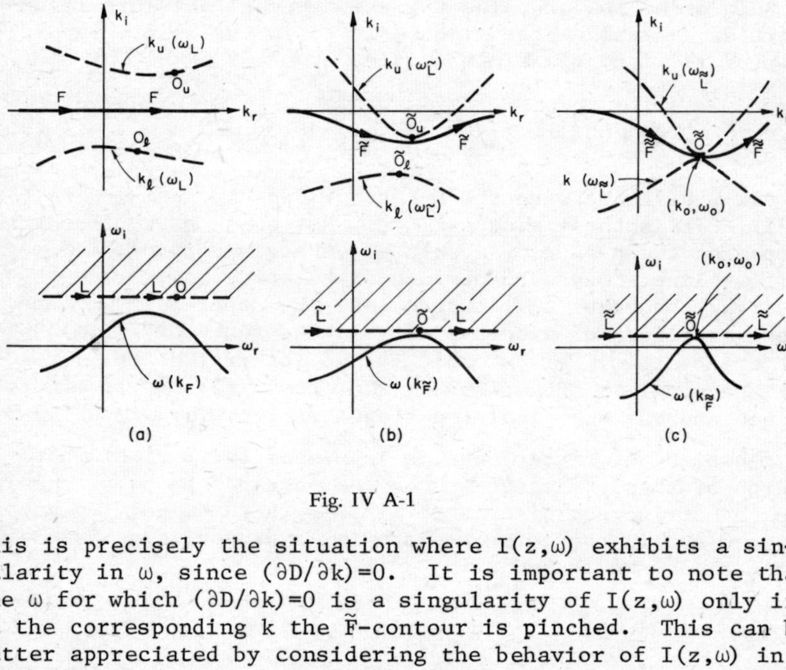


Fig. IV A-1

This is precisely the situation where $I(z,\omega)$ exhibits a singularity in $\omega$, since $(\partial D/\partial k)=0$. It is important to note that the $\omega$ for which $(\partial D/\partial k)=0$ is a singularity of $I(z,\omega)$ only if at the corresponding k the $\tilde{F}$-contour is pinched. This can be better appreciated by considering the behavior of $I(z,\omega)$ in the vicinity of such a singularity. Let such a singularity occur at $\omega=\omega_o$ and its corresponding $k=k_o$, then

$$\left(\frac{\partial D}{\partial k}\right)_{\omega_o,k_o} = 0 \qquad \text{(IVA-2)}$$

and

$$D(k_o, \omega_o) = 0 \qquad \text{(IVA-3)}$$

In the simplest case, the vicinity of this point in $\omega$ and k is characterized by the fact that as $\omega$ approaches $\omega_o$ (from above) two k-roots (from *opposite* sides of $\tilde{F}$) approach each other to form a double root at $k_o$, i.e. (see Fig. IVA-2)

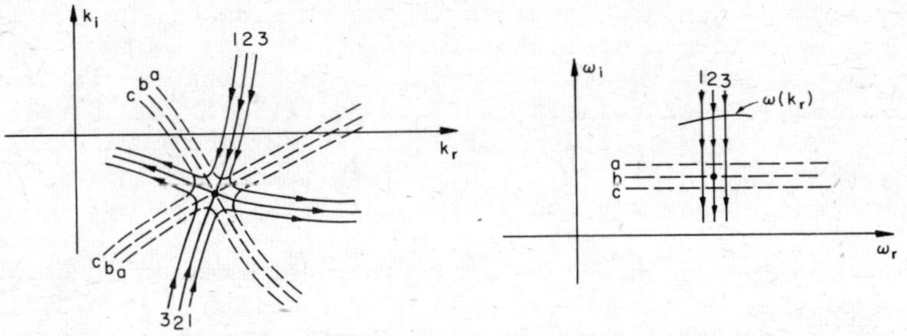


Fig. IV A-2

$$\omega - \omega_o \sim (k - k_o)^2 \qquad \text{(IVA-4)}$$

In this neighborhood we can write the dispersion relation as

$$D(k,\omega) = [k - k_u(\omega)]\,[k - k_\ell(\omega)]\,R(k,\omega) \qquad \text{(IVA-5)}$$

where

$$\left.\begin{aligned} k_u &= k_o + C(\omega - \omega_o)^{1/2} \\ k_\ell &= k_o - C(\omega - \omega_o)^{1/2} \end{aligned}\right\} \qquad \text{(IVA-6)}$$

where C is a complex constant and $R(k,\omega)$ is regular near $(\omega_o, k_o)$ Substituting into Eq. (IVA-1) we find

$$I(z,\omega) = \frac{u(z)e^{ik_u z}}{i2CR(\omega-\omega_o)^{1/2}} + \frac{u(-z)e^{ik_\ell z}}{i2CR(\omega-\omega_o)^{1/2}} + I_R \qquad \text{(IVA-7)}$$

where $I_R$ contains the remaining summations over $k_u$ and $k_\ell$ that are not near $k_o$. We note since $u(-z)=1-u(z)$, in the limit of $k_u \to k_\ell \to k_o$ Eq. (IVA-7) becomes

$$I(z,\omega) = \frac{e^{ik_oz}}{i2CR_o(\omega-\omega_o)^{1/2}} + u(z)\frac{\partial}{\partial k}\left[\frac{e^{ikz}}{iR}\right]_{k_o} + I_{Ro} \quad \text{(IVA-8)}$$

The first term extends over *all* z and has a branch-pole singularity at $\omega=\omega_o$. The second term is just the residue in the second order pole at $k=k_o$, and is regular in $\omega$ as is also the last term. We can now see that if the merging of the two roots in k had been formed by either two $k_u$ or two $k_\ell$ roots, the first two terms in Eq. (IVA-7) would both belong to either u(z) or u(-z) and the singularity term in Eq. (IVA-8) would not exist. Thus the $\omega$-solutions of Eqs. (IVA-2) and (IVA-3) will be singular points of $I(z,\omega)$ only if their corresponding double k values arise from a pinch of the deformed F-contour.

By continuing this type of contour deformation in both complex planes we can uncover all of the "pinch" singularities of $I(z,\omega)$. Assuming that these are found, the evaluation of G(z,t) is given by the integral along the $\tilde{L}$ contour shown in Fig. IVA-3,

$$G(z,t) = \int_{\tilde{L}} \frac{d\omega}{2\pi}\, e^{-i\omega t}\, I(z,\omega) \quad \text{(IVA-9)}$$

Assuming that the highest singularity is of the pinch-type found in Eq. (IVA-8), the time-asymptotic response is

$$\lim_{t\to\infty} G(z,t) \sim \lim_{t\to\infty} \int_{\omega_o} \frac{d\omega}{2\pi}\, e^{-i\omega t}\, \frac{e^{ik_oz}}{(\omega-\omega_o)^{1/2}}$$

$$\sim e^{ik_oz}\,\frac{e^{-i\omega_o t}}{t^{1/2}} \quad \text{(IVA-10)}$$

for *all* z. If $\omega_o$ is in the upper-half $\omega$-plane, the time-asymptotic response found in Eq. (IVA-10) grows in time, at all points in space, i.e. an *absolute instability*.

We can gain further insight into the nature of an absolute instability by examining the manner in which the asymptotic mode of Eq. (IVA-10) is set up. This can be done by tracing the path of the poles that pinch, from their original position when $\omega$ is on L to their pinched position on $\tilde{\tilde{F}}$ when $\omega$ is on $\tilde{\tilde{L}}$,

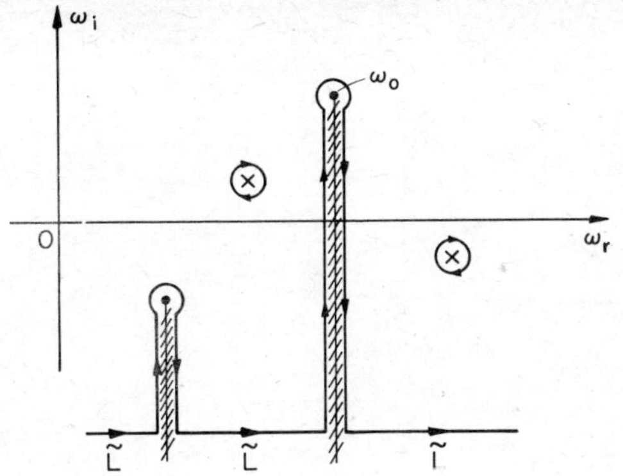


Fig. IV A-3

as shown by the sequence of points labelled O, $\tilde{O}$ and $\tilde{\tilde{O}}$, and their mapped partners, in Fig. IVA-1. Consider the frequencies on L to represent the source excitation of the medium at z=0. Then for $\omega$ on L at O the response is given by $O_u$ and $O_\ell$ both of which are spatially decaying, respectively, for z>0 and z<0, since they are on opposite sides of F (see Fig. IV-2a) The reason that these responses are decaying in space is directly tied to the causality requirement which forces us to place the L-contour above the highest branch of $\omega(k_F)$. Thus even though a medium supports growing (i.e. unstable) responses its localized excitation by a source that grows sufficiently fast in time will produce only decaying responses in space. This is illustrated schematically in Fig. IVA-4, for the z>0 response. In addition we note that the spatial response at z=0 will be discontinuous because $O_u$ and $O_\ell$ are decaying in opposite directions, and their k's have different real parts; this is illustrated in Fig. IVA-5. The discontinuity at the z=0 origin is accommodated by the presence of the source. As we deform F to $\tilde{F}$ and L to $\tilde{L}$ (Fig. IVA-1) the growth rate of the "source" can now be reduced from O to $\tilde{O}$, without violating causality, and we then find the spatial responses at $\tilde{O}_u$ and $O_\ell$, respectively, for z>0 and z<0. Now the response for z>0 is growing in space because $\tilde{O}_u$ is above $\tilde{F}$ and its imaginary part is negative. The appearance of this growing response can again be appreciated from Figs. IVA-1 and 4; since the time-growth rate of the source at $\tilde{O}$ is below that of the highest branch of $\omega(k_F)$ the response $\tilde{O}_u$ is now spatially growing for z>0. The response $\tilde{O}_\ell$ for z<0, not having crossed the real k-axis, is still decaying, and the total spatial response is still discontinuous at z=0, as shown in Fig. IVA-5.

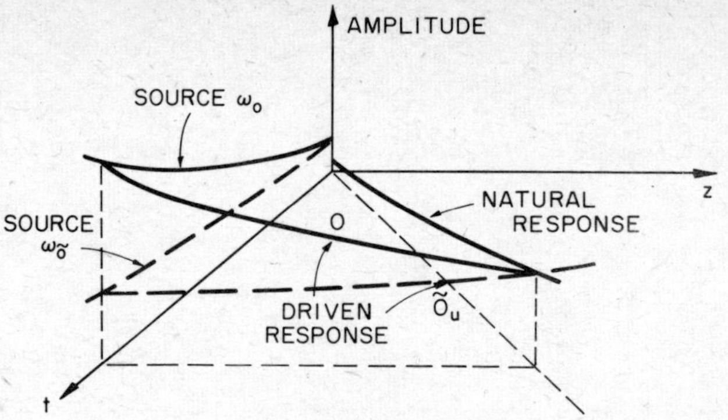


Fig. IV A-4

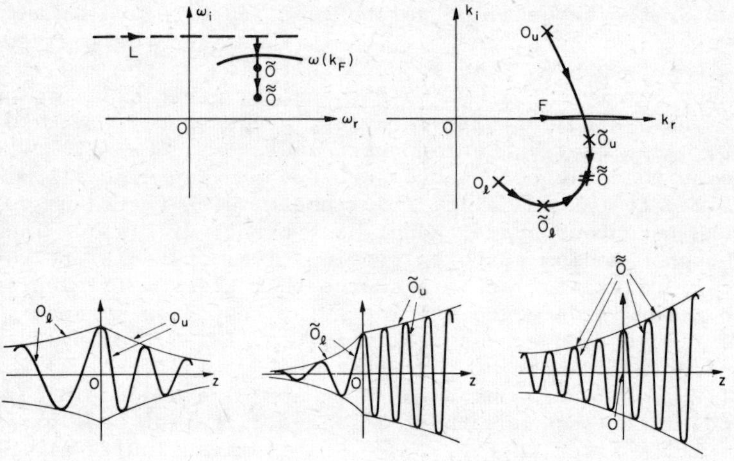


Fig. IVA-5

Finally when the further deformation of $\tilde{F}$ to $\tilde{\tilde{F}}$ and $\tilde{L}$ to $\tilde{\tilde{L}}$ encounters a pinching of $\tilde{\tilde{F}}$ at $\tilde{\tilde{O}}$ the spatial response becomes perfectly continuous at z=0 (see also Eq. (IVA-10)); as $\tilde{O}_\ell$ and $\tilde{O}_u$ approach $\tilde{\tilde{O}}$ the spatial decay rate of the former for z<0 approaches the spatial growth rate for z>0 of the latter and their spatial oscillations coincide (see Fig. IVA-5). Since the discontinuity at z=0 has disappeared there is no longer a necessity for a source at the origin. Hence, the pinch points $\tilde{\tilde{O}}$ represent a *normal mode* of the medium. The normal mode with the largest imaginary part of its frequency will dominate the time-asymptotic response. If the imaginary part of this frequency is positive the normal mode is growing in space and in time, for all points of space, which is what we called an absolute instability. The fact that absolute instabilities are normal modes means that they represent the "breaking into oscillations" of the unstable system, i.e. the build-up of an oscillator; clearly, this can start from noise, or any other initiating excitation, and will reach a final state, perhaps a steady state, that will be determined by nonlinear effects. We remark that by the described contour deformations in the complex k and complex $\omega$ planes all the normal modes given by pinch points can be uncovered. Their number must be equal to the number of independent initial conditions that can be specified, and their superposition represents the general linear response of the medium.

The analytic form of the vicinity of a pinch-type normal mode can now be generalized from that of Eq. (IVA-4) to

$$(\omega - \omega_o)^n \sim (k - k_o)^{p+2} \qquad \text{(IVA-11)}$$

with n = 1, 2 . . . and p = 0, 1 . . ., where at least two of the k-roots approach $k_o$ from opposite sides of the deformed F-contour. The relation of Eq. (IVA-11) implies that in addition to Eqs. (IVA-2) and (IVA-3) other derivatives of D with respect to $\omega$ and/or k may vanish at the pinch point. Following the development of Eqs. (IVA-5 through 8), it is clear that a general pinch-type singularity in $I(z,\omega)$ may be either a simple branch pole, or a simple pole, or a higher-order branch pole, or a higher-order pole, depending on the value of n and p in Eq. (IVA-11). With $\omega_o$ in the upper-half $\omega$-plane these could lead to time asymptotic responses growing even faster than the simplest one of Eq. (IVA-10).

The pinch-type normal mode can also be generalized to the case of a three-dimensional Green's function, as would arise from a point source. In that case the simplest pinch-type singularity in $I(\bar{r},\omega)$ would occur for $\omega=\omega_o$ and $\bar{k}=\bar{k}_o$ satisfying

$$\left(\frac{\partial D}{\partial \bar{k}}\right)_{\omega_o,\bar{k}_o} = 0 \qquad \text{(IVA-12)}$$

and

$$D(\bar{k}_o,\omega_o) = 0 \qquad \text{(IVA-13)}$$

The deformation of contours must now be carried out simultaneously in four complex planes, $\omega$, $k_x$, $k_y$ and $k_z$, with respectively starting contours L, $F_x$, $F_y$, and $F_z$. This is now considerably more complicated since the dispersion relation $D(\overline{k},\omega) = 0$ can only provide the mapping between the $\omega$-plane and one of the k-planes, the other two k-values being fixed. Choosing L to be above the highest branch of $\omega(\overline{k_F})$, so as to satisfy causality, we can establish which of the solutions to Eq. (IVA-12) and (IVA-13) are pinch-type singularities, as follows. Suppose that the four-dimensional point 0 $(\omega_o,\overline{k}_o)$ is a simple pinch-type singularity; then, keeping any two of the three k's real (e.g. $k_x$ and $k_y$) the map of the L-contour into the third k-plane (e.g. $k_z$-plane) must have branches on *both* sides of its real axis; furthermore, deforming the L-contour down to L and passing through $\omega_o$ must result in one of its mapped branches, in the third k-plane (e.g. $k_z$-plane), to partly cross the real k-axis; finally letting each of the two k's that were kept real approach their respective pinch points $k_o$ (e.g. $k_{ox}$ and $k_{oy}$), we must find that two of the mapped branches of $\tilde{L}$ in the third complex k-plane will pinch (at e.g. $k_{oz}$). Figure IVA-6 illustrates such mappings; for simplicity

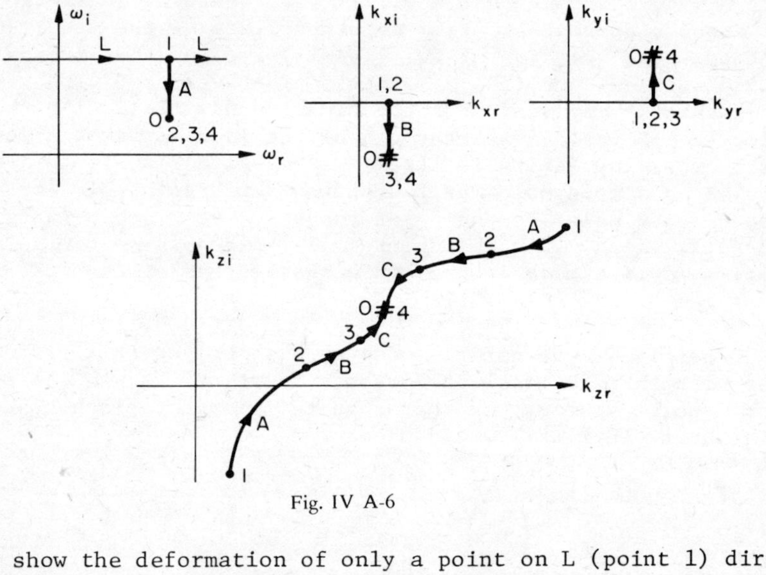


Fig. IV A-6

we show the deformation of only a point on L (point 1) directly to 0 (point 2) through path A whose map is shown in the $k_z$-plan with $k_x$ and $k_y$ remaining real and fixed, etc. Such mapping mus be found for each of the k-planes. If in any one k-plane the described pinching does not occur from L-mapped branches on opposite sides of the real k-axis, the pinch points of Eqs. (IVA-12) and (IVA-13) will not be singular points of $I(\bar{r},\omega)$. For the simplest pinch-type singularity, assuming it is the highest singularity in the ω-plane, the time-asymptotic Green's function is then found to be

$$\lim_{t\to\infty} G(\bar{r},t) \sim e^{i\bar{k}_o\cdot\bar{r}} \frac{e^{-i\omega_o t}}{t^{3/2}} \qquad \text{(IVA-14)}$$

A somewhat simpler situation arises for guided waves. Letting z be the direction of homogeneity and propagation, the transverse wave numbers are then constrained by boundary condition equations

$$k_x = B_x(\omega,k_z) \qquad \text{(IVA-15)}$$

$$k_y = B_y(\omega,k_z) \qquad \text{(IVA-16)}$$

which in general may depend upon ω and $k_z$. The pinch-type singularity is then determined from

$$\left(\frac{\partial D}{\partial k_z}\right)_{\omega_o,k_{zo}} = 0 \qquad \text{(IVA-17)}$$

and

$$D(\omega_o,k_{zo}) = 0 \qquad \text{(IVA-18)}$$

subject to the constraints of Eqs. (IVA-14) and (IVA-15).

So far we have considered pinch-type singularities for which the pinching of the deformed F-contour occurs for finite values of k. This required that the two (or more) k-roots that pinch must originate on opposite sides of the real k-axis when ω is on the L-contour. Pinch-type singularities can also occur at infinite values of k; in this case the pinching of the deformed F-contour can be formed by two (or more) k-roots that originate on the same side of the real k-axis when ω is on the L-contour. Returning to the one-dimensional case and Eq. (IVA-1), we note that these singularities will occur at (finite) values of ω for which at least two $k_u(\omega)$ or two $k_\ell(\omega)$ become infinite.

These are *essential singularities* in $I(z,\omega)$. Although infinite k-values indicate that a further refinement in the model for the medium may be necessary, it is often useful to first study the instability in the model containing the essential singularity. This is best illustrated by an example. Consider the dispersion relation for the one-dimensional, electrostatic, beam-plasma interaction at high-frequencies where the motion of the ions can be neglected:

$$D(k,\omega) = 1 - \frac{\omega_p^2}{\omega^2} - \frac{\omega_b^2}{(\omega - kv_o)^2} = 0 \qquad \text{(IVA-19)}$$

where $\omega_p$ is the plasma frequency of the stationary electron plasma, $\omega_b$ is the plasma frequency of the beam electrons, and $v_o$ is the drift velocity of the beam. Even without solving for the $\omega(k_F)$ branches, we note that when L is taken sufficiently far up above the real $\omega$-axis the roots $k(\omega_L)$ are both in the upper-half of the complex k-plane. Hence at finite k, pinch-type singularities cannot occur in $I(z,\omega)$. However, as F is deformed to $\tilde{F}$ so that L can be brought down to $\tilde{L}$ near the real $\omega$-axis, we find when $\omega \to \omega_p$, from just above the real $\omega$-axis, $k \to \pm i^{\frac{1}{2}}(\infty)$. Hence $\omega = \pm\omega_p$ are essential singularities in $I(z,\omega)$. In fact since both k-roots belong to $k_u$ this singularity also pinches the deformed F-contour, but at infinite k; this can also be established directly from Eqs. (IVA-2) and (IVA-3) using Eq. (IVA-19).

The Green's function for the beam-plasma interaction can be found by using Eq. (IVA-19) in Eq. (IV-1). Carrying out the k integration we find zero for z<0, and for z>0;

$$G(Z,T) = \frac{i\omega_b}{2v_o}\, \omega_p \int_L \frac{d\Omega}{2\pi}\, p^3(\Omega) e^{-i\Omega T} \left(e^{ip(\Omega)Z} - e^{-ip(\Omega)Z}\right) \qquad \text{(IVA-20)}$$

where $Z \triangleq \omega_b z/v_o$, $T = \omega_p(t - z/v_o)$, and $p(\Omega) = \Omega/(\Omega^2-1)^{\frac{1}{2}}$. From this it is also evident that the response is zero for z=0, and for $t<z/v_o$ i.e. before the arrival of the pulse fron at any z>0. The integrand of Eq. (IVA-20) exhibits the essen tial singularities in the exponentials containing $p(\Omega)$ which becomes infinite at $\Omega^2 = 1$. A time-asymptotic form of G can be obtained by the saddle point integration technique. Thus, for example, the saddle point of $[\Omega T - p(\Omega)Z]$ in the upper-half $\Omega$-plane is located for T >> Z at $\Omega \approx 1 + (Z/T)^{\frac{2}{3}} e^{i2\pi/3}/2$, and the L-contour can be deformed to pass through it in the steepest descent direction. As long as Z/T is not too small the contributions from the singularities of $p^3(\Omega)$ can be igno and we obtain

$$G(Z,T) \sim \left\{\exp -iT\left[1 + \frac{3}{4}\left(\frac{Z}{T}\right)^{2/3}\right]\right\}\left[\exp \frac{3\sqrt{3}}{4} T^{1/3} Z^{2/3}\right] \quad \text{(IVA-21)}$$

We note that this is very much different from a finite k-plane pinch singularity. Here, every point for $z<0$ remains unexcited, while all points for $z>0$ once reached by the pulse will grow indefinitely in time. Furthermore, the growth rate in time varies from position to position, being larger just behind the front of the pulse and smaller at the tail end of the pulse. In addition we note that this essential singularity was on the real $\omega$-axis and still gave a considerable time-asymptotic growth. We may therefore expect that small effects such as weak collisions in the plasma or plasma temperature, which will certainly remove the essential singularity from the real $\omega$-axis, will not inhibit the growth in time before the response has reached large amplitudes. To see this consider adding a small, phenomenological collision frequency $\nu$ to the plasma electrons; we then find that the time-asymptotic Green's function amplitude is

$$\sim \exp\left[3\sqrt{3}\, T^{1/3} Z^{2/3}/4 - \nu T/2\omega_p\right]$$

which reaches a maximum amplitude $\sim\exp[(\sqrt{3}/2)^{\frac{3}{2}}(\omega_p/\nu)^{\frac{1}{2}}Z]$ at $T = (\sqrt{3}\omega_p/2\nu)^{\frac{3}{2}}Z$, after which it decays in time. Thus, for $\nu \ll \omega_p$, although the instability is no longer absolute, the growth rate in time persists for a long time and the amplitudes of the response can become very large at almost all points in space.

### *B. Convective Instabilities — Amplification of Signals*

We now turn to the cases where for some $k_r$ Eqs.(IV-3) and (IV-4) give $\omega_i(k_r) > 0$, but the singularities in $I(z,\omega)$ (Eq. IV-10) are all found to lie in the lower-half $\omega$-plane. The F-contour can then be deformed to an F-contour such that all of the branches of $\omega(k_F)$ are in the lower-half $\omega$-plane. Thus the region of absolute convergence in the $\omega$-plane may be extended to at least just below the $\omega_i$-axis (see Fig. IVB-1), and the L-contour deformed to an L-contour that lies entirely in the lower-half $\omega$-plane. On such a contour we have

$$\lim_{t\to\infty} G(z,t) = \lim_{t\to\infty} \int_{\tilde{L}} \frac{d\omega}{2\pi}\, e^{-i\omega t}\, I(z,\omega) = 0 \qquad \text{(IVB-1)}$$

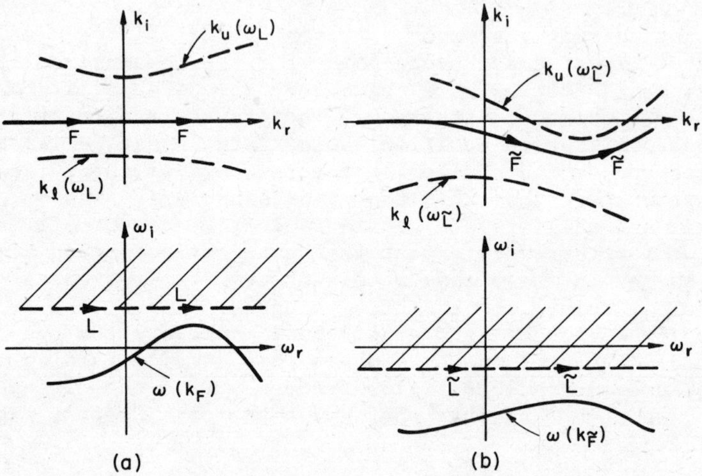


Fig. IV B-1

Hence, every (finite) point in space returns to its original unperturbed state, which implies that there are no absolute instabilities. We note however, from Fig. IVB-1 that since a portion of $\omega(k_F)$ was in the upper-half $\omega$-plane, some part of a branch of $k(\omega_L)$ will end up crossing the real k-axis thus indicating waves growing in space. In the example of Fig. IVB-1 a portion of $k_u(\omega_r)$ will therefore have a negative-imaginary part, indicating spatially growing waves for z>0.

That this represents a convective instability can be further appreciated by considering a Lagrangian version of the Green's function as for an observer moving with a velocity V. For our example in Fig. IVB-1 consider the Green's function for z = Vt > 0. With L on the real $\omega$-axis we then have

$$G(t) = \int_{-\infty}^{\infty} \frac{d\omega_r}{2\pi} \sum_u \frac{iu(Vt)}{\left(\frac{\partial D}{\partial k}\right)_{k_u}}\, e^{-i[\omega_r - Vk_u(\omega_r)]t} \qquad \text{(IVB-2)}$$

wherein the absence of absolute instabilities assures that the integrand is analytic on the $\omega_r$-axis. The time-asymptotic evaluation of G(t) can be obtained by the saddle-point method. Letting $k_u = k_{ur} + i\, k_{ui}$, the saddle points are given by

$$1 - V\frac{\partial k_{ur}}{\partial \omega_r} - i\, V\frac{\partial k_{ui}}{\partial \omega_r} = 0 \qquad \text{(IVB-3)}$$

from which we have

$$\frac{\partial k_{ui}}{\partial \omega_r} = 0 \longrightarrow \omega_r \overset{\Delta}{=} \omega_s \qquad \text{(IVB-4)}$$

as the saddle point (see Fig. IVB-2), when we take

$$V = \left(\frac{\partial k_{ur}}{\partial \omega_r}\right)^{-1}_{\omega=\omega_s} \overset{\Delta}{=} V_s \qquad \text{(IVB-5)}$$

Assuming $\partial^2 k_u/\partial\omega_r^2 \neq 0$ we then find

$$\lim_{t\to\infty} G(t) \sim \frac{1}{t^{1/2}}\, e^{-k_{ui}(\omega_s)V_s t} \qquad \text{(IVB-6)}$$

where $k_{ui}(\omega_s)$ is the maximum negative $k_{ui}$ for real $\omega$. Hence, although for any finite fixed z the time-asymptotic response is zero (Eq. IVB-1), i.e. no absolute instabilities, we find that for a moving observer with velocity $V_s$ the time-asymptotic response can be unstable (Eq. IVB-6). This is precisely the convectively unstable response described in Fig. IV-3.

In a system that is free of absolute instabilities we can also consider its steady-state excitation by an external source at a real frequency $\omega_e(e^{-i\omega_e t})$. The response problem is then formulated in an analogous manner to Eq. IA-11 where $S_{ext} \sim (\omega-\omega_e)^{-1}$. For the one-dimensional case we can write the response as

$$R(z,t) = \int_L \frac{d\omega}{2\pi}\, e^{-i\omega t} \int_F \frac{dk}{2\pi}\, e^{ikz} \cdot \frac{1}{(\omega-\omega_e)D(k,\omega)} \qquad \text{(IVB-7)}$$

If the Green's function returns to rest as in Eq. IVB-1, the time-asymptotic response to the external source will also be at the frequency $\omega_e$, as can be appreciated from the appropriate

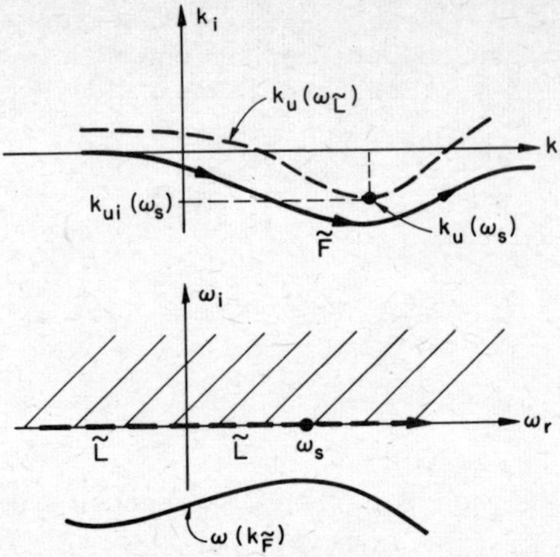


Fig. IV B-2

deformation of the L and F contours, similar to Fig. IVB-1, and shown in Fig. IVB-3. Thus

$$R(z,t) = \int_{\tilde{F}} \frac{dk}{2\pi} \frac{e^{ikz}\, e^{-i\omega_e t}}{D(k,\omega_e)}$$

$$= \sum_u \frac{iu(z)}{\left(\frac{\partial D}{\partial k}\right)_{k_u}} e^{-[\omega_e t - k_u(\omega_e) z]} \tag{IVB-8}$$

$$+ \sum_\ell \frac{-iu(-z)}{\left(\frac{\partial D}{\partial k}\right)_{k_\ell}} e^{-i[\omega_e t - k_\ell(\omega_e) z]}$$

This shows that the excited waves with $k_{ui}(\omega_e) < 0$ are spatially growing for $z>0$, and those with $k_{\ell i}(\omega_e) > 0$ are spatially growing for $z<0$. Such waves can be called amplifying waves since external steady-state signals can be amplified by exciting them at one point and coupling to them at another point where their amplitude is larger.

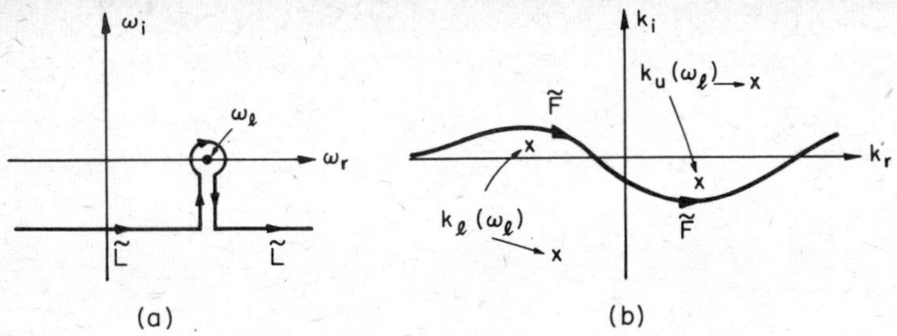


Fig. IV B-3

Summarizing, we note that when the F-contour is deformed to the $\tilde{F}$-contour, such that all branches of $\omega(k_F)$ lie in the lower-half $\omega$-plane, and L may be deformed to $\tilde{L}$ lying entirely in the lower-half $\omega$-plane, the mapping of the real $\omega$-axis into the complex k-plane will exhibit all of the possible spatially amplifying waves: those for $z>0$ having $k_{ui}(\omega_r) < 0$ and lying above $\tilde{F}$; and those for $z<0$ having $k_{\ell i}(\omega_r) > 0$ and lying below $\tilde{F}$. All of these considerations have assumed an infinitely extended homogeneous medium. End effects can give rise to a partial reflection of the amplified signal through the excitation of waves carrying energy in the direction opposite to that of the growing waves. This then would provide a feedback which, if of sufficient strength and of proper phase, can change the amplifying (or convectively unstable) mode into an oscillating (or absolutely unstable) mode. In addition we remark that the spatial amplification is eventually limited by the onset of nonlinear effects that have been neglected so far. These effects can also lead to the excitation of other modes, as we shall show later on, and can also give rise to a feedback that may change a convectively unstable system into an absolutely unstable one.

## C. *Propagation of Unstable Pulses*

The distinction between absolutely and convectively unstable responses can be further clarified by considering the differences in the time-asymptotic pulse shape for the Green's function for these two cases. As is evident from Fig. IV-3, in the case of an absolutely unstable response this pulse shape will encompass the origin, with front and trailing edges of the pulse moving in opposite directions, while for a convectively unstable response the time-asymptotic pulse has moved away from the origin, and its front and trailing edges are moving in the same direction. In order to find this time-asymptotic pulse shape we again consider a Lagrangian description of the Green's function; letting $z = Vt + z_o$ Eq. (IV-1) becomes

$$G(z_o,t) = \int_L \frac{d\omega}{2\pi} e^{-i\omega' t} \int_F \frac{dk}{2\pi} \frac{e^{ikz_o}}{D_V(k,\omega')} \quad \text{(IVC-1)}$$

where

$$\omega' = \omega - kV \quad \text{(IVC-2)}$$

and

$$D_V(k,\omega') = D(k,\omega' + kV) \quad \text{(IVC-3)}$$

As shown in Sec. IVA, the time-asymptotic response will be determined by the highest $\omega$ singularity of the pinch-type, for which now

$$D_V = 0 \quad \text{(IVC-4)}$$

and

$$\frac{\partial D_V}{\partial k} = - \frac{\partial D_V}{\partial \omega'}\left[\frac{d\omega_r}{dk} - V + i \frac{d\omega_i}{dk}\right] = 0 \quad \text{(IVC-5)}$$

For simplicity assume we have only one branch of $\omega(k_F)$ that extends partially into the upper-half $\omega$-plane, as shown in Fig. IVC-1; this will at least be true for any physical

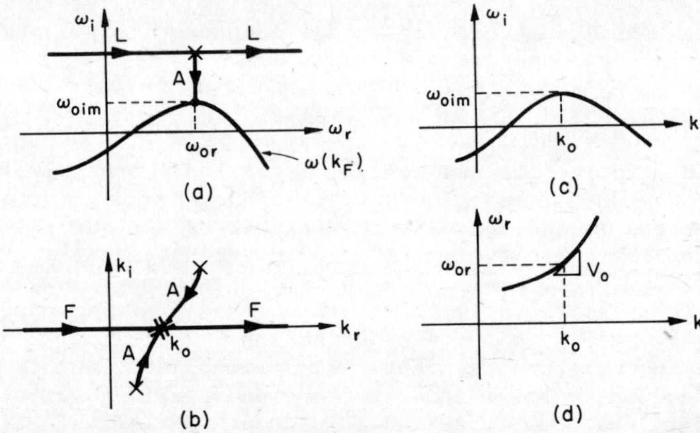


Fig. IV C-1

system in which some set of parameters have just made it unstable. Then it is clear that the maximum growth rate ($\omega_{oim}$) is obtained when the pinch-type singularity occurs where $\omega_i(k_r)$ is a maximum, i.e. at $k_r = k_o$ where $(d\omega_i/dk) = 0$. For this we then require

$$V = \left.\frac{d\omega_r}{dk}\right|_{k=k_o} \triangleq V_o \qquad \text{(IVC-6)}$$

$V_o$ is thus the propagation velocity of the peak of the asymptotic pulse. As V is allowed to vary, the pinch-type singularity growth rates determined from Eqs. (IVC-4) and (IVC-5) will be lower than $\omega_{oim}$, and correspond to positions on the asymptotic pulse that are away from the peak. At the velocities for which the pinch-type singularity growth rate is zero we then have the edges of the asymptotic pulse. A plot of these growth rates times t as a function of the velocity V times t represents the logarithm of the magnitude of the asymptotic Green's function as a function of distance from the origin, as shown in Fig. IVC-2. We note that when the edge velocities have the same sign the asymptotic pulse is convectively unstable, while if the edge velocities are of opposite sign the asymptotic pulse is absolutely unstable. From Fig. IVC-2, for the convectively unstable case, we can also establish the maximum spatial growth rate as the slope of the line from the origin that is tangent to the pulse curve; this must

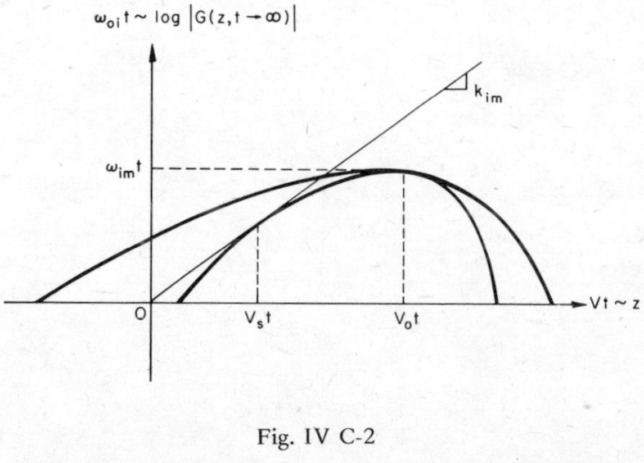


Fig. IV C-2

just be $|k_i(\omega_s)| \triangleq k_{im}$, as determined in Sec. IVB, which is the maximum $k_i$ for real $\omega$ that has crossed the real k-axis in the deformation of L to the real $\omega$-axis. We can also establish bounds on $k_{im}$ with respect to $\omega_{oim} \overset{\Delta}{=} \omega_{im}$, $V_o$, and $V_s$ (Eq. IVB-5).

$$\frac{\omega_{im}}{|V_o|} < |k_{im}| < \frac{\omega_{im}}{|V_s|} \qquad \text{(IVC-7)}$$

the upper bound resulting from the fact that $\omega_{im}$ is the largest growth rate for any velocity. Figure IVC-2 also shows that $|V_o| > |V_s|$

In simple situations, when the instability can be traced to the linear coupling of a positive energy wave and a negative energy wave the edge velocities are just the group velocities of the uncoupled waves at the phase velocity of synchronism. Thus the linear coupling of such a pair of waves having opposite group velocities gives rise to an absolute instability, while if the group velocity of these waves are of the same sign the instability is convective. In the case of a convective instability, the $\omega$ vs. k dispersion diagram exhibits $V_o$ in the plot of $\omega(k_r)$ and $V_s$ in the plot of $k(\omega_r)$, and clearly shows that $|V_o| > |V_s|$. This is illustrated in Fig. IVC-3. In general it should be noted that a convective instability in which the trailing edge of the pulse moves much slower than the front may be difficult to distinguish from an absolute instability since nonlinear effects may become important at almost all points in space before the pulse has had

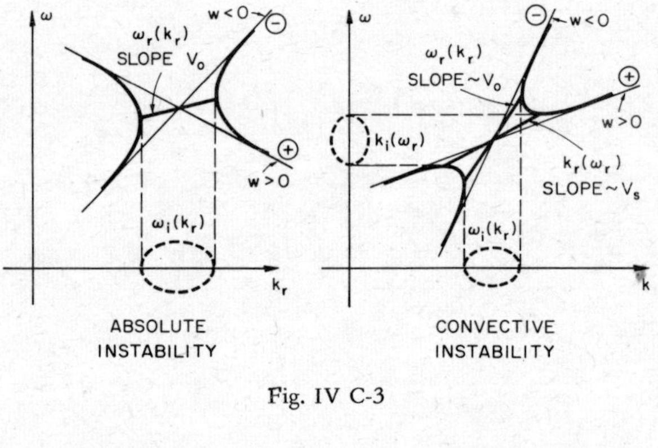


Fig. IV C-3

a chance to convect out of the region of interest. In another extreme, a convective or absolute instability in which both edges of the asymptotic pulse move slowly compared to the time growth rate of the pulse may become highly localized before it has a chance to either convect away or encompass much of the system. (For further details see Appendix 1.)

Another interesting example is afforded by the beam-plasma dispersion relation of Eq. (IVA-19) for which we have also found the time-asymptotic response in Eq. (IVA-21). Using Eqs. (IVC-4) and (IVC-5) with Eq. (IVA-19) we find the pinch-type singularity in $\omega'$ as a function of V:

$$\left(\frac{\omega_o'}{\omega_p}\right)^2 = \left(1 - \frac{V}{v_o}\right)^2 \left[1 + e^{im2\pi/3}\left(\frac{n_b}{n_p}\right)^{1/3} \left(\frac{V/v_o}{1 - V/v_o}\right)^{2/3}\right]^3 \qquad \text{(IVC-8)}$$

where $n_b$ is the unperturbed beam density, $n_p$ is the unperturbed plasma density, and $m = 0,1,2$.

For the growing region of the asymptotic pulse where $\omega_{oi}' > 0$ we can approximate this to

$$\frac{\omega_o'}{\omega_p} \cong 1 - \frac{V}{v_o} - \frac{3}{4}\left(\frac{n_b}{n_p}\right)^{1/3} \left(\frac{V}{v_o}\right)^{2/3} \left(1 - \frac{V}{v_o}\right)^{1/3} + i\,\frac{3\sqrt{3}}{4}\left(\frac{n_b}{n_p}\right)^{1/3} \left(\frac{V}{v_o}\right)^{2/3} \left(1 - \frac{V}{v_o}\right)^{1/3} \qquad \text{(IVC-9)}$$

Figure IVC-4 shows a sketch of $\omega_{oi}'t$ vs. $Vt$, which is proportional to log $|G(z, t \to \infty)|$ vs. z. From this emerge several features which we also obtained from the asymptotic integral evaluation in Sec. IVA: the trailing edge of the asymptotic pulse is pinned at $z = 0$ where it remains of zero amplitude; the front edge moves out at the beam velocity $v_o$; for $n_b \ll n_p$ the vicinity of the front of the pulse grows much faster than the vicinity of the trailing edge.

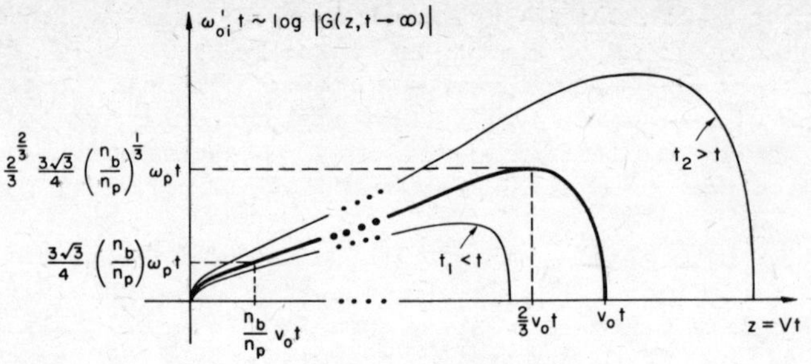


Fig. IV C-4

Finally we note that multiplying Eq. (IVC-9) by t, with V = zt gives

$$\omega_o' t \cong \omega_p\left(t - \frac{z}{v_o}\right) - \frac{3}{4}\left[\omega_p\left(t - \frac{z}{v_o}\right)\right]^{1/3}\left(\frac{\omega_b z}{v_o}\right)^{2/3}$$

$$+ i\,\frac{3\sqrt{3}}{4}\left[\omega_p\left(t - \frac{z}{v_o}\right)\right]^{1/3}\left(\frac{\omega_b z}{v_o}\right)^{2/3} \qquad \text{(IVC-10)}$$

so that $\exp(-i\omega_o' t)$ is seen to be just G(Z,T) as found in Eq. (IVA-21).

### D. *Propagating Waves in an Unstable Medium*

The nature of propagating waves (i.e. those solutions of the dispersion relation which for real k give real ω) in an unstable medium may be surprisingly different from propagating waves in a stable medium. In a medium that is absolutely unstable it is meaningless to speak of the excitation of propagating waves, or even of amplifying waves (Sec. IVB), since the presence of noise (or the transients in turning on of a source), will always drive the medium to its time-asymptotic state of an oscillator characterized by the fastest growing singularity in the ω-plane (and associated k-plane pinch, section IVA). We of course exclude here the situations in which the excitation mechanism may in some sense be so strong as to modify the medium properties and thereby eliminate its absolute instability. On the other hand, in a convectively unstable medium we can consider the excitation of either amplifying waves or propagating waves. The former were discussed in Sec. IVB and their direction of propagation and amplification was determined from the position of the map of the real $\omega$-axis in the k-plane vis-a-vis the deformed F-contour; we must do the same to determine the direction of propagation of propagating waves. For such waves to exist part of the real k-axis will map onto the real $\omega$-axis, and vice versa. However, the deformation of the L-contour onto these portions of the real $\omega$-axis can take place in two distinct ways, depending upon the presence of other branches of $\omega(k_F)$ between it and the branch of $\omega(k_F)$ that maps partly on to the real $\omega$-axis. The former case is illustrated in Fig. IVD-1 where

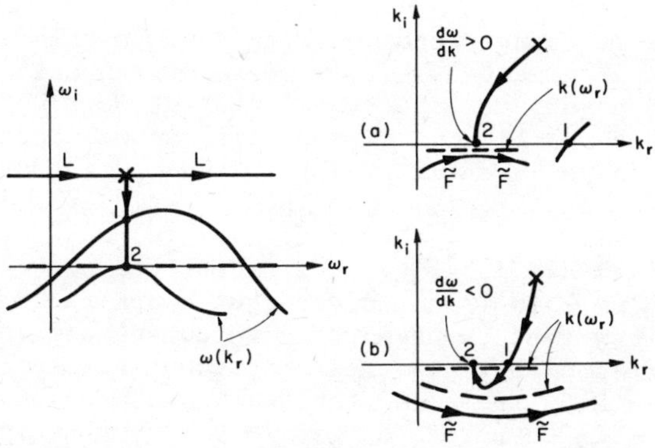


Fig. IV D-1

point 2 represents a possible propagating wave, i.e. a real $\omega$ solution of the dispersion relation for a real k. The deformation of the L-contour to the real $\omega$-axis, and in particular to point 2, is best studied by following the mapping of the line (012) in the complex k-plane. The two possibilities are shown in (a) and (b) of Fig. IVD-1. In case (a), when point 1 is passed the real k-axis is being crossed by a branch which does not include the original point 0, while in case (b) the crossing of the real k-axis at point 1 is by the branch on which point 0 originated. In either case, since point 0 having $\omega$ on L, is in the upper-half k-plane, the excitation of the propagating wave at 2 will exist for $z>0$ only. Suppose however, that we instead ask for the sign of the group velocity for this propagating wave: $(d\omega/dk)_2$. Since $\omega$ and k are both real at point 2 we might have expected that the sign of the group velocity would tell us the direction in which this wave is excited. Thus expanding k near $\omega_2$

$$k(\omega_2 + i\omega_i) = k_2 + i\omega_i\left(\frac{d\omega}{dk}\right)^{-1} + \ldots \qquad \text{(IVD-1)}$$

we find that in case (a) the group velocity is positive and gives the correct direction of excitation, while in case (b) the group velocity is negative and its usual use would erroneously indicate that the wave is excited for $z<0$. We thus find that in a medium that supports convective instabilities, the direction of excitation of even purely propagating waves can no longer be obtained reliably from the sign of the group velocity for such waves.

In order to understand the breakdown in this use of the group velocity we must carefully reexamine the classical derivation which associates the propagation velocity of a pulse with the group velocity. This usually assumes that the asymptotic pulse shape is entirely due to the stationary phase points on the real k-axis. But this is only true for a stable medium, where indeed all the asymptotic responses that are off the real k-axis are decaying (evanescent) responses. In an unstable medium this is no longer true since the deformed F-contour can now lead to asymptotic responses off the real k-axis that are growing, and their contribution may in fact dominate (see Sec. IVB). A stationary phase point on the real k-axis will therefore appear in the asymptotic response only if the F-contour can be made to pass through it; this is so for case (a) in Fig. IVD-1, but not for case (b).

The simplest example that illustrates the above two cases is found in the coupling of modes dispersion relation for a convective instability in Fig. IVC-3. Here we note that at the edges of the coupling region, where the dispersion relation starts to give complex $\omega$ for real k and complex k for real $\omega$, we encounter solutions that have both $\omega$ and k real and group velocities that are negative and rapidly varying in magnitude; clearly these are of the type (b) in Fig. IVD-1 and their group velocities do not characterize their excitation, as for example their excitation is for $z>0$ even when $(d\omega/dk) < 0$. On the other hand, the real $\omega$ and k solutions away from the coupling region are of the type (a) of Fig. IVD-1, and their group velocities do characterize their propagation.

Another example is found in the beam-plasma interaction which can be made convective by the addition of either small (plasma) temperature effects or small (plasma) collisions. For frequencies just above $\omega_p$ and real wavenumbers just above $k_{rc}$ we find propagating waves for which $(d\omega/dk) < 0$, yet we know that for $\omega$ on L (e.g. letting $\omega \to +i\infty$) all the k-roots of the dispersion relation are in the upper-half k-plane indicating that all modes are excited for $z>0$. These are again propagating waves of the type (b) in Fig. IVD-1 for which $(d\omega/dk)$ cannot be taken to have the classical interpretation of group velocity as in a stable lossfree medium. For frequencies much above $\omega_p$ we do however find propagating waves of the type (a) in Fig. IVD-1 for which $(d\omega/dk)$ has the classical meaning of group velocity.

## V. NONLINEAR COUPLING OF WAVES

We shall only touch upon a small but important aspect of the nonlinear dynamics of a plasma. As pointed out earlier, the dispersive nature of wave propagation in a plasma frequently allows one to treat finite amplitude waves by a perturbation scheme. Thus for a large class of waves of finite amplitude the plasma can be considered as only weakly nonlinear and the internal response may be described as a power series in the field amplitudes. In such a scheme we start with an initial unperturbed state and first-order linear perturbations which are described by a set of self-consistent waves (quasi-particles). To second order in the field amplitudes these waves are no longer independent. In a perturbation scheme, these nonlinear effects can be considered to give rise to a coupling between the waves and particles, and between waves themselves. This is the usual development in the study of weak turbulence in plasmas. We shall not pursue this here. Instead we shall focus on *coherent wave interactions*, and furthermore ignore the wave-particle interactions. The latter is certainly legitimate for the fluid modes of a plasma. The nonlinear interaction of coherent waves in a plasma can then be described by a natural extension of the formalism developed in Chapter III. Such interactions are of great interest not only because they describe the modification of the linear waves at finite amplitudes, but more importantly because they lead to new schemes of coupling energy to a plasma, and show new types of instabilities in a plasma, that are not obtainable from the linearized equations.

*A. Nonlinear Coupling of Coherent Waves*

In Sec. IIIA we derived the differential equations for the change in field amplitude and polarization due to a perturbung current density, Eqs. (IIIA-5), (IIIA-6) and (IIIA-9). We now assume that the perturbing current density $\overline{J}_p$ is due to nonlinear effects that were neglected in the linearized analysis. In particular, the nonlinear internal response function for the current density in a homogeneous plasma to second order in the electric field is given by

$$J_i^{(2)}(\bar{r},t) = \int dt' \int dt'' \int d^3r' \int d^3r'' \times$$

$$\hat{\sigma}_{ijk}(t-t', t'-t'', \bar{r}-\bar{r}', \bar{r}-\bar{r}'')E_j(t',\bar{r}')E_k(t'',\bar{r}'') \tag{VA-1}$$

Higher-order currents can be written in a similar manner; we ignore them from here on. In the perturbation scheme outlined above, this current can now be evaluated by assuming that the E-fields in Eq. (VA-1) are given by the slowly-varying linear modes:

$$\overline{E} = \sum_m \overline{E}_{mo} u_m(\bar{r},t) e^{i\bar{k}_m \cdot \bar{r}} e^{-i\omega_m t} \tag{VA-2}$$

Using Eq. (VA-2) in Eq. (VA-1) we arrive at

$$J_i^{(2)} = \sum_{m,p} \sigma_{ijk}(\omega_m,\omega_p,\bar{k}_m,\bar{k}_p)E_{jo}(\omega_m,\bar{k}_m)E_{ko}(\omega_p,k_p) \tag{VA-3}$$

$$u_m(\bar{r},t)u_p(\bar{r},t)e^{i(\bar{k}_m+\bar{k}_p)\cdot\bar{r}} e^{-i(\omega_m+\omega_p)t}$$

where $\sigma_{ijk}$ is the Fourier transform of $\hat{\sigma}_{ijk}$, properly symmetrized.

Substituting Eq. (VA-3) for $\overline{J}_p$ in Eq. (IIIA-9) we obtain the differential equations for coupling of modes due to second-order nonlinearities:

$$\left(\frac{\partial}{\partial t} + \overline{v}_{gn} \cdot \nabla + \gamma_n\right)u_n \triangleq \frac{d}{dt} u_n$$

$$= \frac{-1}{4w_{no}} \cdot \sum_{m;p} \sigma_{ijk}^{mp} E_{io}^{n*} E_{jo}^{m} E_{ko}^{p} u_m u_p e^{i\overline{\delta}_k \cdot \overline{r}} e^{-i\delta_\omega t} \tag{VA-4}$$

Here $\sigma_{ijk}^{mp}$ is the second-order nonlinear conductivity of the medium, the superscripts indicating that it is a function of $(\omega_m, \overline{k}_m)$ and $(\omega_p, \overline{k}_p)$, and

$$\overline{\delta}_k \triangleq \overline{k}_m + \overline{k}_p - \overline{k}_n \tag{VA-5}$$

$$\overline{\delta}_\omega \triangleq \omega_m + \omega_p - \omega_n \tag{VA-6}$$

are small, i.e. of the order of or smaller than the slow variations in d/dt defined in Eq. (VA-4). The superscripts on the fields also indicate the $(\omega, k)$ mode they pertain to.

It is convenient to introduce the following normalizations Let

$$\overline{E}_o^n = \overline{e}_o^n E_{no} \tag{VA-7}$$

where $\overline{e}_o^n$ is the complex polarization vector of the unperturbed wave and has unity magnitude

$$|\overline{e}_o^n| = 1 \tag{VA-8}$$

Let

$$p_n = +1 \text{ or } -1 \tag{VA-9}$$

be the energy parity (i.e. sign) of mode n, so that by Eq. (IIA-19) and with Eqs. (VA-7) and (VA-8) the field amplitude is normalized to the action density (Eq. IIB-21):

$$n_{no} = \frac{w_{no}}{\omega_n} = p_n \left|\frac{\partial D_n}{\partial \omega}\right| \frac{\varepsilon_o}{4} E_{no}^2 \triangleq p_n a_{no}^2 \tag{VA-10}$$

which defines $a_{no}$, and where

$$\frac{\partial D_n}{\partial \omega} \triangleq \bar{e}^*_{no} \cdot \frac{\partial \bar{\bar{D}}^h}{\partial \omega} \cdot \bar{e}_{no} \qquad \text{(VA-11)}$$

Similarly, for the perturbed amplitude we let

$$a_{no} u_n(\bar{r},t) = a_n'(\bar{r},t) \qquad \text{(VA-12)}$$

so that

$$n_n = \frac{w_n}{\omega_n} = p_n |a_n|^2 \qquad \text{(VA-13)}$$

and Eq. (VA-4) becomes

$$\frac{da_n}{dt} = \frac{-p_n}{4\omega_n} \cdot \sum_{m,p} \frac{\sigma^{mp}_{ijk} e^{n*}_{io} e^{m}_{jo} e^{p}_{ko}}{\left(\frac{\varepsilon_o}{4}\right)^{3/2} \left| \frac{\partial D_n}{\partial \omega} \frac{\partial D_m}{\partial \omega} \frac{\partial D_p}{\partial \omega} \right|^{1/2}} a_m a_p e^{i\bar{\delta}_k \cdot \bar{r}} e^{-i\delta_\omega t} \qquad \text{(VA-14)}$$

Equation (VA-14) expresses the coupling of the linear waves due to nonlinear effects up to second-order in the field amplitude. The slowly varying amplitudes that result from these effects are seen to be coupled to all possible pairs of such amplitudes of other waves in the plasma. Because the perturbed field amplitudes are slowly varying the coupling for a particular triplet of waves will be strongest when $\bar{\delta}_k = 0$ and $\delta_\omega = 0$; under these conditions the coupling is called resonant.

## B. *Conservative Coupling of Three Waves*

The simplest case described by Eq. (VA-14) is that for which only three waves satisfy $\bar{\delta}_k$ and $\delta_\omega$ small. From Eq. (VA-14) we then have

$$\frac{da_1}{dt} = K_1 a_2 a_3 e^{i\bar{\delta}_k \cdot \bar{r}} e^{-i\delta_\omega t} \qquad \text{(VB-1)}$$

$$\frac{da_2}{dt} = K_2 a_1 a_3^* e^{-i\bar{\delta}_k \cdot \bar{r}} e^{i\delta_\omega t} \qquad \text{(VB-2)}$$

$$\frac{da_3}{dt} = K_3 a_1 a_2^* e^{-i\bar{\delta}_k \cdot \bar{r}} e^{i\delta_\omega t} \qquad \text{(VB-3)}$$

where

$$\bar{\delta}_k = \bar{k}_2 + \bar{k}_3 - \bar{k}_1 \qquad \text{(VB-4)}$$

$$\delta_\omega = \omega_2 + \omega_3 - \omega_1 \qquad \text{(VB-5)}$$

and the coupling coefficients are

$$K_1 = \frac{-p_1}{4\omega_1}\, \sigma^{23}_{ijk} e^{1*}_{io} e^{2}_{jo} e^{3}_{ko} \Bigg/ \left(\frac{\varepsilon_o}{4}\right)^{3/2} \left|\frac{\partial D_1}{\partial\omega}\,\frac{\partial D_2}{\partial\omega}\,\frac{\partial D_3}{\partial\omega}\right|^{1/2} \qquad \text{(VB-6)}$$

$$K_2 = \frac{-p_2}{4\omega_2}\, \sigma^{1,-3}_{ijk} e^{2*}_{io} e^{1}_{jo} e^{3*}_{ko} \Bigg/ \left(\frac{\varepsilon_o}{4}\right)^{3/2} \left|\frac{\partial D_1}{\partial\omega}\,\frac{\partial D_2}{\partial\omega}\,\frac{\partial D_3}{\partial\omega}\right|^{1/2} \qquad \text{(VB-7)}$$

$$K_3 = \frac{-p_3}{4\omega_3}\, \sigma^{1,-2}_{ijk} e^{3*}_{io} e^{1}_{jo} e^{2*}_{ko} \Bigg/ \left(\frac{\varepsilon_o}{4}\right)^{3/2} \left|\frac{\partial D_1}{\partial\omega}\,\frac{\partial D_2}{\partial\omega}\,\frac{\partial D_3}{\partial\omega}\right|^{1/2} \qquad \text{(VB-8)}$$

In general all three coupling coefficients are independent. However, for a conservative and resonant coupling we shall show that there is only one independent coupling coefficient. Setting $\bar{\delta}_k = 0$ and $\delta_\omega = 0$, and defining a conservative nonlinear perturbation to be such that perturbed waves satisfy both conservation of energy

$$\sum_{i=1}^{3} \frac{d}{dt}\, p_i \omega_i |a_i|^2 = 0 \qquad \text{(VB-9)}$$

and conservation of momentum

$$\sum_{i=1}^{3} \frac{d}{dt}\, p_i \bar{k}_i |a_i|^2 = 0 \qquad \text{(VB-10)}$$

we find that

$$p_1 K_1 = -p_2 K_2^* = -p_3 K_3^* \qquad \text{(VB-11)}$$

This can be shown directly by using Eqs. (VB-1,2, and 3) in Eqs. (VB-9) and (VB-10). Hence under these conditions the three-wave interactions are described by a single coupling coefficient:

$$\frac{da_1}{dt} = p_1 K a_2 a_3 \qquad \text{(VB-12)}$$

$$\frac{da_2}{dt} = -p_2 K^* a_1 a_3^* \qquad \text{(VB-13)}$$

$$\frac{da_3}{dt} = -p_3 K^* a_1 a_2^* \qquad \text{(VB-14)}$$

From Eqs. (VA-12, 13, and 14) it then follows that the action densities $(n_1 + n_2)$ and $(n_1 + n_3)$ remain constant, i.e.

$$-\frac{\dot{w}_1}{\omega_1} = \frac{\dot{w}_2}{\omega_2} = \frac{\dot{w}_3}{\omega_3} \qquad \text{(VB-15)}$$

where the dot stands for $\partial/\partial t$. The relations of Eq. (VB-15) are known as the Manley-Rowe relations. These relations allow us to predict the possible power transfer between the three waves, and show that the ratio of power transfered between any two waves will be like the ratio of their frequencies. Thus, e.g., assuming $\omega_1$, $\omega_2$, and $\omega_3$ are all positive, it is clear that if we are driving wave 1, which is the highest frequency one, the power will transfer to waves 2 and 3 provided they are both positive energy waves. On the other hand, the driving of one of the lower frequency modes will result in the growth of the other two only if one of them is a negative energy mode. Finally, if the highest frequency mode is negative energy and the other two are positive energy, or vice versa, all three modes can grow simultaneously and the three-wave coupling may be spontaneously unstable.

Although the general solution of Equations (VB-12, 13 and 14) may be written in terms of elliptic integrals, this is not too illuminating and will not be done here. Instead, it is convenient to introduce the notion of a "pumped" wave as one that is being driven "externally" so that its amplitude remain constant and unaffected by the nonlinear interaction with the other two waves; under these conditions the nonlinear interaction of the other two waves reduces to a simple coupled set of equations, similar to the linear case in Chapter III except that the coupling coefficient symmetry is different and it is dependent upon the pump amplitude. As in the linear case it is convenient to first consider separately coupling in time and in space.

1. *Temporal Couplings*. Consider the solutions of Eqs. (IVB-1, 2, and 3) for the case when $\overline{\delta}_k = 0$ and $d/dt = \partial/\partial t + \gamma$, i.e. when the perturbation leads to only a slow-time-variation of the field amplitudes. We assume Eq. (VB-11) to hold and allow for a small frequency detuning $\delta_\omega \triangleq \delta$:

$$\left(\frac{\partial}{\partial t} + \gamma_1\right)a_1 = p_1 K a_2 a_3 e^{-i\delta t} \tag{VB-16}$$

$$\left(\frac{\partial}{\partial t} + \gamma_2\right)a_2 = -p_2 K^* a_1 a_3^* e^{i\delta t} \tag{VB-17}$$

$$\left(\frac{\partial}{\partial t} + \gamma_3\right)a_3 = -p_3 K^* a_1 a_2^* e^{i\delta t} \tag{VB-18}$$

*(a) Two Pumps*. The simplest case is obtained by assuming that two of the waves are pumped. Assume that waves 1 and 2 are driven externally so that their amplitude variation due to nonlinear coupling may be ignored. Hence let $a_1$ and $a_2$ be constant in time. Then the nonlinear coupling leads to the excitation of wave 3 as given by Eq. (VB-18):

$$\left(\frac{\partial}{\partial t} + \gamma_3\right)a_3 = -p_3 K^* a_{1o} a_{2o}^* e^{i\delta t}$$

whose solution is

$$a_3 = \frac{-p_3 K^* a_{1o} a_{2o}^*}{\gamma_3 + i\delta}\left(e^{i\delta t} - e^{-\gamma_3 t}\right) \tag{VB-19}$$

and is sketched in Fig. VB-1. We note that the rate at which wave 3 builds up depends upon the product of the coupling coefficient and the two pump amplitudes times the damping of the wave. Furthermore, for $\delta = 0$

$$\frac{W_3}{W_{1o}W_{2o}} = \frac{\omega_3}{\omega_1\omega_2}\frac{|K|^2}{|\gamma_3|^2} \tag{VB-20}$$

and

$$\frac{W_1}{W_{2o}W_{3o}} = \frac{\omega_1}{\omega_2\omega_3}\frac{|K|^2}{|\gamma_1|^2} \tag{VB-21}$$

which shows that for the same pump energies pumping up in frequency leads to a larger energy transfer than pumping down.

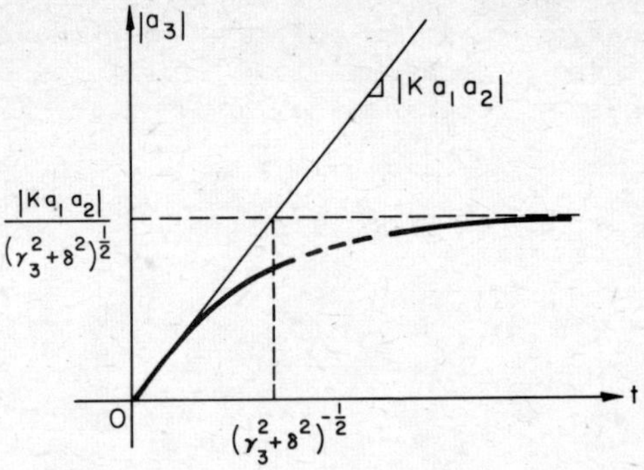


Fig. V B-1

*(b) One Pump.* Now we assume that only one of the waves is driven externally and thus maintained at constant amplitude. Let this be the high-frequency wave at $\omega_1$ ($\omega_1 > \omega_2 > \omega_3$) with amplitude $a_1$. The two remaining waves are thus described by Eqs. (VB-17 and 18):

$$\left(\frac{\partial}{\partial t} + \gamma_2\right)a_2 = -p_2K^* a_{1o}a_3^* e^{i\delta t}$$

$$\left(\frac{\partial}{\partial t} + \gamma_3\right)a_3 = -p_3K^* a_{1o}a_2^* e^{i\delta t}$$

At exact synchronism: $\underline{\delta_\omega = 0}$, and the solutions are of the form

$$a_2 = A_2 e^{\gamma t} \tag{VB-22}$$

$$a_3 = A_3 e^{\gamma^* t} \tag{VB-23}$$

with the dispersion relation

$$(\gamma + \gamma_2)(\gamma + \gamma_3) = p_2p_3|K|^2|a_{1o}|^2 \tag{VB-24}$$

Hence for undamped waves ($\gamma_2 = 0$ and $\gamma_3 = 0$) we obtain

$$\gamma^2 = p_2 p_3 |K|^2 |a_{1o}|^2 \qquad \text{(VB-25)}$$

i.e. growth in time if the two waves have the *same* sign of energy; otherwise the coupling is oscillatory. For waves with finite damping the threshold for growth is $\gamma = 0$ and requires a pump energy which is finite:

$$|a_{1o}|^2 = \frac{\gamma_2 \gamma_3}{|K|^2} \qquad \text{(VB-26)}$$

The complete behavior of $\gamma$ with $|a_1|$ is shown in Fig. VB-2.

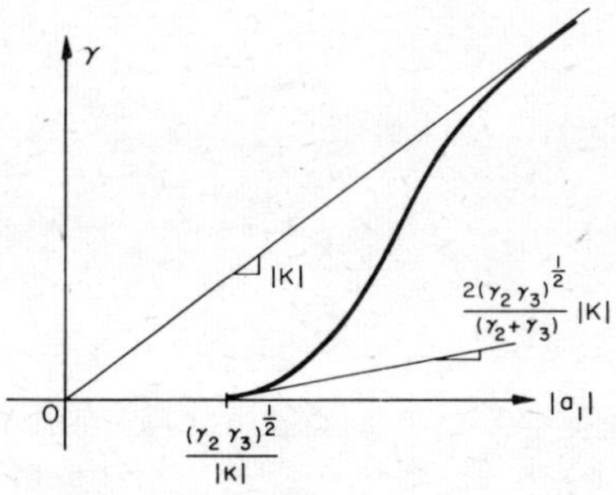


Fig. V B-2

Next consider the case where synchronism is not perfect. $\delta_\omega \neq 0$; for simplicity we neglect wave damping ($\gamma_2 = \gamma_3 = 0$) The solutions are then

$$a_2 = A_2 e^{(\gamma + i\,\delta/2)t} \qquad \text{(VB-27)}$$

$$a_3 = A_3 e^{(\gamma^* + i\,\delta/2)t} \qquad \text{(VB-28)}$$

and

$$\gamma^2 = p_2 p_3 |K|^2 |a_{1o}|^2 - \frac{\delta^2}{4} \qquad \text{(VB-29)}$$

For instability (growth in time) we again require $p_2 p_3 > 0$ i.e. energies of the same sign, and the pump threshold is now

$$|a_{1o}|^2 = \frac{\delta^2}{4|K|^2} \tag{VB-30}$$

Note that by the M-R relations (Eq. VB-15) the power supplied to waves 2 and 3 comes from the pump ($\dot{w}_1 < 0$); the power in waves 2 and 3 divides itself according to the frequencies: $(\dot{w}_2/\dot{w}_3) = (\omega_2/\omega_3)$. Note also that in the case of growth $a_2$ and $a_3$ may reach large enough amplitudes so that Eq. (VB-16) can no longer be ignored; such an effect of *pump depletion* is a possible saturating mechanism for this instability.

Consider next the case where the pump is one of the lower frequency modes, say at $\omega_3 < \omega_2 < \omega_1$, so that $a_3$ is constant. The coupled mode equations are now

$$(\frac{\partial}{\partial t} + \gamma_1)a_1 = p_1 K a_{3o} a_2 e^{-i\delta t}$$

$$(\frac{\partial}{\partial t} + \gamma_2)a_2 = -p_2 K^* a_{3o}^* a_1 e^{i\delta t}$$

The effects of damping ($\gamma_1$, $\gamma_2$) and off-synchronism ($\delta$) are similar to before. With $\gamma_1 = \gamma_2 = \delta = 0$ the solutions to these equations are

$$a_1 = A_1 e^{\gamma t} \tag{VB-31}$$

$$a_2 = A_2 e^{\gamma t} \tag{VB-32}$$

with

$$\gamma^2 = -p_1 p_2 |K|^2 |a_{3o}|^2 \tag{VB-33}$$

Hence for waves having the same sign for their energies the coupling is periodic. The coupling is unstable only if the energies are of opposite sign.

2. *Spatial Coupling.* Other aspects of the possible nonlinear coupling are obtained by considering the waves to have a steady state time dependence at their respective frequencies and the coupling to occur in space. The boundary conditions for wave excitation and propagation must then be carefully associated with the group velocities of the waves,

$\overline{v}_g$. We shall only discuss the results of the one-dimensional case

$$\left(\frac{\partial}{\partial z} + \alpha_1\right)a_1 = \frac{p_1 K}{v_{g1}} a_2 a_3 e^{i\delta_k z} \qquad \text{(VB-34)}$$

$$\left(\frac{\partial}{\partial z} + \alpha_2\right)a_2 = -\frac{p_2 K^*}{v_{g2}} a_1 a_3^* e^{-i\delta_k z} \qquad \text{(VB-35)}$$

$$\left(\frac{\partial}{\partial z} + \alpha_3\right)a_3 = -\frac{p_3 K^*}{v_{g3}} a_1 a_2^* e^{-i\delta_k z} \qquad \text{(VB-36)}$$

where $\alpha = \gamma/v_g$ is the small spatial damping of the wave, $\delta_\omega = 0$ is assumed, and $\delta_k$ is the off-synchronism measure in k. The effects of damping and off-synchronism are again similar to before. The two-pump solution for say wave 3 is similar to Eq. (VB-19) with $t \to z$, $\gamma_3 \to \alpha_3$ and $K \to K/v_{g3}$. The one-pump solution with $\alpha_i = 0$, $\delta_k = 0$, and the pump at the highest frequency $\omega_1$ $(\omega_1 > \omega_2 > \omega_3)$ is

$$a_2 = A_2 e^{\alpha z} \qquad \text{(VB-37)}$$

$$a_3 = A_3 e^{\alpha^* z} \qquad \text{(VB-38)}$$

with

$$\alpha^2 = \frac{p_2 p_3}{v_{g_2} v_{g_3}} |K|^2 |a_{1o}|^2 \qquad \text{(VB-39)}$$

For $p_2 p_3 > 0$ two possible types of unstable solutions exist:

(a) $v_{g2} v_{g3} > 0$ and hence $\alpha^2 > 0$ leading to exponential growth in space.

(b) $v_{g2} v_{g3} < 0$ and hence $\alpha^2 < 0$ leading to periodic backward-wave amplification or oscillation in a critical length

$$L_{min} = \frac{\pi}{2\alpha} \qquad \text{(VB-40)}$$

The detailed solutions for these cases can be adapted from the similar analysis given for the linear coupling of waves. (See Sec. IIIC.)

3. *Coupling in Time and Space — Pulse Response.* We now return to a more complete solution of Eqs. (VB-12,13 and 14) in which the slow variations in both time and space are arbitrary and we ask for the natural response in both time and space of the amplitudes. Let us consider the case of a pump at the highest frequency $\omega_1$, and restrict ourselves to one spatial dimension. Eqs. (VB-13) and (VB-14) become

$$(\frac{\partial}{\partial t} + v_2 \frac{\partial}{\partial z})a_2 = -p_2 K^* a_{10} a_3^* \qquad \text{(VB-41)}$$

$$(\frac{\partial}{\partial t} + v_3 \frac{\partial}{\partial z})a_3 = -p_3 K^* a_{10} a_2^* \qquad \text{(VB-42)}$$

where $v_2$ and $v_3$ are the group velocities, and we first neglect effects due to damping and off-resonance. The solutions to Eqs. (VB-41) and (VB-42) are of the form

$$a_2 = A_2 e^{-ist} e^{i\kappa z} \qquad \text{(VB-43)}$$

$$a_3 = A_3 e^{is^* t} e^{-i\kappa^* z} \qquad \text{(VB-44)}$$

giving a dispersion relation

$$(s - \kappa v_2)(s - \kappa v_3) = -p_2 p_3 |K|^2 |a_{10}|^2 \qquad \text{(VB-45)}$$

The stability analysis of Eq. (VB-45) in the complex s and $\kappa$ planes is straightforward. For real $\kappa$ we obtain growth in time (i.e. Im s>0) only if

$$p_2 p_3 > 0 \qquad \text{(VB-46)}$$

and

$$|K|^2 |a_1|^2 > \frac{\kappa^2}{4} (v_2 - v_3)^2 \qquad \text{(VB-47)}$$

Condition (VB-46) is consistent with what we found before (Eq. VB-25). If in addition:

$$v_2 v_3 < 0 \qquad \text{(VB-48)}$$

the instability is *absolute* with growth rates in time and space

$$s_{oi} = \frac{2|v_2v_3|^{\frac{1}{2}}}{|v_2-v_3|} |K||a_{10}| \tag{VB-49}$$

$$\kappa_{oi} = \frac{-(v_2+v_3)}{|v_2v_3|^{\frac{1}{2}}|v_2-v_3|} |K||a_{10}| \tag{VB-50}$$

On the other hand, if in addition to condition (VB-46)

$$v_2v_3 > 0 \tag{VB-51}$$

the instability is convective and the maximum growth rate in space was found before in Eq. (VB-39). Both the absolute instability or the convective instability develop according to the equations presented in Section III with appropriate modification of the coupling coefficient which for the nonlinear coupling will contain the pump field.

If we include the effects of wave damping, $\gamma_2$ and $\gamma_3$, we find that the absolute instability will require a threshold pump energy given by

$$\begin{aligned} |a_{1o}|^2 &= \frac{|v_2v_3|}{4|K|^2} \left(\frac{\gamma_2}{|v_2|} + \frac{\gamma_3}{|v_3|}\right)^2 \\ &= \frac{|v_2v_3|}{4|K|^2} (\alpha_2 + \alpha_3)^2 \end{aligned} \tag{VB-52}$$

In contrast to Eq. (VB-26), this threshold is finite even if one wave has negligible damping. (See also Appendix 1.)

As an example we shall consider the coupling of Bohm-Gross waves to ion-acoustic waves. The waves satisfying the resonance conditions, assuming for simplicity that the k's are colinear (one space-dimension), can be found from the geometrical construction of a parallelogram with one vertex at the origin and the other three vertices at respectively $(\omega_1, k_1)$, $(\omega_2, k_2$ and $(\omega_3, k_3)$ on the dispersion diagram as shown in Fig. VB-3. With a pump at $(\omega_1, k_1)$ the coupling in Fig. VB-3a is convectively unstable, while the coupling in Fig. VB-3b is absolutely unstable.

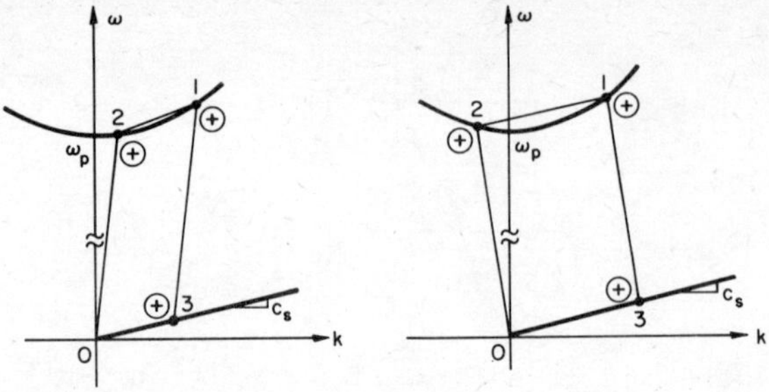


Fig. V B-3

Another interesting example is provided by the possibility of a beam-plasma interaction coupling nonlinearly to ion-acoust waves, as shown in Fig. VB-4. Assume the beam-plasma interaction is sufficiently weak so that it is convective and locali near $(\omega_1, k_1)$. As the amplitude of this wave grows in space along the direction of beam flow (say, along z) it will at some point reach sufficient amplitude to act as a pump and overcome the necessary threshold for nonlinearly exciting waves 2 and 3; this excitation is unstable since 1 is at a frequency higher than either 2 or 3. Since 2 is much closer in frequency to 1 than 3 is to 1, most of the power flow from 1 is to 2 (by the Manley-Rowe relations). Wave 2 on the other hand has a group velocity that is opposite to $v_o$, and since it is strongly excited, as it travels backwards it also can act as a pump to nonlinearly excite 1 and 3; this excitation is also unstable since the wave at 1 is still a negative energy wave. Here

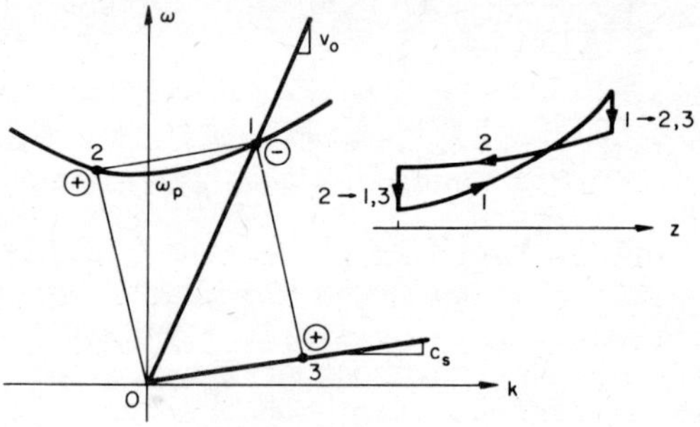


Fig. V B-4

again most of the power transfered from 2 is back to 1 but now in a region where originally the waves were nonlinearly stable. This process can act as a continuous feedback mechanism to drive all three waves into a much more rapidly growing nonlinear absolute instability than the one that may result from just waves 2 and 3 driven by 1 as a pump. Such trhee-wave nonlinear instabilities have been frequently termed as explosive because an idealized analysis of the interaction, that usually ignores the spatial evolution, shows that the growth rate tends to infinity in a finite time. The time and space picture we have presented points up a perhaps more important aspect of such an instability, namely that the "starting length" for such an instability, because of the spatial feedback, may be very short. This would imply that such instabilities can occur even in inhomogeneous media.

### C. *Three Wave Coupling in the Fluid Model of a Plasma*

We shall illustrate the calculation of three wave coupling due to nonlinear effects by using the fluid model of a plasma. The advantage of this model is that the physics of the various nonlinear effects is readily identified, and the nonlinear current can be written down by inspection. In using this model we of course have to be mindful of the fact that the linear waves predicted by this model are not properly described for some regions of $\omega$ and $\overline{k}$ where kinetic effects are important. We must therefore avoid using this model in or near these regions of $\omega$ and $\overline{k}$, except where the kinetic effects just bring in a small damping of the waves that are otherwise basically fluid waves. The nonlinear fluid equations we choose are

$$\left(\frac{\partial}{\partial t} + \overline{v} \cdot \nabla\right)\overline{v} + \gamma v_T^2 \frac{\nabla n}{n} = \frac{q}{m}(\overline{E} + \overline{v} \times \overline{B}) \qquad \text{(VC-1)}$$

$$\frac{\partial n}{\partial t} + \nabla \cdot (n\overline{v}) = 0 \qquad \text{(VC-2)}$$

The set of equations is written for each species (s) of particles in the plasma. The magnetic field is related to the electric field by

$$\frac{\partial \overline{B}}{\partial t} = -\nabla \times \overline{E} \qquad \text{(VC-3)}$$

and an externally applied component $\overline{B}_o$ may also be assumed. The current density is

$$\bar{J} = \sum_s q_s n_s \bar{v}_s \tag{VC-4}$$

Using Eqs. (VC-1-3) the right-hand side of Eq. (VC-4) can be expressed (in principle) as an operator on the electric field. This operator is the nonlinear conductivity. We shall find its form by an iterative solution of Eqs. (VC-1 through 3) in powers of the electric field, up to second order. First we shall assume an unperturbed state of the plasma which is homogeneous in time and space (e.g. $n = n_o$, $B = B_o$, and $T$ = constant, and all other field variables zero). To first-order in the electric field we linearize Eqs. (VC-1 through 4) in the usual manner and find the linear internal response function $\bar{\bar{\sigma}}(k,\omega)$, and the dispersion relation for the linear wave description of these first-order fields. The internal response function to second-order is obtained by assuming that the second order fields are generated by the nonlinear terms evaluated with the first-order fields, and acting as driving forces and flows. The equation for the second-order fields are then again of the same form as the linearized equations, except that the driving forces are changed. The latter arise in the force equation from:

$\bar{v} \times \bar{B}$ (Lorentz nonlinearity),

$(\bar{v} \cdot \nabla)\bar{v}$ (convective nonlinearity),

and $\nabla n/n$ (pressure nonlinearity);

in the continuity equation from

$\nabla \cdot (n\bar{v})$ (continuity nonlinearity);

and in the current desntiy equation from

$n\bar{v}$ (flow nonlinearity).

Each of these can be expressed in terms of the linear electric fields of the possible waves, and when Fourier transformed become convolutions of these fields. Finally the second-order equations of (VC-1) and (VC-2) can be simultaneously solved for the second-order velocity $\bar{v}_2$, and thus the second-order current density

$$\bar{J}_2 = \sum_s q_s (n_o \bar{v}_2 + n_1 \bar{v}_1)_s \tag{VC-5}$$

can be found in terms of the first-order fields. This is precisely what will give us the second-order conductivity as defined earlier. Introducing the resonance conditions

$$\omega_n = \omega_m + \omega_p \qquad \text{(VC-6)}$$

$$\bar{k}_n = \bar{k}_m + \bar{k}_p \qquad \text{(VC-7)}$$

(off-resonance effects can be handled separately, as shown earlier) we obtain the second-order current, for any species

Nonlinearity

$$\bar{J}_2^n = \frac{1}{n_o q} (\bar{E}^p \cdot \bar{\bar{\sigma}}^m \cdot \bar{E}^m) \bar{\bar{\sigma}}^n \cdot \frac{\bar{k}^p}{\omega^p} \qquad \text{Lorentz}$$

$$- \frac{1}{n_o q} \left( \frac{\bar{k}^p}{\omega^p} \cdot \bar{\bar{\sigma}}^m \cdot \bar{E}^m \right) \bar{\bar{\sigma}}^n \cdot \bar{E}^p \qquad \text{Lorentz}$$

$$- \frac{im}{n_o^2 q^3} (\bar{k}^p \cdot \bar{\bar{\sigma}}^m \cdot \bar{E}^m) \bar{\bar{\sigma}}^n \cdot (\bar{\bar{\sigma}}^p \cdot \bar{E}^p) \qquad \text{Convective}$$

$$+ \frac{1}{n_o q} \left( \frac{\bar{\bar{k}}^m}{\omega^m} \cdot \bar{\bar{\sigma}}^m \cdot \bar{E}^m \right) \bar{\bar{\sigma}}^p \cdot \bar{E}^p \qquad \text{Flow}$$

$$- i \frac{\gamma \kappa T}{n_o^2 q^3} \left( \frac{\bar{k}^n}{\omega^n} \cdot \bar{\bar{\sigma}}^p \cdot \bar{E}^p \right) \left( \frac{\bar{k}^m}{\omega^m} \cdot \bar{\bar{\sigma}}^m \cdot \bar{E}^m \right) \bar{\bar{\sigma}}^n \cdot \bar{k}^n \qquad \text{Continuity}$$

$$- i \frac{\gamma(\gamma-2) \kappa T}{n_o^2 q^3} \left( \frac{\bar{k}^p}{\omega^p} \cdot \bar{\bar{\sigma}}^p \cdot \bar{E}^p \right) \left( \frac{\bar{k}^m}{\omega^m} \cdot \bar{\bar{\sigma}}^m \cdot \bar{E}^m \right) \bar{\bar{\sigma}}^n \cdot \bar{k}^p \qquad \text{Pressure}$$

$$+ (m \longleftrightarrow p). \qquad \text{(VC-8)}$$

where again $m \rightleftarrows p$ stands for the same terms with m and p interchanged and $\bar{\bar{\sigma}}^\alpha \triangleq \bar{\bar{\sigma}}(\bar{k}_\alpha, \omega_\alpha)$ is the linear conductivity tensor evaluated for $\omega = \omega_\alpha$ and $\bar{k} = \bar{k}_\alpha$. From Eq. (VC-8) we now find the second-order nonlinear conductivity:

$$\sigma_{ijk} = \sum_s \frac{1}{2} \sum_{m,p} \frac{1}{2qn_o} \left\{ \sigma_{jk}^m \left( \frac{\bar{\bar{\sigma}}^n \cdot \bar{k}_p}{\omega_p} \right)_i - \left( \frac{\bar{\bar{\sigma}}^m \cdot \bar{k}^p}{\omega_p} \right)_j \sigma_{ik}^n \right.$$

$$+ \frac{m}{iq^2 n_o} (\bar{\bar{\sigma}}^m \cdot \bar{k}^p)_j (\bar{\bar{\sigma}}^n \cdot \bar{\bar{\sigma}}^p)_{ik} + \left(\bar{\bar{\sigma}}^m \cdot \frac{\bar{k}_m}{\omega_m}\right)_j \sigma^p_{ik}$$

$$+ \frac{m v_T^2}{iq^2 n_o} \left(\bar{\bar{\sigma}}^m \cdot \frac{\bar{k}_m}{\omega_m}\right)_j \left\{ (\bar{\bar{\sigma}}^p \cdot \bar{k}_n)_i \left(\bar{\bar{\sigma}}^n \cdot \frac{\bar{k}_n}{\omega_n}\right)_k (\gamma - 2) + \right.$$

$$\left. (\bar{\bar{\sigma}}^p \cdot \bar{k}_n)_i \left(\bar{\bar{\sigma}}^n \cdot \frac{\bar{k}_p}{\omega_p}\right)_k \right\} + m \rightleftarrows p$$

(VC-9)

where the summations over species and over all possible pairs appear explicitly.

As an example we evaluate the coupling of two electron-plasma waves (Bohm-Gross) to an ion-acoustic wave as shown in Fig. VB-3. Assuming the ions are cold, $B_o = 0$, and the fields are electrostatic, we find for colinear waves

$$\frac{\bar{e}^*_{10} \cdot (\bar{\mathcal{J}}_{2i} + \bar{\mathcal{J}}_{2e})}{\omega_1} = \frac{-e^3 n_o}{m_i^2} \frac{\frac{k_1}{\omega_1} + \frac{k_2}{\omega_2} + \frac{k_3}{\omega_3}}{\omega_1 \omega_2 \omega_3}$$

$$+ \frac{e^3 n_o}{m_e^2} \frac{\frac{k_1}{\omega_1} + \frac{k_2}{\omega_2} + \frac{k_3}{\omega_3} - \frac{k_1 k_2 k_3}{\omega_1 \omega_2 \omega_3} v_{Te}^2}{\omega_1 \omega_2 \omega_3 \left(1 - \frac{k_1^2 v_{Te}^2}{\omega_1^2}\right)\left(1 - \frac{k_2^2 v_{Te}^2}{\omega_2^2}\right)\left(1 - \frac{k_3^2 v_{Te}^2}{\omega_3^2}\right)}$$

(VC-10)

From this we recognize that to order $m_e/m_i$ and $(k\lambda_D)^2$ the only contributions to Eq. (VC-10) come from the electron current - third term in its numerator and last term in its denominator: hence

$$\frac{\bar{e}_{10}^{*} \cdot (\bar{J}_{2i} + \bar{J}_{2e})}{\omega_1} \cong \frac{\varepsilon_o}{2} \frac{e}{m_e} \frac{\omega_{pi}^2}{c_s \omega_1 \omega_2 \omega_3} \tag{VC-11}$$

is due to the electron convective nonlinearity. From the dispersion relations for these waves we find

$$\frac{\partial D_{1,2}}{\partial \omega} \cong \frac{2\omega_{pe}^2}{\omega_{1,2}^3} \tag{VC-12}$$

$$\frac{\partial D_3}{\partial \omega} \cong \frac{2\omega_{pi}^2}{\omega_3^3} \tag{VC-13}$$

and for the coupling coefficient

$$K \cong \frac{-(e/m_e)}{4\sqrt{2}\,(\varepsilon_o/4)^{1/2}} \frac{\omega_{pi}(\omega_1\omega_2\omega_3)^{1/2}}{c_s \omega_{pe}^2} \tag{VC-14}$$

The growth rate for modes 2 and 3 is then given by

$$\gamma = \frac{1}{4} \frac{v_{10}}{c_s} \frac{\omega_{pi}}{\omega_{pe}} (\omega_2\omega_3)^{1/2} = \frac{1}{4} \frac{v_{10}}{v_{Te}} (\omega_2\omega_3)^{1/2} \tag{VC-15}$$

where we have written

$$v_{10} \triangleq \frac{eE_{10}}{m_e \omega_1} \tag{VC-16}$$

which is the magnitude of the electron velocity induced by the pump field.

In the case where the Bohm-Gross wave (2) and the ion-acoustic wave (3) have small damping rates $\gamma_2$ and $\gamma_3$, respectively, the threshold for instability is given by

$$\left(\frac{v_{10}}{v_{Te}}\right)^2 = 16 \frac{\gamma_2}{\omega_2} \frac{\gamma_3}{\omega_3} \tag{VC-17}$$

In conclusion it should be pointed out that a more complet nonlinear analysis to second-order of a pump driven medium should also include effects of the third order current in whicl the pump enters twice in conjunction with another wave. This then can lead to other effects such as nonoscillatory instabili ties whose threshold is at essentially zero frequency.

## VI. APPENDIX 1

### Unstable Pulse Shape for Coupled-Wave Interactions

In either the linear coupling of a positive and negative energy wave (Sec. IIIC) or the nonlinear (second-order, parametric) coupling of two waves with energies having the same sign (Sec. VB), the slowly varying amplitudes of the coupled waves are described by the dispersion relation (we use the notation of Sec. IB)

$$D(\bar{\kappa},s) = (s - \bar{\kappa}\cdot\bar{v}_1 + i\gamma_1)(s - \bar{\kappa}\cdot\bar{v}_2 + i\gamma_2) + \gamma^2, \qquad (1\text{-}1)$$

where s and $\bar{\kappa}$ characterize the slow variations in time and space, respectively, $\bar{v}_1$ and $\bar{v}_2$ are the group velocities and $\gamma_1$ and $\gamma_2$ the damping rates of the uncoupled waves, and $\gamma^2$ is determined by the coupling mechanism; we shall consider only conservative unstable coupling for which $\gamma^2$ is real and positive.

In the one-dimensional case we can let $\bar{\kappa}\cdot\bar{v} = \kappa v$. First, ignoring the damping rates we find that for $v_1 v_2 < 0$ there is a pinch solution of $D = 0$ and $\partial D/\partial\kappa = 0$ giving an absolute instability growth rate in time

$$s_{oi} = \frac{2|v_1 v_2|^{1/2}}{|v_1| + |v_2|}\gamma, \qquad (1\text{-}2)$$

while for $v_1 v_2 > 0$ the instability is convective with a maximum growth rate in space

$$\kappa_{im} = \frac{\gamma}{(v_1 v_2)^{1/2}}. \qquad (1\text{-}3)$$

For finite damping rates of the unperturbed waves we can find thresholds for these instabilities. The threshold for instability is

$$\gamma^2 = \gamma_1\gamma_2 \triangleq \gamma_c^2. \tag{1-4}$$

If $v_1v_2 < 0$ then there is in addition a threshold for the absolute instability given by

$$\gamma^2 = \frac{\gamma_1|v_2| + \gamma_2|v_1|}{4|v_1v_2|} = \frac{|v_1v_2|}{4}(|\alpha_1| + |\alpha_2|) \triangleq \gamma_a^2, \tag{1-5}$$

where $\alpha_1$ and $\alpha_2$ are the spatial damping rates of the unperturbed waves. For $\gamma_c < \gamma < \gamma_a$ the instability is convective. The unstable pulse shapes, as in Fig. IVC-2, can now be found by forming $D_V \triangleq D(\kappa, s + \kappa V)$, in analogy with Eq. IVC-3, and solving for the pinch root of $D_V = 0$ and $\partial D_V/\partial\kappa = 0$. This gives

$$s_{io}(V) = \frac{2\gamma[(V-v_1)(v_2-V)]^{\frac{1}{2}}}{(v_2-v_1)} - \gamma_1\left(\frac{v_2-V}{v_2-v_1}\right) - \gamma_2\left(\frac{V-v_1}{v_2-v_1}\right), \tag{1-6}$$

from which a plot of $s_{oi}t$ versus $Vt$ gives the shape of the asymptotic pulse. For $\gamma_1 = \gamma_2 = 0$ it is evident that the pulse edge velocities are $v_1$ and $v_2$; the maximum growth rate is $s_{oim} = \gamma$, and it occurs for an observer with velocity $V_o = (v_1+v_2)/2$; and for a convective pulse $V_s = 2v_1v_2/(v_1+v_2)$. In the presence of finite damping of the unperturbed waves these pulse shape properties are:

(a) The velocities at the edges of the unstable pulse

$$V_{\pm} = \frac{V_T + \left(\frac{v_1+v_2}{2}\right)\Gamma^2 \pm \left(\frac{v_2-v_1}{2}\right)\frac{2\gamma\Gamma}{\gamma_1+\gamma_2}}{1+\Gamma^2} \qquad (1\text{-}7)$$

where

$$\Gamma^2 = \frac{4(\gamma^2-\gamma_c^2)}{(\gamma_1+\gamma_2)^2} \qquad (1\text{-}8)$$

and

$$V_T = \frac{\gamma_1 v_2 + \gamma_2 v_1}{\gamma_1+\gamma_2} = (\text{sgn } V_T)\frac{2|v_1 v_2|^{1/2}(\gamma_a^2-\gamma_c^2)^{1/2}}{\gamma_1+\gamma_2} \qquad (1\text{-}9)$$

is the pulse velocity at threshold for instability; i.e., for $\gamma = \gamma_c$ and hence $\Gamma^2 = 0$.

(b) The maximum growth rate of the asymptotic pulse and its velocity,

$$s_{iom} = \gamma\left\{\left[1+\left(\frac{\gamma_1-\gamma_2}{2\gamma}\right)^2\right]^{1/2} - \left(\frac{\gamma_1+\gamma_2}{2\gamma}\right)\right\} = \gamma\left(\frac{\gamma_1+\gamma_2}{2\gamma}\right)\left[(1+\Gamma^2)^{1/2} - 1\right] \qquad (1\text{-}10)$$

$$V_o = \frac{V_T - \left(\frac{v_1+v_2}{2}\right)\left[1-(1+\Gamma^2)^{1/2}\right]}{(1+\Gamma^2)^{1/2}}$$

$$= \left(\frac{v_1+v_2}{2}\right)\left\{1 - \left(\frac{v_1-v_2}{v_1+v_2}\right)\left(\frac{\gamma_1-\gamma_2}{2\gamma}\right)\left[1+\left(\frac{\gamma_1-\gamma_2}{2\gamma}\right)^2\right]^{1/2}\right\}. \qquad (1\text{-}11)$$

(c) For a convectively unstable pulse ($\gamma_c < \gamma < \gamma_a$), the maximum spatial growth rate and its velocity,

$$\kappa_{im} = -(\text{sgn } V_T)\left[\left(\frac{\gamma_a^2-\gamma_c^2}{|v_1v_2|}\right)^{\frac{1}{2}} - \left(\frac{\gamma_a^2-\gamma^2}{|v_1v_2|}\right)^{\frac{1}{2}}\right] \tag{1-12}$$

$$V_s = \frac{2v_1v_2}{(v_1-v_2)} \frac{\left\{\gamma_a^2-\gamma^2\right\}^{\frac{1}{2}}}{(\text{sgn } V_T)\gamma_a + \left(\frac{v_1+v_2}{v_1-v_2}\right)\left\{\gamma_a^2-\gamma^2\right\}^{\frac{1}{2}}} . \tag{1-13}$$

Equations (1-7) through (1-13) describe all the important properties of the unstable asymptotic pulse in one dimension.

For the three dimensional interaction described by the dispersion relation Eq. (1-1), we find for an observer with velocity $\bar{V}$

$$D_{\bar{V}} = D(\bar{\kappa}, s + \bar{\kappa}\cdot\bar{V}) = (s - \bar{\kappa}\cdot\bar{V}_1 + i\gamma_1)(s - \bar{\kappa}\cdot\bar{V}_2 + i\gamma_2) + \gamma^2, \tag{1-14}$$

where $\bar{V}_1 = \bar{v}_1 - \bar{V}$ and $\bar{V}_2 = \bar{v}_2 - \bar{V}$. The pinch condition $\partial D_{\bar{V}}/\partial\bar{\kappa} = 0$ and $D_{\bar{V}} = 0$ then requires (for $\gamma^2 \neq 0$) that $\bar{V}_1$ and $\bar{V}_2$ be colinear, i.e., the unstable pulse appears along the line $V_1,V_2$ joining the tips of $\bar{v}_1$ and $\bar{v}_2$. We can thus let $\bar{\kappa} = \bar{\kappa}_{\parallel} + \bar{\kappa}$ with respect to the directions of $\bar{V}_1$ or $\bar{V}_2$, say $\bar{V}_2$, and obtain an equivalent one-dimensional set of pinch equations

$$D_{V_\parallel} = (s + \kappa_\parallel V_1 + i\gamma_1)(s - \kappa_\parallel V_2 + i\gamma_2) + \gamma^2 = 0 \tag{1-15}$$

and

$$\frac{\partial D_{V_{\parallel}}}{\partial \kappa_{\parallel}} = 0, \tag{1-16}$$

from which $s_{oi}(\overline{V})$ can be found (see Eq. 1-6):

$$s_{oi}(\overline{V}) = \frac{2\gamma(V_1V_2)^{\frac{1}{2}} - (\gamma_1V_2 + \gamma_2V_1)}{V_1 + V_2} \tag{1-17}$$

where $V_1 = \overline{v}_1-\overline{V}$ and $V_2 = \overline{v}_2-\overline{V}$ . We have thus reduced the three-dimensional pulse problem to an equivalent one-dimensional problem which we have already solved (Eqs. 1-7 through 1-13). The three-dimensional pulse shape is thus obtained by solving the equivalent one-dimensional problem (Eq. 1-17) for all possible spatial orientations. In a general anisotropic and dispersive medium the uncoupled waves have different properties in different directions of propagation. Thus both the group velocities $\overline{v}_1$ and $\overline{v}_2$, and the damping rates $\gamma_1$ and $\gamma_2$ may change.

## VII. APPENDIX 2

### Propagation, Conservation, and Coupling in Weakly Inhomogeneous and Time-Varying Plasmas

To extend the linear description (Eq. IA-1) we write the constitutive relation between current and electric field as

$$\overline{J}(\overline{r},t) = \int d^3r' \int dt\ \overline{\overline{\sigma}}(\overline{r}-\overline{r}', t-t', \frac{\overline{r}+\overline{r}'}{2}, \frac{t+t'}{2}) \cdot \overline{E}(\overline{r}',t'), \tag{2-1}$$

and write the electric field (Eq. IIA-5) in the geometrical optics approximation as

$$\overline{E}(\overline{r},t) = \overline{\varepsilon}(\overline{r},t) e^{i\psi(\overline{r},t)} \tag{2-2}$$

so that

$$\overline{k}(\overline{r},t) = \frac{\partial\psi}{\partial\overline{r}} \tag{2-3}$$

and

$$\omega(\overline{r},t) = -\frac{\partial\psi}{\partial t}. \tag{2-4}$$

Proceeding as in Secs. II and III we can expand Eq. (2-1), assuming slow-variations and weak damping, and separate Maxwell's equations for the field amplitudes to zero and first order in the slow-variations. To zero order we obtain the local dispersion relation

$$D(\overline{k},\omega,\overline{r},t) = 0, \tag{2-5}$$

which is just Eq. (IA-20) with $\overline{\overline{K}}(\overline{k},\omega)$ replaced by $\overline{\overline{K}}(\overline{k},\omega,\overline{r},t)$. To first order we obtain the equations for the slowly varying field amplitudes. These lengthy details will be omitted here.

Equation (2-5) together with Eqs. (2-3) and (2-4) is a first order partial differential equation for the phase function $\psi$. Its solution by the method of characteristics is straightforward. Noting that Eqs. (2-3) and (2-4) give

$$\frac{\partial\omega}{\partial\overline{r}} = -\frac{\partial\overline{k}}{\partial t} \tag{2-6}$$

and that Eq. (2-5) can be solved for

$$\omega(\overline{k},\overline{r},t) = \omega_{\overline{k}} \tag{2-7}$$

we find

$$\frac{\partial\overline{k}}{\partial t} + \overline{v}_g\cdot\frac{\partial\overline{k}}{\partial\overline{r}} = \frac{(\partial D/\partial\overline{r})}{(\partial D/\partial\omega)} \tag{2-8}$$

and

$$\frac{\partial\omega}{\partial t} + \overline{v}_g\cdot\frac{\partial\omega}{\partial\overline{r}} = -\frac{(\partial D/\partial t)}{(\partial D/\partial\omega)}\ , \tag{2-9}$$

where

$$\overline{v}_g = \frac{\partial\omega_{\overline{k}}}{\partial\overline{k}} = -\frac{(\partial D/\partial\overline{k})}{(\partial D/\partial\omega)} \tag{2-10}$$

is the local group velocity (viz. Eq. IIA-20). Equations (2-8) and (2-9) are first-order total differential equations, along the ray trajectory $(d\bar{r}/dt) = \bar{v}_g$. Their solution and Eqs. (2-3) and (2-4) give the phase function $\psi$.

The conservation equation can be derived from the first order Maxwell's equations, or more directly from the integral form of the dispersion equation (Eq. IA-7) now extended to

$$\int d^3r' \int dt\, \bar{\bar{D}}(\bar{r}-\bar{r}', t-t', \frac{\bar{r}+\bar{r}'}{2}, \frac{t+t'}{2})\cdot\bar{E}(\bar{r}',t') = 0. \tag{2-11}$$

Using Eqs. (2-2) through (2-4) and expanding as before, for slow-variations and weak damping, we find

$$\frac{\partial n_{\bar{k}}}{\partial t} + \nabla\cdot(\bar{v}_{g\bar{k}} n_{\bar{k}}) + 2\gamma_{\bar{k}} n_{\bar{k}} = 0, \tag{2-12}$$

where

$$n_{\bar{k}}(\bar{r},t) = \left[\frac{1}{4}\,\varepsilon_o \bar{E}^* \frac{\partial \bar{\bar{D}}^h}{\partial \omega}\cdot\bar{E}\right]_{\bar{k}} \tag{2-13}$$

is the action density for the wave (Eqs. IIB-21 and IIA-19) and $\gamma_{\bar{k}}$ its small damping rate (Eq. IIA-21). For a large number of waves we can define a continuum action density $n(\bar{k},\bar{r},t)$ which then obeys

$$\frac{\partial n}{\partial t} + \frac{\partial}{\partial \bar{r}}\cdot\left(\frac{\partial \omega}{\partial \bar{k}}\, n\right) - \frac{\partial}{\partial \bar{k}}\left(\frac{\partial \omega}{\partial \bar{r}}\, n\right) - 2\gamma n = 0, \tag{2-14}$$

where $\gamma = \gamma(\bar{k},\bar{r},t)$ is the continuum weak damping rate. Using the characteristic equations (Eqs. 2-8 and 2-9) we can also establish the fluid equations for the action density

$$\frac{\partial n}{\partial t} + \frac{\partial\omega}{\partial\bar{k}}\cdot\frac{\partial n}{\partial r} - \frac{\partial\omega}{\partial\bar{r}}\cdot\frac{\partial n}{\partial\bar{k}} - 2\gamma n = 0. \tag{2-15}$$

The coupled-wave equations can also be extended by derivations that parallel Secs. IIIA or VA. We shall give the results for the nonlinear, conservative coupling of three waves. Assuming that the uncoupled waves are the WKB solutions described above, the coupling due to the second order nonlinear current gives

$$\frac{da_1}{dt} = p_1\; K(\bar{r},t)a_2a_3\; \exp\{i\delta\psi(\bar{r},t)\} \tag{2-16}$$

$$\frac{da_2}{dt} = -p_2\; K^*(\bar{r},t)a_1a_3^*\; \exp\{-i\delta\psi(\bar{r},t)\} \tag{2-17}$$

$$\frac{da_3}{dt} = -p_3\; K^*(\bar{r},t)a_1a_2^*\; \exp\{-i\delta\psi(\bar{r},t)\} \tag{2-18}$$

where (see Eq. VA-4)

$$\frac{d}{dt} = \left(\frac{\partial}{\partial t} + \bar{v}_g\cdot\nabla + \gamma\right) \tag{2-19}$$

is following the ray trajectory and includes the weak damping, the a's are normalized to action density (Eq. VA-13), the coupling coefficients $K(\bar{r},t)$ are evaluated locally (Eq. VB-6 with $\sigma_{ijk}(\bar{r},t)$ and $D(\bar{r},t)$), and

$$\delta\psi(\bar{r},t) = \psi_2 + \psi_3 - \psi_1 \tag{2-20}$$

is the synchornism dephasing due to the inhomogeneity and time-variation of the medium.

## VIII. REFERENCES

### References for Section I


1. T. H. Stix, The Theory of Plasma Waves, McGraw Hill, New York, 1962.

2. J. F. Denisse and J. L. Delcroix. Plasma Waves, Interscience, New York, 1963.

3. W. P. Allis, S. J. Buchsbaum, and A. Bers, Waves in Anisotropic Plasmas, M.I.T. Press, Cambridge, 1963.

4. V. L. Ginzburg, The Propagation of Electromagnetic Waves in Plasmas, Pergamon Press, London, 1964.

5. V. P. Silin and A. A. Rukhadze, Electromagnetic Properties of Plasmas and Plasma-Like Media, Gordon and Breach, New York, 1965.

6. A. G. Sitenko, Electromagnetic Fluctuations in Plasma, Academic Press, 1967.

7. Yu. L. Klimontovich, The Statistical Theory of Non-Equilibrium Processes in a Plasma, M.I.T. Press, Cambridge, 1967.

8. A. I. Akhiezer, I. A. Akhiezer, R. V. Polovin, A. G. Sitenko, and K. N. Stepanov, Collective Oscillations in a Plasma, M.I.T. Press, Cambridge, 1967.

9. V. D. Shafranov in Reviews of Plasma Physics, Vol. 3 (M. A. Leontovich, Ed.), Consultants Bureau, New York, 1967.

10. D. E. Baldwin, I. B. Bernstein, and M. P. H. Weenink in Advances in Plasma Physics, Vol. 3 (A. Simon and W. B. Thompson, Eds.), Interscience Publications, New York, 1969.


### References for Section II


1. M. E. Gertsenshtein, Zh. Eksp. Theor. Fiz. 26, 680 (1954).

2. P. A. Sturrock, Ann. Phys. 4, 306 (1958).

3. A. Bers and P. Penfield, Jr., IRE Trans. ED-9, 12 (1962).

4. A. Bers in Waves in Anisotropic Plasmas, Sec. 8.5 (W. P. Allis, S. J. Buchsbaum, and A. Bers), M.I.T. Press, Cambridge, 1963.

5. P. A. Sturrock, J. Appl. Phys. 31, 2052 (1961).

6. A. Bers and S. Gruber, Appl. Phys. Letters 6, 27 (1965).

7. B. B. Kadomtsev, A. B. Mikhailovskii, and A. V. Timofeev, Zh. Eksp. Thero. Fiz. 47, 2266 (1965); Sov. Phys.-JETP 20 1517 (1965).

## References for Section III

1. J. R. Pierce, J. Appl. Phys. 25, 179 (1954).

2. W. H. Louisell, Coupled Mode and Parametric Electronics, John Wiley & Sons Inc., New York, 1960.

3. D. L. Bobroff and H. A. Haus, J. Appl. Phys. 38, 390 (1967).

4. Plasma instabilities are treated in many of the references to Section I, viz., Refs. 1, 5, 8, and 10, and in the following.

5. Streaming instabilities are discussed in:

   O. Buneman, Phys. Rev. 115, 503 (1959);

   R. J. Briggs in Advances in Plasma Physics (A. Simon and W. B. Thompson, Eds.), John Wiley & Sons Inc., New York 1971;

   J. A. Davis and A. Bers, Proceedings of the Symposium on Turbulence of Fluids and Plasmas, Polytechnic Press, Brooklyn, 1968.

6. Instabilities due to anisotropic velocity distributions are reviewed in:

   M. N. Rosenbluth in Plasma Physics, IAEA, Vienna, 1965;

   A. V. Timofeev and V. I. Pistunovich in Reviews of Plasma Physics, Vol. 5 (M. A. Leontovich, Ed.), Consultants Bureau New York, 1970.

7. E. G. Harris in Physics of Hot Plasmas (Rye and Taylor, Eds.), Oliver & Boyd, Edinburgh, 1970.

## References for Section IV


1. R. Q. Twiss, Proc. Phys. Soc. (London), B64, 654 (1951); Phys. Rev. 88, 1392 (1952).

2. L. D. Landau and I. M. Lifshitz, Electrodynamics of Continuous Media (in Russian), G.I.T.T.L., Moscow (1953), p. 141; see also their Fluid Mechanics, Pergamon Press, London (1959), p. 113.

3. P. A. Sturrock, Phys. Rev. 112, 1488 (1958).

4. O. Buneman in Plasma Physics, edited by J. E. Drummond, McGraw-Hill Book Co., Inc., New York (1961).

5. Ya. B. Fainberg, V. I. Kurilko, and V. D. Shapiro, Zh. Tekh. Fiz. 31, 633 (1961), Sov. Phys. - Tech. Phys. 6, 459 (1961).

6. R. V. Polovin, Zh. Tekh. Fiz. 31, 1220 (1961), Sov. Phys.-Tech. Phys. 6, 892 (1962).

7. A. Bers and R. J. Briggs, Quarterly Progress Report No. 71, Research Laboratory of Electronics, M.I.T., Cambridge, Mass. (1963), pp. 122-131; Bull. Amer. Phys. Soc., Ser. II, No. 3, 9, 304 (1964).

8. R. J. Briggs, Electron-Stream Interaction with Plasmas, M.I.T. Press, Cambridge, Mass. (1964), Ch. 2.

9. P. Rolland, Phys. Rev. 140, 776 (1965).

10. K. B. Dysthe, Nuclear Fusion 6, 215 (1966).

11. H. Derfler, Phys. Lett. 24A, 763 (1967).

12. L. S. Hall and W. Heckrotte, Phys. Rev. 166, 120 (1968).

13. In mathematics, pinch-type singularities date back to the work of J. Hadamard, Acta Math. 22, 55 (1898); they also play an important role in quantum field theory, see e.g. R. J. Eden, Proc. Roy. Soc. A210, 388 (1952); they were independently discovered in plasma stability theory in reference 7.

## References for Section V

1. R. Z. Sagdeev and A. A. Galeev, Nonlinear Plasma Theory, W. A. Benjamin Inc., New York, 1969.

2. V. N. Tsytovich, Nonlinear Effects in Plasma, Plenum Press, New York, 1970.

3. R. C. Davidson, Methods in Nonlinear Plasma Theory, Academic Press, New York (1972), especially Chapter 6.

4. W. H. Louisel, Coupled Mode and Parametric Electronics, Wiley, New York, 1960.

5. Y. M. Aliev, V. P. Silin, and C. Watson, Zh. Eksp. Teor. Fiz. 50, 943 (1966); (Sov. Phys. JETP 23, 626 (1966).

6. A. Sjolund and L. Stenflo, Physica 35, 499 (1967); and J. Appl. Phys. 38, 2676 (1967).

7. K. Nishikawa, Phys. Soc. Japan 24, 916 and 1152 (1968).

8. D. F. DuBois and M. V. Goldman, Phys. Rev. 164, 207 (1967).

## References for Appendix 2

1. Wave propagation in inhomogeneous plasmas is treated in several of the references of Sec. I, viz. Refs. 1 and 4. (See also the recent review: V. E. Golant and A. D. Piliya, "Linear Transformation of Waves in a Plasma," Soviet Physics Uspekhi 14, 1972.) Instabilities in inhomogeneous plasmas are reviewed in the following references.

2. N. A. Krall in Advances in Plasma Physics, Vol. 1 (A. Simon and W. B. Thompson, Eds.), Interscience Publ., New York, 1968.

3. A. B. Mikhailovskii in Reviews of Plasma Physics, Vol. 3 (M.A.Leontovich, Ed.) Consultants Bureau, New York, 1967.

4. B. B. Kadomtsev and O. P. Pogutse in Reviews of Plasma Physics, Vol. 5 (M.A.Leontovich, Ed.) Consultants Bureau, New York, 1970.

# NON-LINEAR EFFECTS

GUY LAVAL AND RENE PELLAT

*Centre de Physique Théorique de l'Ecole Polytechnique,*
*17 Rue Descartes, 75 Paris V, France*

CONTENTS

# CHAPTER I. QUASI-LINEAR THEORY

## A. THE "OLD" QUASI-LINEAR THEORY

In the linear theory of microinstabilities, the particle distribution function $F(r,v,t)$ is split into two parts, $F_0$ and $F_1$, with $F = F_0 + F_1$. The equilibrium distribution function, $F_0$, usually does not depend on $t$ and is a solution of the Vlasov equation with self-consistent stationary electromagnetic fields $\bar{E}_0$, $\bar{B}_0$. The perturbed distribution function $F_1(r,v,t)$ characterizes the perturbations of the stationary state and is a solution of the linearized Vlasov equation with self-consistent electromagnetic fields $E_1, B_1$. Then we have

$$v \cdot \frac{\partial F_0}{\partial \bar{r}} + \frac{q}{m} (E_0 + \bar{v} \wedge \bar{B}_0) \cdot \frac{\partial F_0}{\partial \bar{v}} = 0 \qquad A(1)$$

$$\frac{\partial F_1}{\partial t} + v \cdot \frac{\partial F_1}{\partial \mathbf{r}} + \frac{q}{m} (\bar{E}_0 + \bar{v} \wedge \bar{B}_0) \frac{\partial F_1}{\partial \bar{v}} = - \frac{q}{m} [\bar{E}_1 + \bar{v} \wedge \bar{B}_1] \cdot \frac{\partial F_0}{\partial \bar{v}} \qquad A(2)$$

Staying in the linearized theory, $F_0$ is assumed to be known and we look for solutions of (2) and the Maxwell equation with $F_1 = \hat{F}_1(r,v,\omega) \exp i\omega t$, $E_1 = \hat{E}_1(r,\omega) \exp i\omega t$ and $B_1 = \hat{B}_1(r,\omega) \exp i\omega t$. The solutions of equation (2) may be written:

$$\hat{F}_1 = - \frac{q}{m} g(\bar{r},\bar{v},\omega)[\hat{E}_1 + \bar{v} \wedge \hat{B}_1] \cdot \frac{\partial F_0}{\partial \bar{v}} \qquad A(3)$$

where $g(r,v,\omega)$ is a linear operator which is defined by:

$$g(\bar{r},\bar{v},\omega)U = \int_{-\infty}^{0} e^{i\omega\tau} U(\bar{r}(\tau),\bar{v}(\tau))d\tau \qquad A(4)$$

where $\bar{r}(\tau)$ and $\bar{v}(\tau)$ are solutions of the unperturbed equation of motion for the particles:

$$\frac{d\bar{r}(\tau)}{d\tau} = \bar{v}(\tau)$$

$$\frac{d\bar{v}(\tau)}{d\tau} = \frac{q}{m} [E_0(\bar{r}(\tau)) + \bar{v}(\tau) \wedge \bar{B}_0(\bar{r}(\tau))] \qquad A(5)$$

such that $\bar{r}(0) = \bar{r}$ and $\bar{v}(0) = \bar{v}$. This way of solving (2) is known as the procedure of integration along unperturbed orbits. Since we are looking for unstable perturbations, $\omega$ is assumed to have a non-zero negative imaginary part so that there is no problem of convergence for $t \to -\infty$. Solution (3) is used in Maxwell's equations to compute the perturbed currents and densities. The dispersion equation is then obtained and the complex roots $\omega_\alpha$ give the unstable modes. The unstable nature of the system depends essentially on the shape of the equilibrium distribution function $F_0$. For several instabilities (weak beam-plasma instability, weak current instability, anisotropy instability with cold plasma, drift waves etc...), the system may become stable by changing the shape of the distribution function in a domain of phase space containing a small fraction of the total number of particles. Drummond, Vedenov, Sagdeev and others noticed that such a change could be provided by the fluctuations themselves and that the stabilization could be achieved with a relatively small change in the particle energy. Then the fluctuation level would remain small during the stabilization process and the linearization of equations could still be used.

Now however, $F_0$ cannot be defined as the initial distribution function of the stationary state but we must consider that $F_0$ is the average value $\langle F \rangle$ of $F$ so that $F_0$ can have any time, space and velocity dependance. In principle, the averaging process should be an averaging on stochastic initial conditions. In fact, the averaging will be defined most of time as an averaging over one or several coordinates in space or over time, when these averaging processes commute with the averaging over initial condition. If the system is initially invariant to space or time translation or to rotation, the averaged quantities will have the same invariance properties and it will often be sufficient to define the averaged quantities as their invariant part under these transformations.

For instance, if we want to study a homogeneous system from time $t = 0$, we shall set $F_0 = \lim_{V \to \infty} (1/V) \int_V F \, d^3r$ and $F_0$ will depend on time and velocity. On the other hand, if the system is stationary, we shall set $F_0 = \lim_{T \to \infty} (1/T) \int F \, dt$ and $F_0$ will depend on space and velocity. As another example, let us consider a plasma in a magnetic mirror where in the absence of fluctuations the particles would experience bouncing motions. These bouncing motions may be described in terms of action and angle conjugate variables $J$, $\theta$. The initial system can be defined by an averaged distribution function $F_0(J)$. Then the averaging is invariant under a change of the origin of the angle $\theta$ and we shall set $F_0 = (1/2\pi) \int F \, d\theta$. We shall denote by brackets the average value of any quantity. In the same way we shall denote $E_0 = \langle E \rangle$, $B_0 = \langle B \rangle$, $\tilde{F} = F - F_0$, $\tilde{B} = B - B_0$, $\tilde{E} = E - E_0$. By averaging the Vlasov equation we get:

$$\frac{\partial F_0}{\partial t} + \bar{v} \cdot \frac{\partial F_0}{\partial \bar{r}} + \frac{q}{m} [\bar{E}_0 + \bar{v} \wedge \bar{B}_0] \frac{\partial F_0}{\partial \bar{v}} = - \frac{q}{m} \left\langle [\tilde{E} + \bar{v} \wedge \tilde{B}] \frac{\partial \tilde{F}}{\partial \bar{v}} \right\rangle \qquad \text{A(6)}$$

By substracting (6) from the exact Vlasov equation we get:

$$\frac{\partial \tilde{F}}{\partial r} + \bar{v} \cdot \frac{\partial \tilde{F}}{\partial \bar{r}} + \frac{q}{m} [\bar{E}_0 + \bar{v} \wedge \bar{B}_0] \frac{\partial \tilde{F}}{\partial \bar{v}} = - \frac{q}{m} [\tilde{E} + \bar{v} \wedge \tilde{B}] \frac{\partial F_0}{\partial \bar{v}} \qquad \text{A(7)}$$

$$- \frac{q}{m} [\tilde{E} + \bar{v} \wedge \tilde{B}] \frac{\partial \tilde{F}}{\partial \bar{v}} + \frac{q}{m} \left\langle [\tilde{E} + \bar{v} \wedge \tilde{B}] \frac{\partial \tilde{F}}{\partial v} \right\rangle$$

In order to prove the self-stabilization, we should solve (6) and (7). In the quasi-linear approximation we replace equation (7) by

$$\tilde{F}(\bar{r},\bar{v},t) = - \sum_{\alpha} \frac{q}{m} g(\bar{r},\bar{v},\omega_\alpha(t)) [\bar{E}_{1\alpha} + \bar{v} \wedge \bar{B}_{1\alpha}] \frac{\partial F_0}{\partial \bar{v}} (t) \qquad \text{A(8)}$$

where $\omega_\alpha(t)$ is the solution of the linear dispersion equation corresponding to the distribution function $F_0(t)$. The fields $\bar{E}_{1\alpha}$ and $\bar{B}_{1\alpha}$ are the values of $E_1$ and $B_1$ which are solutions of Maxwell's equations in the linear approximation with a frequency $\omega_\alpha(t)$. The time dependance of these fields will be given by:

$$\frac{\partial \bar{E}_{1\alpha}}{\partial t} = i\omega_\alpha \bar{E}_{1\alpha} \qquad \text{A(9)}$$

and

$$\tilde{E} = \sum_\alpha \bar{E}_{1\alpha} \quad \tilde{B} = \sum_\alpha \bar{B}_{1\alpha}$$

In other words, we assume that the fluctuating distribution function is well approximated by the instantaneous linearized solution of the Maxwell-Vlasov equations. From (6) and (8) we now get the equation of evolution of $F_0$ in the form:

$$\frac{\partial F_0}{\partial t} + \bar{v} \cdot \frac{\partial F_0}{\partial \bar{r}} + \frac{q}{m} [\bar{E}_0 + \bar{v} \wedge \bar{B}_0] \frac{\partial F_0}{\partial \bar{v}} = \tag{A(10)}$$

$$\frac{q^2}{m^2} < \sum_{\alpha,\beta} [\bar{E}_{1\alpha} + \bar{v} \wedge \bar{B}_{1\alpha}] \frac{\partial}{\partial \bar{v}} g(\bar{r},\bar{v},\omega_\beta(t))[\bar{E}_{1\beta} + \bar{v} \wedge \bar{B}_{1\beta}] > \frac{\partial F_0}{\partial \bar{v}}$$

The set of equations (9) and (10) with the averaged Maxwell's equations is a closed system in $F_0$ and $E_\alpha$ from which the flúctuating distribution functions has been eliminated.

The equations (9) and (10) are the basic equations of the quasilinear theory. In the following we shall establish the conditions of validity of these approximations in several important particular cases. In order that the equation of the quasi-linear theory be valid we must show that

1) in equation (7) the two terms which are quadratic in fluctuating quantities are negligible;

2) we may then neglect the time variation of $F_0$ when we solve (7) though we keep the value of $F_0$ at time $t$ in the solution;

3) we may neglect initial conditions in solving (7) in order to get a markoffian kinetic equation for $F_0$;

4) finally, the time dependence of the electromagnetic fields is correctly given by the quasi-classic approximation (9).

These approximations could seem reasonable provided that the amplitudes of the electromagnetic fields remain small and the characteristic time for the evolution of $F_0$ is much longer than the average period of the fluctuations. In fact, as noticed it is never allowed to write $\tilde{F}_1$ in the form of equation (8), since this expression has poles in the complex $\vec{v}$ plane and we know that for regular initial conditions, this is not possible. Nevertheless we shall show in the following that by considering the problem from another point of view, equations (9) and (10) can be valid in a well-defined range. It is difficult to get these results from a systematic expansion in terms of a unique small parameter because such expansions are not uniformaly convergent in $\vec{v}$ space. The derivation we gave in this section is the traditional one. It may lead to mistakes and we shall give in the following sections a more correct and physical derivation of the quasi-linear equation.

## B. RESONANT PARTICLE QUASI-LINEAR DIFFUSION EQUATION

In this part, we shall see under which conditions it is possible to derive equation A(10) as a good approximation of the system A6, A7. For simplicity we shall begin by treating the simple particular case of one-dimensional electrostatic waves propagating in a homogeneous plasma without average fields ($\tilde{E}_0 = B_0 = 0$). In such a case, $B_1 = 0$ and $\bar{E}_1$ can be taken as a scalar. The velocities in the direction perpendicular to $\bar{E}_1$ may be ignored and equations A(6) and A(7) reduce to

$$\frac{\partial F_0}{\partial t} = -\frac{q}{m} \langle \tilde{E} \frac{\partial \tilde{F}}{\partial \bar{v}} \rangle \qquad \text{B(1)}$$

$$\frac{\partial \tilde{F}}{\partial t} + v \cdot \frac{\partial F}{\partial x} = -\frac{q}{m} \tilde{E} \frac{\partial F_0}{\partial v} - \frac{q}{m} \tilde{E} \frac{\partial \tilde{F}}{\partial v} + \frac{q}{m} \langle \tilde{E} \frac{\partial \tilde{F}}{\partial v} \rangle \qquad \text{B(2)}$$

where we have assumed that the waves propagate along 0 $x$. The averaging is defined by $\langle\rangle = \lim_{L\to\infty} (1/L) \int dx$. $L$ being the length of the system. Following the philosophy of the quasi-linear theory, we shall first assume that the two last terms of the r.h.s. of equation B(2) are negligible and check this approximation a posteriori. Then

$$\frac{\partial \tilde{F}}{\partial r} + v \frac{\partial \tilde{F}}{\partial x} = -\frac{q}{m} \tilde{E} \frac{\partial F_0}{\partial v} \qquad \text{B(3)}$$

We retain in B(3) the time dependance of $F_0$ which is the only difference between B(3) and the linearized equation A(2). We assume periodic boundary condition at the end of a system of length $L$. Then we may write for any fluctuating quantity $\tilde{U}$:

$$\tilde{U} = \sum_k U_k e^{ikx} \qquad U_k = \frac{1}{L} \int \tilde{U} e^{-ikx} dx$$

The Fourier transform of B(3) may be written:

$$\frac{\partial F_k}{\partial t} + ikvF_k = -\frac{q}{m} E_k \frac{\partial F_0}{\partial v} \qquad \text{B(4)}$$

and equation B(1) becomes

$$\frac{\partial F_0}{\partial t} = -\frac{q}{m} \sum_k E_{-k} \frac{\partial F_k}{\partial v} \qquad \text{B(5)}$$

The solution of equation (4) may be written:

$$F_k(t) = F_k(t = 0)e^{-ikvt} - \frac{q}{m}\int_0^t E_k(t')\frac{\partial F_0}{\partial v}(t')e^{ikv(t'-t)}dt' \qquad B(6)$$

Since we know that the theory will be valid only if the frequency of the waves remain close to the linear frequency, let us set:

$$E_k(t) = \hat{E}_k(t)\ \exp\ iS_k(t) \qquad B(7)$$

$$\text{where } S_k(t=0) = 0,\quad \frac{dS_k}{dt} = \omega_k(t) \qquad B(8)$$

$\omega_k(t)$ being the *real* part of the frequency we would get by solving the linear dispersion equations at time $t$ with the distribution function $F_0(t)$. From B(5) we get:

$$\frac{\partial F_0}{\partial t} = -\frac{q}{m}\sum_k \hat{E}_{-k}(t)\frac{\partial}{\partial v}\left(F_k(0)e^{-ikvt-iS_k(t)}\right) \qquad B(9)$$

$$+\frac{q^2}{m^2}\sum_k \hat{E}_{-k}(t)\frac{\partial}{\partial v}\int_0^t \hat{E}_k(t')\frac{\partial F_0}{\partial t}(t')\ \exp[ikv(t'-t)$$

$$+ i(S_k(t') - S_k(t))]dt'$$

where we have used the obvious equality $S_k(t) = -S_{-k}(t)$. The equation B(9) is a non-markoffian kinetic equation for $F_0$ which depends on the initial conditions. In order to go further we have to introduce new approximations.

First, we notice that B(9) may be written:

$$\frac{\partial F_0}{\partial t} = -\frac{q}{m}\sum_k \hat{E}_{-k}(t)\ \frac{\partial}{\partial v}\left(F_k(0)e^{-ikvt-iS_k(t)}\right) \qquad B(10)$$

$$+\frac{q^2}{m^2}\sum_k \hat{E}_{-k}\frac{\partial}{\partial v}\int_0^t d\tau\hat{E}_k(t-\tau)\frac{\partial F_0}{\partial v}(t-\tau)e^{-ikv\tau+i[S_k(t-\tau)-S_k(t)]}$$

Let us set $\bar{\omega}_k(t) = (1/t)S_k(t)$ and consider the first term of the r.h.s. of equation B(10). We may write it:

$$\sum_k A_k(t)\ \exp{-ig(k,t)t} \qquad B(11)$$

where $g(k,t) = kv + \bar{\omega}k$

and $A_k = \hat{\underline{E}}_k(t)\frac{\partial}{\partial v}\ [F_k(0)e^{-igt}]e^{igt}$

The frequency $\omega_k(t)$ depends on the average properties of the plasma such as the density, and temperature. Since we assume that $F_0$ will be only slightly changed, we may consider that $\omega_k$ does not change by an order of magnitude as $t$ varies. Then $g(k,t)$ will be a smoothly varying function of $k$. If the spectrum of excited fluctuations has a half-width $\Delta k$ and is centered at $k_0$, we may introduce the corresponding characteristic variation of $g(k,t)$ over the spectrum:

$$\Delta g(k,t) = g(k_0 + \Delta k,t) - g(k_0 - \Delta k,t) = \frac{1}{T_c}$$

This time $T_c$ may be considered as the coherence time of the fields for resonant particles. It is the characteristic time for the variation of the electric field in a reference frame moving at the velocity $-\bar{\omega}_{k_0/k_0}$. Then when $t \gg T_c$, $\exp ig(K,t)$ is a rapidly oscillating function of $k$. Since $A_k(t)$ is not so, the expression B(11) will go rapidly to zero. This may be seen in another way: if $L \to \propto$, we may set

$$\sum_k A_k \exp{-ig(K,t)t} = L\int dKA(k)\exp{-ig(k,t)t}$$

If $t \gg T_c$ we may apply the saddle point method to the integral and we find that the integral goes to zero except if there is one value of $k$ for which $\partial g/\partial k = 0$, i.e. $U \simeq -\partial\omega_k/\partial k$. For the moment, we shall assume that this is not the case in the range of interest. In the same way, one shows that the integrand of the second term in the r.h.s. of equation B(10) goes to zero when $\tau \gg T_c$. Then if $\tau_E$ and $\tau_F$ are the characteristic times of evolution of $\hat{E}_K$ and $F_0$, we are allowed to replace in B(10) $\hat{E}_K(t-\tau)$ and $(\partial F_0/\partial V)(t-\tau)$ by $\hat{E}_K(t)$ and $(\partial F_0/\partial t)(t)$ provided that $\tau_E \gg T_C$ and $\tau_{F_0} \gg T_C$. By doing this approximation, we modify the integrand only for values of $\tau$ such that it does not contribute to the value of the integral. Then equation B(10) becomes:

$$\frac{\partial F_0}{\partial t} = \frac{q^2}{m^2}\frac{\partial}{\partial v}\sum_K|\hat{E}_K|^2\int_0^t e^{-ikv\tau-i\omega_K(t)\tau}\ d\tau\frac{\partial F_0}{\partial v} \qquad B(12)$$

where we have used also the approximation $S_K(t-\tau) = S_K(t)-\omega_K(t)$ for $\tau \ll \tau_{F_0}$. But as soon as $v\Delta kt \gg 1$, we can use the approximation

$$\int_0^t \exp-i(\omega+kv)\tau \, d\tau = PP \frac{1}{i(\omega+kv)} + \pi\delta(\omega+kv)$$

By retaining only the even term in $K$, we get:

$$\frac{\partial F_0}{\partial t} = \frac{\partial}{\partial v}\left(\sum_K \frac{\pi q^2}{m^2} |E_K|^2 S(\omega_K+kv)\right)\frac{\partial F_0}{\partial v} \qquad \text{B(13)}$$

This is the diffusion equation in velocity space for resonant particles. We note that we have not used any assumption about the stability or instability of the system so that the diffusion equation (13) is valid as well for damped waves. The condition of validity can be written

$$\tau_E \gg T_c \quad \tau_{F_0} \gg T_c \quad \text{with} \quad T_c = [v\Delta k + \Delta\omega_k]^{-1}$$

where $\Delta\omega_k$ is the spread of wave frequencies in the spectrum. Roughly it means that the phase velocity spread must be large.

Let us notice that, if we had used the method of Section A, we would have obtained

$$F_K = -\frac{q}{m} E_K[\omega_k - i\gamma_K + kv]^{-1} \frac{\partial F_0}{\partial v} \qquad \text{B(14)}$$

and

$$\frac{\partial F_0}{\partial t} = \frac{\partial}{\partial v} \sum_K \frac{q^2}{m^2} \left|E_K\right|^2 \frac{\gamma_K}{(\omega_k+kv)^2+|\gamma_K|^2} \frac{\partial F_0}{\partial v} \qquad \text{B(15)}$$

This equation reduces to equation B(13) provided that $\gamma_K > 0$ and $\gamma_K \ll v\Delta k + \Delta\omega_k$. The method used in Section B has two advantages. It makes it possible to determine the domain of validity of the quasi-linear diffusion equation B(15) and shows that in this domain, B(15) is not valid for damped modes since equation (14) is meaningless in such a case for the resonant particles.

At last, we can see that equation B(13) describes an irreversible behaviour for $F_0$. If we multiply B(13) by $F_0$ and integrate over the velocity we get:

$$\frac{\partial}{\partial t}\int F_0^2\, dv = -\frac{2q^2}{m^2}\int dv \sum_K \pi|\hat{E}_K|^2 \delta(\omega_k + kv)\left|\frac{\partial F_0}{\partial v}\right|^2 \leq 0$$

Since $F_0^2$ remains positive, the asymptotic solution for $t \to \infty$ must correspond either to a distribution function $F_0$ such that $\partial F_0/\partial v = 0$ for the resonant particles or to a complete damping of the fluctuations such that $\Sigma_K|E_K|^2 = 0$

## C. NON-RESONANT PARTICLES

We have assumed that the resonant particles represent a small fraction of the total number of particles. Non-resonant particles will denote those particles such that their velocity $v$ cannot fulfill the condition $\omega_K + Kv = 0$ for any value of $K$ such that $E_K \neq 0$. In the lowest order, we find that their distribution function does not vary with time. It is therefore necessary to take higher order terms into account. Then in B(10) we retain one more term in the Taylor expansion of $\hat{E}_K$, $F_0$, $S$. Instead of B(12), we get now:

$$\frac{\partial F_0}{\partial r} = \frac{q^2}{m^2}\sum_K \hat{E}_{-K}(r)\frac{\partial}{\partial v}\int_0^t d\tau \left|\hat{E}_K(r)\frac{\partial F_0}{\partial v}(r) - \tau\frac{\partial}{\partial r}\hat{E}_K(r)\frac{\partial F_0}{\partial v}(r)\right| \times \qquad C(1)$$

$$\exp\{-i[Kv + \omega_K]\tau + i\frac{\tau^2}{2}\frac{\partial \omega_k}{\partial t}\}$$

For simplicity we shall assume that $T_C^2(\partial\omega_k/\partial t) << 1$, though the theory could be worked out without this requirement. Then C(1) becomes:

$$\frac{\partial F_0}{\partial r} = \frac{q^2}{m^2}\frac{\partial}{\partial v}\sum_K \frac{\hat{E}_{-K}(r)\frac{\partial}{\partial r}\left(\hat{E}_K(r)\frac{\partial F_0}{\partial v}(r)\right)}{(\omega_K + Kv)^2}$$

where we have used the relationships:

$$\int_0^t \tau \exp(-i(Kv + \omega K)\tau)d\tau = i\frac{\partial}{\partial \omega K}\int_0^t \exp(-(Kv + \omega_K)\tau)$$

In most cases, for non-resonant particles, we have the inequality:

$$\frac{\partial F_0}{\partial v}\frac{\partial \hat{E}_K}{\partial r} >> \hat{E}_K \frac{\partial^2 F_0}{\partial v \partial r}$$

Then C(2) becomes:

$$\frac{\partial F_0}{\partial t} = \frac{q^2}{2m^2}\frac{\partial}{\partial v}\frac{\sum_K \frac{\partial}{\partial t}\left|E_K\right|^2}{(\omega_k + Kv)^2}\frac{\partial F_0}{\partial v} \qquad C(3)$$

Equation C(3) is not a proper diffusion equation since $(\partial F_0/\partial t)$ is linked to $\partial E_K^2/\partial t$. However, it may be considered as describing the adiabatic part of the distribution function. It accounts for the non-resonant oscillations of particles in the wave fields. These motions cause an apparent increase in velocity dispersion as $E_K^2$ increase and vice-versa.

D. ELECTRIC FIELD SPECTRUM

In order to derive an equation for $\hat{E}_K(t)$ we use the Poisson equation. We obtain $F_K$ from B(6) and then:

$$iKE_K = \frac{q}{\varepsilon_0}\int F_K(t=0)\exp(-ikvt)dv$$

$$- \frac{q^2}{m\varepsilon_0}\int dv\int_0^t E_K(t-\tau)\frac{\partial F_0}{\partial v}(t-\tau)\exp(-ikv\tau)d\tau \qquad D(1)$$

The first term of the r.h.s. is negligible when $t > (k\Delta v)^{-1}$ where $\Delta v$ is the typical scale of variation of $\partial F_0/\partial v$ in velocity space. If we assume that $\tau_E << (k\Delta v)^{-1}$ and $\tau_{F_0} << (k\Delta v)^{-1}$, we may expand the integrand in the vicinity of $\tau = 0$ and we get:

$$iK\hat{E}_K(t) = \frac{iq^2}{m\varepsilon_0}\hat{E}_K(r)\int dv\frac{\partial F_0}{\partial v}(t)\left|pp\frac{1}{\omega_k+Kv} + i\pi\delta(\omega_k + Kv)\right|$$

$$+ \frac{q^2}{m\varepsilon_0}\frac{\partial \hat{E}_K}{\partial t}\int\frac{\frac{\partial F_0}{\partial v}dv}{(\omega_k + Kv)^2} \qquad D(2)$$

where in the last term the integration is restricted to non-resonant particles. Since $\omega_k$ is the solution of the dispersion relation, equation D(2) may be written:

$$\frac{2q^2}{m\varepsilon_0}\left|E_K\right|^2\pi\int dv\frac{\partial F_0}{\partial v}(t)\delta(\omega_k + Kv)$$

$$+ \frac{q^2}{m\varepsilon_0}\frac{\partial}{\partial t}\left|E_K\right|^2\int dv\frac{1}{(\omega_k + Kv)^2}\frac{\partial F_0}{\partial v} = 0 \qquad D(3)$$

or

$$\frac{\partial |E_k|^2}{\partial t} = 2\gamma_k(t)|E_k|^2 \qquad \text{D(4).}$$

where $\gamma_k(t)$ is the value of the growth rate which is obtained from the linear dispersion equation with the distribution function $F_0(t)$.

## E. THE BUMP-IN-TAIL INSTABILITY-LINEAR THEORY

In order to illustrate the quasi-linear stabilizing process we shall apply it to one-dimension electron waves. We assume that the initial distribution function is the sum of a gaussian function at rest and a weak beam such that:

$$F_0(t=0) = \frac{n_0}{v_{th}\sqrt{\pi}} \exp -V^2/v_{th}^2 + \frac{n_1}{\Delta U\sqrt{\pi}} \exp -(V-U)^2/(\Delta U)^2 \qquad \text{E(1)}$$

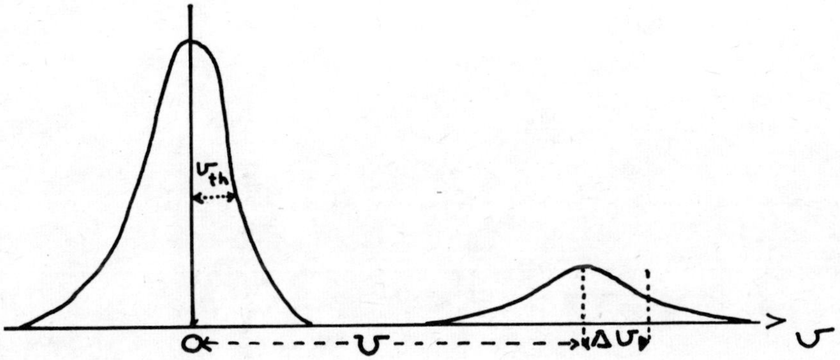


Fig. 1

with $n_1 \ll n_0$, $v_{th} \ll v$, $\Delta v < v$

The unstable waves have frequencies very close to $\omega_p$ and phase velocities larger than $v_{th}$. The growth rate is given by the Landau formula:

$$\frac{\gamma_K}{\omega_k} = - \frac{\pi\omega_p^2}{2n_0 K|K|} \left.\frac{\partial F_0}{\partial v}\right|_{v = -\omega_k/k} \qquad \text{E(2)}$$

which implies that $- \frac{\omega_k}{k} \frac{\partial F_0}{\partial v}\bigg|_{v = -\omega_k/k} > 0$

for the unstable waves. The formula E(2) can be applied only if $[V^3/(\Delta V)^3][n_1/n_0] << 1$. This condition could also be obtained from the inequality $\tau_E << (k\Delta v)^{-1}$ which we had to assume in part D.

## F. THE BUMP-IN-TAIL INSTABILITY-QUASI-LINEAR THEORY

We have to solve the coupled equations B(13) and D(4) for $F_0$ and $E_k^2$. First let us use the transformation $\Sigma_k \rightarrow \frac{L}{2\pi} \int dk$. Then equations B(13) and D(4) become:

$$\frac{\partial F_0}{\partial t} = \frac{q^2}{m^2 \varepsilon_0} \frac{L}{2} \frac{\varepsilon(v)}{v} \frac{\partial F_0}{\partial v} \qquad \text{F(1)}$$

$$\frac{\partial \varepsilon}{\partial t} = 2\gamma(v)\varepsilon(v) \qquad \text{F(2)}$$

$$\text{with } \varepsilon(v) = \frac{\varepsilon_0 |E_k^2|}{2}\bigg|_{k = -\omega_k/v} \qquad \text{F(3)}$$

$$\gamma(v) = \frac{\pi\omega_p |v| v}{2n_0} \frac{\partial F_0}{\partial v} \qquad \text{F(4)}$$

By substituting $\frac{\partial F_0}{\partial v}$ from F(2) into F(1), we obtain

$$\frac{\partial}{\partial t} \left| F_0 - \frac{\omega_p L}{m\pi} \frac{\partial}{\partial v} \left(\frac{1}{v^3} \varepsilon\right) \right| = 0 \qquad \text{F(5)}$$

At $t = 0$, the fluctuations are very small so that:

$$F_0(t) - \frac{\omega_p L}{m\pi} \frac{\partial}{\partial v} \frac{1}{v^3} \varepsilon(t) = F_0(t = 0) \qquad \text{F(6)}$$

or

$$\frac{\omega_p L}{m\pi} \frac{\varepsilon(t)}{v^3} = \int_{v_0}^{v} [F_0(t) - F_0(t = 0)] dv \qquad \text{F(7)}$$

where $v_0$ is the smallest phase velocity of the excited spectrum. We know that the asymptotic state will correspond either to $\partial F_0/\partial v = 0$ or to $\varepsilon(t{=}\infty) = 0$. The second possibility implies obviously, $\partial F_0/\partial v(t{\to}\infty) \leq 0$ in the resonant region which would contradict equation F(6) for $t{\to}\infty$. Then the asymptotic state must correspond to the distribution function which is sketched in Fig. 2.

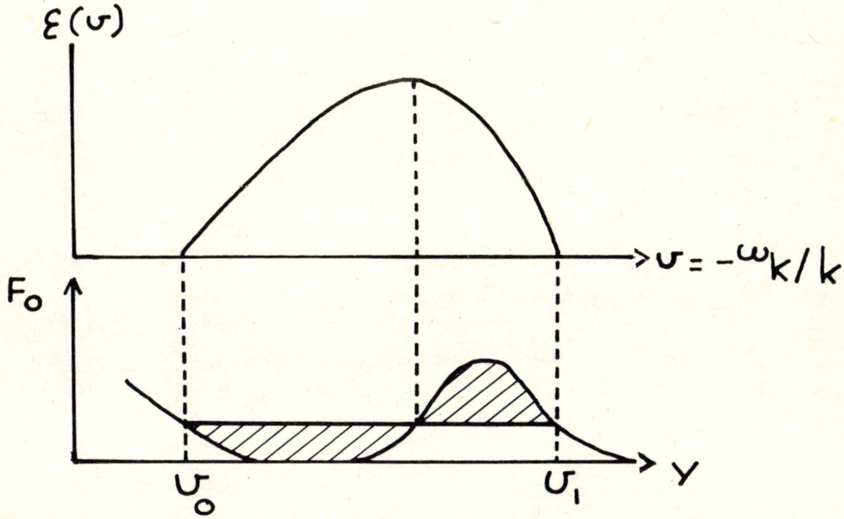


Fig. 2

The horizontal line is chosen so that the areas are equal in order to insure density conservation. From F(7) we can estimate the asymptotic electric field amplitude:

$$\frac{L}{\pi}\,\varepsilon(t{=}\infty) \simeq \frac{n_1 m U^3}{\omega_p} \qquad \text{F(8)}$$

and the electrostatic energy

$$W = \sum_k \frac{\varepsilon_0 |E_k|^2}{2} \simeq \frac{L}{2\pi}\,\varepsilon\Delta k \sim n_1 m U(v_1 - v_0) \qquad \text{F(9)}$$

where $\Delta k$ is the wave spectrum width. It should be noted that if $v_1 - v_0 \ll U$, the electrostatic energy is much less than the beam energy but that the electrostatic energy may be much larger than the thermal energy of the main plasma. The condition $W/\frac{1}{2}\, n_0 m\, Vth^2 < 1$ does not appear as a condition of the quasi-linear theory as we shall see later.

With regards to the non resonant part of the distribution function, we have $\omega_k >> kv$ so that equation C(3) reduces to

$$\frac{\partial F_0}{\partial t} = \frac{q^2}{2m^2\omega_p^2} \frac{\partial}{\partial t} \sum_K |E_k|^2 \frac{\partial^2 F_0}{\partial v^2} \tag{F(10)}$$

or

$$\frac{\partial F_0}{\partial W} = \frac{1}{n_0 m} \frac{\partial^2 F_0}{\partial v^2} \tag{F(11)}$$

the solution is:

$$F_0 = \frac{n_0}{\sqrt{\pi(v_{th}{}^2 + 4W/m)}} \exp{-v^2/(v_{th}{}^2 + 4W/m)} \tag{F(12)}$$

The fluctuations increase the apparent temperature of the plasma but preserve the gaussian shape of the distribution function.

From B(13) and F(10) we obtain:

$$\int_{\text{resonant particles}} \frac{1}{2} mv^2 \frac{\partial F_0}{\partial v} dv = -2 \frac{\partial W}{\partial v}$$

$$\int_{\text{non-resonant particles}} \frac{1}{2} mv^2 \frac{\partial F_0}{\partial v} dv = \frac{\partial W}{\partial v}$$

We see that the beam loses energy, which is transferred in equal part to the fluctuating field and the non-resonant particles.

The beam also loses momentum at a rate which is given by:

$$\frac{\partial}{\partial t} \int_{\text{resonant particles}} mvF_0 \, dv = \sum_K \varepsilon_0 \frac{K}{\omega_k} \frac{\partial E_k^2}{\partial t}$$

This momentum is entirely transferred to the non-resonant particles since the electrostatic wave momentum is zero. In order to find the rate of momentum absorption of non-resonant particles, it is necessary to expand equation C(3) up to the first order in $Kv/\omega_K$. We may say that the beam exerts on the plasma particles a force which is given by

$$\sum_K \varepsilon_0 \frac{K}{\omega_K n_0} \frac{|\partial E_k|^2}{\partial t} \sim \omega_p \left(\frac{n_1}{n_0}\right)^2 m \frac{U^2}{(\Delta U)^2}$$

## G. RANGE OF VALIDITY OF THE QUASI-LINEAR THEORY

For this particular example we can discuss the conditions of validity of the theory. Let us recall that we have already assumed that:

$$\frac{n_1}{n_0} << \left(\frac{\Delta U}{U}\right)^3 \qquad \text{G(1)}$$

Note that we have no trouble with the group velocity in this case since the group velocity of the plasma waves is smaller than $v_{th}$. We must check that:

$$\tau_E >> [v\Delta k + \Delta\omega_k]^{-1} \simeq [v\Delta k]^{-1} \qquad \text{G(2)}$$

$$\tau_{F_0} >> [v\Delta k + \Delta\omega_k]^{-1} \simeq [v\Delta k]^{-1} \qquad \text{G(3)}$$

The first condition G(2) is roughly equivalent to the condition $\gamma << k\Delta v$ that we found earlier and is fulfilled when G(1) is fulfilled. In order to examine G(3), we introduce the time $\tau_p = (m/e\langle k\rangle E)^{1/2}$ where $E^2 = \Sigma_K E_K^2$ and $\langle k\rangle$ is the average value of $k$. This characteristic time is the period of oscillation of a particle in the potential well of a sinusoidal wave of wavenumber $\langle k\rangle$ and amplitude $E$. From F(5) we deduce the relationship

$$\frac{\tau_E}{\tau_{F_0}} = \frac{n_0}{n_1} \frac{1}{\tau_p^4} \frac{1}{k^2 U^2 \Delta k \omega_p}$$

If $\Delta v_b$ is the velocity spread of the beam, we may set

$$\tau_E \simeq \frac{n_0}{n_1} \left(\frac{\Delta v_b}{U}\right)^2 \frac{1}{\omega_p}$$

and

$$\tau_{F_0} \simeq k^2 U^3 \Delta k \left(\frac{\Delta v_b}{U}\right)^2 \tau_p^4 \simeq \tau_p^4 (\langle k\rangle \Delta v_b)^3$$

The condition G(3) may then be written:

$$\tau_p(\langle k\rangle \Delta v_b) >> 1 \qquad G(4)$$

The condition G(4) has a simple interpretation. It means that in the frame which moves at the velocity $-\omega_k/\langle k\rangle$ the electric field varies rapidly enough to make particle trapping impossible. From F(9), we can deduce that at the end of the stabilization process, $\tau_p$ is given by:

$$\tau_p \sim \frac{1}{\langle k\rangle} \left(\frac{n_0}{n_1}\right)^{3/4} \frac{1}{U^{3/4}(\Delta v_b)^{1/4}} \qquad G(5)$$

and the condition G(4) reduces to:

$$\frac{n_1}{n_0} << \left(\frac{\Delta v_b}{U}\right)^3 \qquad G(6)$$

If this last condition is satisfied initially it will remain fulfilled since the velocity spread of the beam increases during the stabilisation process. Then the range of validity of the theory is given by the condition G(1).

Nevertheless it is still necessary to check that non-linear terms are really negligible in equation B(2) for $\tilde{F}$. The exact equation for $F_k$ should be written:

$$\frac{\partial F_k}{\partial t} + v\frac{\partial F_k}{\partial x} = -\frac{q}{m} E_k \frac{\partial F_0}{\partial v} - \frac{q}{m}\sum_{k'} E_{k-k'} \frac{\partial F_{k'}}{\partial v} \qquad G(7)$$

The second term of the r.h.s. must remain negligible when compared to the first term. Using the same method as in B, we may write

$$\sum_{k'} E_{k-k'}F_{k'} = -\frac{q}{m}\frac{\partial F_0}{\partial v} \sum_{k'} E_{k-k'}E_{k'} \exp[i(\omega_{k-k'} + \omega_{k'})t]$$

$$\times \left[pp\,\frac{1}{i(\omega_{k'} + k'v)} + \pi\delta(\omega_{k'} + k'v)\right]$$

$$\simeq \frac{q}{m}\frac{\partial F_0}{\partial v}\frac{W}{\varepsilon_0 \Delta k v}$$

and it extends over a velocity interval

$$\Delta v \sim \Delta\left(\frac{\omega_k}{k}\right) \sim \frac{\Delta k}{k}\frac{\omega_k}{k}$$

Then, it will be negligible provided that

$$1 >> \frac{qk}{m}\frac{E}{(\Delta k)^2 U}$$

or

$$\tau_p(\Delta k U) >> 1$$

which is identical to the condition G(4).

In the non-resonant region, we get:

$$\frac{q}{m}\frac{E}{\omega v_{th}} << 1$$

which is equivalent to the condition $W < n_0 m v^2{}_{th}$.

Then we see that the relative change of the resonant particle distribution function can be large though it has to remain small for the adiabatic part. Nevertheless, the validity conditions are less strict because we need only the first moments of the non-resonant distribution function. It appears that the real condition should be written:

$$(\tau_p)^{-1} \quad << \omega$$

which is always satisfied provided that G(6) is fulfilled.

## APPENDIX

Instead of considering the bump-in-tail instability, we could have assumed that the waves are weakly damped but are excited at a high level at time $t = 0$. This corresponds to the problem of high amplitude wave absorption by the plasma. Then, from F(5), we obtain instead of F(6):

$$F_0(t) - \frac{\omega_p L}{m\pi}\frac{\partial}{\partial v}\frac{1}{v^3}\left[\varepsilon(t) - \varepsilon(t = 0)\right] = F_0(t = 0) \qquad \text{G(8)}$$

Let us assume that the wave packet has a width $\Delta k$ and that the phase velocities extend over a range $\Delta v$.

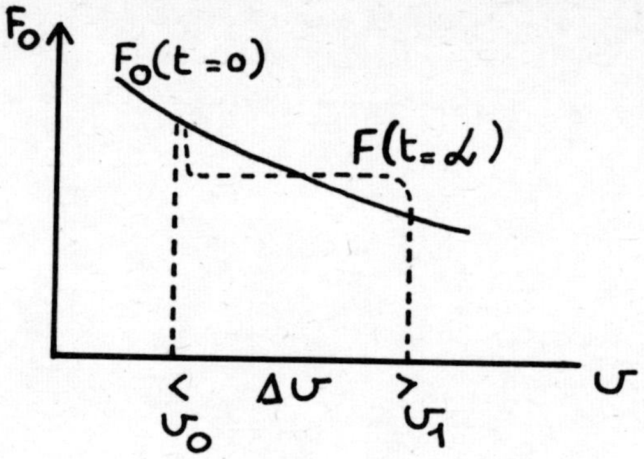


Fig. 3

If the asymptotic state corresponds to non-vanishing electric field amplitudes, then the final distribution function will be constant over the range of resonant velocities as sketched on Fig. 3.

From G(8), we have

$$\frac{\omega_p L}{m\pi v^3}\ [\varepsilon(t) - \varepsilon(0)] = \int_{v_0}^{v} [F_0(t) - F_0(t{=}0)]dv \qquad G(9)$$

Since the r.h.s. of G(9) is negative, in order that $\varepsilon(t)$ be positive, $\varepsilon(0)$ must be larger than a threshold $\varepsilon_T$:

$$L\varepsilon_T = \frac{m\pi v^3}{\omega_p} \int_{v_0}^{v} [F_0(t{=}0) - F_0(t{=}\infty)]dv \qquad G(10)$$

If $\partial F_0/\partial v$ $(t{=}0)$ may be considered as constant in the range of resonant velocities, we obtain:

$$L(\varepsilon_T)_{\text{Max}} = \frac{(\Delta V)^2}{8} \frac{\partial F_0}{\partial v} \frac{\omega_p{}^2}{k^3} m\pi$$

and

$$\left(\sum_k |E_k|^2\right)_T = \frac{(\Delta V)^3}{4V} \frac{m}{\varepsilon_0} \frac{\gamma}{\omega_p} n_0 \qquad G(11)$$

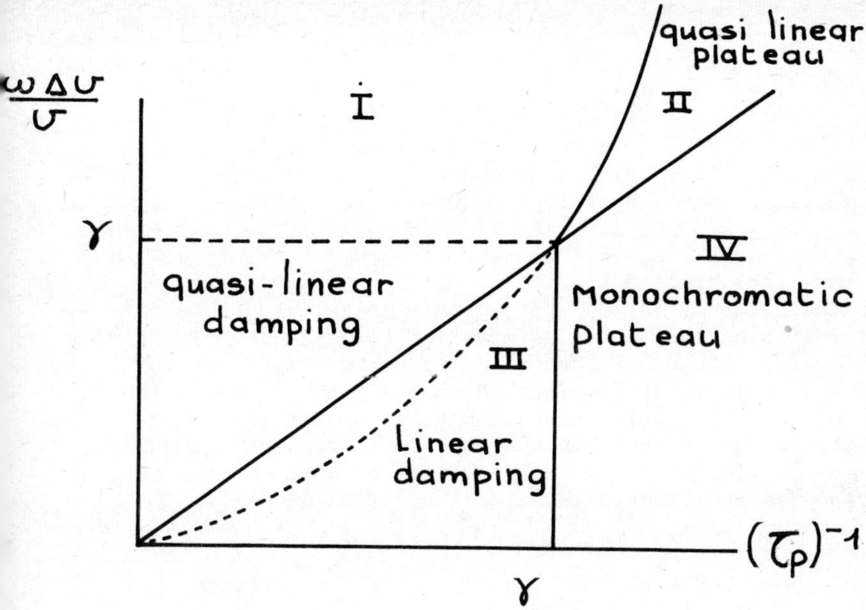


Fig. 4.

where $U$ and $\gamma$ are the initial average damping rate and average phase velocity. This may be expressed in terms of the trapping time, $\tau_p$, as

$$\left(\tau_{p,T}\right)^{-4} = \gamma\omega_p^3\left(\frac{\Delta V}{V}\right)^3 \qquad G(12)$$

If the initial amplitude is smaller than the threshold, the waves damps out before the formation of the plateau. This threshold corresponds to the maximum value of the energy which can be absorbed over the range $\Delta V$ in velocity space.

If $\Delta V$ goes to zero, it appears that $(E_k^2)_T$ will also vanish and this would be in contradiction with the linear theory. However, we must remember that from G(4) we have the condition of validity:

$$\tau_p k\Delta U >> 1$$

i.e.

$$\left[\sum_p\right]^{-1} << \omega\,\frac{\Delta U}{U} \qquad \text{G(13)}$$

The inequality G(13) determines the maximum amplitude which we are allowed to apply the quasi-linear theory. If $(\tau_p^{-1})_T < \omega(\Delta U/U)$ i.e. if $\gamma < \omega(\Delta U/U)$ then the quasi-linear theory may be applied to amplitudes which are large enough to cause the plateau formation. If $\gamma \geq \omega(\Delta U/U)$, then the quasi-linear theory may be applied only to amplitudes which are so small that they decay asymptotically as in the linearized theory. Note that when $\gamma = \omega(\Delta U/U)$, we have $\gamma = (\tau_p)^{-1}$. These results are summarized on the Fig. 4. In regions I and II, the quasi-linear theory may be applied, however the plateau exists only in region II. In regions III and IV, the monochromatic wave theory must be applied. In region IV, the damping disappears.

## H. FOKKER-PLANCK EQUATION FOR RESONANT PARTICLES

The most general form of a Fokker-Planck equation for the evolution of $F_0$ may be written as:*

$$\frac{\partial F_0}{\partial t} = -\frac{\partial}{\partial \xi_i}\,\frac{<\Delta\xi_i>}{\Delta t}\,F_0 + \frac{1}{2}\,\frac{\partial^2}{\partial\xi_i\partial\xi_j}\left[\frac{<\Delta\xi_i\Delta\xi_j>}{\Delta t}\,F_0\right] \qquad \text{H(1)}$$

where the $\xi_i$ are any set of dynamical particle coordinates. The coefficients $<\Delta\xi_i>$ and $<\Delta\xi_i\Delta\xi_j>$ are the average values of the $\xi_i$ fluctuations $\Delta\xi_i = \xi_i(t + \Delta t) - \xi_i(t)$. This equation can be applied to particle-wave interactions if the correlation time, $\tau$, is short enough so that $\delta\xi_i = \xi_i(t + \tau) - \xi(t)$ remains small ($\delta\xi_i << \xi_i$). Then the stochastic process may be considered as markoffian over the time interval $\Delta t >> \tau$ such that $\Delta\xi_i << \xi_i$. These assumptions are precisely the conditions of validity for the quasi-linear theory. Therefore we expect to find the same diffusion equation by applying H(1) to particle wave interactions.

---

*Einstein notation is used in the following.

First, we shall demonstrate an important relationship between the friction coefficient $\langle\Delta\xi_i/\Delta t\rangle$ and the diffusion tensor $\langle\Delta\xi_i\ \Delta\xi_j/\Delta t\rangle$ when the evolution of $F_0$ is due only to the interaction between particles and fluctuating electromagnetic fields.

Let us assume that the unperturbed motion of the particles can be described by a set of variables $J_i$, $\theta_i$ where $J_i$ are constants of the motion and $\theta_i$ are cyclic variables such that:

$$\dot{\theta}_i = \Omega_i(J_1 \ldots J_j \ldots) \quad 0 \leq \theta_i < 2\pi \qquad \text{H(2)}$$

We know that it is possible to find a canonical transformation such that $J_i$ and $\theta_i$ will be canonically conjugate variables. Then, when fluctuations are present we have

$$\dot{\theta}_i = \frac{\partial H}{\partial J_i} \quad \dot{J}_i = -\frac{\partial H}{\partial \theta_i} \qquad \text{H(3)}$$

with

$$H = H_0(J_1 \ldots J_j \ldots) + H_1(J_1 \ldots J_j \ldots,\ \theta_1 \ldots \theta_j \ldots)$$

and

$$\frac{\partial H_0}{\partial J_j} = \Omega_j$$

The hamiltonian $H_1$ takes the perturbation into account, and is small compared to $H_0$. In this set of variables, the Fokker-Planck equation will be written as:

$$\frac{\partial F_0}{\partial t}(J_1 \ldots J_j \ldots) = -\frac{j}{\partial J_i}\left\langle\frac{\Delta J_i}{\Delta t}\right\rangle F_0(J_1 \ldots J_j)$$

$$+\frac{1}{2}\frac{\partial^2}{\partial J_i \partial J_j}\left\langle\frac{\Delta J_i \Delta J_j}{\Delta t}\right\rangle F_0(J_1 \ldots J_j \ldots) \qquad \text{H(4)}$$

In writing H(4) we have assumed that the system was invariant to a transformation that would change the origin of $\theta_i$. We must now compute $\langle\Delta J_i\rangle$ and $\langle\Delta J_i\ \Delta J_j\rangle$. Though we should integrate H(3) from $t - \Delta t$ to $t$ we integrate the equation between $t = 0$ and $t = \Delta t$ for simplicity. In the lowest order we obtain:

$$\theta_{i0} = \theta_i^0 + \Omega_i(..J_j^0...)t \quad J_i^0 = J_{i0} \qquad H(5)$$

where $\theta_i^0$ and $J_i^0$ are initial values of $\theta_i$ and $J_i$. In the next order we have:

$$\dot{\theta}_{i1} = \frac{\partial H_1}{\partial J_i}[...\theta_{j0}...J_{j0}] + \frac{\partial \Omega_i}{\partial J_j}(...J_{j0}...)J_{j1} \qquad H(6)$$

$$\dot{J}_{i1} = -\frac{\partial H_1}{\partial \theta_i}[...\theta_{j0}...,...J_{j0}] \qquad H(7)$$

The averaging process is defined by $\langle A\rangle = \dfrac{\int A\pi_i d\theta_i^0}{\int \pi_i d\theta_i^0}$

and $H_1$ is assumed to be a periodic function of the cyclic variables $\theta_i$. Then from H(5) and H(7) we deduce that $\langle J_{i1}\rangle=0$ and we can compute $\langle J_{i2}\rangle$ from

$$\dot{J}_{i2} = -\frac{\partial^2 H}{\partial\theta_i\partial\theta_j}[...\theta_{j0}...,J_{j0}...]\theta_{j1}$$
$$-\frac{\partial^2 H}{\partial\theta_i\partial J_j}[...\theta_{j0}...,J_{j0}...]J_{j1} \qquad H(8)$$

Then we see that from H(6) and H(7) we can derive the following relationships:

$$\frac{\partial\dot{\theta}_{i1}}{\partial\theta_i^0} = \frac{\partial^2 H_1}{\partial\theta_j^0\partial J_i}[...\theta_{j0}...,J_j^0] + \frac{\partial\Omega_i}{\partial J_k}\frac{\partial J_{k1}}{\partial\theta_j^0} \qquad H(9)$$

$$\frac{\partial\dot{J}_{j1}}{\partial J_{i0}} = -\frac{\partial^2 H_1}{\partial J_i\partial\theta_j^0} - \frac{\partial^2 H_1}{\partial\theta_j^0\partial\theta_k^0}\frac{\partial\Omega k t}{\partial J_i} \qquad H(10)$$

or

$$\frac{\partial\dot{J}_{j1}}{\partial J_{i0}} = -\frac{\partial^2 H_1}{\partial J_{i0}\partial\theta_j^0} + \frac{\partial\dot{J}_{k1}}{\partial\theta_j}\frac{\partial\Omega k t}{\partial J_i} \qquad H(11)$$

$$\frac{\partial\dot{\theta}_{i1}}{\partial\theta_j^0} = -\frac{\partial\dot{J}_{j1}}{\partial J_{i0}} + \frac{d}{dt}\left[\frac{\partial J_k}{\partial\theta_j^0}\frac{d\Omega_i t}{dJ_k}\right] \qquad H(12)$$

Equation H(12) may be written as:

$$\frac{\partial\theta_{i1}}{\partial\theta_j^0} = -\frac{\partial J_{j1}}{\partial J_{i0}} + \frac{\partial J_{k1}}{\partial\theta_j^0}\frac{d\Omega_i}{dJ_{k0}}\,t \qquad \text{H(13)}$$

We also have:

$$\frac{\partial\dot{J}_{i1}}{\partial\theta_j^0} = -\frac{\partial^2 H_1}{\partial\theta_i^0\partial\theta_j^0} = \frac{\partial\dot{J}_{j1}}{\partial\theta_i^0} \qquad \text{H(14)}$$

and

$$\frac{\partial J_{i1}}{\partial\theta_j^0} = \frac{\partial J_{j1}}{\partial\theta_i^0} \qquad \text{H(15)}$$

From equation H(8) we deduce after integration by part, the following:

$$<\dot{J}_{i2}> = <\frac{\partial H_1}{\partial\theta_j^0}\frac{\partial\theta_{j1}}{\partial\theta_i^0}> + <\frac{\partial H_1}{\partial J_{j0}}\frac{\partial J_{j1}}{\partial\theta_i^0}> \qquad \text{H(16)}$$

By using H(13) and H(15), we see that H(16) is equivalent to:

$$<\dot{J}_{i2}> = <\dot{J}_{j1}\left[\frac{\partial J_{i1}}{\partial J_{j0}} - \frac{\partial J_{k1}}{\partial\theta_i^0}\frac{d\Omega_j}{dJ_{k0}}\,t\right]> - <\frac{\partial^2 H}{\partial J_{j0}\,\partial\theta_j^0}J_{i1}> \qquad \text{H(17)}$$

or

$$\begin{aligned}<\dot{J}_{i2}> &= <\dot{J}_{j1}\frac{\partial J_{i1}}{\partial J_{j0}} + J_{i1}\frac{\partial\dot{J}_{j1}}{\partial J_{j0}}> \\ &\quad - <\dot{J}_{j1}\frac{\partial J_{k1}}{\partial\theta_i^0} + J_{i1}\frac{\partial\dot{J}_{j1}}{\partial\theta_k^0}>\frac{d\Omega_j}{dJ_{k0}}\,t\end{aligned} \qquad \text{H(18)}$$

In the second bracket of the r.h.s. of H(18) we may replace $\partial J_{k1}/\partial\theta_i^0$ by $\partial J_{i_1}/\partial\theta_k^0$ and this term cancels out so we have:

$$<\dot{J}_{i2}> = \frac{\partial}{\partial J_{j0}}<\dot{J}_{j_1}\,J_{i1}> \qquad \text{H(19)}$$

Let us go back to equation H(16). We could have written it as:

$$<\dot{J}_{i2}> = \frac{\partial H_1}{\partial \theta_i^0}\frac{\partial \theta_{j1}}{\partial \theta_j^0} > + <J_{j1}\left|\frac{\partial \dot{J}_{i1}}{\partial J_{j0}} - \frac{\partial \dot{J}_{k1}}{\partial \theta_i^0}\frac{d\Omega k}{dJ_0}\, t\right|> \qquad H(20)$$

Using H(13) and H(15) again, equation H(20) becomes

$$<\dot{J}_{i2}> = <J_{i1}\frac{\partial \dot{J}_{j1}}{\partial J_{j0}} + J_{j1}\frac{\partial}{\partial J_{j0}}\dot{J}_{1i}> \qquad H(21)$$

$$- <\dot{J}_{1i}\frac{\partial J_{k1}}{\partial \theta_j^0}\frac{d\Omega j}{dJ_k^0} + J_{j1}\frac{\partial \dot{J}_{k1}}{\partial \theta_i^0}\frac{d\Omega k}{\partial J_j^0}> t$$

The second bracket in the r.h.s. of H(21) cancels out since we may replace $\partial J_{k1}/\partial\theta_i^0$ by $\partial J_{i1}/\partial\theta_k^0$ and interchange $k$ and $j$ in the second term of the bracket. Then,

$$<\dot{J}_{i2}> = \frac{\partial}{\partial J_{j0}} <\dot{J}_{i1}\ J_{j1}> \qquad H(22)$$

From H(19) and H(22) we can deduce immediately

$$<J_{i2}> = \frac{1}{2}\frac{\partial}{\partial J_{j0}} <J_{i1}\ J_{j1}> \qquad H(23)$$

This relationship H(23) is a consequence of the stochastic independance between fluctuating fields and particles which was obviously assumed here. In term of the diffusion and friction coefficients of the Fokker-Planck equation H(4), the relationship H(24) may be written as:

$$<\frac{\Delta J_i}{\Delta t}> = \frac{1}{2}\frac{\partial}{\partial J_j} <\frac{\Delta J_i \Delta J_j}{\Delta t}> \qquad H(24)$$

On the other hand, equation H(4) can also be put under the form

$$\frac{\partial F_0}{\partial t} = -\frac{\partial}{\partial J_i}\left\{<\frac{\Delta J_i}{\Delta t}> F_0 - \frac{1}{2}\frac{\partial}{\partial J_j}\left[<\frac{\Delta J_i \Delta J_j}{\Delta t}> F_0\right]\right\} \qquad H(25)$$

From H(24) we can deduce the general Fokker-Planck equation which describes wave particle interactions:

$$\frac{\partial F_0}{\partial t} = \frac{1}{2} \frac{\partial}{\partial J_i} \left\langle \frac{\Delta J_i \Delta J_j}{\Delta t} \right\rangle \frac{\partial F_0}{\partial J_j} \qquad \text{H(26)}$$

From H(26) we can deduce a $H$- theorem in the following way:

$$\frac{\partial}{\partial t} \int F_0^2 \prod_j dJ_j = -\frac{1}{2} \int \prod_k dJ_k \frac{\partial F_0}{\partial J_i} \left\langle \frac{\Delta J_i \Delta J_j}{\Delta t} \right\rangle \frac{\partial F_0}{\partial J_j} \qquad \text{H(27)}$$

The r.h.s. of equation H(27) is negative definite since the quadratic form associated with the diffusion tensor $\langle \Delta J_i \Delta J_j / \Delta t \rangle$ is positive definite. Then the asymptotic state will correspond either to an absence of fluctuations or to a distribution function which does not depend on the variable $J_j$. Then, equation H(26) describes a purely diffusive process.

An important practical consequence of equation H(26) is that we need only to know $\Delta J_i$ in the lowest order. Second order computations are usually lengthy and cumbersome, therefore the possibility of avoiding them provides a great simplification.

The derivation of equation H(26) from the Fokker-Planck equation provides several features that one could not obtain so easily by the derivation from the Vlasov equation. Such features are the positive definiteness of the diffusion tensor, and the remarkably simple form of the equation when $F_0$ is expressed in terms of the invariant of the unperturbed orbits. Nevertheless, we must emphasize that equation H(26) is unable to describe the adiabatic change of the distribution function of non-resonant particles. This effect is obviously out of the scope of the Fokker-Planck equation since there are no stochastic effects in that case.

If we want $F_0$ to be the distribution function in usual space and momentum coordinates, then equation H(26) may be written as:

$$\frac{\partial F_0}{\partial t} = \frac{1}{2} \frac{1}{D} \frac{\partial}{\partial J_i} \left\langle \frac{\Delta J_i \Delta J_j}{\Delta t} \right\rangle \frac{\partial (F_0 D)}{\partial J_j} \qquad \text{H(28)}$$

where $D = \dfrac{D(..p_i..q_i..)}{D(..\theta_i..J_i..)}$

is the Jacobian of the transformation $p_i, q_i \rightarrow \theta_i, J_i$.

Finally, let us notice that the invariants $J_i$ are defined without ambiguity. They are constants of the particle motion in the average value of the electromagnetic fields $\bar{E}_0$, $\bar{B}_0$. Then in general $J_i$ will be a non-linear function of the coordinate $p_i$, $q_i$. It is not obvious that when $\dot{J}_i$ is zero when computed from $E_0$ and $B_0$, we also have $\langle J_i \rangle = 0$ when computed in the fluctuating fields. As a matter of fact it is a trivial consequence of the hamiltonian form of the motion equation of motion

$$\langle \dot{J}_i \rangle = \langle \frac{\partial H}{\partial \theta_i} \rangle = 0 \qquad \text{H(29)}$$

since, $H$ is a periodic function of $\theta_i$.

Equation H(28) can be applied to many problems. For instance it may be applied to cosmic ray diffusion, replenishment of the earth radiation belts and anomalous transport effects in thermonuclear containment devices. In this chapter we shall treat the very simple case of one dimensional electrostatic waves and compare the resulting diffusion equation to the diffusion equation we obtained previously.

In this case, the average fields are zero and we may choose $v$ as a constant of the motion. Then $x$ will be the conjugate variable and we have:

$$\frac{dv_1}{dt} = \sum_k \frac{q}{m} E_k \exp i(kx_0 + \omega_k t)$$

$$x_0 = x^0 + v_0 t$$

$$v_1 = \sum_k \frac{q}{m} \frac{E_k \exp ikx^0}{i(kv_0 + \omega_k)} [\exp i(kv_0 + \omega_k)t - 1]$$

$$\langle v_1^2 \rangle = \sum_k \frac{q^2}{m^2} \left|E_k\right|^2 \frac{[\cos(kv_0 + \omega_k)t - 1]^2 + \sin^2(kv_0 + \omega_k)t}{|kv_0 + \omega_k|^2}$$

$$\frac{\langle \Delta v^2 \rangle}{\Delta t} = \sum_k \frac{q^2}{m^2} \left|E_k\right|^2 2\pi\delta(\omega_k + kv)$$

The diffusion equation H(26) reduces to:

$$\frac{\partial F_0}{\partial t} = \frac{\partial}{\partial v} \sum_k \frac{q^2}{m^2} \left|E_k\right|^2 \pi\delta(\omega_k + kv) \frac{\partial F_0}{\partial v}$$

which is the same as the diffusion equation for resonant particle of the quasi-linear theory.

*Note*: We have assumed that it was possible to find a cannonical transformation $q_i$, $p_i \rightarrow \theta_i$, $J_i$. Strictly speaking, $\theta_i$ has to be the canonically conjugate variable of $J_i$ such that:

$$\dot{\theta}_i = \frac{\partial H_0}{\partial J_i}$$

Then, we cannot choose any couple $\theta_i$, $J_i$ which fulfills equation H(2). Nevertheless, if we choose as cyclic variables $\theta_i^! = \theta_i \; V_i[\ldots J_j \ldots]$ where $V_i$ are functions of $J_1 \ldots J_j \ldots$, the results are not changed since in the averaging process the the functions $V_i$ will cancel out.

We also emphasize that the functions $J_i$ have to be chosen so that the period of the conjugate cyclic variables $\theta_i$ does not depend on $J_1, \ldots J_j \ldots$ If this condition is not fulfilled, we meet many complications since for instance we have

$$<\frac{\partial A}{\partial J_i}> \neq \frac{\partial}{\partial J_i} <A>$$

In this respect the adiabatic invariants will always be a good choice as variables $J_i$, since in this case $\partial H/\partial J_i = (T_i)^{-1}$ where $T_i$ is the period of the associated motion. Then $\dot{\theta}_i = (T_i)^{-1}$ and the period of $\theta_i$ is automatically equal to $2\pi$. As an important consequence, we find that the diffusion coefficients can be different from zero only if the adiabatic invariance is destroyed by the perturbations. Then the diffusion destroys the gradients of $F_0$ in $J_i$ space.

## I. DIFFUSION BY ELECTROMAGNETIC WAVES

We shall apply the Fokker-Planck diffusion equation to electromagnetic waves which propagate along a uniform magnetic field $Boz$. We know four constants of the unperturbed motion. They are the parallel velocity $v_z$, the perpendicular kinetic energy $w_\perp$ and the transverse coordinates of the guiding center $X$, $Y$. If $x, y, z, v_x, v_y, v_j$ are the particle coordinates we have:

$$w_\perp = \frac{1}{2}(v_x^2 + v_y^2) = \frac{1}{2}\, v_-^2 \qquad X = x + \frac{v_y}{\Omega_c} \quad \Omega_c = \frac{qB_0}{m}$$

$Y = y - v_x/\Omega_c$. The cyclic variables which are associated to $w_\perp$ and $v_z$ are obviously $z$ and $\theta = \text{Arc tg}(-\, v_x/v_y)$. The guiding center coordinate $Y$ will be considered as the cyclic coordinate which is conjugate of $X$. This means that we consider a system which is invariant to translation along $Oy$, so that the perturbed hamiltonian will be periodic function of $y$ and $Y$. This last property is the only one we need.

The perturbed vector potential $A$ is perpendicular to $B_0$ and then the perturbed equation of motion may be written as

$$\frac{dv_x}{dt} = \frac{q}{m}\left[-\left(\frac{\partial A_x}{\partial t} + v_z\frac{\partial A_x}{\partial z}\right) + v_y B_0\right]$$

$$\frac{dv_y}{dt} = \frac{q}{m}\left[-\left(\frac{\partial A_y}{\partial t} + v_z\frac{\partial A_y}{\partial z}\right) - v_x B_0\right] \qquad \text{I(1)}$$

$$\frac{dv_z}{dt} = \frac{q}{m}\left[v_x\frac{\partial A_x}{\partial z} + v_y\frac{\partial A_y}{\partial z}\right]$$

From these equations we deduce:

$$\frac{dw_\perp}{dt} = \frac{iqv_\perp}{m}\left|(\exp i\theta)\left(\frac{\partial A_+}{\partial t} + v_z\frac{\partial A_+}{\partial t}\right) - (\exp -i\theta)\left(\frac{\partial A^-}{\partial t} + v_z\frac{\partial A^-}{\partial z}\right)\right.$$

$$\text{I(2)}$$

$$\frac{dv_z}{dt} = \frac{iqv_\perp}{m}\left|(-\exp i\theta)\frac{\partial A^+}{\partial z} + (\exp -i\theta)\frac{\partial A^-}{\partial z}\right|$$

$$\frac{dx}{dt} = -\frac{1}{B}\left|\frac{\partial A_y}{\partial t} + v_z\frac{\partial A_y}{\partial z}\right|$$

where

$$A^\pm = \frac{A_x \pm iA_y}{2}$$

We know that the solutions of Maxwell's equations split into two modes which have frequencies $\omega_+$ and $\omega_-$ with $A_- = 0$ and $A_+ = 0$ respectively. The sign + or - will be chosen for $k > 0$ since we have

$$A^+_{-k} = (A^-_k)^* \quad A^-_{-k} = A^-_{-k} = (A^+_k)^*$$

where $k$ is the wave vector such that $A_\pm = \sum_k A_k^\pm \exp i(k_z + \omega_\pm t)$

We integrate equations I(2) with $\theta = \theta_0 + \Omega_c t$ and $z = z_0 + v_{z0} t$ and obtain:

$$\Delta w_\perp = \sum_k \{\Delta w^+_{\perp k} + \Delta w^-_{\perp k}\} \qquad \text{I(3)}$$

$$\Delta v_z = \sum_k \{\Delta v^+_{zk} + \Delta v^-_{zk}\}$$

with

$$\Delta w^\pm_{\perp k} = \pm \frac{iq}{m} v^0_\perp \frac{\omega_\pm + kv_{z0}}{\omega_\pm \pm \Omega_c + kv_{z0}} A_k^\pm \exp(\pm i\theta_0 + ikz_0) \qquad \text{I(4)}$$

$$\left[\exp i[\omega_\pm \pm \omega_c + kv_{z0}]t - 1\right]$$

$$\Delta v^\pm_{zk} = \pm \frac{iq}{m} v^0_- k A_k^\pm \exp \pm i\theta_0 + ikz_0 \qquad \text{I(5)}$$

$$\frac{\exp i[\omega_\pm \pm \Omega_c + kv_{z0}]t - 1}{\omega_\pm \pm \Omega_c + kv_{z0}}$$

$$\Delta X = [- A_y(r,t) + A_y(r_0, t = 0)]/B \qquad \text{I(6)}$$

We average on $\theta_0$ and $z_0$ so that

$$<\Delta w_-^2> = \sum_k^{\pm,-} \frac{q^2}{m^2} \left(\omega_\pm + kv_z\right)^2 v_\perp^2 \left|A_k^\pm\right|^2 R$$

$$<\Delta w_\perp \Delta v_z> = \pm \sum_k^{+,-} \frac{q^2}{m^2} k(\omega_\pm + kv_z) v_\perp \left|A_k^\pm\right|^2 R$$

$$<\Delta v_z^2> = = \sum_k^{+,-} \frac{q^2}{m^2} k^2 \left|A_k^\pm\right|^2 v_\perp^2 R$$

with

$$R = \frac{|\exp i[\omega_{\pm}\pm\Omega_c+kv_z]t - 1|^2}{[\omega_{\pm}\pm\Omega_c+kv_z]^2}$$

$$R \simeq 2t\delta[\omega_{\pm}\pm\Omega_c+kv_z]$$

The diffusion equation takes the form

$$\frac{\partial F_0}{\partial t} = \sum_k^{+,-} \{\frac{\partial}{\partial w_\perp}(w_{\pm}+kv_z)^2 D_k^{\pm} \frac{\partial F_0}{\partial w_+} \pm \frac{\partial}{\partial w_+} k(\omega_{\pm}+kv_z)D_k^{\pm} \frac{\partial F_0}{\partial v_z}$$

$$\stackrel{-}{+} \frac{\partial}{\partial v_z} k(\omega_{\pm}+kv_z)D_k^{\pm} \frac{\partial F_0}{\partial w_\perp} + \frac{\partial}{\partial v_z} k^2 D_k^{\pm} \frac{\partial F_0}{\partial v_z} \qquad \text{I(7)}$$

where

$$D_k^{\pm} = v_\perp^2 \frac{q^2}{m^2}\left|A_k^{\pm}\right|^2 \pi\delta[\omega_{\pm}\pm\Omega_c+kv_z]$$

There is no diffusion in $X$ since equations I(6) show that $\Delta X$ is not resonant so that $< \frac{\Delta X^2}{\Delta t} >$, $< \frac{\Delta X\Delta w}{\Delta t} >$, $< \frac{\Delta X\Delta V_z}{\Delta t} >$ will vanish in the limit $\Delta t \to \infty$. Therefore it appears that these electromagnetic waves give rise only to diffusion in $\omega_\perp$ and $v_z$ and not to spatial diffusion. Spatial diffusion arises when they propagate at an angle with the equilibrium magnetic field.

The diffusion equation I(7) could be written as

$$\frac{\partial F_0}{\partial t} = \sum_k^{+,-} \left|\frac{\partial}{\partial w_\perp}(\omega_{\pm}+kv_z)\frac{\partial}{\stackrel{-}{+}\partial v_z} k\right| D_k^{\pm} \left|(\omega_{\pm}+kv_z)\frac{\partial}{\partial w_\perp} \pm k\frac{\partial}{\partial v_z}\right| F_0 \qquad \text{I(8)}$$

In equation I(8), $F$ is considered as the distribution function in $w_\perp$, $w_z$ space. It corresponds to the usual distribution in $v_x v_y v_z$ space since

$$F(\overline{v})\; dv_x dv_y dv_z = F_0(w_\perp, v_z)\; dw_\perp dv_z d\theta$$

Equation I(8) can be put in the form

$$\frac{\partial F_0}{\partial t} = \sum_k^{+,-} \left|\frac{1}{v_\perp}\frac{\partial}{\partial v_\perp}\left(\omega_\pm + kv_z\right)v_\perp \pm kv_\perp \frac{\partial}{\partial v_z}\right| D'_{\pm,k} \qquad \text{I(9)}$$

$$\left|\left(\omega_\pm + ky_z\right)\frac{\partial}{\partial v_\perp} \pm kv_\perp \frac{\partial}{\partial v_z}\right| F_0$$

with

$$D'_{\pm k} = D^\pm_k / v_\perp^2$$

Usually one assumes that the waves are excited by a small population of anisotropic hot particles and propagate in a cold plasma with a density $n_0$ and a plasma frequency $\omega_p$. Then, for the dispersion relation of the electromagnetic waves one writes

$$\frac{k^2c^2}{\omega_\pm{}^2} - 1 = -\sum_{i,e} \frac{\omega_{pj}^2}{\omega_\pm(\omega_\pm \pm \Omega_{cj})}$$

where the sign $\pm$ has the same meaning as before. In the limit where $k^2c^2/\omega^2 >> 1$ and $|\omega\pm| << \Omega_{ce}$ we recover the whistler mode limit. From the resonance condition, we see that for these modes $\omega << kv_z$, so that I(9) can be simplified to yield:

$$\frac{\partial F_0}{\partial t} = \left(\frac{1}{v_\perp}\frac{\partial}{\partial v_\perp} v_z v_\perp - v_\perp \frac{\partial}{\partial v_z}\right)\sum_k \frac{q^2}{m^2}\left|A_k\right|^2 \pi\delta\left(\Omega_{ce} + kv_z\right) \qquad \text{I(10)}$$

$$\left(v_z \frac{\partial}{\partial v_\perp} - v_\perp \frac{\partial}{\partial v_z}\right) F_0$$

This equation is often written in terms of the pitch angle $\alpha = \text{Arc tg}(v\_/v_z)$ and the velocity modulus $v$ as

$$\frac{\partial F_0}{\partial t} = \frac{1}{\sin\alpha}\frac{\partial}{\partial\alpha} \cos\alpha \sum_k \frac{q^2}{m^2}\left|A_k\right|^2 \pi\delta\left(\Omega_{ce} + kv_z\right)\frac{\partial F_0}{\partial\alpha} \qquad \text{I(11)}$$

In this case the diffusion takes place essentially in pitch angle and is much weaker in energy. This is a natural consequence of the conservation of particle energy in the wave frame, though pitch angle can vary. When $\omega << \Omega$, the wave frame nearly coincides with the lab frame so that the energy diffusion is negligible.

As a last remark, it must be emphasized that these modes have a large group velocity and the validity of the quasi-linear theory is not well understood in that case. For instance when the integration on $k$ is carried out, a pole appears in the diffusion tensor when $v_z + \partial\omega(k)/\partial k = 0$. Obviously the theory is not valid in this case.

## J. DIFFUSION ACROSS A MAGNETIC FIELD

In this section we shall derive a general formula relating the diffusion flux of a low β plasma in slab geometry to fluctuations and we study detail diffusion mechanisms.

Let us consider a plasma with a pressure gradient along $ox$ and a confining magnetic field along $oz$. The plasma pressure is assumed to be so small that we may neglect the magnetic intensity gradient. Fluctuations are excited and we want to compute the average density and energy flux across the field lines in the $x$ direction. Let us set:

$$F = F_0 + \tilde{F} \qquad F_0 = \frac{1}{S}\int_S dydzF \tag{J(1)}$$

where $F$ is the ion or electron distribution function, $S$ is the surface across which we compute the diffusion flux. We need only two assumptions:

1) The length $\left|\frac{d}{dx} \log F_0\right|^{-1}$ is much larger than particles Larmor radius and

2) The time $\left|\frac{d}{dt} \log F_0\right|^{-1}$ is much longer than a cyclotron period.

If we set $v_x = v_\perp \cos\theta$ and $v_y = v_\perp \sin\theta$, the Vlasov equation reads

$$\frac{\partial F}{\partial t} + \frac{v\partial F}{\partial r} + \frac{q}{m}\bar{E}\cdot\frac{\partial F}{\partial v} + \Omega_c\frac{\partial F}{\partial\theta} = 0 \tag{J(2)}$$

We shall introduce averaged quantities $\langle A\rangle$ with the same meaning as in equation J(1). Then averaging equation J(2) yields:

$$\frac{\partial F_0}{\partial t} + v_x\frac{\partial F_0}{\partial x} - \Omega_c\frac{\partial F_0}{\partial\theta} = -\frac{q}{m}\left\langle\vec{E}\cdot\frac{\partial\tilde{F}}{\partial v}\right\rangle \tag{J(3)}$$

where we have assumed that $<E>$ was zero. This assumption can easily be removed and does not change the result.

The fluctuations have appeared at a time $t_0$ we can choose equal to $-\infty$. Then the function $F_0$ was an equilibrium distribution function $F_0^0$. We may integrate J(3) along unperturbed orbits by setting

$$\frac{dv'}{dt'} = \frac{q}{m} v' \wedge B_0 \qquad\qquad \text{J(4)}$$

$$\frac{d\bar{r}'}{dt'} = v'$$

The solution of J(4) are given by:

$$v_x' = v_\perp \sin[\Omega_c(t'-t)+\phi] \quad v_\perp \sin\phi = v_x$$

$$v_y' = v_\perp \cos[\Omega_c(t'-t)+\phi] \quad v_\perp \cos\phi = v_y \qquad\qquad \text{J(5)}$$

$$v_z' = v_z$$

$$x'-x = \frac{v_\perp}{\Omega_c}\left[\cos[\Omega_c(t'-t)+\phi] - \cos\phi\right]$$

We find that:

$$F_0(x,\bar{v},t) = F_0^0 - \frac{q}{m}\int_{-\infty}^{t} < \vec{E}(\vec{r}',t')\frac{\partial\tilde{F}}{\partial v'}(r',v',t')\, dt'$$

If $\Gamma_x$ denotes the density flux:

$$\Gamma_x = \int F_0(x,\vec{r},t)d^3vV_x = -\frac{q}{m}\int_{-\infty}^{t} dt'\int d^3vV_x < \tilde{E}\cdot\frac{\partial\tilde{F}}{\partial v'}>_{t',x'} \qquad\qquad \text{J(6)}$$

From J(5) we may compute $\vec{v}'$ in terms of $\vec{v}$ by

$$v_x = v_x' \cos\Omega_c(t'-t) - v_y' \sin\Omega_c(t'-t)$$

$$v_y = v_y' \cos\Omega_c(t'-t) + v_x' \sin\Omega_c(t'-t)$$

$$dv_x dv_y = dv_x' dv_y'$$

In the r.h.s. of J(6) we may take $v'$ as integration variables so that:

$$\Gamma_x = -\frac{q}{m}\int_{-\infty}^{t} dt' \int d^3v'[v_x' \cos \Omega_c(t'-t)$$

$$- v_y' \sin \Omega_c(t'-t)] < E' \cdot \frac{\partial \tilde{F}}{\partial v'} >_{r',x'} \qquad \text{J(7)}$$

$$= \frac{q}{m}\int_{-\infty}^{t} dt'[\cos \Omega_c(t'-t) < E_x \cdot \tilde{n} >_{t'}$$

$$- \sin \Omega_c(t'-t) < E_y \cdot \tilde{n} >_{t'}]$$

Now we use the two assumptions which were made at the beginning of this chapter so that in the r.h.s. of J(7) we may replace $< E\tilde{n} >_{t'}$ by $< E \cdot \tilde{n} >_{x,t}$. The integration on $t'$ is readily performed and we obtain:

$$\Gamma_x = \frac{1}{B} < E_y \tilde{n} >$$

The formula J(8) could be found very easily by noticing that the electric radial drift $E_y/B$ gives rise to the flux $\Gamma_x$. The formula J(8) is a little less trivial since we may apply it to any kind of waves even if their frequency is larger than the cyclotron frequency or if the wave length is smaller than the gyroradius.

We find in the same way:

$$Q_x = \frac{1}{B}\{< E_y \int \tilde{F} \frac{1}{2} mv^2 \, d^3v > + < \int \tilde{F} \frac{1}{2} m \, \overline{E} \cdot \overline{v} \, v_y \, d^3v > \} \qquad \text{J(9)}$$

The formula J(8) shows that the diffusion is ambipolar. If we set:

$$\tilde{n} = \sum_{k_y} n_k \, e^{ik_y y}$$

$$\overline{E}_y = - \frac{\partial \phi}{\partial y} = - i \sum_k k_y \, \phi_k \qquad J(10)$$

$$\Gamma_x = \frac{1}{B} \sum_k i \, k_y \, \phi_k \, n_k$$

from J(10), we deduce that there is no flux diffusion when $\phi_k$ and $n_k$ are symmetric functions of $k_y$. Drift waves are not of this type and then will greatly contribute to diffusion. It does not mean that other instabilities do not give rise to diffusion. But these instabilities are computed in homogeneous plasma so that $n_k$ and $\phi_k$ are symmetric functions of $k_y$. It would be necessary to compute the corrections induced by the inhomogeneity which then would break the symmetry. We shall see that a more practical formula based upon the Fokker-Planck equation may be used.

As regards $Q_x$, we notice that if $\tilde{F}_k$ and $\phi_k$ are symmetric functions of $k_x$ and $k_z$, J(9) reduces to:

$$Q_x = \frac{1}{B} < E_y \int \tilde{F} \frac{1}{2} m(v^2 + 2v_y^2) d^3v >$$

We also note that the diffusion flux $\Gamma_x$ may be zero with $Q_x \neq 0$. This is the case for low frequency waves with parallel phase velocities between the electron and ion thermal velocities. In such cases $\tilde{n}_k{}^i$ is a real function of $\phi$ and $\omega$, though $n_k{}^e$. The frequency becomes real and $\Gamma_x = 0$. However, the energy flux $Q_x$ may remain different from zero.

We can try to estimate the diffusion particle flux $\Gamma_x$ for drift waves by assuming that even in the non-linear regime the relationship between the ion density and $\phi$ will not be very different from the linear expression:

$$\tilde{n}_{ki} = n_0 \, \frac{q\phi_k}{T} \, \frac{\omega^*}{\omega}$$

where $n_0$ is the equilibrium density and $\omega^*$ is the drift frequency

$$\omega^* = k_y\, a_i\, \frac{v_{thi}}{r} \qquad v_{thi} = \sqrt{\frac{2T}{m_i}} \qquad r = \left(\frac{d\log n}{dx}\right)^{-1}$$

We set $\omega = \omega_k - i\gamma_k$ where $\omega_k$ and $\gamma_k$ are the real frequency and growth rate. From J(10), we obtain

$$\Gamma_x = \frac{q}{T}\frac{n_0}{B} \sum_k |\phi_k|^2\, k_y \frac{\omega^*\gamma_k}{|\omega_k|^2+|\gamma_k|^2} \qquad J(11)$$

We must now estimate the amplitudes $|\phi_k|$. This is generally impossible. We shall try to obtain at least an upperbound for $|\Phi_k|$. Let us consider at first the modes with small $k_x$ and $k_y a_i \ll 1$. The lowest order linearized distribution function of a given mode may be written in the form

$$f_k = \frac{n_0 q\phi_k}{T}\left[1-(\omega+\omega^*)\int_{-\infty}^{0} \exp\, i\omega t' + ik[r(t')-r]\, dt' \atop dr'\right] \times$$

$$\exp\, i\omega t + i\bar{k}\cdot\bar{r}$$

where $r(t')$ follows the unperturbed orbits. Since $\omega \ll \Omega_c$ we get in the lowest order:

$$fk = \frac{nq\phi_k}{T}\left|1 - \frac{\omega+\omega^*}{i(\omega+k_z v_z)}\left(1 + \frac{i\bar{k}\cdot\bar{v}\ \bar{e}_z}{\Omega_c}\right)\exp\, i\omega t + i\bar{k}\cdot\bar{r}\right. \qquad J(12)$$

From J(12) we see that $f_k$ has the same dependance in transverse velocity as the equilibrium distribution function. Then if the density gradient of the perturbed distribution function is larger than the equilibrium gradient, the perturbed distribution function will be unstable with respect to higher frequency and shorter wavelengths modes. Therefore we can set an upper bound to $\phi_k$ by requiring that $\Delta\cdot\tilde{n}/n_0 \lesssim 1/r$. Since $\omega \sim \omega^*$, we obtain

$$\frac{q\phi k}{T} \lesssim \frac{1}{kr} \qquad J(13)$$

Under this constraint the maximum spectrum $|\phi_k|^2$ would behave like $\omega_k^{-2}$. In these conditions the electric drift frequency $E/B$ is of the order of the diamagnetic drift frequency which is physically reasonable. If we assume $\gamma_k \sim \omega_k$ we see that the maximum flux is given by:

$$\Gamma_x = \frac{T}{qB} n_0 \sum_k \frac{k_y}{k^2 r^2} \qquad \text{J(14)}$$

The flux is mainly given by the long wavelength modes and the corresponding maximum diffusion coefficient is of the order of $T/qB$ which scales like the Bohm diffusion coefficient.

The preceding argument does not apply to modes with large $k_x$ and small $k_y$. For these modes we notice that the ion guiding center motion equation in the perturbed field is written:

$$\dot{X} = - ik_y\, \phi_k/B$$

$$\dot{Y} = \frac{1}{B} \frac{\partial \phi}{\partial x}$$

The average amplitude of the displacement, $\Delta X$, of the guiding center is then:

$$\Delta X = - \frac{ik_y \phi_k}{\omega_k B}$$

We expect that the non-linear effects will saturate the mode when $\Delta X$ becomes approximately equal to the wavelength $\lambda_x$. Then the saturation limit is given by:

$$\frac{k_y\ \phi_k}{\omega_k B} \lesssim (k_x)^{-1}$$

or J(15)

$$\frac{q\phi_k}{T} \lesssim \frac{1}{k_x r}$$

The mode will saturate at the less severe limit of the two conditions J(13) and J(15). We may summarize the result as follows:

$$\frac{q\phi_k}{T} \lesssim \frac{1}{|k| r} \qquad \Gamma_x \lesssim \frac{n_0}{r} \frac{\gamma_k}{k^2}$$

For modes with $k_y a_i \sim 1$, we may use the same argument since from the preceding considerations we expect the waves to have a more or less isotropic spectrum in this short wavelength region. Then J(14) appears as being something like an upper limit for the diffusion flux.

Now, we shall examine in more detail into the diffusion mechanism by assuming that the quasi-linear or Fokker-Planck equation may be applied. We shall limit our investigation to low frequency electrostatic waves since we have seen that they are mainly responsible for diffusion across the magnetic field. The wave frequencies are smaller than the electron cyclotron frequency and the wavelengths will be larger than the electron Larmor radius. Then the perturbed electron equation of motion can be written:

$$\dot{X} = E_y | B$$

$$v_\perp^2 = 0$$

$$\dot{v}_{\shortparallel} = \frac{q}{m} E_z$$

$$\vec{E} = - \sum_K \vec{\nabla}(\phi_k \exp i\bar{k}\cdot\bar{r}+i\omega_k t)$$

The invariance of $v_\perp^2$ is a consequence of the adiabatic invariance of the magnetic moment. We do not need to consider the ions since we know that the diffusion flux is ambipolar. Then we find immediately:

$$\frac{<\Delta X^2>}{\Delta t} = \sum_k \frac{k_y^2 |\phi k|^2}{B^2} 2\pi\delta(\omega_k+k_z v_z)$$

$$\frac{<\Delta X \cdot \Delta v_z>}{\Delta t} = \frac{q}{m} \sum_k \frac{k_z k_y |\phi k|^2}{B} 2\pi\delta(\omega_k+k_z v_z) \qquad J(16)$$

$$\frac{<\Delta v_z^2>}{\Delta r} = \frac{q^2}{m^2} \sum_K k_z^2 |\Phi_k|^2 \; 2\pi\delta(\omega_k+k_z v_z)$$

The diffusion equation for electrons takes the form:

$$\frac{\partial F}{\partial t} = \frac{q^2}{m^2} \sum_k \left(\frac{k_y}{\Omega_{ce}} \frac{\partial}{\partial x} + k_z \frac{\partial}{\partial v_z}\right) |\phi_k|^2 \delta(\omega_k+k_k v_z)$$

$$\sum_k \left(\frac{k_y}{\Omega_{ce}} \frac{\partial}{\partial x} + k_z \frac{\partial}{\partial v_z}\right) F_0 \left(v_z, \; v_\perp^2, \; X\right)$$

By multiplying this equation by $F$ and integrating on $X$ and $v_z$, we see that $\partial F/\partial t$ will not be zero unless

$$\left(\frac{k_y}{\Omega_{ce}}\frac{\partial}{\partial x} + k_z\frac{\partial}{\partial v_z}\right)F_0 = 0$$

for all excited waves and for $k_z = -\omega_k/v_z$. This would imply that $F_0$ is a functions of $x + (k_y/\Omega_{ce}\omega_k)(v_z{}^2/2)$. If $\omega_k = + k_y V_{de}$ where $V_{de}$ is the diamagnetic electron velocity, we would find

$$x + \frac{k_y}{2\Omega_{ce}\omega_k}v_z^2 = x - \frac{v_z^2}{v_{th}^2\,\frac{1}{n}\frac{dn}{dx}}$$

The Maxwellian equilibrium distribution function is of this type. Unfortunately, for drift waves, $\omega_k$ is not exactly equal to the drift frequency because of the finite Larmor radius effects. These effects are responsible for the growth rate and cannot be neglected. Then the distribution function cannot depend on $X$ in the whole range of parallel resonant velocities. This would imply a non-confined distribution function.

Equations J(15) are easily generalized to cases where the wavelengths are shorter than the electron Larmor radius. They allow to obtain the diffusion coefficient for any kind of low-frequency waves. At this point, it must be stressed that the particle diffusion flux is not determined only by the spatial diffusion coefficient $\langle\Delta X^2/\Delta t\rangle$, since we have:

$$\frac{\partial n}{\partial t} = \frac{\partial}{\partial x}\int d^3v\,\frac{\langle\Delta x^2\rangle}{\Delta t}\frac{\partial F_0}{\partial x} + \frac{\partial}{\partial x}\int d^3v\,\frac{\langle\Delta x\cdot\Delta v_z\rangle}{\Delta t}\frac{\partial F_0}{\partial v_z}$$

We have seen that for drift waves the two terms can nearly compensate each other. The cross diffusion term will be negligible only if $k_y a_e \gg k_z r(V_{the}/V_\phi)$ where $a_e$ is the electron Larmor radius and $V_\phi$ is the parallel phase velocity, or if $|\phi_k|$ and $\omega_k$ are even functions of $k_y$.

# CHAPTER II. MODE-MODE COUPLING

In this chapter, we shall deal with non-linear effects associated with the coupling between different modes of the system.

We shall emphasize a few topics which are not found in the current literature.

At first we treat in details the general relationships which can be obtained from the Vlasov equations in a very formal way. We shall then demonstrate useful symmetry properties for the mode-mode coupling coefficients. Explicit computations of the coefficients are found in the literature and require too much algebra to be usefully reproduced here. We shall only outline the derivation of the non linear Landau damping coefficients for electrons plasma oscillations in a strong magnetic field. The reason of this choice is that the effect has been recently experimentally verified in such a case.

## A. RESONANT MODE-MODE COUPLING

We shall restrict ourselves to electrostatic waves. We assume that the plasma is contained into a box and that the linearized solutions of the wave problem may be reduced to a set of normal modes. Then the Vlasov equation can be written

$$\frac{\partial F}{\partial t} = [H,F] \qquad \text{A(1)}$$

where $F$ is the particle distribution function and $H$ the particle hamiltonian. The hamiltonian is split into an equilibrium hamiltonian $H_0$ and a perturbed hamiltonian $H_1 = q\Phi$ where $\Phi$ is the electrostatic potential. The distribution function is written $F = F_0 + f$ where $F_0$ is the equilibrium distribution functions which fulfill the condition

$$[H_0,F_0] = 0 \qquad \text{A(2)}$$

and where $f$ is obtained from

$$\frac{\partial f}{\partial t} + [H_0, f] = - [H_1, F_0] - [H_1, f]$$

Then the solution of A(2) may also be written

$$f = - \int_{-\infty}^{t} [H_1, F_0] dt' - \int_{-\infty}^{t} [H_1, f] dt' \qquad \text{A(3)}$$

where the integration has to be taken along unperturbed orbits. In the lowest order we obtain:

$$f_1 = - \int_{-\infty}^{t} [H_1, F_0] dt_1 \qquad \text{A(4)}$$

We may set:

$$H_1(t_1) = \int \rho(\bar{r}_1, t_1) \phi(\bar{r}_1, t_1) d\bar{r}_1$$

with

$$\rho(\bar{r}_1, t_1) = q\delta(\bar{r}_1 - \bar{r}(t_1))$$

where $\vec{r}(t_1)$ is the solution of the unperturbed motion equation which is such that $\bar{r}(t) = \bar{r}$. Equation A(4) takes the form:

$$f_1 = - \int d\bar{r}_1 \int_{-\infty}^{t} \phi(\bar{r}_1, t_1) [\rho(\bar{r}_1, t_1), F_0] dt_1$$

Let us set $\bar{r}(t = t_0) = \bar{r}_0, \bar{p}(t = t_0) = \bar{p}_0$ and:

$$\bar{f}_1(\bar{r}_0, \bar{p}_0, t) = f_1[\bar{r}(\bar{r}_0, \bar{p}_0, t), \bar{p}(\bar{r}_0, \bar{p}_0, t), t]$$

We may write the perturbed density

$$n_1(\bar{r}, t) = \int f_1 d^3p = \int \bar{f}_1 \, d^3p$$

and

$$n_1(r', t) = \int f_1 d^3\bar{p} d^3\bar{r} \delta(\bar{r} - \bar{r}') = \int \bar{f}_1 d^3p_0 d^3r_0 \delta[r(t) - r']$$

since $r$ is the position at time $t$ of the particle which was at $r_0$, $p_0$ at time $t = t_0$. Then:

$$n_1(\bar{r},t) = -\int d\bar{r}_1 \int_{-\infty}^{t} \phi(\bar{r}_1,t_1)dt_1 \int dr_0 dp_0 \delta[\bar{r}(t)-\bar{r}] [\rho(\bar{r}_1,t_1),F_0] \qquad A(5)$$

where the Poisson bracket has to be taken on variables $r_0$, $p_0$ since we know that it is invariant with respect to dynamical variables when they are followed along the particle motion. The Poisson bracket may now be considered as an antisymmetric linear operator in $r_0$, $p_0$ space acting on the function $F_0$. Then:

$$\int d\bar{r}_0 d\bar{p}_0 \delta[\bar{r}(t)-\bar{r}][\rho(\bar{r}_1,t_1),F_0] = \int d\bar{r}_0 d\bar{p}_0 F_0 [\delta(\bar{r}-\bar{r}(t)),\rho(\bar{r}_1,t_1)]$$

The perturbed density becomes:

$$n_1(\bar{r},t) = -\int d\bar{r}_1 \int_{-\infty}^{t} dt_1 \phi(\bar{r}_1,t_1) < [\delta(\bar{r}-\bar{r}(r)),p(\bar{r}_1,t_1)]> \qquad A(6)$$

where the brackets mean that we multiply by $F_0(r_0,p_0)$ and integrate over $\bar{r}_0$ and $\bar{p}_0$. If we introduce A(6) into the Poisson equation we obtain:

$$\Delta\phi = -\sum_{i,e} \frac{q}{\varepsilon_0} n_1(\bar{r},t) \qquad A(7)$$

Let us notice an invariance property of the bracket in A(6). The equilibrium fields are stationary. Then:

$$\bar{r}(r) = \bar{R}[\bar{r}_0,\bar{\phi}_0,t-t_0]$$

$$\bar{r}(t_1) = \bar{R}[\bar{r}_0,\bar{p}_0,t_1-t_0]$$

If $r_0'$, $p_0'$ are the particle positions at $t_0 + \tau$ we have obviously

$$\bar{R}[\bar{r}_0',\bar{p}_0',t-t_0-\tau] = R[\bar{r}_0,\bar{p}_0,t-t_0]$$

$$\bar{R}[\bar{r}_0',\bar{p}_0',t_1-t_0-r] = R[\bar{r}_0,p_0,t_1-t]$$

Since $F_0$ does not depend on time and $d^3r_0d^3p_0 = d^3r_0'd^3p_0'$ we find that the brackets are invariant with respect to time translation

$$<[\delta(\bar{r}-\bar{r}(t-\tau)),\rho(\bar{r}_1,t_1-\tau)]> = <[\delta(\bar{r}-t)),\rho(\bar{r}_1,t)]>$$

We look for solutions of A(7) under the form of normal modes of the form $\phi = \Phi(r,\omega)\exp i\omega t$. Then equation A(7) becomes

$$\Delta\Phi = \sum_{i,e}\frac{q}{\varepsilon_0}\int d\bar{r}_1\Phi(\bar{r}_1,\omega)\int_{-\infty}^{t}\exp(i\omega t_1)dt_1<[\delta(\bar{r}-\bar{r}(t)],\ \rho(\bar{r}_1,t_1)]>$$

$$= \sum_{i,e}\frac{q}{\varepsilon_0}\int d\bar{r}_1\Phi(\bar{r}_1,\omega)\int_{-\infty}^{0}\exp(i\omega t_1)dr_1<[\delta(\bar{r}-\bar{r}(0)),\ \rho(\bar{r}_1,t_1)]> \qquad A(8)$$

If we multiply A(8) by any function $\Psi^*(r)$ which fulfill boundary conditions and integrate, we obtain

$$\int\Psi^*\Delta\Phi dr + \sum_{i,e}\frac{1}{\varepsilon_0}\int d\bar{r}_1d\bar{r}\Phi(\bar{r}_1)\Psi(\bar{r})\int_{-\infty}^{0}\exp(i\omega t_1)dt_1<\rho(\bar{r},t=0)\ \rho(\bar{r}_1,t_1)>$$

$$= \int\Phi^*\Delta\Psi dr + \sum_{i,e}\frac{1}{\varepsilon_0}\int d\bar{r}_1,dr\Phi(\bar{r}_1)\Psi(\bar{r})\int_{0}^{\infty}\exp(-i\omega t_1)\ dt_1<[\rho(r,t_1),\rho(r_1,0)]>$$

Let us set:

$$\nu(\bar{r},\bar{r}_1,\omega) = \int_{-\infty}^{+\infty}\exp(i\omega t_1)dt_1<[\rho(\bar{r},t=0),\rho(\bar{r}_1,t_1)]$$

We deduce from the invariance property of the bracket:

$$\nu(\bar{r},\bar{r}_1,\omega) = \int_{-\infty}^{+\infty}\exp(-i\omega t_1)<[\rho(\bar{r}_1,t_1),\rho(\bar{r},t=0)]>$$

$$= -\nu[\bar{r}_1,\bar{r},-\omega]$$

On the other hand

$$\int_{-\infty}^{0} \exp(i\omega t_1)\, dt_1 < [\rho(\bar{r},t=0),\rho(\bar{r}_1,t_1)] >$$

$$= \frac{1}{2i\pi} \int \frac{\nu(\bar{r},\bar{r}_1,\omega')d\omega'}{\omega + \omega'} = -\frac{1}{2i\pi} \int \frac{\nu(\bar{r}_1,\bar{r},-\omega')d\omega'}{\omega + \omega'}$$

$$= \int_{0}^{\infty} \exp(-i\omega t_1)\, dt_1 < [\rho(\bar{r},t_1),\rho(\bar{r}_1,t=0)] >$$

We are interested here in modes with small growth or damping rate. Then, we may consider that the imaginary part of $\omega$ is negative and very small. All the following computations will be done in this limit. Moreover, for the moment, we neglect particle resonant effects so that we are allowed to consider that when the imaginary part of $\omega$ goes to zero, the integrals over $t_1$ become real. We find from A(8) bis that equation A(8) can be formally written:

$$H(\omega)\phi = 0 \qquad \text{A(9)}$$

where $H(\omega)$ is an hermitian operator. Generally, we shall find a set of real values $\omega_\alpha$ and of functions $\phi_\alpha$ which are solution of A(9).

Let us now compute higher order term. We find in the next order:

$$f_2 = \int d\bar{r}_1 \int_{-\infty}^{t} dt_1 \int_{-\infty}^{t_1} dt_2 \phi(\bar{r}_1,t_1)\phi(\bar{r}_2,t_2) \Big[\rho(\bar{r}_1,t_1), [\rho(\bar{r}_2,t_2),F_0]\Big]$$

By using the same trick as in the first order we obtain:

$$qn_2 = \int d\bar{r}_1 d\bar{r}_2 \int_{-\infty}^{t} dt_1 \int_{-\infty}^{t_1} dt_2 \phi(\bar{r}_1,t_1)\phi(\bar{r}_1,t_2) < \Big[[\rho(\bar{r},t),\rho(\bar{r}_1,t)], \rho(\bar{r}_2,t_2)\Big] >$$

If we set

$$\phi = \Sigma\, C_\alpha(r) \exp i\omega_\alpha t\ \Phi_\alpha(\bar{r}) + \Phi_2$$

we know that $C_\alpha$ will be a constant in the lowest order and $\Phi_2$ will be of second order. The Fourier transform of $\Phi$ will be written:

$$\Phi(\omega) = \sum_\alpha C_\alpha(\omega-\omega_\alpha)\ \Phi_\alpha + \Phi_\alpha(\omega)$$

where $C_\alpha(\omega-\omega_\alpha)$ is proportional to $\delta(\omega-\omega_\alpha)$ in the lowest order and is a function of $\omega$ which is peaked around $\omega = \omega_\alpha$ in the second order. Then we have from the Poisson equation:

$$\sum_\alpha C_\alpha(\omega-\omega_\alpha)\ H(\omega)\ \Phi_\alpha + H(\omega)\ \Phi_2(\omega) = -\frac{1}{\varepsilon_0} q \int_{-\infty}^{+\infty} n_2 \exp^{-i\omega t}\ dt \qquad \text{A(10)}$$

In order to solve A(10), we develop $\Phi_2$ on the eigenfunctions of $H(\omega)$. For $\omega \neq \omega_\alpha$, we may neglect the time dependence of $C_\alpha$ and solve A(10) by neglecting the first term of the l.h.s. In the vicinity of $\omega = \omega_\alpha$ we would obtain diverging solutions as $\omega$ approaches $\omega_\alpha$. Then we use the Bogoliubov method by taking into account the time dependence of $C_\alpha$. Then let us multiply A(10) by $\Phi_\alpha^*$ and integrate over $r$. Since $C_\alpha$ will be large only for $\omega \sim \omega_\alpha$, we may write the result:

$$(\omega-\omega_\alpha)\ C_\alpha(\omega-\omega_\alpha) \left[\int \Phi_\alpha^* \frac{\partial H}{\partial \omega} \Phi_\alpha\ d\bar{r}\right]_{\omega=\omega_\alpha}$$

$$= -\frac{q}{\varepsilon_0} \int d\bar{r}dt\ \Phi_\alpha^*(\bar{r}) n_2(\bar{r},t) \exp{-i\omega t}$$

or

$$\omega C_\alpha(\omega) \left[\int \Phi_\alpha^* \frac{\partial H}{\partial \omega} \Phi_\alpha\ d\bar{r}\right]_{\omega=\omega_\alpha}$$

$$= -\frac{1}{\varepsilon_0} \int drdr\ \Phi_\alpha^*(\bar{r}) n_2(\bar{r},t) \exp -i(\omega+\omega_\alpha)t$$

The r.h.s. of A(11) writes:

$$-\frac{1}{\varepsilon_0} \sum_{\beta,\gamma} \int d\bar{r}d\bar{r}_1 d\bar{r}_2 \int_{-\infty}^{+\infty} dt\ \exp i[\omega-\omega_\alpha+\omega_\beta+\omega_\gamma]t$$

$$\int_{-\infty}^{0} dt_1 \int_{-\infty}^{t_1} dt_2\ \Phi_\alpha^*(\bar{r}) \Phi_\beta(\bar{r}_1) \Phi_\gamma(\bar{r}_2)$$

$$\times\ C_\beta C_\gamma < \left[ [\rho(\bar{r}, t = 0),\ \rho(r_1, t_1)],\ \rho(\bar{r}_2, t_2) \right] >$$

$$\exp\ i[\omega_\beta t_1 + \omega_\gamma t_2]$$

We have used once more the time invariant property of the bracket. Since we have assumed that $C(\omega)$ was peaked close to $\omega = 0$, we shall retain only the triplets $\alpha\ \beta\ \gamma$ such that

$$\omega_\beta + \omega_\gamma = \omega_\alpha + \Delta\omega \quad \text{with} \quad \Delta\omega << \omega_\alpha \qquad \text{A(12)}$$

Then the time evolution of $C_\alpha$ will be given by:

$$i \left[ \int \Phi_\alpha^* \frac{\partial H}{\partial \omega} \Phi_\alpha\, d\bar{r} \right]_{\omega=\omega_\alpha} \frac{dC_\alpha}{dt} = \sum_{\substack{\beta,\gamma \\ \omega_\beta+\omega_\gamma=\omega_\alpha}} C_\alpha C_\beta \mu_\alpha[\omega_\beta, \omega_\gamma] \exp\ i\Delta\omega t \qquad \text{A(13)}$$

with

$$\mu_\alpha(\omega_\beta, \omega_\gamma) = \frac{1}{\varepsilon_0} \int d\bar{r} d\bar{r}_1 d\bar{r}_2\ \Phi_\alpha^*(\bar{r}) \Phi_\beta(\bar{r}_1) \Phi_\gamma(\bar{r}_2) \int_{-\infty}^{0} dt_1 \int_{-\infty}^{t_1} dt_2$$

$$\exp\ i[\omega_\beta t_1 + \omega_\gamma t_2]$$

$$\times < \left| [\rho(\bar{r}, t=0), \rho(\bar{r}_1, t_1)], \rho(\bar{r}_2, t_2)] \right| > + (\beta \leftrightarrow \gamma) \qquad \text{A(14)}$$

In principle we cannot neglect $\exp\ i\Delta\omega t$ in the r.h.s. of A(13) because for large value of time it will lead to large change of the coupling term. On the other hand, when we compute A(14) we may neglect it and set $\omega_\alpha = \omega_\beta + \omega_\gamma$ in the integral.

We notice that if condition A(12) is satisfied for $\alpha = \alpha_0$ and $\beta = \beta_0$, then the amplitude of mode $\beta_0$ and $\gamma_0$ will also have a secular behavior and we shall have for instance:

$$i \left| \Phi_{\beta_0} \frac{\partial H}{\partial \omega} \Phi_{\beta_0} \right|_{\omega=\omega_{\beta_0}} \frac{\partial C_{\beta_0}}{\partial t} = C_{\alpha_0} C_{\gamma_0}^* \mu_{\beta_0} [\omega_{\alpha_0}, -\omega_{\gamma_0} \Big| \exp\ -i\Delta\omega t \qquad \text{A(15)}$$

if we agree that in $\mu_{\beta_0}[\omega_{\alpha_0}, -\omega_{\gamma_0}]$, the eigenfunction $\Phi_{\gamma_0}$ is replaced by $\Phi_{\gamma_0}^*$

## B. SYMMETRY OF RESONANT MODE-MODE COUPLING COEFFICIENTS

We want to show that there exists a simple symmetry relationship between $\mu_\alpha[\omega_\beta,\omega_\gamma]$ and $\mu_\beta[\omega_\alpha,-\omega_\gamma]$. At first, let us set:

$$\nu(\bar{r},\bar{r}_1,\omega',\omega'') = \int_{-\infty}^{+\infty} dt_1 \int_{-\infty}^{+\infty} dt_2 \exp -i[\omega' t_1+\omega'' t_2]$$

$$<\Big[[\rho(\bar{r},t=0),\rho(\bar{r}_1,t_1)],\rho(\bar{r}_2,t_1)\Big]>$$

We have:

$$\nu[\bar{r},\bar{r}_1,\bar{r}_2,-\omega'-\omega'',\omega''] = \int_{-\infty}^{+\infty} dr_1 \int_{-\infty}^{+\infty} dr_2 \exp i\omega' t_1-\omega''(t_2-t_1) \quad < \quad >$$

$$= -\int_{-\infty}^{+\infty} dt_1 \int_{-\infty}^{+\infty} d\tau \exp i[\omega' t_1-\omega''\tau]$$

$$<\Big[[\rho(\bar{r}_1,0),\rho(\bar{r},-t_1)],\rho(\bar{r}_2,\tau)\Big]>$$

$$= -\nu[\bar{r}_1,\bar{r},\bar{r}_2,\omega',\omega''] \qquad \text{B(1)}$$

Let us recall the Jacobi identity of Poisson brackets

$$[[f,g],h] + [[g,h],f] + [[h,f],g] = 0$$

Then we may write:

$$\nu(\bar{r},\bar{r}_1,\bar{r}_2,\omega',\omega'') + \nu(\bar{r}_1,\bar{r}_2,\bar{r},\omega'',-\omega'-\omega'')$$
$$+ \nu(\bar{r}_2,\bar{r},\bar{r}_1,-\omega'-\omega'',\omega') = 0 \qquad \text{B(2)}$$

Now we use $\nu[r,r_1,r_2]$ to compute $\mu_\alpha[\omega_\beta,\omega_\gamma]$. As a matter of fact we have:

$$\int_{-\infty}^{0} dt_1 \int_{-\infty}^{t_1} dt_2 \exp i[\omega_\beta t_1+\omega_\gamma t_2]$$

$$<\Big[[\rho(\bar{r},t=0),\rho(\bar{r}_1,t_1)],\rho(\bar{r}_2,t_2)\Big]> = \qquad \text{B(3)}$$

$$= -\frac{1}{(2\pi i)^2}\int d\omega' d\omega'' \frac{\nu[\bar{r},\bar{r}_1,\bar{r}_2,\omega',\omega'']}{(\omega_\gamma-\omega'')[\omega_\beta+\omega_\gamma-\omega'-\omega'']}$$

Then we may write:

$$\mu_\alpha(\omega_\beta,\omega_\gamma) = -\frac{1}{(2i\pi)^2\varepsilon_0}\int d\bar{r}d\bar{r}_1 d\bar{r}_2 \Phi_\alpha^*(\bar{r})\Phi_\gamma(\bar{r}_2)$$

$$\times \frac{d\omega' d\omega''}{\omega_\alpha-\omega'-\omega''}\left[\frac{\nu(\bar{r},\bar{r}_1,\bar{r}_2,\omega',\omega'')}{\omega_\gamma-\omega''} + \frac{\nu(\bar{r},\bar{r}_2,\bar{r}_1,\omega',\omega'')}{\omega_\beta-\omega''}\right] \qquad \text{B}(4)$$

By using B(1) and B(2) we establish easily the following identity:

$$\frac{1}{\omega_\alpha-\omega'-\omega''}\left[\frac{\nu(\bar{r},\bar{r}_1,\bar{r}_2,\omega',\omega'')}{\omega_\gamma-\omega''} + \frac{\nu(\bar{r},\bar{r}_2,\bar{r}_1,\omega',\omega'')}{\omega_\beta-\omega''}\right] =$$

$$-\frac{\nu(r_1,r,r_2,-\omega'-\omega'',\omega'')}{(\omega_\gamma-\omega'')(\omega_\beta-\omega')} + \frac{\nu(\bar{r}_1,\bar{r}_2,\bar{r},\omega',-\omega'-\omega'')}{(\omega_\beta-\omega'')(\omega_\alpha-\omega'-\omega'')}$$

By a change of the integration variables $\omega'$, $\omega''$ we get immediately

$$\int \frac{d\omega' d\omega''}{\omega_\alpha-\omega'-\omega''}\left[\frac{\nu(\bar{r},\bar{r}_1,\bar{r}_2,\omega',\omega'')}{\omega_\gamma-\omega''} + \frac{\nu(\bar{r},\bar{r}_2,\bar{r}_1,\omega',\omega'')}{\omega_\beta-\omega''}\right]$$

$$= \int \frac{d\omega' d\omega''}{\omega_\beta+\omega'+\omega''}\left[\frac{\nu(\bar{r}_1,\bar{r}_2,\bar{r},\omega',\omega'')}{\omega_\alpha+\omega''} + \frac{\nu(\bar{r}_1,\bar{r},\bar{r}_2,\omega',\omega'')}{\omega''-\omega_\gamma}\right]$$

By comparing with B(4), we obtain the symmetry relationship

$$\mu_\alpha(\omega_\beta,\omega_\gamma) = [\mu_\beta(\omega_\alpha,-\omega_\gamma)]^*$$

Then the three coupled equations for the amplitude of the triplet $\alpha_0,\beta_0,\gamma_0$ will be given by

$$i\left[\Phi_\alpha^* \frac{\partial H}{\partial \omega}\Phi_\alpha\right]_{\omega_\alpha}\frac{dC_\alpha}{dt} = C_\beta C_\gamma \mu \exp i\Delta\omega t$$

$$i\left|\Phi_\beta^* \frac{\partial H}{\partial \omega} \Phi_\beta\right|_{\omega_\beta} \frac{dC_\beta}{dt} = C_\alpha C_\gamma^* \mu^* \exp -i\Delta\omega t$$

$$i\left|\Phi_\gamma^* \frac{\partial H}{\partial \omega} \Phi_\gamma\right|_{\omega_\gamma} \frac{dC_\gamma}{dt} = C_\alpha C_\beta^* \mu^* \exp -i\Delta\omega t$$

By an appropriate choice of the origin of the $C_\alpha$ phase we may always write the system:

$$s_\alpha i \frac{dA_\alpha}{dr} = A_\beta A_\gamma V \exp i\Delta\omega t$$

$$s_\beta i \frac{dA_\beta}{dr} = A_\alpha A_\gamma^* V \exp -i\Delta\omega t$$

$$s_\gamma i \frac{dA_\gamma}{dr} = A_\alpha A_\beta^* V \exp -i\Delta\omega t$$

with

$$V = \frac{|\mu|}{\left[\left(\Phi_\alpha^* \frac{\partial H}{\partial \omega_\alpha} \Phi_\alpha\right)\left(\Phi_\beta^* \frac{\partial H}{\partial \omega_\beta} \Phi_\beta\right)\left(\frac{\partial H}{\partial \omega_\gamma} \Phi_\gamma\right)\right]^{1/2}}$$

and

$$A_\lambda = C_\lambda\left|\Phi_\lambda^* \frac{\partial H}{\partial \omega_\lambda} \Phi_\lambda\right|^{1/2} \qquad s_\alpha = \operatorname{sign}\left(\Phi_\alpha^* \frac{\partial H}{\partial \omega_\alpha} \Phi_\alpha\right)$$

Then we see that the symmetries of the mode-mode coupling coefficients are only a consequence of the time invariance of the equilibrium as well as of the non dissipative characters of the waves. These symmetries are preserved even in an inhomogeneous system with boundary conditions.

## C. THE RANDOM-PHASE-APPROXIMATION

In the preceding section, we were interested in a situation where a finite number of well separated modes were nonlinearly coupled. Then we found that among the finite number of coupling terms, the largest ones were the terms corresponding to $\Delta\omega$ as small as possible. Then we kept these terms and discarded the others which were assumed to be much smaller. We remained then with a finite number of coupled equations for the amplitude. Let us now assume that the dimension of the system increases in such a way that the number of excited modes becomes infinite in a finite frequency interval. Then we cannot really distinguish the terms correspondent to small values of $\Delta\omega$ since $\Delta\omega$ becomes a nearly continuous function of the labels $\alpha$, $\beta$, $\gamma$. We are obliged to retain all the contributions to the mode-mode coupling terms in equation A(10). Since we work now in a large system where the boundary conditions do not play any role, we may assume that locally the system is uniform and that the modes are characterized by a wave vector $k$ and that we shall have:

$$\Phi_\alpha = \exp ikr$$

Then equation A(10) becomes:

$$C_k(\omega-\omega_k)H(\omega,k) + H(\omega,k)\Phi_2(k,\omega) = \frac{q}{\varepsilon_0}\int_{-\infty}^{+\infty} n_{2,k} \exp^{-i\omega t}\, dt$$

with

$$H(\omega,k) = \frac{1}{V}\int \exp -i\bar{k}\bar{r}\; H(\omega)\exp ikr\; d^3r$$

$$n_{2k} = \frac{1}{V}\int n_2(\bar{r}) \exp -ikr\; d^3r \qquad \text{C(1)}$$

$$\Phi_{2k} = \frac{1}{V}\int \Phi_2(r,\omega) \exp -ikr\; d^3r$$

where $V$ is the volume of the system. In the r.h.s. of C(1) we find series of the form:

$$\sum_{k',k''}\left\{\frac{1}{\varepsilon_0}\int dr dr_1 dr_2 \exp -i\bar{k}_0\bar{r} + i(\bar{k}_0'\bar{r}_1 + \bar{k}_0'\bar{r}_2)\right.$$

$$\int_{-\infty}^{0} dt_1 \int_{-\infty}^{t_1} dt_2 \exp i(\omega_{k'}t_1 + \omega_{k''}t_2) \qquad \text{C(2)}$$

$$\left.\left\langle\left[[\rho(\bar{r},t=0),\rho(\bar{r}_1,t_1)],\rho(\bar{r}_2,t_2)\right)\right]\right\rangle + \langle k'\leftrightarrow k''\rangle\right\}$$

Now we have a new invariant property of the bracket since the system is also uniform in space. Then if we set $r_1 = r_1' + r$, $r_2 = r_2' + r$ and use the invariance of the bracket, equation

C(1) becomes:

$$C_k(\omega-\omega_k)H(\omega,k) + H(\omega,k)\Phi_2(k,\omega) =$$

$$\sum_{k'} \mu_{k,k'}[\omega_{k'},\omega_{k-k'}]\ C_{k'}C_{k-k'}\delta(\omega + \omega_{k'} + \omega_{k,k})$$

with

$$\mu_{k,k'}[\omega_{k'},\omega_{k-k'}] = \frac{1}{\varepsilon_0}\int dr_1 dr_2\ \{\exp i[\bar{k}'\bar{r}_1+(\bar{k}-\bar{k}')\bar{r}_2]$$

$$\int_{-\infty}^{0} dt_1 \int_{-\infty}^{t_1} dt_2 \exp i(\omega_{k'}t_1+\omega_{k,k'}t_2) \qquad C(3)$$

$$+ \exp i[(\bar{k}-\bar{k}')\bar{r}_1+\bar{k}'\bar{r}_2] \int_{-\infty}^{0} dt_1 \int_{\infty}^{t_1} dt_2 \exp i(\omega_{k-k'}t_1+\omega_{k'}t_2)\}$$

$$<[[(\rho(r=0,t=0),\rho(r_1,t_1)],\rho(r_2,t_2))]>$$

We solve the equation C(3) as in part A when $k'$ is such that $\Delta\omega_k = \omega_{k'} + \omega_{k-k'} - \omega_k << \omega_k$ and we obtain:

$$i\,\frac{\partial C_k}{\partial t}\,\frac{\partial H}{\partial \omega}(\omega_k,k) = \sum_{k'} \mu_{k,k'}\ (\omega_{k'}+\omega_{k-k'})C_{k'}C_{k-k'} \exp i\Delta\omega_k t \qquad C(4)$$

In the domain of variation of $k'$ where $\Delta_{\omega_k}$ is not much smaller than $\omega_k$, we may retain only the term $\Phi_2$ and we obtain

$$\Phi_{2k}(t) = \exp i\omega_k t\ V^{-1} \sum_{k'} \frac{\mu_{k,k'}[\omega_{k'},\omega_{k-k'}]C_{k'}C_{k-k'}}{H[\omega_{k'}+\omega_{k-k'},k]} \exp i\Delta\omega_k t \qquad C(5)$$

If we set now:

$$\Phi_k = C_k(t) + \Phi_{2k}(t) \exp -i\omega_k r$$

the results of C(4) and C(5) may be summarized by the equation:

$$\frac{\partial \Phi_k}{\partial t} = \sum_{k'} \frac{\mu_{k,k'}[\omega_{k'},\omega_{k-k'}]i\Delta\omega_k\Phi_{k'}\Phi_{k-k'}\exp i\Delta\omega_k t}{H[\omega_{k'}+\omega_{k-k'},k]} \qquad C(6)$$

In equation C(6) we may replace the summation by an integral if we consider $\Phi_k$ as a continuous function of $k$. It implies that the mode interaction time $T_i$ is such that $\delta\omega_k T_i << 1$ where $\delta\omega_k$ is the frequency interval between two neighboring modes. In such conditions we may use the transformation $\sum_k \to V\int dk$ where $V$ is the volume of the system and C(6) becomes:

$$\frac{\partial \Phi_k}{\partial t} = V\int dk' \frac{\mu_{k,k'}[\omega_{k'},\omega_{k-k'}]i\Delta\omega_k\Phi_{k'}\Phi_{k-k'}\exp i\Delta\omega_k t}{H(\omega_{k'}+\omega_{kk'},k)} \qquad C(7)$$

In the integrand of the r.h.s. integral, the factor of $\exp i\Delta\omega_k t$ is a well behaved function of $k'$ which has no singularity and goes to zero at the edge of the spectrum. On the other hand $\Delta\omega_k$ is a function of $k'$ and for large $t$, the exponential becomes a very rapidly varying function of $k'$ so that we may once more apply the arguments which were used in the quasi-linear theory. Then we conclude that the r.h.s. of equation C(7) will be zero after a time of the order of $[\Delta(\Delta\omega_k)]^{-1}$ where $\Delta(\Delta\omega_k)$ is the average variation of $\Delta\omega_k$ over the whole spectrum and provided that the condition $(\partial/\partial k')\Delta\omega = 0$ cannot be met over the spectrum. If $T_i >> [\Delta(\Delta\omega_k)]^{-1}$, the equation C(7) does not describe a secular variation of the amplitude $C_k$ and we conclude that at this order the mode-mode coupling terms can be neglected. If we want to find a secular behavior we have to go to next order.

This result is usually interpreted as resulting from an averaging of equation A(13) over the phases of $C_\beta C_\gamma$. An assumption on the randomness of this phase makes it possible to demonstrate that the r.h.s. is zero. This method is known as the random phase approximation and is used as a trick to get rid elegantly of the coherent three waves interaction terms. However, it appears to be a purely formal process which does not define the validity conditions.

Since we are interested in the secular behaviour of the mode amplitudes, we have to go one order further into the solution of the Maxwell-Vlasov equation. We set $\Phi = \Phi_1 + \Phi_2 + \Phi_3$ and we obtain:

$$f_3 = \int dr_1 dr_2 \int_{-\infty}^{t} dt_1 \int_{-\infty}^{t_1} dt_2 [\Phi_1(\bar{r}_1,t_1)\Phi_2(\bar{r}_2,t_2) + \Phi_2(\bar{r}_1,t_1)\Phi_1(\bar{r}_2,t_2)]$$

$$[\rho(\bar{r}_1,t_1),[\rho(\bar{r}_2,t_2,F_0]\Big]-\int dr_1dr_2dr_3\int_{-\infty}^{t}dt_1\int_{-\infty}^{t_1}dt_2\int^{t_2}dt_3$$

$$\Phi_1(\bar{r}_1,t_2)\Phi_1(\bar{r}_2,t_2)\Phi_3(\bar{r}_3,t_3)\Big[\rho(\bar{r}_1,t_1),[\rho(\bar{r}_2,t_2),[\rho(\bar{r}_3,t_3),F_0]\Big]\Big]$$

In the same way as for lower orders we get:

$$qn_3 = \int d\bar{r}_1,d\bar{r}_2\int_{-\infty}^{t}dt_1\int_{-\infty}^{t_1}dt_2[\Phi_1(\bar{r}_1,t_1)\Phi_2(\bar{r}_2,t_2)$$

$$+\Phi_2(\bar{r}_1,t_1)\Phi_1(\bar{r}_1,t_1)]$$

$$<\Big[[\rho(\bar{r},t),\rho(\bar{r}_1,t_1)],\rho(\bar{r}_2,t_2)\Big]>$$

$$-\int dr_1dr_2dr_3\int_{-\infty}^{t}dt_1\int_{-\infty}^{t_1}dt_2\int_{-\infty}^{t_2}dt_3[\Phi_1(r_1,t_1)\Phi_1(r_2,t_2)\Phi_1(r_3,t_3)]$$

$$<\Big[\Big[\big[\rho(\bar{r},t),\rho(\bar{r}_1,t_1)\big],\rho(\bar{r}_2,t_2)\Big],\rho(r_3,t_3)\Big]>$$

We shall set $qn_3 = qn_3^{(1)} + qn_3^{(2)}$ where $qn_3^{(1)}$ stands for the first multiple integral and $qn_3^{(2)}$ for the second one. For the moment we shall consider only $qn_3^{(1)}$ and we shall see later that $qn_3^{(2)}$ does not contribute.

We deduce from C(6) that we may write after integration on time:

$$\Phi_{k,2} = V\exp i\omega_k t\int dk'\mu_{k,k'}[\omega_{k'},\omega_{k-k'}]\Phi_{k'}\Phi_{k-k'}$$

$$\left[pp\frac{1}{H(\omega_{k'}\omega_{k+k'},k)}+i\pi\delta[H(\omega_{k'}+\omega_{k-k'},k)\right]$$

or, in a more compact form:

$$\phi_{k,2}(t) = \sum_{k'}\frac{\mu_{k,k'}[\omega_{k'},\omega_{k-k'}]\Phi_{k'}\Phi_{k-k'}}{H[\omega_{k'}+\omega_{k-k'}-i\varepsilon,k]}\exp i[\omega_{k'}+\omega_{k-k'}]t$$

Then, from C(8), we obtain immediately:

$$\frac{q}{\varepsilon_0} n_3^{(1)}(k,t) = \sum_{k',k''} \frac{\mu_{k,k'}[\omega_{k'},\omega_{k''}+\omega_{k-k'-k''}]\mu_{k-k',k''}[\omega_{k''},\omega_{k-k'-k}]}{H[\omega_{k''}+\omega_{k-k'-k''}-i\varepsilon,k-k']}$$

$$\times\ \Phi_{k'}\Phi_{k''}\Phi_{k-k'-k''} \exp i[\omega_{k'}+\omega_{k''}+\omega_{k-k'-k''}]t$$

$$+ \sum_{k',k''} \frac{\mu_{k,k-k'}[\omega_{k-k'},\omega_{k''}+\omega_{k'-k''}]\mu_{k',k''}[\omega_{k''},\omega_{k'-k''}]}{H[\omega_{k''}+\omega_{k'-k''}-i\varepsilon,k']}$$

$$\times\ \Phi_{k-k'}\Phi_{k''}\Phi_{k'-k''} \exp i[\omega_{k-k'}+\omega_{k''}+\omega_{k'-k''}]t$$

As a matter of fact we are essentially interested in the time evolutions of the amplitudes $|\Phi_k|$ and we obtain it from the equation:

$$\frac{\partial|\Phi_k|^2}{\partial t}\ \frac{\partial}{\partial\omega} H(k,\omega_k) = 2\mathrm{Im}[\Phi_{-k_1}\ n_{k_3} + n_{k_2}\Phi_{-k_2}] \qquad \mathrm{C}(8)$$

since we have found previously that the products of first order terms do not contribute to the secular behaviour of $|\Phi_k|$. Let us see now how we obtain the kinetic equation by applying the random phase approximation. Let us consider for instance a term of the following type which appears in the product $\Phi_{-k_1}\cdot n_{k_3}$:

$$\sum_{k',k''} \mu_{k'k'}[\omega_{k'},\omega_{k''}+\omega_{k-k'-k''}]\mu_{k-k',k''}[\omega_{k''},\omega_{k-k'-k''}]$$
$$\Phi_{-k}\Phi_{k'}\Phi_{k''}\Phi_{k-k'-k''} \times \frac{\exp i[\omega_{-k}+\omega_{k'}+\omega_{k''}+\omega_{k-k'-k''}]t}{H[\omega_{k''}+\omega_{k-k'-k''}-i\varepsilon,k-k']}$$

If $k''$ is kept fixed, by applying the same arguments as in the preceding order, we conclude that the summation over $k'$ gives a negligibly small result except if $k'' = k$ since in this case the whole argument of the exponential cancels out. The case where $k' = 0$ has not to be considered since it would not correspond to a wave. Then when we integrate over $k''$ we have an integrand which is zero every where except in the vicinity of $k'' = k$. The usual approximation is equivalent to considering that in the series only the term $k'' = k$ is different from zero.* Applying the same argument to the terms of equation C(8) we obtain the usual kinetic equation for resonant mode-mode coupling:

$$\frac{dN_k}{dt} = \sum_{k'} |V_{k,k'}|^2 \delta(\omega_k - \omega_{k'} - \omega_{k-k'}) \Big[ N_{k'} N_{k-k'}$$

$$- N_k N_{k'} \operatorname{sign}\left( \frac{\partial H(\omega_k, k)}{\partial \omega_k} \frac{\partial H(\omega_{k-k'}, k-k')}{\partial \omega_{k-k'}} \right)$$

$$- N_k N_{k-k'} \operatorname{sign}\left( \frac{\partial H(\omega_k, k)}{\partial \omega_k} \frac{\partial H(\omega_{k'}, k')}{\partial \omega_{k'}} \right) \Big]$$

where we have used the symmetry rules:

$$\mu_{k,k'}[\omega_{k'}, \omega_k - \omega_{k'}] = \{\mu_{k-k', +k}[\omega_k, -\omega_{k'}]\}^*$$

$$\mu_{k,k-k'}[\omega_{k-k'}, \omega_k + \omega_{k'-k}] = \{\mu_{k', k'-k}[\omega_{k'-k}, \omega_k]\}^*$$

and where:

$$\left| V_{k,k'} \right|^2 = \frac{|\mu_{k,k-k'}[\omega_{k'}, \omega_{k-k'}]|^2}{\left| \frac{\partial}{\partial \omega_k} H[\omega_k, +k] \frac{\partial}{\partial \omega_{k'}} H[\omega_{k'}, k'] \frac{\partial}{\partial \omega_{k-k'}} H[\omega_{k-k'}, k-k'] \right|}$$

with $N_k = |\Phi_k|^2 \left|\frac{\partial H}{\partial \omega_k}\right|$.

The number $N_k$ may be interpreted as the number of plasmons associated with the mode $k$ since the energy of the mode $k$ is given by $\omega_k N_k$. For systems in static or thermodynamic equilibrium the signs of $\partial H/\partial \omega$ are positive for positive frequencies. When it is not the case peculiar phenomena as explosive instabilities may arise. It is easy to check that such a mode coupling equation conserves the total wave energy since:

---

*This phase mixing argument is valid only if the functions of $k$ are smooth enough. It implies, of course, that a stochastic averaging has been done in such a way that all functions of $k$ can be considered as continuous.

$$\Sigma \omega_k N_k \text{ sign } \frac{\partial}{\partial \omega_k} H(\omega_k, k) = 0$$

The contribution of $N_{k_3}{}^{(2)}$ may be ignored because it gives no imaginary term in C.8 in the absence of particle wave interaction. The general method used here to find the symmetry properties of the coefficient is not the good one to obtain completely neglected the resonant terms so that it will be often possible to use the moment equations to compute these coefficients and so to take advantage of large algebraic simplifications.

## D. NON-LINEAR LANDAU DAMPING

In the preceding sections we have neglected all the resonant effects. In the lowest order they give rise to the linear damping or growth rate. In all problems of interest, these rates are assumed to be small enough. They must be of the order of the mode-mode coupling terms without going out of the frame of the weak turbulence theory. Now, we are going to see an example where non-linear effects induce new resonance effects which play an important role in turbulent processes.

Let us consider for simplicity the motion of a charged particle in the field of two waves of frequencies $\omega_1, \omega_2$. The equation of motion is written

$$m \frac{dv}{dt} = qE_1 \cos(\omega_1 t + k_1 x) + E_2 \cos(\omega_2 t + k_2 x)$$

In the zero order approximation we shall have:

$$x^\circ = x_0 + v_0 t$$

$$v = v_0$$

In the first order we obtain:

$$x_1 = -\frac{q}{m}\left\{\frac{E_1 \cos(\omega_1 t_1 + k_1 x^\circ)}{(\omega_1 + k_1 v_0)^2} + \frac{E_2 \cos(\omega_2 t + k_2 x^\circ)}{(\omega_2 + k_2 v_0)^2}\right\} \qquad D(1)$$

$$v_1 = \frac{q}{m}\left\{\frac{E_1 \sin(\omega_1 t + k_1 x^\circ)}{\omega_1 + k v_0} + \frac{E_2 \sin(\omega_2 t + k_2 x^\circ)}{\omega_2 + k_2 v_0}\right\}$$

Let us assume that we are interested in plasma waves for which $\omega_1/k_1 \sim \omega_2/k_2 >> V_{th}$ where $V_{th}$ is the average thermal velocity. Then the resonance at $\omega_1 + k_1v_1 = 0$ and $\omega_2 + k_2v_2 = 0$ will interest a small fraction of particles which will give rise to the small growth or damping rate. If the particle has a velocity of the order of the thermal velocity, we can expect a resonant effect only in the non-linear higher order terms. These effects will be small for one particle but as they are much more numerous that the high velocity resonant particles, the resulting effect can be of the same order as the linear effect. Then since $\omega_1 >> kv_0$ and $\omega_2 >> kv_0$ we have in the next order:

$$\frac{dv_2}{dt} = \frac{q^2}{m^2}\left|E_1k_1\ \sin(\omega_1t+k_1x°) + E_2k_2\ \sin(\omega_2t+k_2x°)\right| \times \left|\frac{E_1\ \cos(\omega_1t+k_1x°)}{\omega_1^2} + \frac{E_2\ \cos(\omega_2t+k_2x°)}{\omega_2^2}\right| \qquad D(2)$$

We see that we shall have resonance if $(\omega_1 \pm \omega_2)/k_1 \pm k_2$ may be of the order of the thermal velocity. This is possible provided that waves with neighbouring frequencies and opposite phase velocities exists or if the wave frequency depends weakly on $k$. The second order electric field induced by these velocity perturbations will also interact resonantly with the particles. Since the corresponding fluctuations will oscillates at frequencies $\omega_1 \pm \omega_2$ we have to couple them again with the waves in order to recover a field fluctuation at frequencies $\omega_1$ and $\omega_2$ which will take part to the build up or to the damping of the wave. Then we shall have to go to third order in the expansion of the Valsov equation as we shall see now.

We shall limit ourselves to the computation of the non-linear Landau damping of electron waves in a strong magnetic field such that the plasma frequency $\omega_p$ is smaller than the cyclotron frequency $\omega_{ce}$. The perpendicular velocities are ignorable and we solve a one-dimension Vlasov equation.

Using a Fourier transform in $x$ and $t$, we set

$$F = f_0 + \sum_{k,\omega} f_{k,\omega}\ \exp(ikx + i\omega t)$$

The equation for the fluctuating distribution function $f_{k,\omega}$ is written

$$i(\omega+kv)f_{k,\omega} = \frac{q}{m} ik\Phi_{k,\omega} \frac{\partial f_0}{\partial v} + \frac{q}{m} \sum_{k',\omega'} i(k-k')\Phi_{\substack{k-k' \\ \omega-\omega'}} \frac{\partial f_{k',\omega'}}{\partial v} \qquad D(3)$$

where $\Phi_{k,\omega}$ is the Fourier transform of the perturbed potential. If we set $F(k,\omega) = \frac{q}{m} \frac{k\partial f_0/\partial_v}{\omega + k_v}$ and $f_{k,\omega} = f^1_{k,\omega} + f^2_{k,\omega} + \ldots,$

$$\Phi_{k,\omega} = \Phi_k + \Phi^{(2)}_{k,\omega} + \ldots$$

we obtain

$$f^1_{k,\omega} = F(k,\omega)\Phi_k$$

$$f^2_{k,\omega} = F(k,\omega)\Phi^2_{k,\omega} + \frac{q}{m}\sum_{k'+k''=k} \frac{\Phi_{k'}\Phi_{k''}}{\omega+kv} \frac{\partial}{\partial v}\left[k''F(k',\omega_{k'})+k'F(k'',\omega k'')\right]$$

$$f^3_{k,\omega} = F(k,\omega)\Phi^3_{k,\omega} + \frac{q}{m} \sum_{\substack{k'+k''=k \\ \omega'+\omega''=\omega}} \frac{k''}{\omega+kv}\left[\Phi^{(2)}_{k'',\omega''} \frac{\partial f^{(1)}_{k',\omega'}}{\partial v} + \Phi^{(1)}_{k''} \frac{\partial f^{(2)}_{k',\omega'}}{\partial v}\right]$$

where $\Phi_k$ stands for $\Phi^1_{k,\omega}$. If now $N(k,\omega)$ denotes $\int F(k,\omega)dv$, we obtain the perturbed densities:

$$n^{(2)}_{k,\omega} = N(k,\omega)\Phi^{(2)}_{k,\omega} + \frac{q}{m} \sum_{k'+k''=k} \int \frac{\Phi_{k'}\Phi_{k''}dv}{\omega+kv} \left[\frac{\partial}{\partial v} (k''F(k',\omega_{k'}) + k'F(k'',\omega_{k''}))\right]$$

Using

$$\int \frac{dv}{\omega kv} \frac{\partial}{\partial v} F(k',\omega') = k \int \frac{F(k,\omega')dv}{(\omega+kv)^2}$$

and

$$\int \frac{F(k',\omega')dv}{(\omega+kv)^2} = - \frac{\partial}{\partial\omega} \int \frac{F(k',\omega')}{(\omega+kv)}$$

We get, from the definition of $F(k,\omega)$:

$$\int \frac{F(k',\omega')}{(\omega+kv)^2}\,dv = k'\,\frac{\partial}{\partial\omega}\left[\frac{N(k,\omega)-N(k',\omega')}{\omega'k-\omega k'}\right]$$

and after some algebra:

$$n^{(2)}_{k,\omega} = N(k,\omega)\Phi^{(2)}_{k,\omega} - \frac{q}{m}\sum_{k'+k''=k} kk'k''\Phi_{k'}\Phi_{k''}\,\frac{\partial}{\partial\omega}\left[\frac{N(k',\omega')-N(k,\omega)}{k'\omega-k'\omega'} + \frac{N(k'',\omega'')-N(k,\omega)}{k''\omega-k\omega''}\right] \qquad D(4)$$

In the same way and after some more algebra:

$$n^{(3)}_{k,\omega} = N(k,\omega)\Phi^{(3)}_{k,\omega} - \frac{q}{m}\sum_{\substack{k'+k''=k\\ \omega'+\omega''=\omega}} \frac{\partial}{\partial\omega}\left[\frac{N(k',\omega')-N(k,\omega)}{k'\omega-k\omega'}\left(\Phi^{(2)}_{k''}\Phi_{k'} + \Phi_{k''}\Phi^{(2)}_{k'}\right) + \begin{pmatrix} h' \leftrightarrow k'' \\ \omega' \leftrightarrow \omega''\end{pmatrix}\right] + \frac{q^2}{m^2}\sum_{\substack{k'+k''=k\\ \omega'+\omega''=\omega}}$$

$$\left[\sum_{\substack{q'+q''=k'\\ \omega_{q'}+\omega_{q''}=\omega'}} k'k''q'q''\,\Phi_{k''}\Phi_{q'}\Phi_{q''}\left(k\frac{\partial^2}{\partial\omega^2} + k'\frac{\partial^2}{\partial\omega\partial\omega'}\right)\right.$$

$$\{(k\omega'-k'\omega)^{-1}\,(k\omega_{q'}-q'\omega)^{-1}(k'\omega_{q'}-q'\omega')^{-1}$$

$$\times\,[(k'\omega_{q'}-q'\omega')kN(k,\omega) - (k\omega_{q'}-q'\omega)k'N(k',\omega')$$

$$\left. + (k\omega'-k'\omega)q'N(q',\omega_{q'})] + (q' \leftrightarrow q'')\} + \begin{pmatrix} k' \leftrightarrow k'' \\ \omega' \leftrightarrow \omega''\end{pmatrix}\right] \qquad D(5)$$

In equation D(5), the second term of the r.h.s represents the contribution of induced non-linear scattering. The mixing of two waves gives rise to a non-linear virtual wave $\Phi^{(2)}_k$ which induces a new wave by coupling with a third one.

The third and fourth terms may be considered as four plasmons interactions since it is the direct coupling of three waves which induces a fourth one by the non-linear motion of the particles.

Let us restrict ourselves to two initial waves which are defined by their wave numbers $k_1$ and $k_2$ and the frequencies $\omega_1$ and $\omega_2$. The second order terms will contain perturbations with wave numbers and frequencies

$k_1 + k_2,\ \omega_1 + \omega_2$

$k_2 + k_1,\ \omega_2 - \omega_1$

$k_1 - k_2,\ \omega_1 - \omega_2$

$-k_1 - k_2, -\omega_1 - \omega_2$

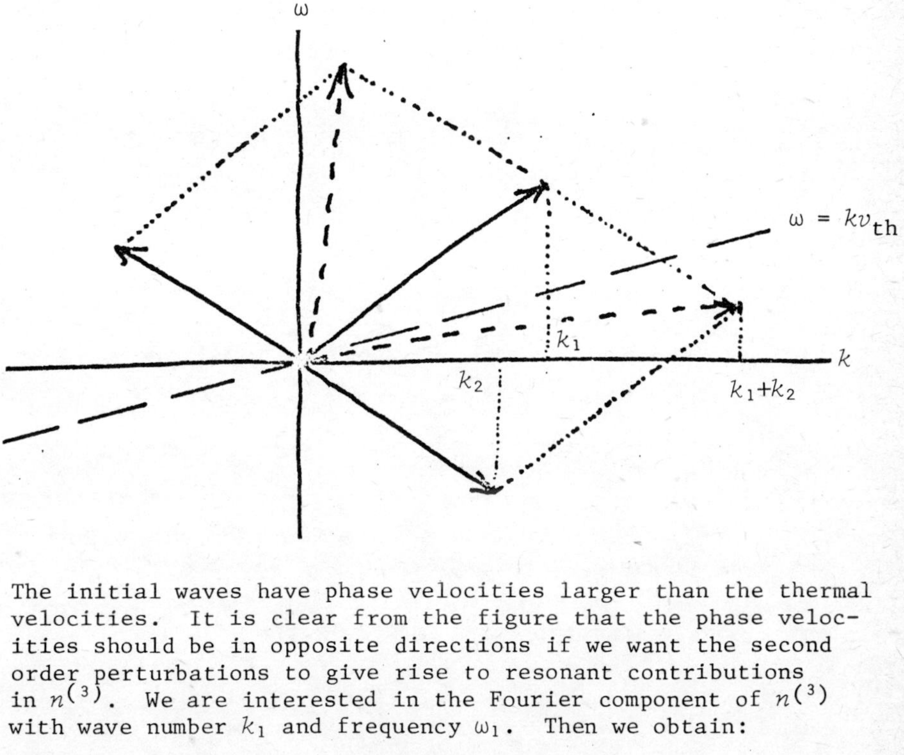


The initial waves have phase velocities larger than the thermal velocities. It is clear from the figure that the phase velocities should be in opposite directions if we want the second order perturbations to give rise to resonant contributions in $n^{(3)}$. We are interested in the Fourier component of $n^{(3)}$ with wave number $k_1$ and frequency $\omega_1$. Then we obtain:

$$n^{(3)}(k_1,\omega) = N(k,\omega)\Phi^{(3)}_{k,\omega} + \frac{q}{m}\,\frac{k_1k_2(k_1+k_2)}{[k_1\omega_2-k_2\omega_1]^2}\,[k_2N(-k_2,-\omega_2)$$

$$+\,k_1N(k_1,\omega_1) - (k_1+k_2)N(k_1+k_2,\omega_{k_1}+\omega_{k_2})]\Phi^{(2)}_{k_1+k_2}\ \Phi_{-k_2}$$

$$+ \frac{q^2}{m^2} \frac{k_1{}^2 k_2{}^2 |\Phi_{k_2}|^2 \Phi_{k_1} (k_1+k_2)^4}{(k_1\omega_2-k_2\omega_1)^4} N(k_1+k_2,\omega_1+\omega_2) \qquad D(6)$$

The second order potentials $\Phi_k^{(2)}$ have to be computed from the Poisson equation in which we introduce the second order perturbed density $n_k^{(2)}$ of equation D(4). We obtain:

$$n_{k_1+k_2}^{(2)} = N[k_1+k_2,\omega_1+\omega_2]\Phi_{k_1+k_2}^{(2)} + \frac{q}{m} \frac{k_1 k_2 (k_1+k_2)\Phi_{k_2}\Phi_{k_1}}{(k_1\omega_2-k_2\omega_1)^2}$$

$$[k_1 N(k_1,\omega_1) + k_2 N(k_2,\omega_2) - (k_1+k_2)N(k_1+k_2,\omega_1+\omega_2)]$$

If $K$ denotes now the total wave vector, the second order potential $\Phi_{k_1+k_2}^{(2)}$ given by the Poisson equation is:

$$\Phi_{k_1+k_2}^{(2)} = \frac{q^2}{m\varepsilon_0} \frac{(k_1+k_2)k_1 k_2 \Phi_{k_1}\Phi_{k_2}}{(k_1\omega_1-k_2\omega_2)^2[(k_1+k_2)^2-(q/\varepsilon_0)N(k_1+k_2)]} \times$$

$$[k_1 N(k_1) + k_2 N(k_2) - (k_1+k_2)N(k_1+k_2)]$$

Then the third order perturbed density reduces to:

$$n_{k_1}^{(3)} = N(k_1,\omega)\Phi_{k_1}^{(3)}(\omega) + \frac{q^2}{m^2} \frac{k_1{}^2 k_2{}^2 (k_1+k_2)^2}{(k_1\omega_2-k_2\omega_1)^4} \times$$

$$\frac{\Phi_{k_1}|\Phi_{k_2}|^2}{[(k_1+k_2)^2-\frac{q}{\varepsilon_0}N(k_1+k_2)]} \times$$

$$\{\frac{q}{\varepsilon_0}[k_1 N(k_1)+k_2 N(k_2)-(k_1+k_2)N(k_1+k_2)]^2 +$$

$$+ (k_1+k_2)^2 N(k_1+k_2)[(k_1+k_2)^2-\frac{q}{\varepsilon_0}N(k_1+k_2)]\}$$

In the curly bracket of the r.h.s of equation D(7), the last term is the four plasmons interaction term and the other ones are the induced nonlinear scattering terms. Since $k_1\omega_1$ are solutions of the dispersion relation we have:

$$N(k_1) = \frac{k_1{}^2\varepsilon_0}{q} \qquad N(k_2) = \frac{k_2{}^2\varepsilon_0}{q}$$

and

$$N(k_1+k_2) = \frac{q}{m}\int pp\,\frac{(k_1+k_2)\partial f_0/\partial v}{\omega_1+\omega_2+(k_1+k_2)v}\,dv + \frac{q}{m}\,i(k_1+k_2)n$$

$$\int dv\delta\,[\omega_1+\omega_2+(k_1+k_2)v]\frac{\partial f_0}{\partial v}$$

Then $N(k_1 + k_2)$ is of the order of $\frac{q}{m}\frac{n_0}{V_{th}^2} \sim \frac{\varepsilon_0}{q}\frac{\omega^2 p}{V_{th}^2}$, and is much larger than $N(k)$. We see that in the lowest order the non-linear induced scattering cancels exactly the four plasmas term. This feature occurs frequently in non-linear Landau damping conputations.

We remain with:

$$n_{k_1}^{(3)} = N(k,\omega)\Phi_{k,\omega}^{(3)} + \frac{q^2}{m^2}\,\frac{k_1^2k_2^2(k_1+k_2)^2\;\Phi_{k_1}|\Phi_{k_2}|^2}{(k_1\omega_2-k_2\omega_1)^4[(k_1+k_2)^2-\frac{q}{\varepsilon_0}N(k_1+k_2)]}\times$$

$$N(k_1+k_2)(k_1+k_2)[2(k_1+k_2)k_1k_2-k_1k_2{}^2-k_2k_1{}^2]$$

In the lowest order, we can neglect $(k_1+k_2)^2$ in the denominator provided that $|k_1+k_2|^2\lambda_D^2 << 1$ where $\lambda_D$ is the Debye length. Then, we have a further reduction in the contribution since $N(k_1+k_2)$ cancels and all the resonant terms have disappeared. Finally we obtain:

$$n_{k_1}^{(3)} = N(k,\omega)\Phi_{k,\omega}^{(3)} - \frac{q}{m^2}\,\varepsilon_0\,\frac{k_1{}^2k_2{}^2(k_1+k_2)^2\;\Phi_{k_1}|k_2|^2}{(k_1\omega_2-k_2\omega_1)^4}$$

$$\left[1-\frac{\varepsilon_0(k_1+k_2)^2}{qN(k_1+k_2)}\right](k_1+k_2)[2(k_1+k_2)k_1k_2-k_1k_1{}^2-k_2k_2{}^2]$$

The kinetic equation for the wave amplitudes is readily obtained from

$$\frac{\partial}{\partial t}(\Phi_{k_1})^2 = \Phi_{k_1}^*\frac{\partial\Phi_{k_1}}{\partial t} + \Phi_{k_1}\frac{\partial\Phi_{k_1}^*}{\partial t}$$

Writing $\Phi_{k_1} = \int \Phi_{k_1}(\omega)\exp i\omega t\,d\omega$ and linearizing the Poisson equation around $(\omega_{k_1},k_1)$ we get:

$$\Phi_{k_1,\omega_1}(\omega_1-\omega_{k_1})\frac{\partial}{\partial\omega_1}[k_1{}^2-\frac{q}{\varepsilon_0}N(k_1,\omega_1)] = \frac{q}{\varepsilon_0}n_3(k_1,\omega_1)$$

and after some algebra:

$$\frac{\partial}{\partial \omega_1} [k_1{}^2 - \frac{q}{\varepsilon_0} N(k_1,\omega_1)] \frac{\partial}{\partial t} \left|\Phi_{k_1}\right|^2 = + i \frac{q}{\varepsilon_0}\left[\Phi^*_{k_1} n_{k_1} - \Phi_{k_1} n^*_{k_1}\right]$$

It reads:

$$\frac{\partial}{\partial \omega_1} \left[k_1{}^2 - \frac{q}{\varepsilon_0} N(k_1\omega_1)\right] \frac{\partial}{\partial t}\left|\Phi_k\right|^2 = \frac{2q^2\varepsilon_0}{m^3}$$

$$\frac{k_1{}^2 k_2{}^2 (k_1+k_2)^3 |\Phi_{k1}|^2 |\Phi_{k2}|^2}{|k_1\omega_1 - k_2\omega_2|^4}$$

$$[2(k_1+k_2)k_1k_2 - k_1k_1{}^2 - k_2k_2{}^2] \frac{[k_1+k_2]^2}{|N(k_1+k_2)|^2} \int \pi dv \delta[\omega_1+\omega_2+(k_1+k_2)v] \frac{\partial f_0}{\partial v}$$

If we introduce the number of plasmons:

$$N_1 = \left|\frac{\partial}{\partial \omega_1} [k_1{}^2 - \frac{q}{\varepsilon_0} N(k_1,\omega_1)]\right| |\Phi_{k_1}|^2$$

$$N_2 = \left|\frac{\partial}{\partial \omega_2} [k_2{}^2 - \frac{q}{\varepsilon_0} N(k_2,\omega_2)]\right| |\Phi_{k2}|^2$$

the kinetic equation for waves may also be written if $\omega_1 > 0$ and $\omega_2 < 0$:

$$\frac{dN_1}{dt} = W_{k_1 k_2} N_1 N_2$$

$$\frac{dN_2}{dt} = - W_{k_1 k_2} N_1 N_2$$

where

$$W_{k_1,k_2} = \frac{2\varepsilon_0 q^2}{m^3} \frac{k_1{}^2 k_2{}^2 (k_1+k_2)^3}{|k_1\omega_1 - k_2\omega_2|^4} (k_1+k_2)^2 [2(k_1+k_2)k_1k_2 -$$

$$k_1 k_1{}^2 - k_2 k_2{}^2] \left|\frac{\partial}{\partial\omega_1}[k_1{}^2 - \frac{q}{\varepsilon_0} N(k_1,\omega_1)]\frac{\partial}{\partial\omega_2}[k_2{}^2 - \frac{q}{\varepsilon_0} N(k_2,\omega_2)]\right|^{-1}$$

D(7)

$$|N(k_1+k_2)|^{-2} \int \pi dv \delta[(\omega_1+\omega_2) + (k_1+k_2)v]\frac{\partial f_0}{\partial v}$$

The equation D(7) appeals to the following remarks. We see that the time derivatives of $N_1$ and $N_2$ are obtained at the same order than the resonant mode-mode coupling term in the random phase approximation. We should have needed the R.P.A. only in order to eliminate the four mode resonant coupling terms. The energy is not conserved in the interaction process but the total number of plasmons $N_1 + N_2$ is conserved. The symmetry rules that we have found are general and can be obtained by the same formal computations as for resonant mode-mode coupling. It will be noticed also that the sign of $V_{k_1 k_2}$ changes with the sign of $\omega_1 + \omega_2$.

## E. NON-LINEAR STABILIZATION

The main problem of weak turbulence is to understand the mechanism of non-linear stabilization of weak instabilities. The quasi-linear mechanism has been shown to be an efficient one. But often it cannot be applied because of the non-resonant nature of the particle wave interaction, because of its own inefficiency like in the case of drift waves or because collisions destroy the good shape of the distribution function. It is then necessary to take into account other non-linear effects to have an idea of the mechanisms of wave saturation.

At first we shall consider the case of coherent modes which are coupled by the equations we found at the end of §B. We must at first include the damping coefficients into those equations which become:

$$is_\alpha[\frac{dA_\alpha}{dt} - \gamma_\alpha] = \sum_{\omega_\alpha \simeq \omega_\beta + \omega_\gamma} A_\beta A_\gamma V \exp i\Delta\omega t \qquad E(1)$$

In general the equations E(1) cannot be solved if a great number of modes are coupled. Hence, we shall study the simple case of three coupled modes. We shall assume that they are positive energy modes and we shall choose positive signs for the three frequencies $\omega_0, \omega_1, \omega_2$ with $\omega_0 = \omega_1 + \omega_2 + \Delta\omega$. We shall set:

$$A_0 = a_0 \exp i[\Delta\omega t + \phi_0]$$

$$A_1 = a_1 \exp i[\Delta\omega t + \phi_1]$$

$$A_2 = a_2 \exp i[\Delta\omega t + \phi_2]$$

where $a_0, a_1, a_2, \phi_0, \phi_1, \phi_2$ are real. The evolution of $a_0, a_1, a_2, \phi_0, \phi_1, \phi_2$ are ruled by the equations:

$$\frac{da_0}{dt} = \gamma_0 a_0 - V \text{ Sin } \theta \, a_1 a_2 \qquad E(2)$$

$$\frac{da_1}{dt} = \gamma_1 a_1 + V \text{ Sin } \theta \, a_0 a_2 \qquad E(3)$$

$$\frac{da_2}{dE} = \gamma_2 a_2 + V \text{ Sin } \theta \, a_0 a_1 \qquad E(4)$$

$$\frac{d\theta}{dr} = V\left[- \frac{a_1 a_2}{a_0} + \frac{a_0 a_2}{a_1} + \frac{a_0 a_1}{a_2}\right] \text{Cos } \theta + \Delta\omega \qquad E(5)$$

where $\theta = \phi_0 - \phi_1 - \phi_2$. We want to see if by coupling those modes it is possible to limit the growth of the modes which have a positive $\gamma$. Let us notice, at first, that such a stabilization is impossible if the three $\gamma$ are positive or zero. If we multiply E(1) by $a_0$, E(2) by $a_1$ and E(3) by $a_2$, we obtain:

$$\frac{d}{dt}(a_0^2 + a_1^2) = 2\gamma_0 a_0^2 + 2\gamma_1 a_1^2$$

$$\frac{d}{dt}(a_0^2 + a_2^2) = 2\gamma_0 a_0^2 + 2\gamma_2 a_2^2$$

These equations show that we must choose negative values for one or two $\gamma$. Let us assume, at first, that $\Delta\omega$ is small when compared with the growth or damping rates so that we may neglect it in E(5). We can find an equilibrium situation for the phases and amplitudes with:

$$\text{Cos } \theta = 0; \; a_0^2 = \frac{\gamma_1\gamma_2}{V^2}; \; a_1^2 = -\frac{\gamma_0\gamma_2}{V^2}; \; a_2^2 = -\frac{\gamma_0\gamma_1}{V^2} \qquad E(6)$$

The equilibrium is possible if $\gamma_0 > 0$, $\gamma_1 < 0$, $\gamma_2 < 0$ or if $\gamma_0 < 0$, $\gamma_1 > 0$, $\gamma_2 > 0$. If one refers to the decay instability we may say that the unstable wave decays into two lower frequency waves or the two unstable waves generate a higher frequency damped wave by mode-mode coupling. We must see if the equilibrium E(6) is stable. Then, we perturb the phases and the amplitudes and we linearize the equations E(2), (3), (4), (5). The perturbed quantities are proportional to exp $st$ where $s$ is given by the secular equation:

$$s^3 - [\gamma_0 + \gamma_1 + \gamma_2]s^2 + 4\gamma_0\gamma_1\gamma_2 = 0 \qquad E(7)$$

for the perturbed amplitudes

$$s = +(\gamma_0 + \gamma_1 + \gamma_2) \qquad E(8)$$

for the perturbed phases which are decoupled from the perturbed amplitudes. We must have $\gamma_0 + \gamma_1 + \gamma_2 < 0$ for the stability of the phases and when this inequality is fulfilled the equations E(7) has always two roots with positive real parts. Then the equilibrium is always unstable and numerical computations have shown that the amplitudes grow without limitations. The only possible stable state is reached if $\omega_1 = \omega_2 = \omega_0/2$ and if the unstable wave can generate a damped resonant harmonic ($\gamma_0 < 0$, $\gamma_1 > 0$) then, there is a degeneracy and the equilibrium is stable provided that $\gamma_0 + 2\gamma_1 < 0$. Numerical computations have proved that in such a case there was a rapid convergence towards the equilibrium state.

If now the frequency matching is such that $\Delta\omega$ cannot be neglected, the same kind of computations can be done. The results are more easily understood if we assume that the unstable wave decays into a damped subharmonic ($\gamma_0 > 0$, $\gamma_1 < 0$, $\omega_1 = \omega_2$). This case is unstable when $\Delta\omega$ is neglected. The equilibrium becomes stable if

$$2\gamma_1 < -\gamma_0[1+\sqrt{3}]; \quad \left(\frac{\Delta\omega}{2\gamma_1+\gamma_0}\right)^2 > \frac{\gamma_0^2+2\gamma_0\gamma_1+\gamma_1^2}{2\gamma_1^2+2\gamma_0\gamma_1-\gamma_0^2}$$

Then, stabilization is possible in this case provided that the subharmonic is damped enough and that $\Delta\omega$ is large enough. Nevertheless, if $\Delta\omega$ is too large there is no stabilization because in this case the equilibrium is given by:

$$a_0 = \frac{-\gamma_1}{V\sin\theta_0}; \quad a_1^2 = -\frac{\gamma_0\gamma_1}{V^2\sin^2\theta_0}; \quad \mathrm{tg}\,\theta_0 = \frac{2\gamma_1+\gamma_0}{\Delta\omega}$$

If $\Delta\omega \gg 2\gamma_1 + \gamma_0$, $\theta_0 \to 0$ and the amplitudes $a_0$, $a_1$ become infinite. Then coherent mode-mode coupling can be an efficient stabilization mechanism.

Let us assume now that we may apply the RPA. Once more, we need a mixture of damped and unstable wave since we immediately deduce from C(9):

$$\frac{\partial}{\partial t}\sum_{k} |\omega_k| N_k = \sum_{k} \gamma_k N_k |\omega_k|$$

To prove the possibility of stabilization, we assume that we may apply the RPA to three wave packets. We neglect the frequency and wave number spreading inside each wave packet and we assume that they are resonantly coupled with $k_0 = k_1 + k_2, \omega_0 = \omega_1 + \omega_2$, $\omega_1 > 0$, $\omega_2 > 0$, $\omega_3 > 0$, then:

$$\frac{dN_0}{dt} = 2\gamma_0 N_0 + V^2[N_1N_2 - N_0N_1 - N_0N_2]$$

$$\frac{dN_1}{dt} = 2\gamma_1 N_1 + V^2[N_2N_0 + N_0N_1 - N_1N_2]$$

$$\frac{dN_2}{dt} = 2\gamma_2 N_2 + V^2[N_1N_0 + N_0N_2 - N_1N_2]$$

The equilibrium is defined by:

$$N_0 = -\ 2\gamma_1\gamma_2/V^2(\gamma_0 + \gamma_1 + \gamma_2)$$

$$N_1 = 2\gamma_0\gamma_2/V^2(\gamma_0 + \gamma_1 + \gamma_2) \qquad \text{E(9)}$$

$$N_2 = 2\gamma_0\gamma_1/V^2(\gamma_0 + \gamma_1 + \gamma_2)$$

The amplitudes $N_0$, $N_1$, $N_2$ are of the order of $\gamma$ though in the coherent case they were of the order of $\gamma^2$. Then the RPA gives much larger saturation levels than the coherent wave interaction. It can be checked as in the preceding case that the equilibrium E(9) is stable and exists only if $\gamma_0 > 0$, $\gamma_1 < 0$, $\gamma_2 < 0$, $\gamma_0 + \gamma_1 + \gamma_2 < 0$. In the same way we may hope that non-linear Landau damping can be a stabilizing mechanism. In this case the equations of evolution for the wave amplitudes are given by:

$$\frac{dN_k}{dt} = 2\gamma_k N_k + \sum_{k'} W_{kk'} N_k N_{k'} \qquad \text{E(10)}$$

with $W_{k,k'} = -\ W_{k',k}$ if we assume that the waves are positive energy waves with large phase velocities. From the relationship

$$\frac{d}{dt}\sum_k N_k = 2\sum_k \gamma_k N_k$$

we deduce that we need once more damped waves. The equilibrium will be given by:

$$2\gamma_k + \sum_{k'} W_{kk'} N^0_{k'} = 0$$

If we look for the stability of this equilibrium, we set $N_k = N_k{}^0 + \delta N_k$ and we have

$$\frac{1}{N_k{}^0}\frac{d}{dt}\delta N_k = \sum_{k'} W_{k,k'}\,\delta N_{k'}$$

If we multiply this equation by $\delta N_k$ and sum over $k$ we find:

$$\frac{d}{dt}\sum_k \frac{1}{N_k{}^0}\left(\delta N_k{}^0\right)^2 = 0$$

Then, it appears that any departure from the equilibrium will not relax to the equilibrium. It means that the equilibrium state will not be reached and the solution of E(10) will be oscillating in time or unstable.*

ACKNOWLEDGEMENTS

We gratefully acknowledge the help from the scientists who attended the Ecole d'Eté des Houches. We want to thank also Mrs. Andalot from the Ecole Polytechnique and Miss Pophillat who efficiently took care of the manuscript.

REFERENCES

*Books*

DAVIDSON, R.C., *Methods in Non-linear Plasma Theory*, Academic Press, N.Y., 1972.

KADOMITSEV, B.B., *Plasma Turbulence*, Academic Press, N.Y., 1965.

SAGDEEV, R.Z. and GALEEV, A.A., *Non-linear Plasma Theory*, Benjamin, N.Y., 1969.

---

*This property of non-linear Landau damping has been pointed out to us by J. Coste, C. Montes, and J. Peyraud.

A very complete list of references will be found in R.C. Davidson's book. We give here references which are more closely related to the subject of the lectures.

*Chapter I*

AAMODT, R.E. and DRUMMOND, W.E., *Phys. Fluids, 7*, 1816 (1964).

AL'TSHUL, L.M. and KARPMAN, V.I., *J.E.T.P., 20*, 1043 (1965).

BERNSTEIN, I.B. and ENGELMANN, F., *Phys. Fluids, 9*, 937 (1966).

BURNS, T. and KNORR, G., *Phys. Fluids, 15*, 610 (1972).

DRUMMOND, W.E. and PINES, D., *Nucl. Fusion Suppl., Pt 3*, 1049 (1962).

GALEEV, A.A. and RUDAKOV, L.I., *J.E.T.P., 18*, 444 (1964).

KENNEL, C.F. and ENGELMANN, F., *Phys. Fluids, 9*, 2377 (1966).

KENNEL, C.F. and PETSCHEK, H.E., *J. Geophys. Res., 71*, 1 (1966).

KLOZENBERG, J.P. and BERNSTEIN, I.B., *J. Plasma Physics, 4*, 595 (1970).

MORSE, R.L. and NIELSON, C.W., *Phys. Fluids, 12*, 2418 (1969).

O"NEIL, *Phys. Fluids, 8*, 2255 (1965).

ROBERSON, C., GENTLE, K.W. and NIELSEN, P., *Phys. Rev. Letters, 26*, 226 (1971).

SHAPIRO, V.D. and SHEVCHENKO, V.I., *J.E.T.P., 18*, 1109 (1964).

VAHALA, G. and MONTGOMERY, D., *J. Plasma Physics, 4*, 677 (1970).

VEDENOV, A.A., VELIKHOV, E.P. and SAGDEEV, R.Z., *Nucl. Fusion, 1*, 82 (1961); *Nucl. Fusion Suppl., Pt. 2*, 465 (1962).

*Chapter II*

AKHIEZER, I.A., *J.E.T.P., 20*, 617 (1965); *20*, 1519 (1965).

COPPI, B., ROSENBLUTH, M.N. and SUDAN, R.N., *Ann. Phys., 55*, 207 (1969).

GALEEV, A.A., KARPMAN, V.I. and SAGDEEV, R.Z., *Nucl. Fusion, 1*, 20 (1965), AEC tr. 6669.

KADOMTSEV, B.B., *J.E.T.P., 16*, 1578 (1963).

KADOMTSEV, B.B., *J.E.T.P., 18*, 847 (1964).

KADOMTSEV, B.B., MIKHAILOVSKI, A.B. and TIMOFEEV, A.V., *J.E.T.P., 20*, 1517 (1965).

KADOMTSEV, B.B. and PETVIASHVILI, V.I., *J.E.T.P., 16*, 1578 (1963).

KOVRIZHNIKH, L.M. and TSYTOVITCH, V.N., *J.E.T.P., 20*, 978 (1965).

SILIN, V.P., *J.E.T.P., 21*, 1127 (1965).

SLOAN, M.L. and DRUMMOND, W.E., *Phys. Fluids, 13*, 2554 (1970).

# ATOMIC AND MOLECULAR PROCESSES IN IONIZED GASES

J.L. DELCROIX

*Laboratoire de Physique Des Plasmas,*
*91405 Orsay, France*

CONTENTS

# CHAPTER 1: ELASTIC AND INELASTIC COLLISIONS IN AN IONIZED GAS

## 1.1 *Introduction*

In its classical approach, plasma theory deals with a fully ionized gas in which the atomic and molecular processes can be neglected. Usually, it concentrates on a gas of electrons and protons which interact only through Coulomb forces. This pure plasma physics description is essentially correct for controlled fusion research and many astrophysical problems.

On the other hand, most of the laboratory plasmas are not fully ionized, and in this case the atomic and molecular processes may play an important role in the dynamics of the medium. This is even more pronounced in the study of weakly ionized gases that are encountered in such fields of research as gas discharges, ionospheric physics, plasma chemistry, ... The aim of this series of lectures is to provide some general information on atomic and molecular processes which may be of importance in such fields, and to deal with a few specific processes for which important progress has been made during recent years.

The interest will be focused not on the acquisition of atomic and molecular data which would include for example the discussion of quantum theory of atomic collisions [1] as well as many experimental techniques like crossing and merging beams [2] but rather on the evaluation of these data and of their significance for studies of ionized gases. For this work, there are two main difficulties; first the knowledge of these phenomena is changing quickly at the present time, and second, as is often the case in chemical physics, it is difficult to find general rules, or theories which apply to a large class of systems. On the contrary many gases have specific properties related to their particular chemical composition. In such a situation the need for a system of automatic information including a bank of basic data of gas physics is felt by many people and such systems have been started in some laboratories [3].

## 1.2 *Atomic and Molecular Species in an Ionized Gas*

The particles which may be present in an ionized gas are the following:

| | |
|---|---|
| electrons | e |
| atoms in the ground level | A |
| excited atoms | A* |
| positive and negative atomic ions | $A^+$, $A^-$ |
| molecules in the ground level | $A_2$ |
| rotationnally and vibrationnally excited molecules | $A_2{}^r$, $A_2{}^v$ |
| electronically excited molecules | $A_2{}^*$ |
| positive and negative molecular ions | $A_2{}^+$, $A_2{}^-$ |

If a is one of these species, the state of this component can be described by:

$f(\vec{r},\vec{w}_a,t)$ the distribution function of velocities $\vec{w}_a$ or by:
$n_a(\vec{r},t)$ the number density
$\vec{v}_a(\vec{r},t)$ the fluid velocity of species a
$T_a(r,t)$ the kinetic temperature

The evolution of these quantities is determined by applied external conditions (applied fields, injected fluxes, ...) and by collision phenomena.

## 1.3 *Collisions Between Particles*

1.3.1 *Binary and ternary collisions*. The collisions may involve two or three particles. The process occurring can be specified by writing it as a chemical reaction:

$P + Q \rightarrow S + T$ Binary collisions
$P + Q + R \rightarrow S + T + U$ Ternary collisions

Alternatively, these reactions can be written in a more condensed manner as:

P, Q/S, T
P, Q, R/S, T, U

1.3.2 *Binary elastic collisions*. An elastic collision is one in which the final products are identical to the initial. It means that in such a collision there has been no change in the internal state of the colliding particles but only an exchange of momentum and kinetic energy. The probability of elastic collisions between two

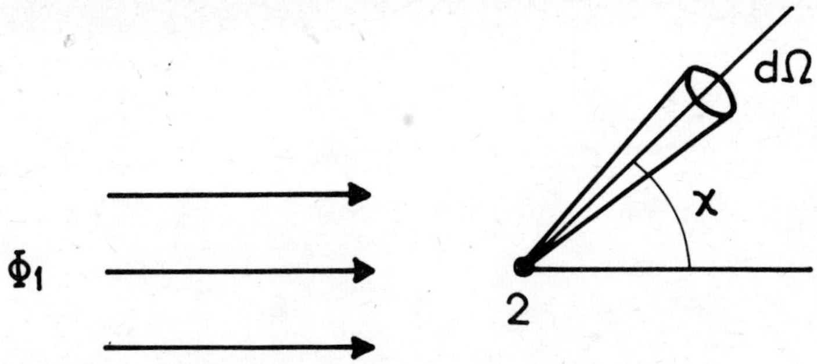


Fig. 1 Elastic differential cross-section

species of particles is described by the differential cross section $\sigma(\chi)$ (cm$^2$ · steradian $^{-1}$). This quantity is defined by the idealized experiment shown in Fig. 1, where a beam of particles carrying a flux $\Phi_1$ (cm$^{-2}$ sec$^{-1}$) of particles 1 irradiates a particle 2. The figure is drawn assuming that 2 is an infinite mass particle but the definition applies also to finite mass particles by considering the center of mass system (see for instance ref. [4] for a more complete discussion). Assuming that the interaction between particles 1 and 2 is described by a central potential, the space distribution of scattered particles depends only on the scattering angle $\chi$. If $dN_1$ is the number of particles scattered during time interval dt into a small solid angle $d\Omega$ around $\chi$, then by definition

$$\frac{dN_1}{dt} = \Phi_1 \sigma(\chi, g_{12}) d\Omega \tag{1}$$

where $g_{12}$ is the relative velocity of particles 1 and 2.

In gas physics one must often consider collisions at low (thermal) energies. If these energies are low enough, that the de Broglie wavelength of the particles is larger than the range of the interacting potential, then the scattering becomes isotropic:

$$\sigma(\chi, g_{12}) \rightarrow \sigma(g_{12})$$

Moreover, $\sigma(g_{12})$ usually converges to a finite limit when $g_{12} \rightarrow 0$ (at least for short range interactions which occur in electron-atom, electron-molecule, atom-atom and molecule-molecule collisions).

Starting from $\sigma(\chi)$ one may obtain, by integrating over $\chi$, a series of integrals called transport cross sections:

$$\sigma_o = \int \sigma(\chi)\, d\Omega \tag{2}$$

$$\sigma_1 = \int (1 - \cos\chi)\sigma(\chi)\, d\Omega \tag{3}$$

$$\sigma_n = \int (1 - P_n(\cos\chi)\sigma(\chi)\, d\Omega \tag{4}$$

$\sigma_0$, the so-called "total cross-section" is measured in scattering experiments using monoenergetic, collimated beams of particles. It does not enter into the hydrodynamic equations of the medium. $\sigma_1$, is the "momentum transfer cross-section" which appears in the hydrodynamic equations, where it serves to express the friction term between the components 1 and 2. It can be deduced from a set of measurement of the appropriate transport coefficients.

The detailed properties of $\sigma(\chi)$, $\sigma_0$, $\sigma_1$, ... can be deduced in a rather general way from the quantum theory of collisions. In its most standard form, this theory expands the incoming plane wave of particles into spherical waves [5]. These partial waves are characterized by an angular dependance expressed by a factor $P_\ell(\cos\chi)$. The partial wave of order $\ell$ suffers a phase shift $\delta_\ell$ when going through its focus at particle 2. By solving the Schrödinger equation of the problem, one gets the value of the phase shifts $\delta_\ell$ as a function of relative energy (or wave number k).

When the $\delta_\ell$ are known one obtains $\sigma(\chi)$, $\sigma_0$ and $\sigma_1$ through the following formulae:

$$\sigma(\chi) = \frac{1}{k^2} \left| \sum_{\ell=o}^{\infty} (2\ell + 1)\, e^{i\delta_\ell} \sin\delta_\ell\, P_\ell(\cos\chi) \right|^2 \tag{5}$$

$$\sigma_o = \frac{4\pi}{k^2} \sum_{\ell=o}^{\infty} (2\ell + 1) \sin^2 \delta_\ell \tag{6}$$

$$\sigma_1 = \frac{4}{k^2} \sum_{\ell=o}^{\infty} (\ell + 1) \sin^2 (\delta_\ell - \delta_{\ell+1}) \tag{7}$$

The behaviour of these quantities when the energy goes to zero ($k \to 0$) has been studied in detail in the simplest case of electron-atom collisions by many authors.

The most used formulae are the so-called modified effective range theory formulae obtained by O'Malley [6] in the following form:

$$\text{tang}\ \delta_0 = -Ak - \frac{\pi}{3a_0}\alpha k^2 - \frac{4}{3a_0}\alpha A k^3 \ln(ka_0) + O(k^3) \quad (8)$$

$$\text{tang}\ \delta_1 = \frac{\pi}{15a_0}\alpha k^2 - A_1 k^3 + O(k^4) \quad (9)$$

$$\text{tang}\ \delta_\ell = \pi\left[(2\ell+3)(2\ell+1)(2\ell-1)a_0\right]^{-1} \times \alpha k^2 + O(k^4) \quad \ell > 1 \quad (10)$$

from which one gets:

$$\sigma(\chi) = A^2 + \frac{\pi}{a_0}\alpha A k \sin\frac{\chi}{2} + \frac{8}{3a_0}\alpha A^2 k^2 \ln(ka_0) + O(k^2) \quad (11)$$

$$\sigma_0 = 4\pi\left[A^2 + \frac{\pi}{3a_0}\alpha A k + \frac{8}{3a_0}\alpha A^2 k^2 \ln(ka_0) + Bk^2 + \ldots\right] \quad (12)$$

$$\sigma_1 = 4\pi\left[A^2 + \frac{\pi}{3a_0}\alpha A k + \frac{8}{3a_0}\alpha A^2 k^2 \ln(ka_0) + Ck^2 + \ldots\right] \quad (13)$$

In these formulae [1] $a_0$ is the electron Bohr radius, and $\alpha$ the polarizability of the atom.** The constants A, B, C, $A_1$ may, in principle, be evaluated theoretically, but they are usually considered as parameters to be determined by experiment. A is the so-called scattering length. It has the meaning that

---

*It is claimed that these formulae are valid up to around 3 eV for He and 0.6 eV for Ar [6][7].

**Formulae (8) to (13) apply to any neutral atom not possessing a permanent quadrupole moment.

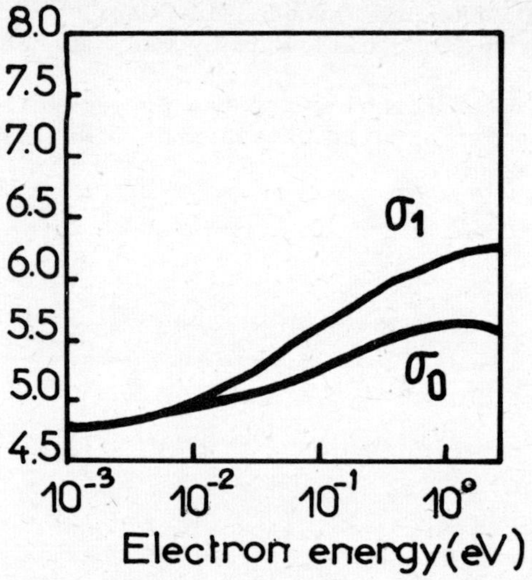


Fig. 2 Total cross section $\sigma_0$ and momentum transfer cross-section $\sigma_1$ for electrons in *He* (after Golden [8]).

$4\pi A^2$ is the limit towards which $\sigma_0$, $\sigma_1$ (and all the $\sigma_n$) tend when the energy of the electron goes to zero. Moreover, it can be shown that $\sigma_0$ and $\sigma_1$ remain very close to each other up to $10^{-1}$ eV about. As an example Fig. 2 shows compared values of $\sigma_0$ and $\sigma_1$ obtained by Golden [7] in He. In this work, the author has measured $\sigma_0$ in a beam experiment and deduced $\sigma_1$ from $\sigma_0$ by using the O'Malleys formulae.

1.3.3 *Binary inelastic collisions*. An inelastic binary collision is one in which at least one of the interacting particles changes its internal state. For example, the processes

$$e, N_2 \; / \; e, N_2^{V}$$

$$e, N_2 \; / \; e, e, N_2^{+}$$

$$e, N_2 \; / \; e, N, N$$

are inelastic collisions. It will be convenient to consider the new state of this particle (such as $N_2^v$, $N_2^+$ in the first two examples above) as a new chemical species. Then one can say that an inelastic collision is a process through which at least one species is transformed into one or several others.

For most of the inelastic reactions one does not know the probability that the new particles will be formed in a given configuration of velocities and energy. It means that one does not know the related differential cross-sections. Moreover, this information is not very useful for studying the dynamics of a reacting medium. For these reasons one usually is interested only in total cross-sections. For a given process say:

1, 2/3, 4, 5

the total cross-section $\sigma$ is defined by the experiment sketched on Fig. 3, where a total number $dN_1$ of particles 1 of

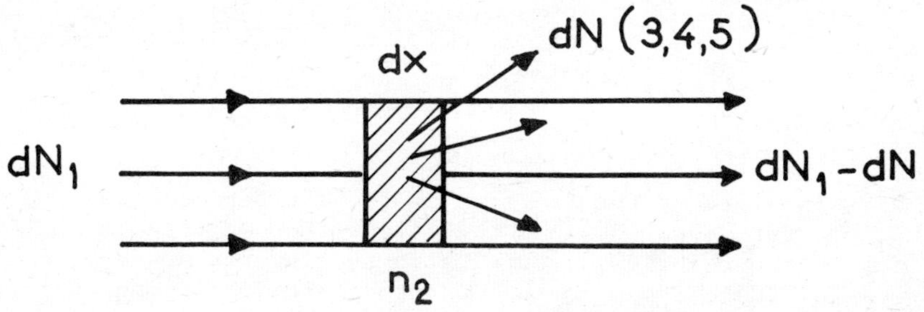


Fig. 3 Total inelastic cross section of process 1, 2/3, 4, 5.

given velocity $g_{12}$ traverses a target containing particles 2 with number density $n_2$. If the target thickness is dx and the number of reactions 1, 2/3, 4, 5 produced is $dN \ll dN_1$ (single collision condition) then by definition the cross-section $\sigma(g_{12})$ for this process is related to the probability of reaction $dN/dN_1$ by the relation:

$$\frac{dN}{dN_1} = n_2 \, \sigma(g_{12}) \, dx \tag{14}$$

For most of the important inelastic processes in gases these total inelastic cross-sections have been measured using beam techniques. Unfortunately, this information is not always available in the low (thermal) energy domain due to the difficulty of producing and detecting accurately enough, beams of slow particles.

If instead of a beam experiment one considers now a gas where the particles 1 and 2 have velocity distribution functions $f_1(\vec{w}_1)$ and $f_2(\vec{w}_2)$ then the rate of the reaction (number of reactions per $cm^3$ and sec) is related to the corresponding cross section by the relation:

$$\frac{dn}{dt} = \int \sigma(g_{12}) g_{12}\, f_1(\vec{w}_1) f_2(\vec{w}_2) \underline{dw_1}\ \underline{dw_2} \tag{15}$$

$$\text{with } g_{12} = [\overrightarrow{w_1 - w_2}] \tag{16}$$

or introducing the number densities of particles 1 and 2:

$$\frac{dn}{dt} = k\, n_1 n_2 \tag{17}$$

$$\text{with } k = \langle \sigma_{12}\, g_{12} \rangle = \frac{\int \langle \sigma_{12}\, g_{12} \rangle f_1 f_2 \underline{dw_1}\ \underline{dw_2}}{\int f_1 f_2 \underline{dw_1}\ \underline{dw_2}} \tag{18}$$

k is the rate of the reaction in the "forward" sense 1, 2 → 3, 4, 5. For given physical conditions the net rate is smaller due to the reactions occurring in the reverse sense 3, 4, 5 → 1, 2. In a gas in thermodynamic equilibrium the net rate is indeed zero because forward and backward ractions balance each other exactly.

The dimension of k is $cm^3\ sec^{-1}$. A typical order of magnitude for a reaction between atoms of molecules may be obtained by setting:

$$g_{12} = 3.10^4 \text{ cm sec}^{-1}$$

$$\sigma_{12} = 3.10^{-16} \text{ cm}^2$$

$$\text{whence } k \sim 10^{-11} \text{ cm}^3 \text{ sec}^{-1} \tag{19}$$

1.3.4 *Ternary inelastic collisions.* Even in gases at low pressures ternary inelastic collisions are often important. The most usual are the reactions:

P, Q, R/PQ, R'

where two of the particles P, and Q unite to produce a new compound PQ through an exoenergetic reaction.

Examples of this type are:

$He^+$, He, He/$He_2^+$, He molecular ion formation
$He^+$, e, e/He, e three body recombination i; e; e

These reactions would not occur if only the first two particles P, Q were present, because the compound would be formed in a highly excited PQ** state which would decay back into the initial P and Q particles. The third body R is said to stabilize the reaction. Before PQ** decays it reacts with R, PQ is stabilized in a lower energy state PQ*, and R' takes away some part of the reaction energy. If R and R' are electrons the quasi-totality of the reaction energy is transformed into kinetic energy of the electron. The "recoil" energy of PQ is small, due to the large mass ratio between PQ and the electron.

The microscopic mechanisms of ternary collisions are difficult to describe and not well known. One may try to define a generalised cross-section for a reaction

1, 2, 3/4, 5

For this let us consider a gas volume where the particles 1, 2, 3 have distribution functions $f_1(\vec{w}_1)$, $f_2(\vec{w}_2)$, $f_3(\vec{w}_3)$. We can write an equation similar to (15) for the total rate:

$$\frac{dn}{dt} = \int k(\vec{w}_1, \vec{w}_2, \vec{w}_3)\, f_1\, f_2\, f_3\, \underline{dw_1}\, \underline{dw_2}\, \underline{dw_3} \qquad (20)$$

$k(\vec{w}_1, \vec{w}_2, \vec{w}_3)$ is the differential rate due to particles of specified velocities $\vec{w}_1$, $\vec{w}_2$, $\vec{w}_3$. For binary collisions, the analogous quantity was written as the product $\sigma_{12}g_{12}$ of a cross-section by the relative velocity. In the case of ternary collisions one sees easily that $k(\vec{w}_1, \vec{w}_2, \vec{w}_3)$ has dimension $L^6\ T^{-1}$, but there is no unequivocal and generally accepted way of analysing this quantity. One can, however, use the scheme of the stabilisation by the third body to write this as

$$k(\vec{w}_1, \vec{w}_2, \vec{w}_3) = \sigma_{12}\, g_{12}\, V_{123} \qquad (21)$$

where $V_{123}$ is a volume of interaction of the transient compound with the third body.

Introducing the number densities $n_1$, $n_2$, $n_3$ one finally writes

$$\frac{dn}{dt} = k\, n_1\, n_2\, n_3 \tag{22}$$

where $k = \langle k(\vec{w}_1, \vec{w}_2, \vec{w}_3)\rangle$ (23)

A typical order of magnitude estimate of the rate k for a ternary reaction between atoms and molecules can again be obtained by putting

$$\sigma_{12}\, g_{12} = 10^{-11}\ cm^3\ sec^{-1}$$

$$V_{123} = (10^{-7})^3\ cm^3$$

whence $k \sim 10^{-32}\ cm^6\ sec^{-1}$ (24)

1.3.5 *Effect of elastic and inelastic collisions.* The effects produced by elastic and inelastic collisions in the dynamics of a medium are usually of different kinds:

-elastic collisions are the main mechanism of momentum transfer between particles. They are important for determining the fluid velocities $\vec{v}_a$ of the different species. Consequently they determine such transport coefficients as the electric mobility of charges particles, the diffusion coefficients of electrons, ions or excited species, ...

- inelastic collisions create and destroy the different chemical species.

They are important for determining the various numerical densities $n_a$. In an elastic collision there is a large exchange of kinetic and potential energy of the particles. Consequently, inelastic collisions play an important role in the balance of energy and the associated determination of the kinetic temperature $T_a$ of the different species.

## 1.4 *Main Collision Processes in an Ionized Gas*

The most important collision processes occuring in a monoatomic gas like helium, mercury vapour, ... are listed in Table I. The usual name of these processes and their significance in the dynamics of the gas are given in the second and third column of this table. In a diatomic gas ($H_2$, $N_2$, $O_2$, ...) one has a very similar list of processes with the few additional ones listed in Table 2 which are essentially related to the vibrational excitation and dissociation of molecules. Finally, some gases like $O_2$ can form negative ions. In these gases one has to consider the additional processes listed in Table 3.

TABLE 1. Collision Processes in A Monoatomic Gas

| Process | Name | Macroscopic effects |
|---|---|---|
| e, e/EL | Coulomb interaction | Maxwellisation of electrons |
| e, A/EL | Elastic electron-atom collision | Electron mobility |
| e, A/e, A* | Excitation by an electron | Metastables production |
| e, A/e, e, $A^+$ | Direct ionisation by an electron | Electron multiplication |
| e, A*/e, A | Deexcitation (quenching) | Destruction of metastables |
| e, A*/e, e, $A^+$ | Two-step ionization | Electron multiplication |
| e, $A^+$/EL | Coulomb interaction | Ambipolar diffusion |
| e, e, $A^+$/e, A* | Three-body e-e-i recombination | Strongly ionized plasmas decay |
| e, $A^+$/$h\nu$, A* | Radiative recombination | Strongly ionized plasmas decay |
| e, $A_2^+$/EL | Coulomb interaction | Ambipolar diffusion |
| e, $A_2^+$/A*, A | Dissociative recombination | Weakly ionized plasmas decay |
| e, e, $A_2^+$/e, $A_2$* | Three body e-e-i recombination | Weakly ionized plasmas decay |
| A, A/EL | Elastic atom-atom collision | Diffusion, viscosity, thermal conductivity of neutral gases |
| A*, A/EL | Elastic atom-metastable collision | Diffusion of metastables |
| A*, B/e, A, $B^+$ | Penning reaction | Ionisation in gas mixtures |

TABLE 1. Continued

| Process | Name | Macroscopic effects |
|---|---|---|
| $A^+$, A/EL | Elastic ion-atom collision | Ionic mobility |
| $A^+$, A/A, $A^+$ | Resonant charge transfer | Ionic mobility |
| $A^+$, A, $A/A_2^+$, A | Three body $A_2^+$ formation | Production of molecular ions |
| $A_2^+$, A/EL | Elastic ion-atom collision | Ionic mobility |
| A*, A*/e, $A^+$, A | "Crossed" Penning reaction | Matastables destruction |
| A*, A/e, $A_2^+$ | Hornbeck-Molnar process | Molecular ion formation |
| $A^+$, $A^+$/EL | Coulomb interaction | Maxwellisation of ions |
| $A_2^+$, $A/A^+$, A, A | Collisional destruction of $A_2^+$. | Destruction of $A_2^+$ |
| $A_2^+$, $A_2^+$/EL | Coulomb interaction | Maxwellisation of ions |

TABLE 2. Additional Processes in diatomic gases

| Process | Name | Macroscopic effects |
|---|---|---|
| e, $A_2$/e, $A_2^v$ | Vibrational excitation by electron | Vibrational temperature enhancement |
| e, $A_2$/e, A*, A | Dissociation by electron | Production of atoms |
| e, $A_2$/e, $A^+$, A | Dissociative ionization | Production of atomic ions |
| $A_2^v$, $A_2/A_2^{v'}$, $A_2$ | Transfer of vibration | Vibration energy relaxation |

TABLE 3. Some Additional Processes in electronegative gases

| Process | Name | Macroscopic effects |
|---|---|---|
| e, $A_2/A^-$, A | Dissociative attachment | Production of $A^-$ |
| e, $A_2$, $A_2/A_2^-$, $A_2$ | Three-body attachment | Production of $A_2^-$ |
| e, $A^-$/e, e, A | Detachment by electron collision | Destruction of $A^-$ |

TABLE 3. Continued

| Process | Name | Macroscopic effects |
|---|---|---|
| $A^-$, $A_2/A$, $A_2^-$ | Charge exchange | Transformation of $A^-$ into $A_2^-$ |
| $A^-$, $A_2$, $A_2/A_3^-$, $A_2$ | | Production of $A_3^-$ |
| $A_2^-$, $A_2^+/A_2$, $A_2$ | Ion-ion recombination | Destruction of $A_2^-$ |

## CHAPTER 2: ROTATIONAL AND VIBRATIONAL EXCITATION OF MOLECULES BY LOW ENERGY ELECTRONS

### 2.1 *Introduction*

In this chapter we deal with the two processes

$$e,\ A_2/e,\ A_2{}^{r} \tag{25}$$

$$e,\ A_2/e,\ A_2{}^{v} \tag{26}$$

where $A_2{}^{r}$ and $A_2{}^{v}$ are rotationally and vibrationally excited molecules. The importance of these phenomena in the description of motion of low energy electrons through molecular gases has been recognized for many years. It is essentially due to the rather low value of the quanta associated with these two modes of excitations of the molecules. The energy levels of the first rotational ($J = 1,2,3$; $v = o$) and vibrational levels ($J = 0$, $v = 1,2$) above ground state ($J = 0$, $v = 0$) are given in Table 4 for $H_2$, $N_2$ and $O_2$.

TABLE 4. Rotation and Vibration Energy Levels (eV) of $H_2$, $N_2$ and $O_2$

| | Levels | $H_2$ | $N_2$ | $O_2$ |
|---|---|---|---|---|
| | J = 1 | $15.10^{-3}$ | $0.50\ 10^{-3}$ | $0.36\ 10^{-3}$ |
| v = 0 | J = 2 | $44.10^{-3}$ | $1.50\ 10^{-3}$ | $1.07\ 10^{-3}$ |
| | J = 3 | $88.10^{-3}$ | $3.0\ 10^{-3}$ | $2.15\ 10^{-3}$ |
| J = 0 | v = 1 | 0.516 | 0.29 | 0.19 |
| | v = 2 | 1.002 | 0.57 | 0.38 |

To discuss the importance of processes (25) and (26) let us first recall the values of the thermal energies kT:

$$T = 300\ ^{o}K \qquad kT = 26 \times 10^{-3}\ eV$$

$$T = 77\ ^{o}K \qquad kT = 6.6 \times 10^{-3}\ eV$$

$$T = 20\ ^{o}K \qquad kT = 1.7 \times 10^{-3}\ eV$$

Comparing these figures with the values in Table 4, one sees that one has:

$$kT \ll E_v \text{ (for temperatures below 2000° K)} \tag{27}$$

$$kT \gg E_r \text{ (except for } H_2 \text{ at low temperatures)} \tag{28}$$

where $E_v$ and $E_r$ are the energies of the first vibrational and rotational levels.

The relation (27) means that most of the molecules of a gas in thermal equilibrium are in the ground vibrational level (v = 0) except for high temperatures whereas relation (28) shows that for all gases, except $H_2$ at low temperatures, the molecules in thermal equilibrium occupy a rather large number of rotational states.

Considering now the interactions of electrons and molecules in a weakly ionized gas the importance of processes (25) and (26) depends on the value of the thermal energy $kT_e$ of the electrons. When the electrons are in thermal equilibrium with the neutrals ($kT_e = kT_0$), it follows from (27) that they are unable to excite the vibrational levels of the molecules. But it follows from (28) that they can excite the rotational levels of the molecules. Except for $H_2$ at low temperatures, this situation is rather complex. The excitation collisions

$$e, A_2^{r}/e, A_2^{r'} \qquad J' > J$$

are balanced by the reverse collisions

$$e, A_2^{r'}/e, A_2^{r} \qquad J < J'$$

Now in many cases the electron temperature $T_e$ is larger than the neutral temperature $T_0$. In the simplest case this is due to the existence of a DC electric field E applied to the gas, and then it is well known that

$$T_e = f\left(T_o, \frac{E}{n_o}\right)$$

If $E/n_0$ is small enough then $T_e \sim T_0$ and the situation is essentially one of thermal equilibrium. The main interactions between electrons and ions are the elastic collisions. The rotational excitation and deexcitation are nearly balanced and there is essentially no vibrational excitation.

Then if $E/n_0$ increases from its equilibrium value, and the exchange of energy between electrons and molecules by rotation starts to be significant (excitation > deexcitation). When $kT_e$ becomes of the order of $E_v$ the excitation of vibrations starts. Usually the reverse process

$$e, A_2^v / e, A_2$$

is not important because the population of vibrationally excited molecules $A_2^v$ saturates at low relative values* (due to relaxation processes which will be discussed in next chapter). For higher values of $kT_e$ other more energetic processes (dissociation, electronic excitation, ionisation...) may occur, but we shall not consider this field in the present chapter.

## 2.2 *Theoretical Analysis of "Electron Swarms" in Weakly Ionised Gases*

2.2.1 *Transport coefficients of electrons.* The transport coefficients of electrons in a weakly ionized gas are essentially the electric mobility $\mu_e$ and the diffusion coefficient $D_e$. They are defined by the relation:

$$n_e \vec{v}_e = n_e \mu_e \vec{E} - D_e \vec{\nabla} n_e \qquad (29)$$

where $\vec{v}_e$ is the drift velocity of electrons and it is assumed for simplicity that $T_e$ does not vary spatially.

The measurements of $\mu_e$ and $D_e$ are usually made by the "electron swarm" technique. In this kind of experiments, electrons are emitted from a cathode into a region of uniform electric field filled with gas at an appreciable fraction of atmospheris pressure. The electrons quickly come to an equilibrium in which the rate of gain of energy from the electric field and the rate of loss of energy through collisions with the gas molecules are equal. One determines the electron drift velocity $v_e$ from the time required to cross the drift region.** One can also measure the $D_e/\mu_e$ ratio from the lateral spread of the electrons as they drift along the tube. As an example of such measurements Fig. 4 shows the results obtained in $H_2$ as a function of $E/n_0$ for two values of $T_0$ (77° K and 300° K).

---

*This is not true in molecular gas lasers where one aims precisely to obtain large concentrations of $A_2^v$ by passing large currents through the gas.

**This drift velocity is related to $\mu_e$ by $\vec{v}_e = \mu_e \vec{E}$ because the longitudinal gradient of $n_e$ is negligible.

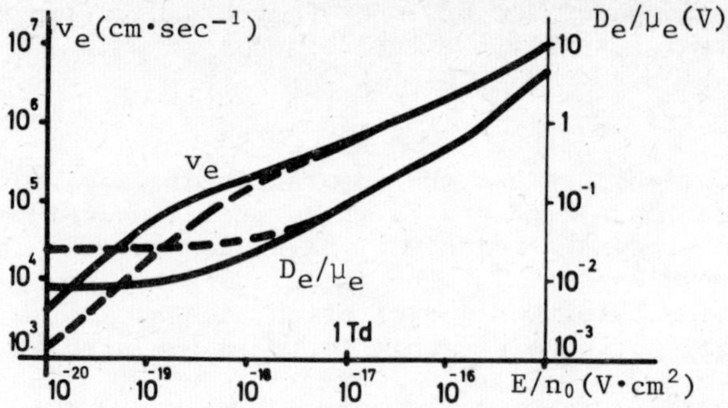


Fig. 4 Drift velocity and $D_e/\mu_e$ for $H_2$ at 77°K (——) and room temperature (– – –).

2.2.2 *Theory including only elastic collisions.* The theoretical analysis of these experiments is based on the kinetic theory of weakly ionized gases which has been developed by many authors [9][10]. The first approximation to this theory is made by assuming that the only interactions between electrons and molecules are elastic collisions described by a differential cross section $\sigma(\chi)$ defined as in chapter 1. This theory is based on the so-called Lorentz model which takes into account the two conditions:

$\frac{n_e}{n_o} \ll 1$ ($\rightarrow$ no interactions between electrons)

$\frac{m_e}{m_o} \ll 1$ ($\rightarrow$ exchange of energy between an electron and a molecule very small in an elastic collision)

Then if one considers the electric field and the gradients of $n_e$ as small perturbation terms one can write the distribution function $f_e(\vec{r},\vec{w}_e,t)$ of the electrons as

$$f_e(\vec{r},\vec{w}_e,t) = f_o(\vec{r},w_e,t) + f_1(\vec{r},\vec{w}_e,t) \tag{30}$$

where $f_0$ is a dominant term isotropic ($f_0$ is function only of the modulus $w_e$ of $\vec{w}_e$) and $f_1$ is a small anisotropy describing the macroscopic current produced by $\vec{E}$ and $\nabla\vec{n}_e$. Then by expanding $f_e$ in spherical harmonics in velocity space and neglecting second order terms one gets (cf. [10] pp. 00, 00 and 00).

$$\frac{\partial f_o}{\partial t} = J + D + K \tag{31}$$

with
$$J = \frac{1}{3}\frac{e^2}{m^2} E^2 \frac{1}{w^2}\frac{\partial}{\partial w}\left(\frac{w^2}{\nu_1}\frac{\partial f_o}{\partial w}\right) \tag{31a}$$

$$D = \frac{1}{3}\frac{w^2}{\nu_1}\nabla^2 f_o \tag{31b}$$

$$K = \frac{m}{m_o}\frac{1}{w^2}\frac{\partial}{\partial w}\left[\nu_1 w^3 \left(f_o + \frac{kT_o}{mw}\frac{\partial f_o}{\partial w}\right)\right] \tag{31c}$$

and
$$f_1 = -\frac{e}{m\nu_1}\frac{\vec{w}}{w}\frac{\partial f_o}{\partial w} \cdot \vec{E} - \frac{\vec{w}}{\nu_1} \cdot \vec{\nabla} f_o \tag{32}$$

with
$$\nu_1(w) = n_o \sigma_1(w) w \tag{33}$$

In all these formulae we have for simplicity written w and m instead of $w_e$ and $m_e$.

The three terms J, D, K in equation (31) represent respectively the heating of the electrons by the applied electric field (Joule effect), the so-called "diffusion cooling," and the cooling by elastic collisions with the molecules. For "average" electrons ($w = (2kT_e/m)^{1/2}$) the orders of magnitude of J, D, K in a system of characteristic length L ($L \sim$ diffusion length) are

$$J \sim \frac{1}{3}\frac{e^2}{m^2}\frac{E^2}{n_o \sigma_1 w^3} f_o \tag{34}$$

$$D \sim \frac{w}{3 n_o \sigma_1 L^2} f_o \tag{35}$$

$$K \sim \frac{m}{m_o} n_o \sigma_1 w \left(1 - \frac{T_o}{T_e}\right) f_o \tag{36}$$

Usually, the diffusion cooling D is negligible with respect to the collision cooling K; the condition for this is

$$n_o \sigma_1 L \left(1 - \frac{T_o}{T_e}\right)^{1/2} \gg \left(\frac{m_o}{3m}\right)^{1/2} \tag{37}$$

Then the isotropic part $f_0$ of the electron distribution is given in stationary state by:

$$J + K = 0 \tag{38}$$

This equation, where the momentum transfer cross section $\sigma_1(w)$ is a given function of w, must be solved numerically. But one can get an order of magnitude of the electron temperature $T_e$ by using the formulae (34) and (36); one gets:

$$\left(\frac{m_o}{12m}\right)^{1/2} \frac{eE}{n_o \bar{\sigma}_1} \sim kT_e \left(1 - \frac{T_o}{T_e}\right)^{1/2} \tag{39}$$

where $\bar{\sigma}_1 = \sigma_1(\sqrt{2kT_e/m})$.

Once $f_0$ is known, one gets $f_1$ from equation (32). The two terms in this equation represent the small anisotropies produced respectively by the electric field and the spatial gradient of $f_0$. Multiplying $f_1$ by $\vec{w}$ and integrating over velocity space one gets the resulting electron flux:

$$n\vec{v} = \frac{e}{m} \int_{w=o}^{\infty} f_o \, d\left[\frac{4\pi w^3}{3\nu_1}\right] \vec{E} - \vec{\nabla}\left[\frac{4\pi}{3} \int_{w=o}^{\infty} f_o \frac{w^4}{\nu_1} dw\right] \tag{40}$$

from which one gets:

$$\mu_e = \frac{e}{n_e m} \int_{w=o}^{\infty} f_o \, d\left[\frac{4\pi w^3}{3\nu_1}\right] \tag{41}$$

and assuming that only the number density $n_e$ contained in $f_0$ (but not the shape of $f_0$) is varying spatially

$$D_e = \frac{4\pi}{3} \int_{w=o}^{\infty} \frac{f_o}{n_e} \frac{w^4}{\nu_1} dw \tag{42}$$

Since the two quantities which are easy to measure are $\mu_e$ and $D_e/\mu_e$ it is convenient to write these quantities in the simple form:

$$\mu_e = \frac{e}{m \nu_m} \tag{43}$$

$$\frac{D_e}{\mu_e} = \frac{\varepsilon_k}{e} \tag{44}$$

where $\nu_m$ is the macroscopic collision frequency for momentum transfer. From (41) one finds that $\nu_m$ is related to the corresponding microscopic frequency $\nu_1(w)$ by:

$$\frac{1}{\nu_m} = \left\langle \frac{1}{w^2} \frac{d}{dw} \left(\frac{w^3}{3\nu_1}\right) \right\rangle \tag{45}$$

where the symbol <> represents an average over velocity space weighted by $f_0$.

$\varepsilon_k$ has the dimensions of an energy. It is of the same order of magnitude as the electron thermal energy $kT_e$. Thus, it is an experimentally significant measure of the average energy of the electrons and it has been called by Frost and Phelps [12] the characteristic energy. More precisely one gets from (41) and (42):

$$\varepsilon_k = \frac{\langle mw^2/\nu_1 \rangle}{\left\langle \frac{1}{w^2} \frac{d}{dw} \left(\frac{w^3}{\nu_1}\right) \right\rangle} \tag{46}$$

Finally the rate of exchange of energy by collisions between electrons and molecules can be obtained from (31c). Multiplying this equation by $1/2\ mw^2$, integrating over velocity space and dividing by $n_e$ one gets for the power loss per electron:

$$P_K = \nu_\varepsilon (\varepsilon - kT_o) \tag{47}$$

with $$\nu_\varepsilon = \frac{m}{m_o} \left\langle \frac{1}{w^2} \frac{d}{dw} (\nu_1 w^3) \right\rangle \tag{48}$$

$$\varepsilon = \frac{\langle mw^2 \nu_1 \rangle}{\left\langle \frac{1}{w^2} \frac{d}{dw} (w^3 \nu_1) \right\rangle} \tag{49}$$

$\nu_\varepsilon$ is a macroscopic collision frequency for energy exchange between electrons and neutrals; and $\varepsilon$ is a characteristic energy governing this exchange.

2.2.3 *Theory including inelastic collisions.*

a) *Boltzmann equation.* The cross sections for inelastic collisions of electrons with molecules are usually about two orders of magnitude smaller than the one for elastic collisions. Since the elastic collisions are very efficient for the exchange of momentum, one can neglect inelastic collisions in the calculation of the anisotropic part of the electron distribution function: equation (32) is practically unaffected by inelastic collisions.

On the other hand inelastic collisions may play an important role in the determination of the isotropic part $f_0$, because the elastic collision term K acting on $f_0$ contains the very small factor $m/m_0$. To take inelastic collisions into account the equation for $f_0$ is changed to:

$$\frac{\partial f_o}{\partial t} = J + D + K + K' \tag{50}$$

The inelastic term K' is expressed [10][12] by the formulae:

$$\frac{mw}{2} K' = \sum_{i<j} \frac{mw}{2} K'_{ij} = \sum_{i<j} (A_{ij} + B_{ji}) \tag{51}$$

$$A_{ij} = n_i \left[ - u\, \sigma_{ij}(u) f(u) + (u+u_{ij}) \sigma_{ij}(u+u_{ij}) f(u+u_{ij}) \right] \tag{52}$$

$$B_{ji} = n_j \left[ - u\, \sigma_{ji}(u) f(u) + (u-u_{ij}) \sigma_{ji}(u-u_{ij}) f(u-u_{ij}) \right] \tag{53}$$

In these formulae $u = 1/2\ m\ w^2$ is the energy of the group of electrons considered. i and j represent two states of the molecules with which the electrons are making inelastic collisions. $u_{ij}$ is the difference in energy between these two states with energy of state i assumed smaller than the one of state j. $\sigma_{ij}$ and $\sigma_{ji}$ are respectively the cross sections for the transition i → j (collisions of first kind) where the electrons loose energy, and for the inverse transition j → i (collisions of second kind, also called superelastic) where they gain energy. $n_i$ and $n_j$ are the number densities of molecules in states i and j.

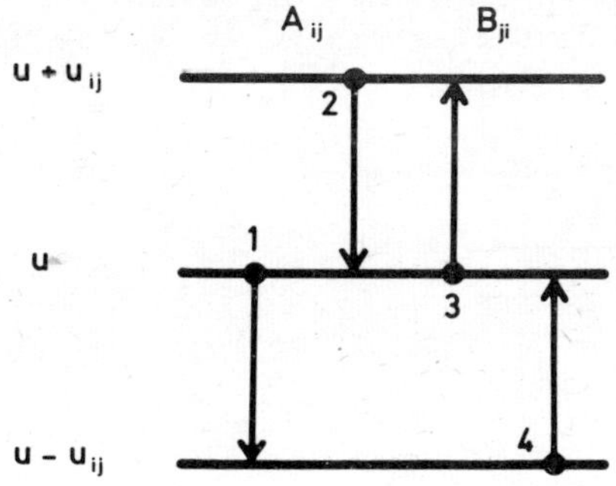


Fig. 5 The four classes of inelastic collisions connecting states *i* and *j* of molecules ($E_i < E_j$). The arrows represent change in energy of the electrons.

The significance of the four terms included in (52) and (53) is visualised on Fig. 5 where the arrows show the change in energy of the electrons. The terms 1 and 2 included in $A_{ij}$ represent collisions of first kind. Term 1 depopulates the group of electrons of energy u. Term 2 creates new electrons in this group coming from energy $u + u_{ij}$. Similarly terms 3 and 4 included in $B_{ji}$ represent superelastic collisions with again a depopulating term 3 and a source term 4 due to electrons of initial energy $u - u_{ij}$.

The cross sections $\sigma_{ij}$ and $\sigma_{ji}$ for two inverse processes are related by the Klein-Rosseland formula:

$$g_i\ u\ \sigma_{ij}(u) = g_j(u - u_{ij})\sigma_{ji}(u - u_{ij}) \tag{54}$$

where $g_i$ and $g_j$ are the statistical weights of states i and j. Using this formula and grouping now together the reverse terms, 1, 4 and 2, 3 we get:

$$\frac{mw}{2} K'_{ij} = - g_i \, u \, \sigma_{ij}(u) \left[\frac{n_i}{g_i} f(u) - \frac{n_j}{g_j} f(u - u_{ij})\right]$$

$$- g_j \, u \, \sigma_{ji}(u) \left[\frac{n_j}{g_j} f(u) - \frac{n_i}{g_i} f(u + u_{ij})\right] \tag{55}$$

One sees that to calculate the inelastic term K' one needs to know the number densities of various states i, j ... If one *assumes that the molecules are in thermal equilibrium* these densities are given by the usual Gibbs formula

$$n_i = n_o \frac{g_i \, e^{-E_i/kT_o}}{Z} \tag{56}$$

$$\text{with } Z = \sum_i g_i \, e^{-E_i/kT_o} \tag{57}$$

from which one gets:

$$\frac{mw}{2} \frac{Z}{n_o} K'_{ij} = - g_i \, u \, \sigma_{ij}(u) e^{-E_i/kT} \left[f(u) - e^{-u_{ij}/kT} f(u - u_{ij})\right]$$

$$- g_j \, u \, \sigma_{ji}(u) e^{-E_j/kT} \left[f(u) - e^{u_{ij}/kT} f(u + u_{ij})\right] \tag{58}$$

b) *Power loss of electrons by collisions*. One verifies on formula (58) that K' cancels when the distribution function f is a maxwellian of temperature $T_e = T_0$.

But, due to the action of the electric field, f has not this equilibrium expression and K' is non-zero. Then K' produces an energy exchange between electrons and molecules. The *power loss per electron* is:

$$P_{K'} = - \frac{1}{n_e} \int_o^{\infty} \frac{1}{2} mw^2 \, K' \, 4 \pi w^2 \, dw = \sum_{i<j} P_{ij} \tag{59}$$

From formulae (51), (52) and (53) one gets:

$$P_{ij} = (\nu_{ij} - \nu_{ji})\, u_{ij} \tag{60}$$

with $$\nu_{ij} = \frac{n_i}{g_i} a_{ij} \tag{61}$$

$$\nu_{ji} = \frac{n_j}{g_j} a_{ji} \tag{62}$$

$$a_{ij} = \frac{1}{n_e} \frac{8\pi}{m^2} \int_{u_{ij}}^{\infty} g_i\, u\, \sigma_{ij}(u) f(u) du \tag{63}$$

$$a_{ji} = \frac{1}{n_e} \frac{8\pi}{m^2} \int_{o}^{\infty} g_j\, u\, \sigma_{ji}(u) f(u) du \tag{64}$$

$\nu_{ij}$ and $\nu_{ji}$ are two collision frequencies for energy exchange associated with collisions of first and second kind respectively. By using Klein-Rosseland formula $a_{ji}$ can be re-written as:

$$a_{ji} = \frac{1}{n_e} \frac{8\pi}{m^2} \int_{u_{ij}}^{\infty} g_i\, u\, \sigma_{ij}(u) f(u - u_{ij}) du \tag{65}$$

Comparing (63) and (65) one sees that the expressions for $a_{ij}$ and $a_{ji}$ are very similar, the only difference being a translation in the electron distribution function used for computing the integral.

If one assumes that the molecules are in thermal equilibrium one gets for the power loss per electron due to transitions $i \overset{\leftarrow}{\rightarrow} j$:

$$P_{ij} = u_{ij}\, \nu_{ij} \left[1 - \frac{a_{ji}}{a_{ij}} e^{-u_{ij}/kT_o}\right] \tag{66}$$

The second term in the bracket expresses the ratio of power gain by superelastic collisions to power loss by collisions of first kind. The resulting effect may even be a net gain if $a_{ji}/A_{ij}$ is large enough. One may introduce for the transitions $i \leftrightarrows j$ an effective electron temperature $T_{eij}$ by writing

$$\frac{a_{ji}}{a_{ij}} = e^{u_{ij}/T_{eij}} \tag{67}$$

$$P_{ij} = u_{ij}\, \nu_{ij} \left[1 - e^{u_{ij}\left(\frac{1}{kT_{eij}} - \frac{1}{kT_o}\right)}\right] \tag{68}$$

The condition for having a net gain by transitions $i \leftrightarrows j$ is:

$$k\, T_{eij} < k\, T_o \tag{69}$$

Finally one can approximate these formulae by considering two classes of inelastic collisions:

- "weak" collisions for which $u_{ij} \ll kT_0$
- "strong" collisions for which $u_{ij} \gg kT_0$

As explained in the introduction excitation of rotations correspond usually to weak collisions while excitation of vibrations correspond to strong collisions.

- For *weak collisions* one may write:

$$e^{-u_{ij}/k\, T_o} = 1 - u_{ij}/k\, T_o \tag{70}$$

from which one gets

$$P_{ij} = \nu_{uij}\, (\varepsilon_{ij} - k\, T_o) \tag{71}$$

$$\text{with } \varepsilon_{ij} = u_{ij} \frac{e^{u_{ij}/kT_{eij}}}{e^{u_{ij}/kT_{eij}} - 1} \tag{72}$$

$$\nu_{uij} = \nu_{ij} \frac{u_{ij}}{kT_o} \left(e^{u_{ij}/kT_{eij}} - 1\right) \tag{73}$$

Formula (71) is very similar to the one obtained for elastic collisions. $\nu_{uij}$ is the effective frequency for energy exchange due to transitions $i \leftrightarrows j$ and $\varepsilon_{ij}$ the corresponding characteristic energy. Usually $T_{eij}$ is not much smaller than $T_0$ and one has:

$$u_{ij} << kT_e \tag{74}$$

from which one gets

$$\varepsilon_{ij} = kT_{eij} \tag{75}$$

$$\nu_{uij} = \nu_{ij} \left(\frac{u_{ij}}{kT_{eij}}\right)\left(\frac{u_{ij}}{kT_o}\right) \tag{76}$$

For *strong collisions* one may neglect the contribution of superelastic collisions* and write simply:

$$P_{ij} = \nu_{ij} u_{ij} \tag{77}$$

But one can also with no significant error write it as:

$$P_{ij} = \nu_{ij}(u_{ij} - kT_o) \tag{78}$$

Then by *adding the contributions of elastic, weak and strong inelastic collisions* one finally writes the power loss per electron as:

$$P = \nu_u(u_K - kT_o) \tag{79}$$

with: (80)

$$\nu_u = \nu_\varepsilon + \sum_{i<j}^{W} \nu_{uij} + \sum_{i<j}^{S} \nu_{ij}$$

$$u_K = \frac{1}{\nu_u}\left(\varepsilon\nu_\varepsilon + \sum_{i<j}^{W} \varepsilon_{ij}\, \nu_{uij} + \sum_{i<j}^{S} u_{ij}\, \nu_{ij}\right) \tag{81}$$

---

*This assumption is valid only if the molecules are in thermal equilibrium; it may be wrong in discharges where the applied electric power is large enough to produce large quantities of excited molecules.

where $\Sigma^W$ and $\Sigma^S$ mean summations over weak and strong collisions respectively. $\nu_u$ is the total effective frequency for energy exchange and $u_K$ the corresponding characteristic energy.

## 2.3 *Populations of Rotational States in* $H_2$ *and* $N_2$

Before discussing the application of the preceding theory to determination of cross sections for excitation of rotations and vibrations, it is necessary to recall the statistical properties of homonuclear diatomic molecules with non-zero nuclear spin such as $H_2$ and $N_2$.

It is well known (cf. [11] p. 139) that for these molecules the rotational states of even or odd J have different symmetries with respect to the exchange of nuclei, and that it may take months, if not years before a molecule goes from an even-numbered to an odd-numbered rotational level. Gases such as $H_2$ and $N_2$ may then be regarded as mixtures of two modifications: a symmetrical(s) with only even numbered rotational levels and an antisymmetrical(a) with only odd-numbered rotational levels. If the nuclear spin is I, the molecules can have a resultant nuclear spin T which takes the values:

$$T = 2I, 2I - 1, \dots 0 \tag{82}$$

It may be shown that for the symmetric levels only the even T values and for the antissymetric levels only the odd T values are possible. When the nuclear spin I = 1/2, as in spins (T = 0) and the antisymmetrical modification has parallel nuclear spins (T = 1). For a gas in thermal equilibrium at a temperature $kT_0 << E_v$ all the molecules are in the ground vibrational level; their distribution among rotational levels is given by Gibbs formula (56) with the statistical weights

$$g_J = (2J + 1)\, g_n \tag{83}$$

where $g_n$ is the factor due to nuclear spins. It is obtained by summing the statistical weights 2T + 1 of a configuration of resultant spin T over even values of T for a symmetric level and odd values of T for an antisymmetric. Depending on the values of I one finds in this way the values of $g_n$ shown in Table 5.

TABLE 5. Factor $g_n$ in the Statistical Weight of Rotational Levels

| J \ I | n | $n + \frac{1}{2}$ |
|---|---|---|
| even ( $s$) | (2I + 1) (I + 1) | (2I + 1) I |
| odd ( $a$) | (2I + 1) I | (2I + 1) (I + 1) |

The modification with the greater statistical weight is usually called the ortho modification and the other the para modification.

In $H_2$ the $a$ modification (J = 1, 3 ...) is ortho-hydrogen with $g_n = 3$, the $s$ modification (J = 0,2, ...) is para-hydrogen with $g_n = 1$. At high temperatures where many rotational levels are populated one sees easily that the thermal equilibrium mixture contains a ratio of ortho-$H^2$ to para-$H_2$ which is just the ratio 3/1 of the values of $g_n$. This is still approximately valid at 293° K as shown in first line of Table 6 which gives the populations of rotational levels. This mixture will be called normal-$H_2$. If it is cooled to 77° K without any exchange line of this table; in true thermal equilibrium one would have different populations as shown in third line. Finally by cooling normal-$H_2$ to very low temperature and catalysing by active charcoal the transformation ortho para one gets pure para-$H_2$; when this is warmed up to 77° K one gets the populations shown in the last line.

TABLE 6. Relative Populations (%) of Rotational Levels in $H_2$

| Experimental conditions | | J = 0 | J = 1 | J = 2 | J = 3 |
|---|---|---|---|---|---|
| normal $H_2$ | 293° K | 13.2 | 66.2 | 11.6 | 8.7 |
| normal $H_2$ | 77° K | 24.8 | 75.0 | 0 | - |
| thermal equlibrium | 77° K | 65.6 | 33.7 | 0.7 | - |
| para-$H_2$ | 77° K | 99.3 | - | 0.7 | - |

In $N_2$, the nuclear spin is I = 1; ortho-$N_2$ is the $s$ modification (J = 0, 2, ...) with $g_n = 6$; para-$N_2$ is the $a$ modification (J = 1, 3, ...) with $g_n = 3$. But due to the small value of the rotational quanta, there are some 20 rotational levels that must be taken into account even at liquid nitrogen temperature; consequently one is always dealing with normal-$N_2$ which is a mixture of ortho-$N_2$ and para-$N_2$ with a concentration ratio 2/1 given by the values of $g_n$.

## 2.4 *Determination of Cross-Sections from Swarm Analysis*

2.4.1 *Principle of the method.* The analysis of section 2.2 has been used as a basis to determine the momentum cross-section $\sigma_1$ and various inelastic cross sections $\sigma_{ij}$. The procedure is to assume a set of approximate cross sections based on any available information; to substitute these cross sections in the Boltzmann equation; and to solve this equation on a high speed digital computer for the distribution function $f_0$. This $f_0$ is then used together with $\sigma_1$ to calculate the transport coefficients given in (41), (42) or some combination of them. These theoretical results are then compared with values obtained from experiment. On the basis of discrepancies between these two set of values, the elastic and inelastic cross sections are adjusted and the procedure is repeated until one obtains agreement between calculated and experimental transport coefficients which is comparable with the accuracy of the experimental data.

Application of this method, to the determination of electron cross-sections for excitations of rotations and vibrations of molecules was developed by two groups. First Phelps and co-workers [12,13,14,18] analysed the method in detail and applied it to $H_2$ and $N_2$. More recently, Crompton and co-workers [15,17,21] discussed the accuracy and uniqueness of the method and obtained better results in $H_2$.

The interest of swarm technique for determining these types of cross-sections is obvious because the electron energies involved are very low of the order of 1 eV or less and the direct measurements of these cross sections by beam techniques are very difficult. But on the other hand the swarm technique can produce unique results only when the complexity of processes involved is not too great.

Lack of uniqueness is an inherent problem when, cross sections for a number of competing processes must be determined from relatively limited experimental data, i.e. data for two or three transport coefficients. To analyse this problem one must first remember that the electron energies in a swarm have a mean value which is a function of $E/n_0$. In calculating the transport integral, there is an effective range extending from zero to about twice this mean energy, within which nearly all the integral is contained.

If elastic scattering is the only process involved, then one may get the variation with energy of the momentum transfer cross section $\sigma_1$ from data of one transport coefficient like taken at a series of values of $E/n_0$.

The situation is not as favourable when more than one inelastic process is present to control the energy distribution of the swarm. If the two processes with cross-sections $\sigma_2(u)$, $\sigma_3(u)$ and thresholds $u_2$, $u_3$ are reasonably separate so that $u_2 << u_3$ then one still can get definite results in the following way:

- $\sigma_1(u)$ and $\sigma_2(u)$ can be determined in the range $0 < u < u_2/2$
- $\sigma_3(u)$ can be determined with relatively good accuracy above the threshold $u_3$ if extrapolations for $u \sim u_3$ of $\sigma_1(u)$ and $\sigma_2(u)$ are known even with relatively poor accuracy
- $\sigma_3(u)$ for $u >> u_3$ can be determined together with the momentum cross section $\sigma_1$.

The most delicate domain of energy is the threshold of the second inelastic process around which both inelastic processes may be of comparable importance. This analysis shows clearly that it is usually impossible to determine the cross sections for excitations of rotations because, as we have seen in preceding section many rotational levels are populated, and many processes are competing. Even in normal hydrogen at low temperatures the $J = 0$ and $J = 1$ are both involved and the two transitions $J = 0 \rightarrow 2$ and $J = 1 \rightarrow 3$ shown on Fig. 6 have not their thresholds enough separated to allow an accurate determination of the corresponding cross-sections.

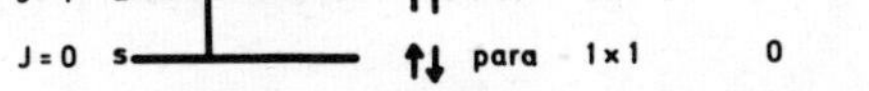


Fig. 6 *Rotational levels of hydrogen* (values of *J*, symmetries, nuclear spin-orientation ortho and parahydrogen, statistical weights, and energies).

The only simple case is the one of parahydrogen at low temperature. In this gas 99.5% of the molecules are in the v = 0, J = 0 ground state and the only two processes involved are the transitions J = 0 → 2 and the transition* v = 0 → 1 with well separated thresholds at 44 meV and 520 meV. Taking advantage of this situation Crompton and co-workers [15] were able to determine the cross section for the process J = 0 → 2. Then in a subsequent work on normal hydrogen Gibson [17] could determine the cross section for J = 1 → 3 using the known value J = 0 → 2.

2.4.2 *Fitting parameters*. To improve the accuracy of the unfolding method that we have described it is possible to make the adjustments not directly on the transport coefficients but on some combination of them chosen to be sensitive to one of the processes studied. For this purpose Phelps and co-workers introduced to effective frequencies $\nu_m$ and $\nu_u$ deduced from the experimental results by the formulae

$$\mu_e = \frac{e}{m\,\nu_m} \tag{84}$$

$$|e\,E\,v_e| = \nu_u(\varepsilon_K - kT_o) \tag{85}$$

The variations of $\nu_m$ and $\nu_u$ in normal $H_2$ at 77° K are shown on Fig. 7 as functions of $\varepsilon_K$. Also shown on this figure in dashed line is $\nu_\varepsilon$ the energy exchange frequency computed from elastic collisions only. It is clear on this diagram that $\nu_m$ is essentially governed by elastic collisions while $\nu_u$ is sensitive to inelastic collisions.

Crompton and co-workers [15] introduced instead of $\nu_m$ and $\nu_u$ two effective cross sections $\sigma_m$* and $\sigma_i$* defined by the formulae:

$$\sigma_m^* = A\,(\frac{E}{n_o})/v_e(\varepsilon_K/e)^{1/2} \tag{86}$$

---

*As we shall see further this last is indeed complex because it may involve simultaneous excitation of rotation in such processes as v = 0 → 1, J = 0 → 2.

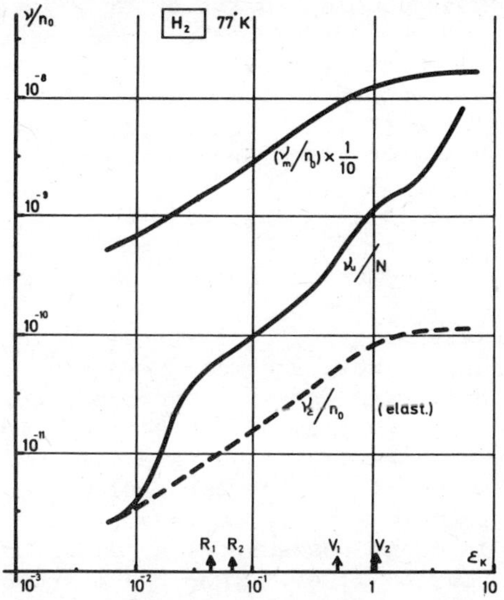


Fig. 7 Momentum transfer and energy exchange collision frequencies for $H_2$ at 77°K [11]. $R_1$, $R_2$, $V_1$, $V_2$ are the threshold energies of processes $J = 0 \rightarrow J = 2$; $J = 1 \rightarrow J = 3$; $v = 0 \rightarrow v = 1$; $v = 0 \rightarrow v = 2$.

$$\sigma_i^* = B \frac{v_e\, E/n_o}{(\varepsilon_K/e)^{1/2}} \left[1 - 2.47 \frac{\varepsilon_K - 1.07\, kT_o}{m_o v_e^2}\right] \tag{87}$$

where A and B are constants. They obtain these expressions by assuming that elastic collisions and one type of inelastic collisions are involved and by introducing the model of constant cross sections into the exact equations. The numerical constants include integrals over the energy distribution function. The authors show that they vary little with the shape of this function and they choose the value shown in the formulae. Then they claim that $\sigma_m$* and $\sigma_i$* which can be obtained from experiment as functions of E/n (or $\varepsilon_K$) provide two good fitting parameters closely related to the momentum transfer cross section $\sigma_1(u)$ and the inelastic cross section $\sigma_i(u)$.

2.4.3 *Conclusion on swarm analysis.* In spite of the difficulties mentioned above the swarm analysis is still a powerful tool to determine low energy electron-molecule cross sections. It is especially useful to determine threshold behaviour of vibrational cross-sections at least when reasonably accurate estimation of rotational excitation cross sections are available either from experiment or from theory.

But it may be considered the other way: if one assumes that all the involved cross sections are known the swarm analysis allows to calculate the various excitation rates and provides then a detailed description of the mechanisms through which the energy is transferred from the applied electric field via the electrons to the various degrees of freedom of the molecules. We shall give in section 2.6 some of the results obtained in this way.

### 2.5 *Determination of Cross-Sections by Beam Techniques*

Direct measurements of cross-sections for collisions between electron and molecules can be performed by shooting a beam of monoenergetic electrons across a collision chamber filled with gas at low pressure or across a molecular beam of low density. These techniques are difficult to apply to the vibrational and rotational excitation cross-sections because these phenomena occur at very low electronic energies and electrons in beams possess energy distributions which can frequently extend to $\sim 1$ eV. Then one must obtain in the collision region a narrower energy distribution and this can only be achieved by some method of velocity selection. The available methods are based on:

- retarding methods of analysis
- deflection in electric and magnetic fields
- time of flight analysis

These methods were discussed recently by Hasted ([2] chapter 3). Using these techniques the vibrational excitation cross-sections have been measured for the most usual gases [18]. Until recently there were no similar measurements for rotational i excitation. But there is now some recents measurements in hydrogen [14].

### 2.6 *Experimental Results in* $H_2$

2.6.1 *Rotational excitation.* The present situation is summarized on Figs. 8,9 and 10. Figure 8 shows for the transition ($J = 1 \rightarrow 3$) at moderate energies the comparison of cross-sections obtained by different authors. The values obtained by Ehrhard and Linder [13] by beam techniques are in fair agreement with the theoretical values of values of Henry and Lane [16].

The early values of Engelhardt and Phelps are now considered to be in error. They were obtained by using in the swarm analysis some early theories [19,20,21] which gave too small values for the cross section and not the correct

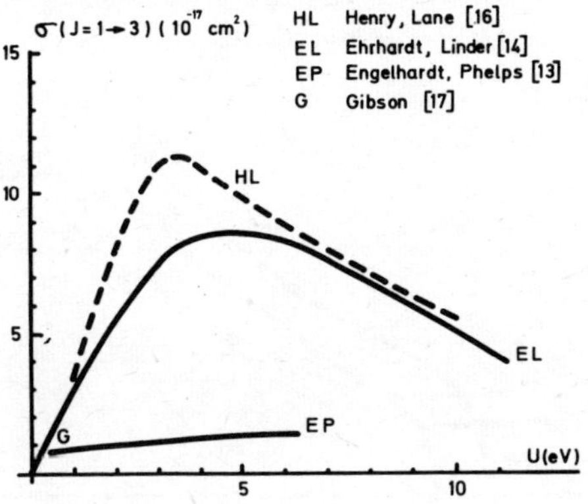


Fig. 8 Experimental (——) and theoretical (– – –) values of the corss section for the transition $J = 1 \rightarrow 3$ in hydrogen.

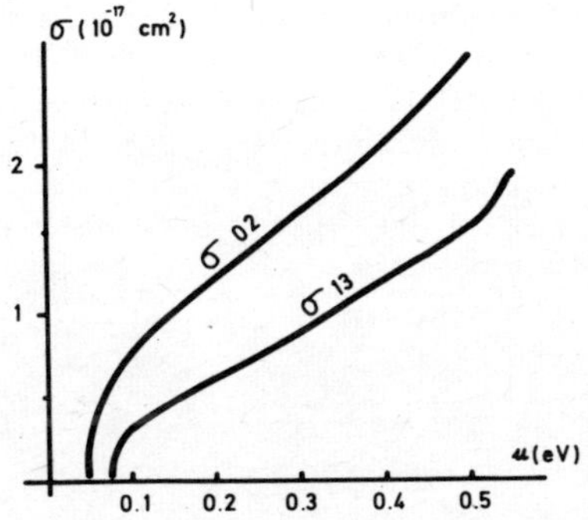


Fig. 9 Cross sections for the excitation of rotations in $H_2$ at low electron energy.
$\sigma_{02}$ transition $J = 0 \rightarrow 2$ after Crompton et al [15].
$\sigma_{13}$ transition $J = 1 \rightarrow 3$ after Gibson [17].

energy dependence. At low energies (u < 0.5 eV) the only values are those of Gibson obtained by swarm analysis [17]. Also shown on the same figure are the values for the transition J = 1 → 3 obtained by Crompton and co-workers [15]. Both are in very good agreement with the theoretical values of Henry and Lane.

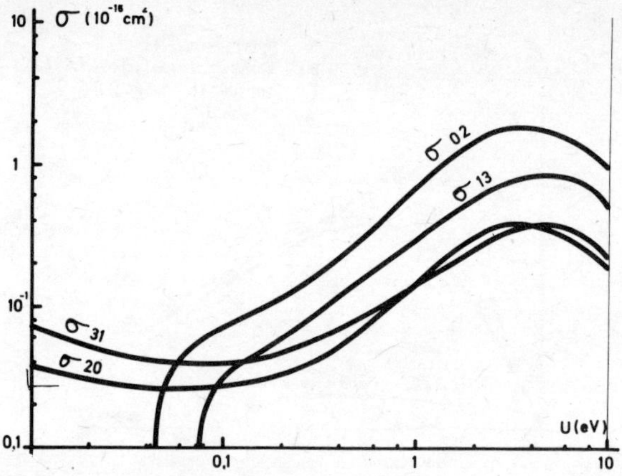


Fig. 10 Suggested cross sections for excitation and deexcitation of rotations in $H_2$. $\sigma_{02}$ $(J = 0 \to 2)$, $\sigma_{20}$ $(J = 2 \to 0)$, $\sigma_{13}$ $(J = 1 \to 3)$, $\sigma_{31}$ $(J = 3 \to 1)$.

Finally we have shown on Fig. 9 the cross-sections for the superelastic processes J = 2 → 0 and J = 3 → 1. They were computed with the Klein-Rosseland starting from the values of direct cross-sections shown on the same figure which are the best compromise between the works of Crompton et al. [15] Gibson [17] Ehrhardt et al. [14] and Henry and Lane [16].

For u > 5eV the four cross sections are related by the formulae

$$\sigma_{13} = \frac{3}{5}\sigma_{02} \tag{88}$$

$$\sigma_{20} = \frac{1}{5}\sigma_{02} \tag{89}$$

$$\sigma_{31} = \frac{3}{7}\sigma_{13} = \frac{9}{35}\sigma_{02} \tag{90}$$

The last two, result directly from the Klein-Rosseland formula which says that when $u \gg u_{ij}$ the cross sections $\sigma_{ji}$ and $\sigma_{ij}$ are in an inverse ratio with the statistical weights $g_j$ and $g_i$

Formula (88) has been found both experimentally and theoretically; it shows that the two excitation cross sections $\sigma_{02}$ and $\sigma_{13}$ are in an inverse ratio with the corresponding threshold energies. At intermediate energies (0.5 eV $< u <$ 5 eV) formulae (88) to (90) are a poorer approximation as seen on Fig. 9. Below 0.5 eV they become completely wrong.

The increasing of the superelastic cross-sections at very low energies is related to the very large slope of the excitation cross sections near the thresholds. If, as suggested by the results of Crompton and Gibson shown on Fig. 9 these slopes are infinite, then it results from the general properties of the Klein-Rosseland formula discussed in Appendix 2A that the superelastic cross-sections $\sigma_{20}$ and $\sigma_{31}$ tend to infinity for $u \to 0$.

### 2.6.2 *Vibrational excitation*

a) *Excitation of first vibrational level.* For the excitation of vibrations the situation is complicated by the possible simultaneous excitation of rotations. Let us for instance consider the excitation of one vibrational quantum. Since the only states populated in $H_2$ at moderate and low temperatures are $v = 0$, $J = 0,1$ one has to take into account the following processes with the related cross sections:

| | | |
|---|---|---|
| $v = 0 \to 1$ | $J = 0 \to 0$ | $\sigma_{v1,00}$ |
| $v = 0 \to 1$ | $J = 0 \to 2$ | $\sigma_{v1,02}$ |
| $v = 0 \to 1$ | $J = 1 \to 1$ | $\sigma_{v1,11}$ |
| $v = 0 \to 1$ | $J = 1 \to 3$ | $\sigma_{v1,13}$ |

In high resolution recent beam experiments one could in principle measure directly any of these elementary cross-sections. But usually one determines some combination of them.

The simplest case is again the one of *parahydrogen at low temperature* where in beam experiments one could determine the total $v = 0 \to 1$ cross section:

$$\sigma_{v,1}\,(\text{total}, J = 0) = \sigma_{v1,00} + \sigma_{v1,02} \tag{91}$$

While in the swarm analysis one introduces an effective cross section $\sigma_{v1}$ (swarm) for which Crompton and co-workers [24] have given the approximate expression

$$\sigma_{v1} \text{ (swarm, } J = 0) = \sigma_{v1,00} + 1.08\, \sigma_{v1,02} \tag{92}$$

The difference between the two expressions (91) and (92) is not significant within the present accuracy of experiments and theories.

In the case of *normal hydrogen* at moderate and low temperature one may write the total v = 0 → cross section per hydrogen molecule as:

$$\sigma_{v1} \text{ (total, n } H_2) = \frac{1}{4}\, \sigma_{v1} \text{ (total, } J = 0) + \frac{3}{4}\, \sigma_{v1} \text{(total, } J = 1) \tag{93}$$

with $\sigma_{v1}\text{(total, } J = 1) = \sigma_{v1,11} + \sigma_{v1,13}$ (94)

But it has been shown by Crompton and co-workers [24], that, between threshold and at least 1.5 eV one has:

$$\sigma_{v1}\text{(total, } J = 0) = \sigma_{v1}\text{(total, } J = 1) = \sigma_{v1} \tag{95}$$

This means that the total cross-section (93) for normal hydrogen has also the same value (95).

We arrive then at the conclusion that one can in principle define for the excitation of v = 0 → 1 transition a unique cross section $\sigma_{v1}$ which should be approximately independent of the way it is measured (beam or swarm technique), of the precise modification of $H_2$ used (para, normal) and of the gas temperature.

The experimental values of this total cross-section $\sigma_{v1}$ are shown on Fig. 11 for electron energies E > 1 eV. These results are in good agreement with previous measurements, but the situation near threshold is not at all clear as can be seen on Fig. 12.

Considering the partial cross sections the results obtained by Linder are shown on Fig. 11 reproduced from [25]; they show that one has approximately

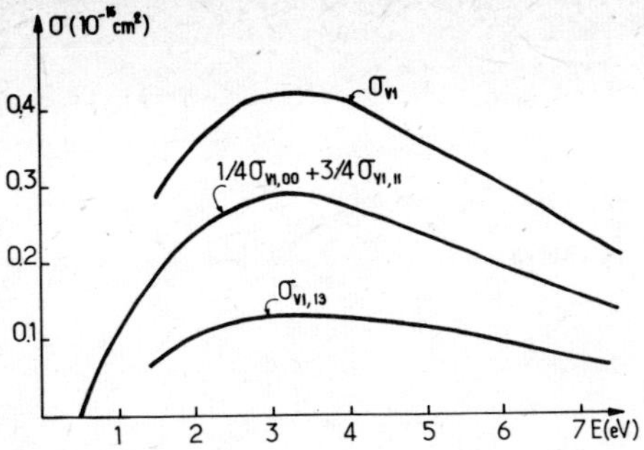


Fig. 11 Cross sections for excitation of $v = 0 \rightarrow 1$ transitions in $H_2$ ([25] after Linder).

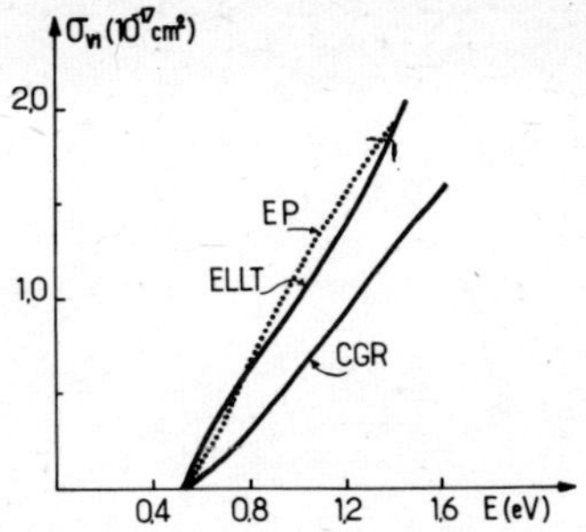


Fig. 12 Total cross section for $v = 0 \rightarrow 1$ excitation in $H_2$ near threshold (EP [27], ELLT [22], CGR [24]).

$$\sigma_{v1,13} \sim \frac{1}{3}\sigma_{v1} \tag{96}$$

from which one gets:

$$\sigma_{v1,11} \sim \frac{2}{3}\sigma_{v1} \tag{97}$$

On the other hand it has been shown theoretically [25, 26] that

$$\frac{\sigma_{v1,02}}{\sigma_{v1,13}} \sim \frac{5}{3} \tag{98}$$

from which one gets:

$$\sigma_{v1,02} \sim \frac{5}{9}\sigma_{v1} \tag{99}$$

$$\sigma_{v1,00} \sim \frac{4}{9}\sigma_{v1} \tag{100}$$

b) *Excitation of higher vibrational levels.* The excitation of two or three vibrational quanta by electron collisions have relatively small cross-sections as shown on Fig. 13 taken from reference [22]. The measurement was made

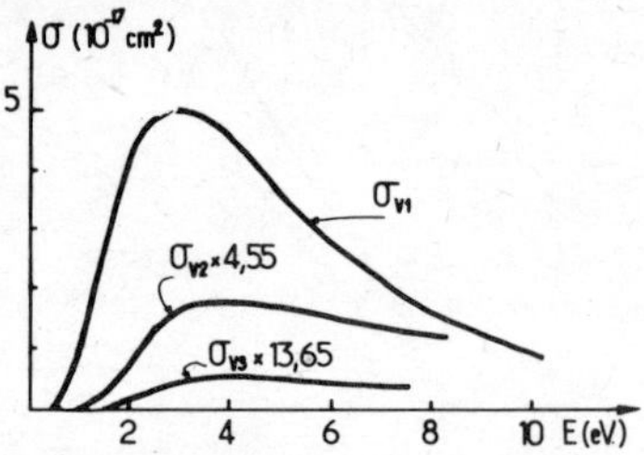


Fig. 13 Cross sections $\sigma_{v_1}, \sigma_{v_2}, v_3$ for excitation of one, two three vibrational quanta by electron collisions in $H_2$ [22].

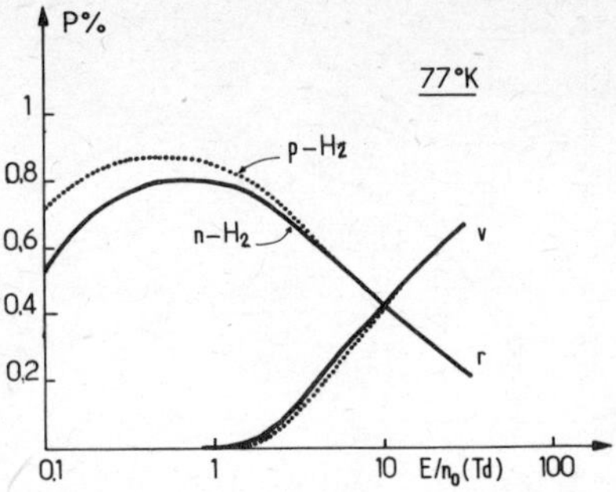


Fig. 14 Power absorbed by rotational and vibrational excitation in normal and para-hydrogen as a function of $E/n_0$ (Td). *1 Townsent (Td) = $10^{-17}$ V.cm²*.

for an electron scattering angle of 20 deg. but it was shown that the cross-sections for v = 0 → 2 and v = 0 → 3 excitations are approximately isotropics.

2.6.3 *Power absorbed by rotational and vibrational excitation.* An important by-product of swarm analysis is the calculation of the repartition of the absorbed electric power between different types of collision processes. The results obtained by Crompton and co-workers [24] are shown on Fig. 14. It is seen that the fraction of the power absorbed by rotation is practically the same for normal and parahydrogen as soon as $E/n_0$ is larger than a few townsends (1 townsend = $10^{-17}$ V cm$^2$); this result is related to the equality (95).

## 2.7 *Experimental and Theoretical Results in* $N_2$

2.7.1 *Rotational excitation.* As we have seen in section 2.3, in nitrogen one must take into account some 20 rotational levels even at liquid nitrogen temperature. Consequently it was impossible up to now to perform beam measurements of cross-sections for rotational excitation. The swarm analysis is also unable to derive the full set of cross-sections involved. It is then necessary to rely on theoretical calculations. Figures 15 and 16 show the results obtained by Geltman and Takayanagi [28] and Chen [29]. The values of Fig. 15 are probably overestimated for E > 0.2 eV, but the exact accuracy of these theoretical values is not well known at present time.

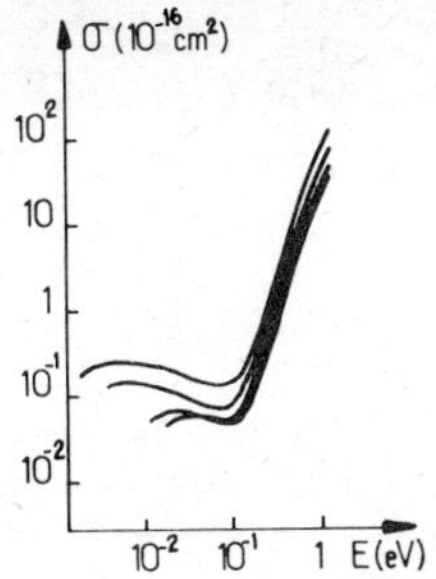


Fig. 15 Cross-sections for rotational excitation $(J \rightarrow J + 2)$ of $N_2$ [28].

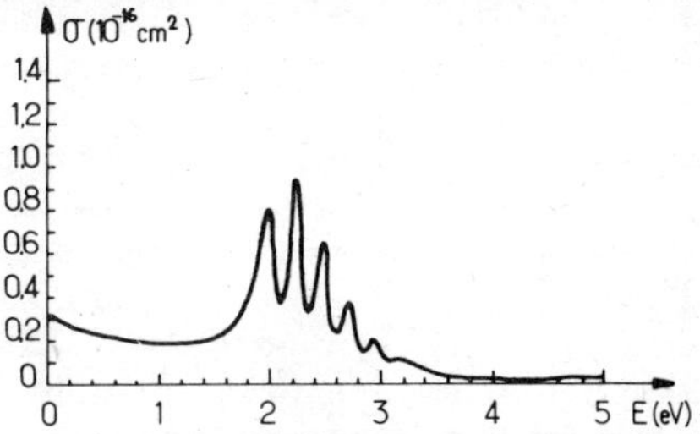


Fig. 16 Energy dependence of the cross sections for two-quantum rotational excitation of $N_2$ molecules by slow electron impact [29].

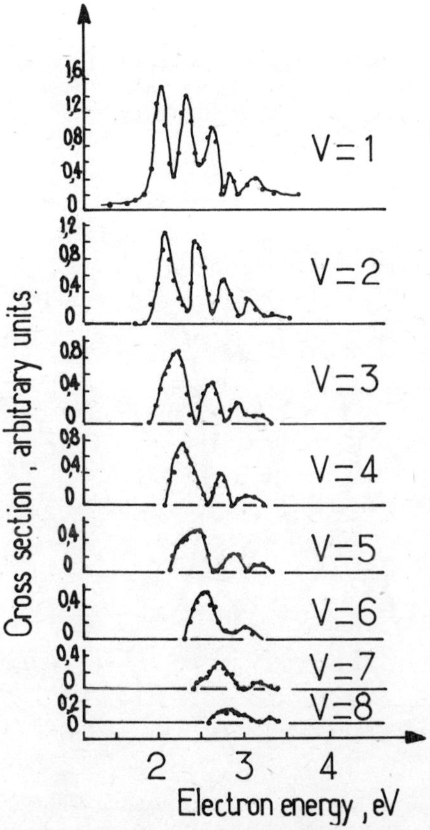


Fig. 17 Vibrational excitation of $N_2$ in the resonance region near 2 $eV$ as measured by Schulz [30]. The units of cross-section are approximately $10^{-16}$ cm$^2$.

2.7.2 *Vibrational excitation.* Figure 17 shows cross-sections for excitation of the first eight vibrational levels in $N_2$ as measured by Schulz using electron beam techniques [30]. Although the cross sections given are in arbitrary units, Schulz concludes that the units are approximately $10^{-16}$ cm$^2$. This means that the total vibration excitation cross section reaches approximately $5 \times 10^{-16}$ cm$^2$ at about 2.2 eV. This result is consistent with the results of analyses of transport data [31]. In addition to the very large cross sections shown by this data, each of the cross-section curves contains a great deal of structure. This structure and the large cross section have led to the postulate that this is a good example of a case in which the electron molecule collisions should be described in terms of the resonance picture that we shall discuss in the next section.

2.7.3 *Power absorbed by rotational and vibrational excitation.* The fractional power loss due to excitation of vibrations and rotations has been calculated by Engelhardt and co-workers [31]; their results are shown on Fig. 18.

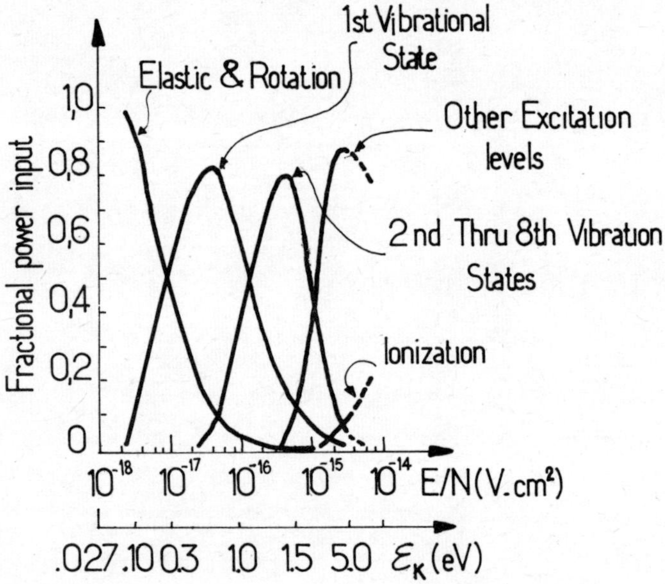


Fig. 18 Fractional power input to elastic and inelastic collisions for electrons in $N_2$ at 77°$K$ as a function of $E/N$. An ε K scale has been inserted under the $E/N$ scale for convenience [31].

## 2.8 *Theoretical Discussion of Excitation Mechanisms*

2.8.1 *Direct and resonance excitation.* The theory of electron scattering by diatomic molecules with excitation of rotations or vibrations has been developed along two different approaches which leads to distinguish two types of scattering.

In the direct scattering electron propagates mainly in the unexcited field of the target molecules, tolerating comparatively smaller distortion. The main mechanism of interaction are due to the permanent quadrupole [19] and to the (anisotropic) polarisability of the molecule [32]. The calculation may be done following Born approximation or a better distorted wave formalism.

In the resonance scattering, the projectile electron propagates in the excited field of the target molecule, suffering considerable distortion in forming compound states. In other words one may assume that the excitation takes place in two steps:

$$e + A_2 \rightarrow A_2^- \rightarrow e + A_2^* \tag{101}$$

In the first step a compound negative ion state is formed. But this negative ion has a short life time and decays by autoionization into an electron and an excited milecule $A_2$*.

The direct excitation mechanism results into a rather small cross-section which has a smooth variation with electron energy while the resonance mechanism produces a large enhancement of the cross section in an energy band determined by the energy of formation of $A_2^-$ compound state. It then appears indeed on the cross sections curves as a broad resonance. This resonance has a fine structure due to some interference effects that we shall describe in the next section.

2.8.2 *Resonance excitation of vibration in* $H_2$, $N_2$, $O_2$ The resonance mechanism has been analysed by many authors. We analyse here the recent paper of Birtwistle and Herzenberg [33] which studies in detail the excitation of vibration in $N_2$ through the process:

$$e + N_2 \rightarrow N_2^- \, (^2\Pi_g) \rightarrow e + N_2^v$$

Their method can be explained with the help of the diagram of Fig. 19 which shows the potential curves of the ground state ($X^1\Sigma_g^+$) of $N_2$ and of the $N_2^-$ ($^2\Pi_g$) state. This compound state is characterized by a complex energy.

$$E(R) = V(R) - \frac{1}{2} i \, \Gamma(R) \tag{102}$$

R being the distance between the two nitrogen nuclei, the real part V(R) is the electronic energy of $N_2^-$. The complex part Γ(R) characterizes the probability of autoionisation with the usual relation

$$\Gamma \, \tau = \hbar \tag{103}$$

where τ is the life time of $N_2^-$ state. V(R) and Γ(R) have been determined in reference [33] and independently from ab initio calculations by Krauss and Mies [34]. The results of Birtwistle and Herzenberg are shown on Fig. 19 for V(R) and Fig. 20 for Γ(R). The curves V(R) has a minimum corresponding to an equilibrium distance $R_0^-$ of $N_2^-$ slightly larger than the corresponding value $R_0$ in $N_2$ ground state. The vibrational levels associated with $N_2^-$ correspond to an energy

$$\hbar\omega_v^- \sim 0.24 \text{ eV} \tag{104}$$

but they have no true physical meaning because as shown in Fig. 20 one has

$$\Gamma \gtrsim \hbar\omega_v$$

which means that the autoionization time is shorter than the vibration period.

The $N_2^-(^2\Pi_g)$ state is the results of the addition of an extra electron in a $\Pi_g$ orbital, to the ground state $X^1 \Sigma_g^+$ of $N_2$. When this electron is nea by the nuclei one can consequently write the wave function of the system as a product of an electronic and nuclear wave function:

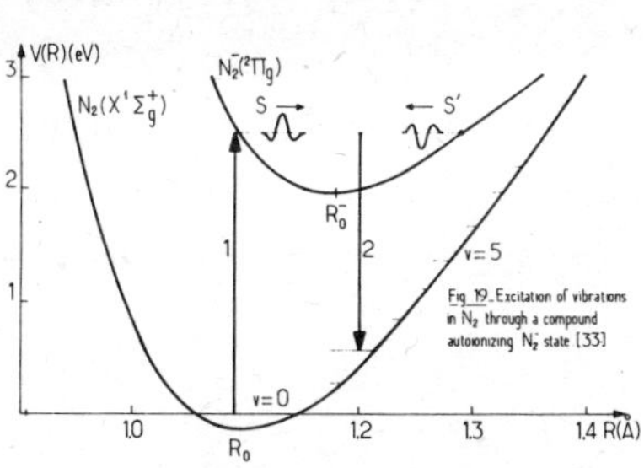


Fig. 19 Excitation of vibrations in $N_2$ through a compound autoionizing $N_2^-$ state [33].

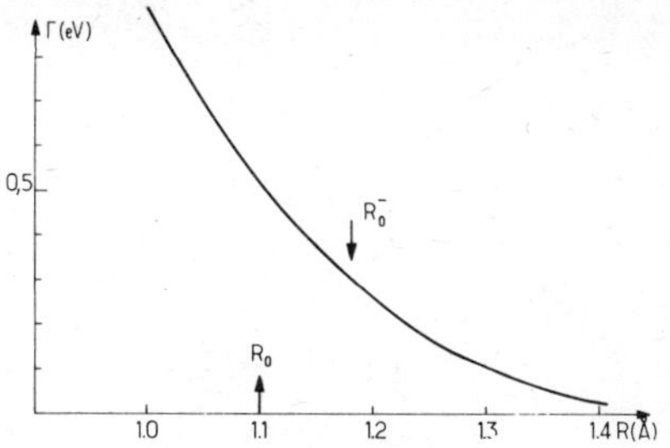


Fig. 20 Autoionization of $N_2^-(^2TT_g)$ state [33].

$$\psi = \left[\Sigma_{m=\pm 1} (4\pi)^{1/2}\, Y^{*}_{2m}(\vec{k}_0)\, \psi_m(q, \vec{r}, \vec{R})\right] \xi(R, \varepsilon) \qquad (105)$$

where $\vec{k}_0$ and $\varepsilon$ are the initial wave vector and energy of the electron, $\vec{R}$ the vector joining the two nuclei, $\vec{r}$ the vector position of the extra electron and q the others electrons coordinates. The first bracket is the electronic part of the wave function, and $\xi$ is the nuclear part. There are two coupled equations governing those two parts.

Now the resonance mechanism is analysed as shown in Fig. 19. Starting from $N_2$ in the ground state with v = 0 the electron induces a transition to the $N_2^-$ state. This produces in the equation for the nuclear wave function of the system a source term around $R = R_0$; an outgoing nuclear wave S propagates from this distance toward larger values of R; it is eventually reflected on the other side of the potential curve; the resulting S' wave then propagates toward $R = R_0$. But along this pro propagation the waves S and S' are strongly damped by the autoionization process. In $N_2$ this damping is strong enough for the reflected wave S' to be almost annihilated when arriving back at $R = R_0$.

The cross section for the excitation of a given vibrational level of $N_2$ is then obtained by an integration over the autoionization probability taking into account the nuclear wave function (S + S'). The interferences between S and S' are responsible for the fine structure of the cross-sections. To perform the detailed calculations Birtwistle and Herzenberg assume given functional variations of V(R) (Morse potential) and Γ(R) (ascribing the variations of Γ with R entirely to the variation of the penetrability of the centrifugal barrier). The functions V(R) and Γ(R) are then completely determined by

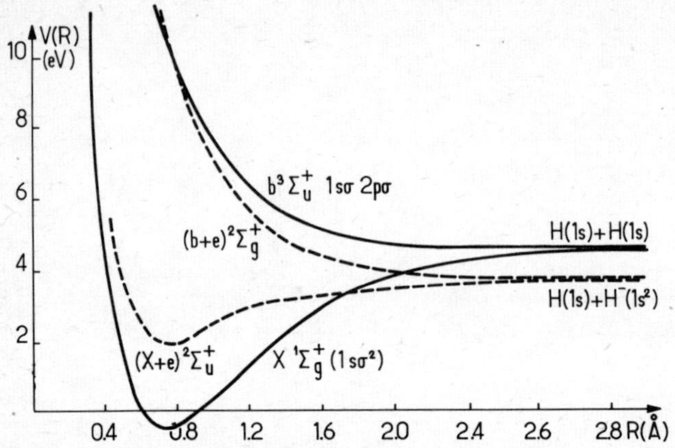


Fig. 21 Lower states of $H_2$ and $H_2^-$ [35].

a few parameters. The values of those are adjusted to give a good fit to the shape of the cross-section curves for the excitation of the levels v = 1; 3; 5; 7. Then one finds that these parameters give also very good agreement for the other levels excitation. This theory could still be improved by (i) calculating the distribution of final rotational states (ii) including direct scattering (iii) including exchange.

It is interesting now to compare the situation in the three diatomic gases $H_2$, $N_2$, $O_2$. The data on negative ion compound state are shown in Fig. 21 for $H_2$ [35], Fig. 22 for $O_2$ [36] and Table 7 for the three gases.

In $H_2$ the negative ion state hardly exists because $\Gamma$ is not only much larger than $h\omega_v$ but even of the same order as the energy of formation of the state. Consequently there is no reflected wave S' and only a strongly damped outgoing wave S. The resonance which appears in the cross-sections for vibration excitation is just a broad maximum with no fine structure. Also as $R_0$ and $R_0^-$ are about equal the cross section for v > 1 are very small. Finally one may question whether the resonance description of equation (101) is a useful concept and we shall see indeed in next section that for $H_2$ it is possible to make a direct calculation giving the proper results.

In $O_2$ the situation is more complicated because the negative ion $O_2^-$ is a stable state, and the phenomena were

TABLE 7. Properties of Neutral Molecule Ground State and Negative ion Compound State in $H_2$, $N_2$, $O_2$.

| | $H_2$ | $N_2$ | $O_2$ |
|---|---|---|---|
| *Ground state* $A_2$ | | | |
| Structure | $^1\Sigma_g^+$ | $^1\Sigma_g^+$ | $^3\Sigma_g^-$ |
| $R_0$ (Å) | 1.3 | 1.1 | 1.2 |
| $h\omega_v$(eV) | 0.52 | 0.29 | 0.19 |
| *Compound state* $A_2^-$ | | | |
| Structure | $^2\Sigma_u^+$ | $^2\Pi_g$ | $^2\Pi_g$ |
| $R_0^-$ (Å) | 1.3 [35] | 1.2 [33] | 1.4 [36] |
| $E^-(R_0^-)-E(R_0)$ | 2 [35] | 2 [33] | -0.43[36] |
| $h\omega_v^-$(eV) | ? | 0.24[33] | 0.11[36] |
| $\Gamma(R_0)$(eV) | $\sim$ 2 [ ] | 0.52[33] | $\sim 10^{-3}$[36] |

not well understood until the recent work of Schulz and coworkers [36]. The electron affinity of $O_2$ is positive ($E(R_0) - E^-(R_0^-) = 0.43$ eV) but the potential curve of $O_2^-$ crosses the one of $O_2$ (point K on Fig. 22). On the right of the crossing point, $O_2^-$ is stable. On the left in can autoionize, but the life time for autoionisation is much longer than in $N_2$ and $H_2$. The value of $\Gamma$ is not well known but much smaller than $h\omega_v^-$. This means that the vibration states of $O_2^-$ do exist and that, in the resonance process similar to the one of Fig. 19, the S wave can reflect many times before being damped by autoionization. Consequently it has been shown both theoretically [33] and experimentally [37] that the cross section for excitation of vibration has sharp maxima. Each of these maxima corresponds to the excitation of a given vibrational state of $O_2^-$, and indeed the distance between two maxima is about 0.11 eV = $h\omega_v^-$.

2.8.3 *Phase shift analysis of* e, $H_2$ *collisions*. A theory which do not postulate at the beginning the formation of a compound state has been developed for e, $H_2$ collisions by Henry and Chang [25]. It shows however as we shall see that the incoming electron is trapped for some time in a $p\sigma$ orbital which corresponds to the compound $(1s\sigma_g^2\ 2p\sigma_u)\,^2\Sigma_u^+ H_2^-$ state.

To make this study one first need to choose the proper interaction potential between the incoming electron and the $H_2$

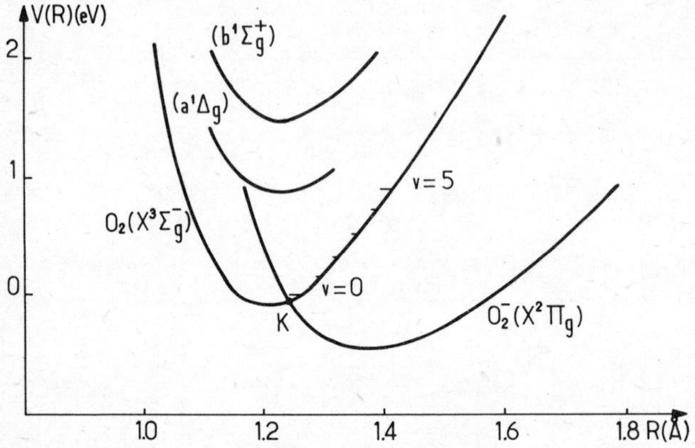


Fig. 22 Lower states of $O_2$ and $O_2^-$ [36].

molecule. Henry and Chang take ac ount of a short range direct and exchange potential, and of a long range potential including a polarization and an electron-quadrupole interaction. A cutoff radius $r_c$ is introduced in the long range potential because for distances shorter than $r_c$ the classical electrostatic description of the interaction breaks down.

Now, to analyse the wave function of the e, $H_2$ system one may use two different sets of base functions: the laboratory set where the state of the molecule is defined by the two quantum numbers J and v for rotation and vibration and the molecular frame set where one considers the molecule as non rotating and having a parametrically fixed internuclear distance R, the electron state being specified by the quantum numbers $\ell$ and $\lambda$ ($\lambda$ is the projection of $\ell$ along the internuclear axis). When the electron is far away from the molecule the electron-molecule interaction is weaker than the rotational and vibrational Hamiltonians and the system is conveniently described by the quantum numbers J and v (laboratory frame). On the other hand, when the electron is close in, the system is more appropriately described by the quantum numbers $\lambda$ and R (molecular frame). For $H_2$ this short distance region extends to something between 5 $a_0$ and 10 $a_0$. Now it can be shown that for electrons of energy larger than 1 eV most of the accumulation in the phase shift arises from within 5 $a_0$. Thus the molecular frame is the appropriate one for calculations of phase shifts. Another way to look at it is to say that

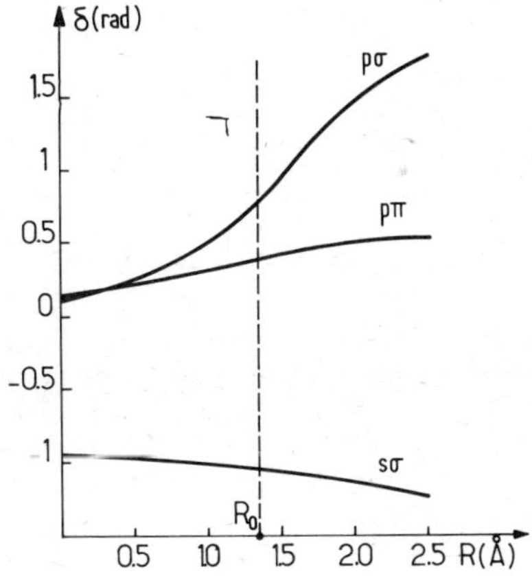


Fig. 23 Phase shifts of the three orbitals $s\sigma$, $p\sigma$ and $p\pi$ for an electron energy of 4.5 *eV* ($R_0$ *= equilibrium distance in* $H_2)^-$

the time of interaction at short distance is much shorter than the rotation period of the molecule

$(\omega_R v_e/5 a_0 << 1$ or using the compound state description $\Gamma >> h\omega_R)$

Following these remarks the procedure to solve the e, $H_2$ problem at energies between 1 eV and 10 eV is first to compute the electron phase shifts in the molecular frame (this is a relatively simple problem) and then to compute the cross sections by using a frame transformation.

For the first step Henry and Chang take into account the three orbitals s $\sigma$, p $\sigma$ and p$\pi$. The corresponding phase shifts are shown as a function of R on Fig. 23 for an electron energy of 4.5 eV. Considering now the cross sections one has to distinguish between elastic collisions and vibrational excitation. For elastic collisions it is shown that the cross sections are determined by the values of $\delta(R_0)$ and the contribution of one partial orbital would vanish is $\delta(R_0) = k\pi$. Looking at Fig. 23 it is concluded that the three orbitals s$\sigma$, p$\sigma$ and p$\pi$ contribute to the elastic cross section. On the other hand, for the excitation of vibration it is shown that the cross section is determined by the values of $\delta(R_0)$ and $\frac{d\delta}{dR}\Big|_{R=R_0}$ and would vanish if $\frac{d\delta}{dR}\Big|_{R=R_0} = 0$. Looking at Fig. 23

it is seen that the p$\sigma$ curve has a larger slope than the two others and one concludes that this orbital, gives the main contribution to the vibrational excitation. As we have yet said the large influence of this p$\sigma$ orbital is equivalent to the formation of $(^2\Sigma_u{}^+)$ $H_2{}^-$ state. But the present formulation is clearly better in $H_2$ due to the short autoionization life time of this state.

Finally it has been shown by Henry and Chang that this theory give very good agreement with the experimental values quoted in section 2.6 both for the pure vibrational excitation and the simultaneous vibrational and rotational excitation.

## REFERENCES

*Chapter 1*

1. MASSEY, H.S.W. and BURHOP, E.H.S., *Electronic and Ionic Impact Phenomena*, 2nd edition, London: Oxford University Press, 1969.

2. HASTED, J.B., *Physics of Atomic Collisions*, 2nd edition, London: Butterworth, 1972.

3. DELCROIX, J.L., *Rapport Interne L.P. 140*, Laboratoire de Physique des Plasmas, Orsay, 1972.

4. DELCROIX, J.L., *Physique des Plasmas, Vol. 1*, Monographie Dunod (Paris 1963), English Trans., Wiley: New York, 1965.

5. MESSIAH, A., *Mecanique Quantique, Tome 1*, Chapter 10, Dunod Paris, 1960.

6. O'MALLEY, T.F., *Phys. Rev., 130*, 1020 (1963).

7. GOLDEN, D.E., *Phys. Rev., 151*, 48 (1966).

*Chapter 2*

8. KIHARA, T., *Rev. Mod. Phys., 24*, 45 (1952); *25*, 844 (1953).

9. ALLIS, W.P., *Handbuch der Physik, Vol. 21*, p. 383-444, Springer: Berlin, 1956.

10. DELCROIX, J.L., *Physique des Plasmas, Vol. 2*, Monographie Dunod. (Paris 1966), English Trans., Wiley: New York, 1967.

11. HERZBERG, G., *Spectra of Diatomic Molecules*, 2nd edition, Van Nostrand, Princeton, 1950.

12. FROST, L.S. and PHELPS, A.V., *Phys. Rev.*, *127*, 1621 (1962).

13. ENGELHARDT, A.G. and PHELPS, A.V., *Phys. Rev.*, *131*, 2115 (1963).

14. EHRHARDT, H. and LINDER, F., *Phys. Rev. Letters*, *21*, 419 (1968).

15. CROMPTON, R.W., GIBSON, D.K., and Mc INTOSH, A.I., *Austral. J. Phys.*, *22*, 715 (1969).

16. HENRY, R.J.W. and LANE, N.F., *Phys. Rev.*, *183*, 221 (1969).

17. GIBSON, D.K., *Austral. J. Phys.*, *23*, 683 (1970).

18. PHELPS, A.V., *Rev. Mod. Phys.*, *40*, 399 (1968).

19. GERJUOY, E. and STEIN, S., *Phys. Rev.*, *97*, 1671 (1955); *98*, 1848 (1955).

20. SAMPSON, D.H. and MJOLNESS, R.C., *Phys. Rev.*, *140*, A1466 (1965).

21. GELTMAN, S. and TAKAYANAGI, K., *Phys. Rev.*, *143*, 25 (1966).

22. EHRHARDT, H., LANGHANS, L., LINDER, F. and TAYLOR, H.S., *Phys. Rev.*, *173*, 222 (1968).

23. LINDER, F., *Int. Conf. Phys. Elec. Atom. Coll.*, Cambridge, 1969, p. 141.

24. CROMPTON, R.W., GIBSON, D.K. and ROBERTSON, A.G., *Phys. Rev.*, *A2*, 1386 (1970).

25. HENRY, R.J.W. and CHANG, E.S., *Phys. Rev.*, *A5*, 276 (1972).

26. ABRAM, R.A. and HERZENBERG, A., *Int. Conf. Phys. Elec. Atom. Coll.*, Cambridge, 1969, p. 100.

27. ENGELHARDT, A.G., PHELPS, A.V. and RISK, C.G., *Phys. Rev.*, *135*, A1566 (1964).

28. GELTMAN, S. and TAKAYANAGI, K., *Phys. Rev.*, *143*, 25 (1966).

29. CHEN, J.C.Y., *Phys. Rev., 146*, 61 (1966).

30. SCHULZ, G.J., *Phys. Rev., 135A*, 988 (1964).

31. ENGELHARDT, A.G., PHELPS, A.V. and RISK, C.G., *Phys. Rev., 135*, A1566 (1964).

32. DALGARNO, A. and MOFFETT, R.J., *Proc. Nat. Acad. Sc. India, A33*, 511 (1963).

33. BIRTWISTLE, D.T. and HERZENBERG, A., *J. Phys., B4*, 53 (1971).

34. KRAUSS, M. and MIES, F.H., *Phys. Rev., A1*, 1592 (1970).

35. SHARP, T.E., *Atomic Data, 2*, 119 (1971).

36. BONESS, M.J.W. and SCHULZ, G.J., *Phys. Rev., A2*, 2182 (1970).

37. SPENCE, D. and SCHULZ, G.J. SCHULZ, *Phys. Rev., A2*, 1802 (1970).

# TOPICS ON PLASMA RESPONSE FUNCTIONS*

G. KALMAN

*Department of Physics, Boston College, Chestnut Hill, Massachusetts 02167, U.S.A.*

---

*Partial support from Contract F 19628-71-C-0079 in the preparation of these notes is gratefully acknowledged.

*Mutato nomine de te*
*fabula narratur*

HORATIUS

CONTENTS

## I. PLASMA RESPONSE FUNCTIONS

The properties of a plasma can conveniently be described in terms of "response functions" which describe the response of the system to an externally imposed driving force. The response functions, apart from their obvious property of describing how the plasma behaves under the influence of the perturbation, contain a great deal of additional information. Most of this information stems from the relation between the response functions and the two-, three-, etc. point density and current correlation functions of the *unperturbed* system, known as the Fluctuation Dissipation Theorem [1,2].

### I-1. *Definitions*

We will consider an infinite homogeneous stationary classical plasma. Consider further an external electric (or magnetic) field (i.e., generated by external sources) imposed on the plasma. Such a field will give rise to charge separations and currents and, therefore, the actual field will be different from the external one. We adopt the notation*

| | |
|---|---|
| External field | $\hat{\mathbf{E}}, \hat{\mathbf{B}}$ |
| Plasma field | $\boldsymbol{E}, \boldsymbol{B}$ |
| Total field | $\mathbf{E}, \mathbf{B}$ |

Obviously

$$\mathbf{E} = \hat{\mathbf{E}} + \boldsymbol{E}\,, \qquad \mathbf{B} = \hat{\mathbf{B}} + \boldsymbol{B} \tag{1}$$

In general the plasma current j can be expressed as a function of either the external or the actual field. In the sequel we will mostly consider electric perturbation only

---

*To make contact with the language of conventional electrodynamics [3] the following correspondence can be made:

$$\mathbf{E} \rightarrow \mathbf{E},\ \hat{\mathbf{E}} \rightarrow \mathbf{D},\ \boldsymbol{E} \rightarrow -4\pi\mathbf{P}$$

$$\mathbf{B} \rightarrow \mathbf{B},\ \hat{\mathbf{B}} \rightarrow \mathbf{H},\ \boldsymbol{B} \rightarrow +4\pi\mathrm{M}$$

(a static magnetic field must be included in description of the unperturbed state; a time dependent magnetic field is not independent of the accompanying electric field). Then, provided the field is "weak" enough to permit an expansion, one can represent $\underset{\sim}{j}$ as

$$j(\hat{E}) = \underset{1}{\hat{\sigma}}\hat{E} + \underset{2}{\hat{\sigma}}\hat{E}\hat{E} + \ldots$$

or (2)

$$j(E) = \underset{1}{\sigma}E + \underset{2}{\sigma}EE + \ldots$$

in a rather symbolic notation (since the $\sigma$-s are, in general, operators). The $\hat{\sigma}$-s will be referred to as *external* conductivities while $\sigma$-s are the *ordinary* conductivities. In a similar fashion the (external and ordinary) polarizabilities can be introduced*

$$\mathcal{E}(\hat{E}) = -\underset{1}{\hat{\alpha}}\hat{E} - \underset{2}{\hat{\alpha}}\hat{E}\hat{E} = -\underset{1}{\alpha}E - \underset{2}{\alpha}EE$$

$$\mathcal{B}(\hat{B}) = -\underset{1}{\hat{\xi}}\hat{B} - \underset{2}{\hat{\xi}}\hat{B}\hat{B} = -\underset{1}{\xi}B - \underset{2}{\xi}BB \qquad (3)$$

which lead to the definitions of the dielectric and diamagnetic response functions, $\varepsilon$ and $\nu$

$$\varepsilon = 1 + \alpha$$

$$\nu = 1 + \xi \qquad (4)$$

It should be noted that although these definitions are simple, they are not operationally very useful because $\alpha$ and $\xi$ appear as functions of the unknown E and B. For the time being we concentrate on the linear terms in the expansion, ignoring higher order responses. The response functions are non local operations in space and time, but because of the space and time translation invariance, they depend on the difference coordinates only. Therefore a Fourier transformation will carry them into multiplicative functions [3]. Thus we are led to consider

$$\underset{\approx}{\hat{\sigma}}(\underset{\sim}{k}\omega)\ ,\ \underset{\approx}{\sigma}(\underset{\sim}{k}\omega)\ ,\ \underset{\approx}{\alpha}(\underset{\sim}{k}\omega)\ ,\ \underset{\approx}{\xi}(\underset{\sim}{k}\omega)$$

---

*See, however, Eq. (47), and (8) and (14) for more precise definitions.

where we now emphasize the three dimensional tensor character of each of these functions.

There are simple relations between these response functions. First we note that it follows from the Maxwell equations, written down* first with external sources ($\hat{j}$, $\rho$) and then with the external and plasma sources ($\underset{\sim}{j} + \hat{\underset{\sim}{j}}, \rho + \hat{\rho}$)

$$(\underset{\sim}{k} \times \underset{\approx}{\nu} \times \underset{\sim}{k})\cdot\underset{\sim}{E} + \frac{\omega^2}{c^2}\underset{\approx}{\varepsilon}\cdot\underset{\sim}{E} = -4\pi i\frac{\omega}{c^2}\hat{\underset{\sim}{j}} \qquad (5)$$

$$\underset{\sim}{k}\cdot\underset{\approx}{\varepsilon}\cdot\underset{\sim}{E} = -4\pi i\hat{\rho}$$

$$\underset{\sim}{k} \times \underset{\sim}{k} \times \underset{\sim}{E} + \frac{\omega^2}{c^2}\underset{\sim}{E} = -4\pi i\frac{\omega}{c^2}(\underset{\sim}{j}+\hat{\underset{\sim}{j}}) \qquad (6)$$

$$\underset{\sim}{k}\cdot\underset{\sim}{E} = -4\pi i\ (\rho+\hat{\rho})$$

implies that

$$(\underset{\sim}{k} \times \underset{\approx}{\xi} \times \underset{\sim}{k})\cdot\underset{\sim}{E} + \frac{\omega^2}{c^2}\underset{\approx}{\alpha}\cdot\underset{\sim}{E} = +4\pi i\frac{\omega}{c^2}\underset{\sim}{j}$$

$$\underset{\sim}{k}\cdot\underset{\approx}{\alpha}\cdot\underset{\sim}{E} = +4\pi i\rho \qquad (7)$$

which can be regarded as a more operational definition of $\underset{\approx}{\alpha}$ and $\underset{\approx}{\xi}$. However, it is obvious that the equations (7) are not sufficient to determine *both* $\underset{\approx}{\xi}$ and $\underset{\approx}{\alpha}$. Indeed, it is redundant to use both of them for the description of a classical plasma. (For a quantum plasma, the situation is somewhat different.) It is a matter of convenience to adopt some subsidiary condition between $\underset{\approx}{\xi}$ and $\underset{\approx}{\alpha}$: the scheme usually followed in plasma physics is to set $\underset{\approx}{\xi} = 0$. This of course does not mean that the plasma does not possess magnetic properties; merely that the responsibility of describing the magnetic properties is shifted to the transverse element of $\underset{\approx}{\alpha}$. However, it should be kept in mind that one always can revert to the more physical diamagnetic polarizability by

---

* $(\underset{\sim}{k} \times \underset{\approx}{\nu} \times \underset{\sim}{k})_{mn} = \varepsilon_{mrs}k_r\,\nu_{su}\,\varepsilon_{nuv}\,k_v$

$$\underset{\approx}{\xi} = \frac{c^2}{\omega^2}\underset{\sim}{k} \times \underset{\approx}{\alpha} \times \underset{\sim}{k}$$

With this choice, we introduce:

$$\underset{\approx}{\Delta}_\circ = n^2\underset{\approx}{T} - \underset{\approx}{1}$$
$$\underset{\approx}{\Delta} = \underset{\approx}{\Delta}_\circ - \underset{\approx}{\alpha} \tag{8}$$

$$n \equiv \frac{kc}{\omega} \qquad \text{(index of refraction)}$$

$$\underset{\approx}{T} \equiv \underset{\approx}{1} - \frac{\underset{\sim}{k}\underset{\sim}{k}}{k^2} \qquad \text{(transverse projection tensor)}$$

and thus:

$$\underset{\approx}{\Delta}_\circ \cdot \widehat{\underset{\sim}{E}} = +\frac{4\pi i}{\omega}\hat{\underset{\sim}{j}}$$
$$\underset{\approx}{\Delta} \cdot \underset{\sim}{E} = +\frac{4\pi i}{\omega}\hat{\underset{\sim}{j}} \tag{9}$$

Therefore

$$\underset{\sim}{E} = \underset{\approx}{\Delta}^{-1} \cdot \underset{\approx}{\Delta}_\circ \cdot \hat{\underset{\sim}{E}} \tag{10}$$

which can be regarded as another operational definition of $\underset{\approx}{\alpha}$. Since

$$\underset{\approx}{\Delta}_\circ^{-1} = \frac{n^2}{n^2-1}\underset{\approx}{T} - \underset{\approx}{1} \equiv \underset{\approx}{P} - 1 \tag{11}$$

an alternative form of (10) is

$$\hat{\underset{\sim}{E}} = \underset{\approx}{\Delta}_\circ^{-1} \cdot \underset{\approx}{\Delta} \cdot \underset{\sim}{E} = \{\underset{\approx}{1} + (\underset{\approx}{1} - \underset{\approx}{P}) \cdot \underset{\approx}{\alpha}\} \cdot \underset{\sim}{E} \tag{12}$$

Finally, we note the following relations

$$\underset{\approx}{\alpha} = \frac{4\pi i}{\omega}\underset{\approx}{\sigma}$$
$$\underset{\approx}{\sigma} = \hat{\underset{\approx}{\sigma}} \cdot \{\underset{\approx}{1} + (\underset{\approx}{1} - \underset{\approx}{P}) \cdot \underset{\approx}{\alpha}\} \tag{13}$$

The conceivable driving or perturbing quantities for a plasma can be listed as $\hat{\underset{\sim}{j}}$, $\underset{\sim}{E}$, $\hat{\underset{\sim}{E}}$, which, in turn, excite a response characterized by $\underset{\sim}{j}$, $\underset{\sim}{\mathcal{E}}$, or $\underset{\sim}{E}$. The following table gives a summary of the respective response functions:

| | $\hat{\underset{\sim}{j}}$ | $\hat{\underset{\sim}{E}}$ | $\underset{\sim}{E}$ |
|---|---|---|---|
| $\underset{\sim}{j}$ | $\underset{\approx}{\alpha}\cdot\underset{\approx}{\Delta}^{-1}$ | $\hat{\underset{\approx}{\sigma}}$ | $\underset{\approx}{\sigma}$ |
| $\underset{\sim}{\mathcal{E}}$ | $\frac{4\pi i}{\omega}\underset{\approx}{\Delta}_o^{-1}\cdot\underset{\approx}{\alpha}\cdot\underset{\approx}{\Delta}^{-1}$ | $\underset{\approx}{\Delta}^{-1}\cdot\underset{\approx}{\alpha}$ | $\underset{\approx}{\Delta}_o^{-1}\underset{\approx}{\alpha}$ |
| $\underset{\sim}{E}$ | $\frac{4\pi i}{\omega}\underset{\approx}{\Delta}^{-1}$ | $\underset{\approx}{\Delta}^{-1}\cdot\underset{\approx}{\Delta}_o$ | $\underset{\approx}{1}$ |

(14)

When the plasma is anisotropic (either because of an anisotropic velocity distribution or because it is in an external magnetic field) all the tensor elements of the response function are, in general, non-vanishing (although not all of them are independent). In an isotropic system, however, they become diagonal in the coordinate system where k points along one of the coordinate axes. Setting $\underset{\sim}{k} = (ook)$, one finds

$$\underset{\approx}{\alpha} = \begin{pmatrix} \alpha^T & & \\ & \alpha^T & \\ & & \alpha^L \end{pmatrix} \tag{15}$$

etc.; the superscripts indicating transverse and longitudinal elements. Our main interest in this case will lie with $\alpha^L$ and the related other longitudinal response functions since these are responsible for the effects associated with the Coulomb interaction.

The transverse and longitudinal components of the electric fields and currents are now decoupled from each other and the previous table (14) simplifies to:

| | $\hat{\rho}$ | $\hat{E}^L$ | $E^L$ |
|---|---|---|---|
| $\rho$ | $-\frac{\alpha}{\varepsilon}$ | $\frac{k}{\omega}\hat{\sigma}$ | $\frac{k}{\omega}\sigma$ |
| $\mathcal{E}^L$ | $+\frac{4\pi i}{k}\frac{\alpha}{\varepsilon}$ | $-\frac{\alpha}{\varepsilon}$ | $-\alpha$ |
| $E^L$ | $\frac{4\pi i}{k}\frac{1}{\varepsilon}$ | $\frac{1}{\varepsilon}$ | $1$ |

(16)

and relations (13) become

$$\alpha = \frac{4\pi i}{\omega}\,\sigma \qquad \hat{\sigma} = \frac{\sigma}{\varepsilon} \tag{17}$$

The superscript $^L$ has been and henceforth will be omitted (whenever no confusion can arise).

### I-2. *Kramers-Kronig Relations*

The response functions, when they are represented in the time domain, connect the response and the perturbation at different times, e.g.

$$j(t) = \int_{-\infty}^{+\infty} \sigma(t-t')\,E(t')\,dt' \tag{18}$$

However, causality requires that

$$\sigma(t) = 0 \qquad \text{for} \qquad t < 0 \tag{19}$$

This, in turn, leads to the condition that the Fourier-transform $\sigma(\omega)$ is analytic in the upper $\omega$ half plane, or, in other words, $\sigma(\omega)$ is a plus-function [4]. A simple theorem [4] states that any - sufficiently well behaved - function of the complex variable $\omega$ can uniquely be decomposed into plus- and minus- functions with the aid of the projection operations $\delta_{\pm}$:

$$F_{\pm}(\omega) = \int_{-\infty}^{+\infty} \delta_{\pm}(\omega-\omega')\,F(\omega')\,d\omega' \tag{20}$$

which are, respectively, analytic in the upper or lower $\omega$ half planes; the $\delta_{\pm}$ functions (distributions, to be precise) are given by

$$\delta_+(x) = \frac{1}{2\pi}\int_0^{\infty} e^{ixt}\,dt = \lim_{o\to 0}\frac{i}{2\pi}\,\frac{1}{x+io}$$

$$= \frac{1}{2}\,\delta(x) + \frac{i}{2\pi}\,P\frac{1}{x} \tag{21}$$

$$\delta_-(x) = \frac{1}{2}\int_{-\infty}^{0} e^{ixt}\,dt = \lim_{o\to 0}\frac{-i}{2\pi}\,\frac{1}{x-io}$$

$$= \frac{1}{2}\,\delta(x) - \frac{i}{2\pi}\,P\frac{1}{x}$$

($P$ stands for principal value)

Since $\sigma(\omega)$ is already a plus function

$$\sigma(\omega) = \int_{-\infty}^{+\infty} \delta_+(\omega-\omega')\;\sigma\;(\omega')\;d\omega' \tag{22}$$

$$= \frac{1}{2}\,\sigma(\omega) + \frac{i}{2\pi}\,P\!\int \frac{d\omega'}{\omega-\omega'}\,\sigma(\omega')$$

or, separating out the real and imaginary parts of $\sigma$

$$\sigma = \sigma' + i\sigma'' \tag{23}$$

one arrives at the famous Kramers-Kronig relations [1,3]

$$\sigma'(\omega) = -\frac{1}{\pi}\,P\!\int_{-\infty}^{+\infty} \frac{d\omega'}{\omega-\omega'}\,\sigma''(\omega') \tag{24}$$

$$\sigma''(\omega) = +\frac{1}{\pi}\,P\!\int_{-\infty}^{+\infty} \frac{d\omega'}{\omega-\omega'}\,\sigma'(\omega')$$

Similar relations must exist for the other response functions figuring in (14) or (16) with the qualification that their behaviour at $\omega$=0 or at $\omega\to\infty$ might be pathologic. (The longitudinal conductivity must remain finite at $\omega=0$ and vanishes as $1/\omega$ for $\omega\to\infty$ on physical grounds). Therefore, some care has to be taken by subtracting the infinite part, so that the plus-ness (including the real axis) is not violated.

## I-3. *Limiting Behaviors*

The limiting behavior of $\alpha$ for $\omega\to\infty$ and for $\omega\to 0$ *and* $k\to 0$ (in this order) can be discussed on physical grounds and is of importance [5]. We consider a single species (electron) plasma here, with a neutralizing background. The generalization to multicomponent case is obvious.

(i) For $\omega\to\infty$ both the interaction between and the thermal motion of the plasma particles become negligible and therefore $\underset{\approx}{\alpha}$ goes to its cold plasma value

$$\underset{\approx}{\alpha} = -\frac{\omega_o^2}{\omega^2}\underset{\approx}{1} \qquad \text{for } \omega\to\infty \tag{25}$$

$$\omega_o^2 = 4\pi\frac{e^2 n}{m}$$

(ii) The above implies that

$$\underset{\approx}{\xi} = \frac{\omega^2}{k^2c^2}\underset{\approx}{T} \qquad \text{for } \omega\to\infty \tag{26}$$

This is the consequence of the perfect diamagnetic screening property of the plasma for high frequency magnetic disturbances.

(iii) The following limit applies to thermal equilibrium. For $\omega\to\infty$ (static perturbation) and $k\to 0$ (long wavelength limit) one would expect that a simple hydrodynamic description holds. If the system is now perturbed by an external field $E_{\underset{\sim}{k}}e^{i\underset{\sim}{k}\cdot\underset{\sim}{x}}$ and finds its stationary equilibrium, the pressure force and the electric force have to balance each other. In the linear approximation (V is the volume)

$$n = n_\circ + n_1 = n_\circ + \frac{1}{V}\Sigma n_{\underset{\sim}{k}}\, e^{i\underset{\sim}{k}\cdot\underset{\sim}{x}}$$

$$P = P_\circ + P_1 = P_\circ + \frac{1}{V}\Sigma P_{\underset{\sim}{k}} e^{i\underset{\sim}{k}\cdot\underset{\sim}{x}} \tag{27}$$

$$-\frac{1}{n_\circ}\,\mathrm{grad}P_1 = e\underset{\sim}{E}\ .$$

However, while on the one hand

$$\underset{\sim}{E}_{\underset{\sim}{k}} = \frac{\hat{\underset{\sim}{E}}_{\underset{\sim}{k}}}{\varepsilon^L(\underset{\sim}{k}0)} \tag{28}$$

on the other hand P can be related to n with the aid of the isothermal compressibility K:

$$K = \frac{1}{n}\frac{\partial n}{\partial P} \tag{29}$$

Thus

$$P_{\underset{\sim}{k}} = \frac{n_{\underset{\sim}{k}}}{Kn_\circ} \tag{30}$$

while

$$-i\underset{\sim}{k}\phi(k)\, n_{\underset{\sim}{k}} = e\mathcal{E}_{\underset{\sim}{k}} = -e\frac{\alpha^L(\underset{\sim}{k}0)}{\varepsilon^L(\underset{\sim}{k}0)}\,\hat{E}_{\underset{\sim}{k}} \tag{31}$$

with the notation (to be amply used in the sequel)

$$\phi(k) = 4\pi\frac{e^2}{k^2}$$

Combination of equations (27) through (30) leads to

$$ik\,\frac{1}{Kn_\circ^{\,2}}\,\frac{1}{ik\phi(k)}\quad e\,\frac{\alpha^L(\underset{\sim}{k}0)}{\varepsilon^L(\underset{\sim}{k}0)}\quad \hat{E}_{\underset{\sim}{k}} = e\,\frac{\hat{E}_{\underset{\sim}{k}}}{\varepsilon^L(k0)} \tag{32}$$

Finally, one obtains

$$\lim_{k\to 0} \alpha^L(\underset{\sim}{k}0) = Kn_\circ{}^2\phi(k) = \frac{K}{K_\circ}\frac{\kappa^2}{k^2} \tag{33}$$

where $K_0 = \frac{\beta}{n_0}$ (compressibility of a perfect gas)

$$\beta = \frac{1}{kT} \tag{34}$$

$$\kappa = (4\pi e^2 n_\circ \beta)^{1/2} \quad \text{(inverse Debye length)}$$

The importance of this limiting behavior is that the simple screening property of the plasma derived in the Vlasov approximation (where $\alpha^L = \kappa^2/k^2$) is not affected (apart from a modification of the Debye length) by higher order effects.

I-4. *Sum Rules*

Sum rules are direct consequences of the Kramers-Kronig relations. In general, sum rules can be derived [1] for all the tensor elements of the various response functions. Here we consider longitudinal rules only for isotropic plasmas [5].

Since simple limits exist for the real part of $\alpha$, one considers the Kramers-Kronig relation which connects $\alpha'$ with the integral of $\alpha''$. (Alternatively, the same relations can be derived from the K-K relations for $\sigma$)

$$\alpha'(\underset{\sim}{k}\omega) = -\frac{1}{\pi} P \int_{-\infty}^{+\infty} \frac{d\omega'}{\omega-\omega'} \alpha''(\underset{\sim}{k}\omega') \tag{35}$$

Letting $\omega\to\infty$ on both sides of (35) one arrives at

$$-\frac{\omega_\circ{}^2}{\omega^2} = -\frac{1}{\pi}\int_{-\infty}^{+\infty} \frac{d\omega'}{\omega}\left(1+\frac{\omega'}{\omega}\right) \alpha''(\underset{\sim}{k}\omega') \tag{36}$$

or (since $\alpha''(\omega)$ is an odd function)

$$\int_0^\infty d\omega\,\omega\alpha''(\underset{\sim}{k}\omega) = \frac{\pi}{2}\,\omega_0^{\,2} \tag{37}$$

(37) is the *conductivity* or *polarizability* sum rule.

The next sum rule is derived by setting $\omega=0$ in (35). This leads to

$$\int_0^\infty \frac{d\omega}{\omega}\,\alpha''(\underset{\sim}{k}\omega) = \frac{\pi}{2}\,\alpha'(\underset{\sim}{k}0) \tag{38}$$

which combined with (33) yields

$$\lim_{k\to 0}\int_0^\infty \frac{d\omega}{\omega}\,\alpha''(\underset{\sim}{k}\omega) = \frac{\pi}{2}\,\frac{K}{K_o}\,\frac{\kappa^2}{k^2} \tag{39}$$

known as the *compressibility sum rules*.

A further set of sum rules results from considering the response function $\alpha/\varepsilon$ (or alternatively, $\hat{\theta}$). The associated K-K relation is

$$\mathrm{Re}\,\frac{\alpha(\underset{\sim}{k}\omega)}{\varepsilon(\underset{\sim}{k}\omega)} = -\frac{1}{\pi}\,P\!\int \frac{d\omega'}{\omega-\omega'}\;\mathrm{Im}\frac{\alpha(\underset{\sim}{k}\omega')}{\varepsilon(\underset{\sim}{k}\omega)} \tag{40}$$

The high frequency expansion yields

$$\int_0^\infty d\omega\,\omega\frac{\alpha''(\underset{\sim}{k}\omega)}{|\varepsilon(\underset{\sim}{k}\omega)|^2} = \frac{\pi}{2}\,\omega_o^{\,2} \tag{41}$$

or

$$\int_0^\infty d\omega\,\omega\,\mathrm{Im}\frac{1}{\varepsilon(\underset{\sim}{k}\omega)} = -\frac{\pi}{2}\,\omega_o^{\,2} \tag{42}$$

These sum rules have become known as the *f-sum rules*. The comparison of (41) with (37) is noteworthy.

The $\omega \to 0$ limit in (40) finally leads to

$$\int_0^\infty \frac{d\omega}{\omega} \frac{\alpha''(k\omega)}{|\varepsilon(\underset{\sim}{k}\omega)|^2} = \frac{\pi}{2}\left(1 - \mathrm{Re}\frac{1}{\varepsilon(\underset{\sim}{k}0)}\right) \tag{43}$$

In virtue of (33) this means that

$$\lim_{k\to 0} \int_0^\infty \frac{d\omega}{\omega} \frac{\alpha''(\underset{\sim}{k}\omega)}{|\varepsilon(\underset{\sim}{k}\omega)|^2} = \frac{\pi}{2} \tag{44}$$

In contrast to the previous case the great difference between (44) and (39) should be stressed; (44) is the *perfect screening sum rule*.

I-5. *Quadratic Response Functions*

We now concentrate on the second order terms in the expression (2) [6]. The quantities of interest are

$$\sigma_{\mu\nu\sigma}\ (\underset{\sim}{k}_1\omega_1\ ;\ \underset{\sim}{k}_2\omega_2) \qquad \alpha_{\mu\nu\sigma}\ (\underset{\sim}{k}_1\omega_1\ ;\ \underset{\sim}{k}_2\omega_2)$$
$$\varepsilon_{\mu\nu\sigma}\ (\underset{\sim}{k}_1\omega_1\ ;\ \underset{\sim}{k}_2\omega_2)$$

and the corresponding external quantities

$$\hat{\sigma}_{\mu\nu\sigma}\ (\underset{\sim}{k}_1\omega_1\ ;\ \underset{\sim}{k}_2\omega_2) \qquad \hat{\alpha}_{\mu\nu\sigma}\ (\underset{\sim}{k}_1\omega_1\ ;\ \underset{\sim}{k}_2\omega_2)$$

It is understood that $k = k_1 + k_2$ and $\omega = \omega_1 + \omega_2$ thus the symbolic notation, say $\tilde{j} = \tilde{\sigma}_2 EE\tilde{}$ stands for

$$j_\mu(\underset{\sim}{k}\omega) = \int \frac{d\omega_1}{2\pi} \sum_{\underset{\sim}{k}_1} \sigma_{\mu\nu\sigma}\ (\underset{\sim}{k}_1\omega_1\ ;\ \underset{\sim}{k}-\underset{\sim}{k}_1\ \ \omega-\omega_1)$$
$$E_\nu(\underset{\sim}{k}_1\omega_1)E_\sigma(\underset{\sim}{k}-\underset{\sim}{k}_1\ \ \omega-\omega_1) \tag{45}$$

In an isotropic system there are, at most 7 independent elements of the polarizability, etc. tensors. In the following, we concentrate on the longitudinal-longitudinal-longitudinal projection only

$$j = k_\mu \sigma_{\mu\nu\sigma} k_{1\nu} k_{2\sigma} / k\, k_1 k_2 E_1 E_2 \tag{46}$$

etc.

In deriving the connection between external and ordinary conductivities and polarizabilities one has to realize that in the expansion of $E$ or j in terms of the external field, there are contributions of different order. Considering, for example the polarizabilities, one can write

$$\begin{aligned} E^{(1)} &= -\hat{\underset{1}{\alpha}}\, \hat{E} = -\underset{1}{\alpha}\, E^{(1)} \\ E^{(2)} &= -\hat{\underset{2}{\alpha}}\hat{E}\hat{E} = -\underset{2}{\alpha}\, E^{(1)} E^{(1)} - \underset{1}{\alpha} E^{(2)} \end{aligned} \tag{47}$$

From this, it follows that

$$\hat{\underset{2}{\alpha}}(\underset{\sim}{k}_1\omega_1 \,;\, \underset{\sim}{k}_2\omega_2) = \frac{\underset{2}{\alpha}\,(\underset{\sim}{k}_1\omega_1 \,;\, \underset{\sim}{k}_2\omega_2)}{\varepsilon(\underset{\sim}{k}_1\omega_1)\,\varepsilon(\underset{\sim}{k}_2\omega_2)\,\varepsilon(\underset{\sim}{k}\omega)} \tag{48}$$

$$\varepsilon = 1 + \underset{1}{\alpha}$$

The physical interpretation of this result is obvious. While in linear situation an excitation can produce very large response when its k and $\omega$ approximately satisfy the linear dispersion relation $\varepsilon(k\omega)=0$, in the case of quadratic response the response will be enhanced when $k_1$ and $\omega_1$ or $k_2$ or $\omega_2$ *or, in addition*, their combinations $\underset{\sim}{k}$ and $\omega$ satisfy the same linear dispersion relation.

Although this result is correct to second order in the excitation amplitude and formally there is no change in the linear dispersion relation within this perturbation scheme, this approach fails to give an account of the important situation when one of the excitation amplitudes is much higher than any of the others and (a) and expansion to second order is warranted only because of the high intensity of this single mode, and (b) the change in the structure of the linear dispersion relation cannot be neglected.

Consider now the case when there is an externally imposed intense field* $\hat{E}_0$ corresponding to the total field

*Since now the quantities under consideration are longitudinal, we omit vector notation most of the time.

$E_0$($E_n$, etc. is an abbreviation for $E(k_n \omega_n)$) which interacts strongly with the other modes, say $E_1$, $E_2$,...$E_i$.

Then, to second order in $\hat{E}_0 \hat{E}_i$, but to first order in $\hat{E}_i$

$$\begin{aligned} E_i &= -\hat{\alpha}_i \hat{E}_i - \sum_j \hat{\alpha}_{\circ j} \hat{E}_\circ \hat{E}_j \\ E_\circ &= -\hat{\alpha}_\circ \hat{E}_\circ \end{aligned} \tag{49}$$

while, considering the total fields as driving quantities

$$\begin{aligned} E_i &= -\alpha_i E_i - \sum_j \alpha_{\circ j} E_\circ E_j \\ E_\circ &= -\alpha_\circ E_\circ \end{aligned} \tag{50}$$

The equations pertaining to $E_0$ are obviously uncoupled to the others, while the ones determining the $E_i$-s can be regarded as a linear relation for the set $\underline{E} \equiv \{E_i\}$, in terms of the generalized polarizabilities $a(E_0)$

$$\underline{\underline{E}} = - \hat{a}\hat{\underline{\underline{E}}} = - a\underline{\underline{E}} \tag{51}$$

Then it follows immediately that

$$\hat{a} = (1+a)^{-1} a \tag{52}$$

The equivalent of the linear dispersion relation is now

$$| \; 1 + \; a(E_\circ) \; | = 0 \tag{53}$$

The above formalism is the basis of the description of parametric amplification (see, e.g. [8]). What is important from our point of view is that the relations between external and ordinary response functions will be affected by $E_0$, but the external response functions themselves (to be related to fluctuations by the FDT) will not.

To see this in detail, consider the practically important case when only two modes are coupled to $E_0$($i,j$ = 1,2). Then

$$a = \begin{pmatrix} \alpha_1 & \alpha_{o1}E_o \\ \alpha_{o2}E_o & \alpha_2 \end{pmatrix} \tag{54}$$

and one finds

$$\hat{a}_1 = 1 - \frac{\varepsilon_2}{\varepsilon_1\varepsilon_2 - \alpha_{o1}\alpha_{o2}E_o^2} \simeq \frac{\alpha_1}{\varepsilon_1(1 - \Lambda E_o^2)}$$

$$\hat{a}_2 = 1 - \frac{\varepsilon_1}{\varepsilon_1\varepsilon_2 - \alpha_{o1}\alpha_{o2}E_o^2} \simeq \frac{\alpha_2}{\varepsilon_2(1 - \Lambda E_o^2)} \tag{55}$$

$$\hat{a}_{o1} = \frac{\alpha_{o1}}{\varepsilon_1\varepsilon_2(1 - \Lambda E_o^2)} E_o \qquad \hat{a}_{o2} = \frac{\alpha_{o2}}{\varepsilon_1\varepsilon_2(1 - \Lambda E_o^2)} E_o$$

$$\Lambda = \frac{\alpha_{o1}\alpha_{o2}}{\varepsilon_1\varepsilon_2}$$

Since our approximation is an expansion up to second order in the amplitude, formally it would be necessary to expand the denominators in $E_0^2$; this, however, would not be correct, since the denominator exhibits a resonance behavior (the zeros depend on $E_0^2$) which would never be recovered by any expansion technique.

Finally we comment on the K-K relations satisfied by the non-linear response functions. $\sigma(\omega_1,\omega_2)$ is obviously a plus-function of $\omega_1$ and $\omega_2$ and therefore of $\omega$ also. The real and imaginary parts of $\sigma$ satisfy the double Hilbert transform relations

$$\sigma''(\omega_1, \omega_2) = -\frac{1}{4\pi^2} \, PP \int_{-\infty}^{+\infty} d\mu \int_{-\infty}^{+\infty} d\nu \, \frac{\sigma''(\mu, \nu)}{(\omega_1-\mu)(\omega_2-\nu)}$$

$$\sigma'(\omega_1, \omega_2) = -\frac{1}{4\pi^2} \, PP \int_{-\infty}^{+\infty} d\mu \int_{-\infty}^{+\infty} d\nu \, \frac{\sigma'(\mu, \nu)}{(\omega_1-\mu)(\omega_2-\nu)} \tag{56}$$

It should be noted that for the quadratic response functions the physical significance of the real and imaginary parts is different from the corresponding relations for the linear response functions and no simple identification of σ' with dissipative processes can be made.

## II. EQUILIBRIUM FLUCTUATION DISSIPATION THEOREM

The linear fluctuation dissipation theorem (FDT) establishes a relation between the two-point unperturbed correlation functions of the plasma and the linear response functions characterizing the response of the plasma to a perturbation [1,2]. The equilibrium theorem (pertaining to the case where the unperturbed state is a state of thermal equilibrium) is *exact* and will be discussed below; there exists also a non-equilibrium theorem, which is based, however, on perturbation expansion. This is discussed in Appendix A.

### II-1. *Basic Relations*

In this derivation we consider an electron plasma with infinitely heavy ions. The result will be extended later to a two-component plasma.

The equilibrium phase space distribution function of a plasma-electromagnetic radiation system is given by

$$\Omega^\circ\ (x,p\ ;\ A,E) = Z^{-1} e^{-\beta H^\circ}$$
$$Z = \int d\Gamma e^{-\beta H^\circ} \tag{1}$$

$\{\underset{\sim}{x}_i\}$ $\{\underset{\sim}{p}_i\}$ and $\{\underset{\sim}{A}_{\underset{\sim}{k}}\}$, $\{\underset{\sim}{E}_{\underset{\sim}{k}}\}$ are the coordinates and momenta of the particles and the field, respectively. We use a longitudinal (Bohm) gauge and thus

$$\underset{\sim}{E}^L_{\underset{\sim}{k}} = -\dot{\underset{\sim}{A}}_{\underset{\sim}{k}} \qquad \Phi = 0 \tag{2}$$

The longitudinal electric field is not an independent coordinate and therefore the phase-space element is

$$d\Gamma = d\left\{\underset{\sim}{x}_i\right\} d\left\{\underset{\sim}{p}_i\right\} d\left\{\underset{\sim}{A}_{\underset{\sim}{k}}\right\} d\left\{\underset{\sim}{E}_{\underset{\sim}{k}}\right\} \delta(\underset{\sim}{k}\cdot\underset{\sim}{E}_{\underset{\sim}{k}} + \frac{1}{V}4\pi i e \sum_i e^{-i\underset{\sim}{k}\cdot\underset{\sim}{x}_i}) \tag{3}$$

The hamiltonian* is

---

*Volume factors will be omitted ("unit volume"), but will be restored when necessary.

$$H^\circ = \sum_i \frac{1}{2}mv_i^2 + \frac{1}{8\pi}\sum_{\underset{\sim}{k}} \left\{\underset{\sim}{E}_{\underset{\sim}{k}}\cdot\underset{\sim}{E}_{-\underset{\sim}{k}} + k^2c^2\ \underset{\sim}{A}_{\underset{\sim}{k}}\cdot\underset{\approx}{T}\cdot\underset{\sim}{A}_{-\underset{\sim}{k}}\right\}$$

$$v_i = \frac{1}{m}\left\{p_i + e\left[\sum_{\underset{\sim}{q}} A_{\underset{\sim}{q}}\, e^{i\underset{\sim}{q}\cdot\underset{\sim}{x}_i} + \underset{\sim}{A}_\circ\ (x_i)\right]\right\} \tag{4}$$

$\underset{\sim}{A}_0$ is associated with the eventual uniform external field. The equilibrium two-point correlations of interest are defined by [9]

$$P(\underset{\sim}{\xi}\tau) = \langle\rho(\underset{\sim}{r}t)\ \rho(\underset{\sim}{r}+\underset{\sim}{\xi}\ ,\ \ t+\tau)\rangle^\circ$$

$$Q_{\mu\nu}(\underset{\sim}{\xi}\ ,\ \tau) = \langle\,\underset{\sim}{j}_\nu\ (\underset{\sim}{r}t)\ \underset{\sim}{j}_\mu(\underset{\sim}{r}+\underset{\sim}{\xi}\ ,\ \ t+\tau)\rangle^\circ \tag{5}$$

$$\langle.\ .\ .\rangle^\circ = \int d\Gamma\ .\ .\ .\ .\ .\ \Omega^\circ$$

where $\rho,\underset{\sim}{j}$ are the exact microscopic charge and current densities. The appropriate Fourier transform relations are of the type*

$$Q_{\mu\nu}(\underset{\sim}{k}\tau) = \langle j_{-\underset{\sim}{k}\nu}\ (0)\ j_{\underset{\sim}{k}\mu}\ (\tau)\rangle^{\circ *}$$

$$Q_{\mu\nu}(\underset{\sim}{k}\omega)\ \delta(\omega-\omega') = \frac{1}{2\pi}\langle j_{\underset{\sim}{k}\mu}(\omega)\ j_{-\underset{\sim}{k}\nu}\ (\omega')\rangle^{\circ *}$$

$$P(\underset{\sim}{k}\omega) = \frac{\underset{\sim}{k}\underset{\approx}{Q}\cdot\underset{\sim}{k}}{\omega^2} \tag{6}$$

$$\rho_{\underset{\sim}{k}}(t) = -e\sum_i e^{i\underset{\sim}{k}\cdot\underset{\sim}{x}_i(t)}$$

$$j_{\underset{\sim}{k}}(t) = -e\sum_i \underset{\sim}{v}_i(t)\ e^{-i\underset{\sim}{k}\cdot\underset{\sim}{x}_i(t)}$$

It will be of interest to consider the equal time charge density correlation $P(\underset{\sim}{k},\tau = 0)$:

---

* Volume factors are omitted ("unit volume"), but will be restored when necessary.

$$P(\underset{\sim}{k}\ ,\ \tau=0) = \frac{e^2}{V} \sum_{i,j} \langle e^{-i\underset{\sim}{k}\cdot(\underset{\sim}{x}_i-\underset{\sim}{x}_j)}\rangle^\circ =$$

$$= \frac{e^2}{V} N\ (N-1)\ \langle e^{i\underset{\sim}{k}\cdot(\underset{\sim}{x}_1-\underset{\sim}{x}_2)}\rangle^\circ + e^2\frac{N}{V} \tag{7}$$

The first term in the second line originates from the $i\neq j$, while the second from the $i=j$ contributions. The former can be evaluated with the aid of the one-particle (F) and two-particle (G) distribution functions:

$$G(12) = F\ (1)\ F(2)\ \{1 + g\ (12)\} \tag{8}$$

(8) defines the equilibrium pair correlation function $g(12) \equiv g(|\underset{\sim}{x}_1-\underset{\sim}{x}_2|)$.

Thus

$$P(\underset{\sim}{k}\ ,\ \tau=0) = e^2\left\{\frac{N^2}{V^3}\int d\underset{\sim}{x}_1 d\underset{\sim}{x}_2 e^{-i\underset{\sim}{k}\cdot(\underset{\sim}{x}_1-x_2)}\cdot \left(1+g\ (\underset{\sim}{x}_1-\underset{\sim}{x}_2)\right) + n\right\} \tag{9}$$

$$= e^2 n\left\{1 + ng(\underset{\sim}{k})\right\} + e^2 n^2 V\delta_{\underset{\sim}{k}o}$$

This is the correlation function due to the electronic charges. The contribution from the ion background actually compensates the $\delta_{\underset{\sim}{k}0}$ term, which thus will be omitted.

The equal time correlation for $\underset{\approx}{Q}(\underset{\sim}{k}\tau)$ is simpler because the velocities of different particles are not correlated in equilibrium.

$$Q_{\mu\nu}(\underset{\sim}{k}\ ,\ \tau=0) = \frac{e^2}{V}\sum_{ij} \langle v_{i\mu}v_{j\nu}\rangle^\circ \cdot \langle e^{-ik\cdot(\underset{\sim}{x}_i-\underset{\sim}{x}_j)}\rangle^\circ$$

$$= e^2\sum_{ij} \delta_{\mu\nu}\delta_{ij}\frac{1}{3}\langle v^2\rangle\frac{1}{V} \tag{10}$$

$$= e^2\delta_{\mu\nu}\frac{n}{\beta m}$$

II-2. *Dynamical FDT*

We now introduce a small perturbing field (given by its vector potential) $-e\underset{\sim}{A}_q \exp(i\underset{\sim}{q}\cdot\underset{\sim}{x}_i)$ which modifies $H, \Omega, \underset{\sim}{j}$:

$$H = H° + H'$$

$$\Omega = \Omega° + \Omega'$$

$$\underset{\sim}{j} = \underset{\sim}{j}° + \underset{\sim}{j}'$$

The perturbing terms are

$$H' = m\sum_i \underset{\sim}{v}_i \cdot \delta\underset{\sim}{v}_i = +e\sum_i \underset{\sim}{v}_i \cdot \hat{\underset{\sim}{A}}_{\underset{\sim}{q}}\, e^{i\underset{\sim}{q}\cdot\underset{\sim}{x}_i} =$$

$$= -\hat{\underset{\sim}{A}}_{\underset{\sim}{q}} \cdot \underset{\sim}{j}^{\circ}_{-\underset{\sim}{q}} \tag{11}$$

$$\underset{\sim}{j}'_{\underset{\sim}{k}} = -e\sum_i \delta\underset{\sim}{v}_i\, e^{-i\underset{\sim}{k}\cdot\underset{\sim}{x}_i} = -\frac{e^2}{m}\sum_i \hat{A}_{\underset{\sim}{q}}\, e^{i(\underset{\sim}{q}-\underset{\sim}{k})\cdot\underset{\sim}{x}_i}$$

while $\Omega'$ will be determined from the Liouville equation:

$$\frac{\partial\Omega}{\partial t} = -iL\Omega \qquad L = -i\,[H, \ .\ .\ .] \tag{12}$$

[ ] representing the Poisson bracket.

With the splitting of the Hamiltonian into equilibrium and perturbation part a similar splitting of $L$ follows:

$$L = L° + L' \tag{13}$$

To first order the Liouville equation becomes

$$\frac{\partial\Omega'}{\partial t} = -i\,L°\,\Omega' - i\,L'\,\Omega° \tag{14}$$

which leads to the immediate solution

$$\Omega' = -i\int_{-\infty}^{t} e^{-iL°(t-t')}\, L'\,(t')\,\Omega°\,dt'$$

$$= -i\int_{0}^{\infty} e^{-iL°\tau} L'(t-\tau)\,\Omega^{\circ} d\tau \tag{15}$$

$$= \int_0^\infty e^{-iL^\circ\tau}\hat{A}_{\underset{\sim}{q}\mu}(t-\tau)\ [j^\circ_{-\underset{\sim}{q}\mu}(t)\ ,\ \Omega^\circ]\ d\tau$$

$$= \beta\int_0^\infty \hat{A}_{\underset{\sim}{q}\mu}(t-\tau)\ e^{-iL^\circ\tau}\ \frac{dj^\circ_{-\underset{\sim}{q}\mu}}{dt}(t)\ \Omega^\circ\ d\tau \tag{15}$$

$$= \beta\Omega^\circ\int_0^\infty \hat{A}_{\underset{\sim}{q}\mu}\ (t-\tau)\ \frac{d}{dt}\ \underset{\sim}{j}^\circ_{-\underset{\sim}{q}\mu}\ (t-\tau)\quad d\tau$$

In arriving at the final expression $\exp(-iL^\circ\tau) = U(\tau)$ has been recognized as the time evolution operator which shifts the time argument of any dynamical variable by $-\tau$ (in terms of the evolution of the unperturbed system).

Having obtained the perturbed distribution function, we calculate the expectation value of the current in the perturbed ensemble

$$\begin{aligned}(\langle j\rangle)' &= \langle j^\circ\rangle' + \langle j'\rangle^\circ \\ \langle\ldots\rangle' &= \int d\Gamma\ .\ .\ .\ \Omega'\end{aligned} \tag{16}$$

First we calculate the first term:

$$\langle j^\circ_{\underset{\sim}{k}\mu}\rangle' = \beta\int_0^\infty d\tau\hat{A}_{\underset{\sim}{k}\nu}(t-\tau)\int d\Gamma\Omega^\circ j^\circ_{\underset{\sim}{k}\mu}(t)\ \frac{d}{dt}\ j^\circ_{-\underset{\sim}{k}\nu}(t-\tau)$$

$$= -\beta\int_0^\infty d\tau\hat{A}_{\underset{\sim}{k}\nu}(t-\tau)\ \frac{d}{d\tau}\ \langle j_{\underset{\sim}{k}\mu}\ (t)\ j_{-\underset{\sim}{k}\nu}\ (t-\tau)\rangle^\circ \tag{17}$$

$$= -\beta\int_0^\infty d\tau\hat{A}_{\underset{\sim}{k}\nu}(t-\tau)\ \frac{d}{d\tau}\ Q_{\mu\nu}\ (\underset{\sim}{k}\tau\ )$$

$$= \beta\int_0^\infty d\tau\hat{E}_{\underset{\sim}{k}\nu}\ (t-\tau)\ Q_{\mu\nu}\ (\underset{\sim}{k}\tau)\ -\beta\hat{A}_{\underset{\sim}{k}\mu}(t)\ Q_{\mu\nu}(\underset{\sim}{k}\ ,\ \tau=0$$

assuming that $\underset{\approx}{Q}(\underset{\sim}{k}\tau\to\infty) \to 0$.

The value of the second term in (16) can easily be obtained. From (11) one has

$$<j'_{\underset{\sim}{k}\mu}>^{\circ} = \frac{-e^2}{m} \hat{A}_{\underset{\sim}{q}} (t) \sum_i \int d\Gamma \, e^{i(\underset{\sim}{q}-\underset{\sim}{k})\cdot\underset{\sim}{x}_i} \, \Omega_{\circ} \tag{18}$$

$$= \frac{e^2}{m} \hat{A}_{\underset{\sim}{k}} (t) \, n$$

The value of $Q_{\mu\nu}(k,\tau=0)$ is given by (10) as $\delta_{\mu\nu}e^2 \frac{n}{\beta}$. Thus (18) and the last term in (17) cancel each other. Then the final result is (19):

$$<j_{\underset{\sim}{k}\mu}(\omega)>' = \beta\hat{E}_{\underset{\sim}{k}\nu} \int_{-\infty}^{+\infty} d\omega' \delta_+ (\omega-\omega') \, Q_{\mu\nu}(\underset{\sim}{k}\omega') \tag{19}$$

which implies that

$$\hat{\sigma}_{\mu\nu} (\underset{\sim}{k}\omega) = \beta \int_{-\infty}^{+\infty} d\omega' \delta_+ (\omega-\omega') \, Q_{\mu\nu} (\underset{\sim}{k}\omega') \tag{20}$$

(63) already provides the required connection between response function and equilibrium correlation. It can be cast into a more attractive form by observing that $\underset{\approx}{Q}$ is hermitian

$$Q_{\mu\nu} (\underset{\sim}{k}\omega) = Q^*_{\nu\mu}(\underset{\sim}{k}\omega) \tag{21}$$

which follows from the defining equation (5) and from the reality of $\underset{\approx}{Q}(\underset{\sim}{\xi}\tau)$. Taking the hermitian part of (20) one finds

$$\frac{1}{2} (\underset{\approx}{\hat{\sigma}} + \underset{\approx}{\hat{\sigma}}^+) = \frac{\beta}{2} \underset{\approx}{Q} \tag{22}$$

Thus the equilibrium correlation can be expressed in terms of the dissipative part of the conductivity.

Taking the longitudinal projection of (22) one has

$$\hat{\sigma}^{L'} (\underset{\sim}{k}\omega) = \frac{\beta}{2} \frac{\omega^2}{k^2} P(\underset{\sim}{k}\omega) \tag{23}$$

or

$$P(\underset{\sim}{k}\omega) = \frac{k^2}{2\pi\beta\omega} \frac{\alpha^{L''}(\underset{\sim}{k}\omega)}{|\varepsilon^L(\underset{\sim}{k}\omega)|^2} \tag{24}$$

(24) provides the contact with the idea of dressed particles. In the Vlasov approximation (and only there)

$$\alpha^L(\underset{\sim}{k}\omega) = \frac{\omega_o{}^2}{k^2}\int \frac{1}{\omega-\underset{\sim}{k}\cdot\underset{\sim}{v}+io} \frac{\partial F(\underset{\sim}{v})}{\partial \underset{\sim}{v}} \cdot \underset{\sim}{k}\, d\underset{\sim}{v} \tag{25}$$

and

$$\alpha^{L''}(\underset{\sim}{k}\omega) = -\pi \frac{\omega_o{}^2}{k^2}\int \delta(\omega-\underset{\sim}{k}\cdot\underset{\sim}{v})\, \underset{\sim}{k}\cdot\frac{\partial F(\underset{\sim}{v})}{\partial \underset{\sim}{v}}\, d\underset{\sim}{v}$$

$$= \pi\beta m\omega_o{}^2 \frac{\omega}{k^2}\int \delta(\omega-\underset{\sim}{k}\cdot\underset{\sim}{v})\, F(\underset{\sim}{v})\, d\underset{\sim}{v} \tag{26}$$

(26) combined with (24) yields the two-point charge density correlation in terms of uncorrelated dressed particles [9, 22]

$$P(\underset{\sim}{k}\omega) = 2\pi e^2 n \int F(\underset{\sim}{v}) \frac{\delta(\omega-\underset{\sim}{k}\cdot\underset{\sim}{v})}{|\varepsilon(\underset{\sim}{k}\ \underset{\sim}{k}\cdot\underset{\sim}{v})|^2}\, d\underset{\sim}{v} \tag{27}$$

It should be emphasized that on the one hand (27) is easily generalized to non-equilibrium situations (see Appendix A), while (24) is not; on the other hand (24) is *exact*, while (27) is valid to lowest order in the plasma parameter only.

II-3. *Static FDT*

The importance of the static FDT is that it provides a relation between the equilibrium pair correlation function, whose knowledge is sufficient to determine all the equilibrium properties of the system, and the dielectric response function. Dropping the L superscript again, (24) can be restated as

$$\mathrm{Im}\,\frac{1}{\varepsilon(\underset{\sim}{k}\omega)} = -2\pi\beta\frac{\omega}{k^2}P(\underset{\sim}{k}\omega) \tag{28}$$

Dividing through by and integrating over ω, the left-hand-side can be evaluated with the aid of the sum rule (I.43) while the right-hand-side by

$$\int P(\underset{\sim}{k}\omega)\,d\omega = 2\pi\ P(\underset{\sim}{k}\ ,\ \tau=0) \tag{29}$$

The latter can be related to the pair correlation function through (8) to obtain

$$g(k) = \frac{1}{n}\left\{\frac{1}{\beta n\phi(k)}\left[1 - \mathrm{Re}\,\frac{1}{\varepsilon(\underset{\sim}{k}0)}\right] -1\right\} \tag{30}$$

If $\varepsilon(k0)$ is replaced by its Vlasov value ($\varepsilon(k0) = 1 + (\kappa^2/k^2)$) one finds the well-known Debye correlation function

$$g_D(k) = -\frac{1}{n}\ \frac{\kappa^2}{k^2+\kappa^2} \tag{31}$$

II-4. *Two Component Plasmas*

For an electron-ion plasma the currents and thus the conductivities have contributions both from the electrons and ions

$$\underset{\approx}{\hat{\sigma}} = \underset{\approx}{\hat{\sigma}}^{(e)} + \underset{\approx}{\hat{\sigma}}^{(i)} \tag{32}$$

etc. The FDT as a relation between the *total* conductivity and the *total* current - or charge-density correlations is unchanged. However, it is more problematic to establish relations between the response functions and correlations pertaining to the individual species. The source of the difficulty is that there are *two* conductivities, but *three* possible correlations (ee, ei=ie, ii).

We quote here the result for the static FDT, for Z-times ionized ions

$$\sum_{B=e,i} \frac{e_A e_B}{e^2} g_{AB}(k) = \frac{1}{n}\left\{\frac{1}{\beta n\phi(k)}\right.$$

$$\left. \cdot \operatorname{Re} \frac{\alpha^{(A)}(\underset{\sim}{k}0)}{\varepsilon(\underset{\sim}{k}0)} - Z_A \right\} \tag{33}$$

$$A = e\ ,\ i$$

$$Z_e = +1 \qquad e_e = -e$$

$$Z_i = +Z \qquad e_i = +e$$

*Appendix* II-A: *Non-equilibrium FDT*

In this Appendix we demonstrate the validity of a non-equilibrium fluctuation dissipation theorem [22,23]. This theorem - in contrast to the equilibrium one - is valid to first order in the plasma coupling parameter $\gamma = \kappa^3/n$ (inverse number of particles within the Debye sphere) only.

The derivation is greatly facilitated by the introduction of the Vlasov-Dupree operator, $V(\underset{\sim}{k}\omega;\underset{\sim}{v}\underset{\sim}{v}')$ which is defined by writing the Fourier transform of the linearized Vlasov equation (where $F(\underset{\sim}{v})$ is any stationary, homogeneous velocity distribution and $f(\underset{\sim}{k}\omega;\underset{\sim}{v})$ is the small deviation from it)

$$-i(\omega-\underset{\sim}{k}\cdot\underset{\sim}{v})f - \frac{i}{m}\phi(k)\underset{\sim}{k}\cdot\frac{\partial F}{\partial \underset{\sim}{v}}\int f\, d\underset{\sim}{v}' = f(t=0) \tag{34}$$

in the form

$$-i\{\omega - V(\underset{\sim}{k}\omega)\}\ f = f(t=0) \tag{35}$$

The significance of the second term is

$$V(\underset{\sim}{k}\omega)f \equiv \int V(\underset{\sim}{k}\omega;\underset{\sim}{v}\underset{\sim}{v}')\ f(\underset{\sim}{v}')d\underset{\sim}{v}'$$
$$\equiv V(\underset{\sim}{k}\omega;\underset{\sim}{v}\underset{\sim}{v}')f(\underset{\sim}{v}') \tag{36}$$

In the second line we introduced the convention that integration over repeated velocity arguments is always understood.

Inversion of (35) yields the "solution" of the initial value problem

$$f(\omega) = \frac{i}{\omega - V(\underset{\sim}{k}\omega)} f(t=0) \tag{37}$$

or, introducing the time evolution operator $W(\underset{\sim}{k}t)$ and its Fourier Transform as

$$\begin{aligned} f(\omega) &= W(\underset{\sim}{k}\omega) f(t=0) \\ &\equiv W(\underset{\sim}{k}\omega;\underset{\sim}{v}\ \underset{\sim}{v}') f(t=0;\underset{\sim}{v}') \end{aligned}$$

or

$$\begin{aligned} f(t) &= W(\underset{\sim}{k}t) f(t=0) \\ &\equiv W(\underset{\sim}{k}t;\underset{\sim}{v}\ \underset{\sim}{v}') f(t=0;\underset{\sim}{v}') \end{aligned} \tag{38}$$

for $t > 0$. Obviously

$$W(\underset{\sim}{k}\omega) = \frac{i}{\omega - V(\underset{\sim}{k}\omega)} \tag{39}$$

Since the closed solution of the Vlasov equation is readily obtained, an explicit representation of $W(k\omega)$ can be exhibited:*

$$W(\underset{\sim}{k}\omega;\underset{\sim}{v}\ \underset{\sim}{v}') = \frac{i}{\omega - \underset{\sim}{k}\cdot\underset{\sim}{v} + io} \left\{ \delta(\underset{\sim}{v}-\underset{\sim}{v}') - \frac{\partial F}{\partial \underset{\sim}{v}} \cdot \underset{\sim}{k} \frac{\phi(k)}{\varepsilon(\underset{\sim}{k}\omega)} \frac{1}{\omega - \underset{\sim}{k}\cdot\underset{\sim}{v}' + io} \right\} \tag{40}$$

The io ($o \to 0$) term in the singular denominators will be omitted but understood in the sequel. We emphasize that $F(\underset{\sim}{v})$ can be any stationary distribution, but obviously $V$ and $W$ are funcionals of F.

We now consider the microscopic distribution function in the one particle phase space

---

*We set $m = 1$ in the sequel.

$$N(\underset{\sim}{x}\underset{\sim}{v};t) = \sum_i \delta(\underset{\sim}{x}-\underset{\sim}{x}_i[t])\,\delta(\underset{\sim}{v}-\underset{\sim}{v}_i[t]) \tag{41}$$

The ensemble average of this is the customary one-particle distribution function

$$\langle N \rangle = F(\underset{\sim}{v};t)$$

Our interest is in the fluctuations of N around F, which will be denoted by $\nu$

$$\nu = N - \langle N \rangle \tag{42}$$

N exactly satisfies the (non-linear) Vlasov equation (in contrast to F for which the Vlasov equation is an approximation). Assuming now that the fluctuations are small, $\nu$ will satisfy the linearized Vlasov equation, i.e.

$$-i(\omega - V(F))\,\nu = \nu(t=0)$$

or (43)

$$\nu(t) = W(F;t)\quad \nu(0)$$

Consider now the correlation of phase space density fluctuations $\Pi$:

$$\Pi(\underset{\sim}{\xi}\tau;\ \underset{\sim}{v}_1\underset{\sim}{v}_2) = \langle \nu(\underset{\sim}{x}+\underset{\sim}{\xi},\underset{\sim}{v}_1;\ t+\tau)\ \nu(\underset{\sim}{x}\ \underset{\sim}{v}_2;t)\rangle$$

$$= \langle \nu(\underset{\sim}{\xi},\ \underset{\sim}{v}_1;\tau)\,\nu(0,\underset{\sim}{v}_2;0)\rangle \tag{44}$$

Now, in view of (43), for $\tau > 0$

$$\Pi(\underset{\sim}{\xi}\tau;\underset{\sim}{v}_1\underset{\sim}{v}_2) = \theta(\tau)\ \Pi(\underset{\sim}{\xi}\tau;\underset{\sim}{v}_1\underset{\sim}{v}_2) \tag{45}$$

$$\theta(\tau)\Pi(\underset{\sim}{\xi}\tau;\underset{\sim}{v}_1\underset{\sim}{v}_2) = W(\tau;\underset{\sim}{v}_1\underset{\sim}{v}_1')\quad \langle \nu(\underset{\sim}{\xi},\underset{\sim}{v}_1';0)\ \nu(0,\underset{\sim}{v}_2;0$$

or

$$\Pi_+(\underset{\sim}{k}\omega;\underset{\sim}{v}_1\underset{\sim}{v}_2) = W(\underset{\sim}{k}\omega;\underset{\sim}{v}_1\underset{\sim}{v}_1')\quad \langle \nu(\underset{\sim}{k}\underset{\sim}{v}_1';0)\ \nu^*(\underset{\sim}{k}\underset{\sim}{v}_2;0)\rangle \tag{46}$$

The equal-time correlation can be expressed in terms of the one-particle and two-particle distribution functions, in a manner similar to what has been done for P (cf. Eq. (8)):

$$\langle \nu(\underset{\sim}{k},\underset{\sim}{v}_1) \quad \nu^*(\underset{\sim}{k},\underset{\sim}{v}_2)\rangle = = F(\underset{\sim}{v}_1) \quad \delta(\underset{\sim}{v}_1-\underset{\sim}{v}_2)+\mathcal{G}(\underset{\sim}{k};\underset{\sim}{v}_1\underset{\sim}{v}_2) \tag{47}$$

where

$$G(12) = F(1)F(2) + \mathcal{G}(12) \tag{48}$$

$\mathcal{G}(k;\underset{\sim}{v}_1\underset{\sim}{v}_2)$ can be determined from the second equation of the BBGKY hierarchy. The derivation is given in Appendix IV-A, with the result

$$\mathcal{G}(\underset{\sim}{k};\underset{\sim}{v}_1\underset{\sim}{v}_2) = \int \frac{d\omega}{2\pi} W\ (k\omega;\underset{\sim}{v}_1\underset{\sim}{v}_1') \ W^*(k\omega;\underset{\sim}{v}_2\underset{\sim}{v}_2') \ I(\underset{\sim}{k};\underset{\sim}{v}_1'\underset{\sim}{v}_2')$$

$$= \int d\tau \ W(\underset{\sim}{k}\tau;\underset{\sim}{v}_1'\underset{\sim}{v}_1') \ W(\underset{\sim}{k}\tau;v_2 v_2') \ I(\underset{\sim}{k};\underset{\sim}{v}_1'\underset{\sim}{v}_2')$$

$$I(\underset{\sim}{k};\underset{\sim}{v}_1\underset{\sim}{v}_2) = i\phi(k)\underset{\sim}{k}\cdot\left(\frac{\partial F}{\partial \underset{\sim}{v}_1} F(\underset{\sim}{v}_2) - \frac{\partial F}{\partial \underset{\sim}{v}_2} F(\underset{\sim}{v}_1)\right) \tag{49}$$

(49) then combined with (47) and (46) leads to an expression for $\Pi$ in terms of the one particle distribution function

$$\theta(\tau)\Pi(\underset{\sim}{k}\tau;\underset{\sim}{v}_1\underset{\sim}{v}_2) = W\ (\underset{\sim}{k},\tau;\underset{\sim}{v}_1\underset{\sim}{v}_1') \ F(\underset{\sim}{v}_1') \ \delta(\underset{\sim}{v}_1'-\underset{\sim}{v}_2)$$

$$+ \ \theta(\tau) \int dt \ W(\underset{\sim}{k},t;\underset{\sim}{v}_1\underset{\sim}{v}_1')W^*(\underset{\sim}{k},t;\underset{\sim}{v}_2\underset{\sim}{v}_2') \ I(\underset{\sim}{k};\underset{\sim}{v}_1'\underset{\sim}{v}_2') \tag{50}$$

from which it follows that

$$\Pi_+(\underset{\sim}{k}\omega;\ \underset{\sim}{v}_1\underset{\sim}{v}_2) = W(\underset{\sim}{k}\omega;\underset{\sim}{v}_1\underset{\sim}{v}_1') \quad F(\underset{\sim}{v}_1')\,\delta(\underset{\sim}{v}_1'-\underset{\sim}{v}_2)$$

$$+\int d\omega' \ W(\underset{\sim}{k}\omega';\underset{\sim}{v}_1\underset{\sim}{v}_1') \ W^*(\underset{\sim}{k}\omega';\underset{\sim}{v}_2\underset{\sim}{v}_2')\,I(\underset{\sim}{k};\underset{\sim}{v}_1'\underset{\sim}{v}_2')$$

$$\bullet\ \delta_+ \ (\omega-\omega') \tag{51}$$

P can be obtained from $\Pi$ by integrating over the velocities. To evaluate this we note that

$$\int \underset{\sim}{dv} \quad W(\underset{\sim}{k}\omega;\underset{\sim}{v}\ \underset{\sim}{v}') = \frac{i}{\omega-\underset{\sim}{k}\cdot\underset{\sim}{v}'} \quad \frac{1}{\varepsilon(\underset{\sim}{k}\omega)} \tag{52}$$

Thus (for $\eta(\underset{\sim}{k}\omega)$ see (4.48))

$$P_{+}(\underset{\sim}{k}\omega) = e^2\int d\underset{\sim}{v}_1\int d\underset{\sim}{v}_2 \ \Pi_{+}(\underset{\sim}{k}\omega;\underset{\sim}{v}_1\underset{\sim}{v}_2)$$
$$= \frac{e^2}{\phi(k)} \left\{\frac{\eta(k\omega)}{\varepsilon(\underset{\sim}{k}\omega)} - \int d\omega' \frac{\alpha(\underset{\sim}{k}\omega')\eta^*(\underset{\sim}{k}\omega') + \alpha^*(\underset{\sim}{k}\omega')\eta(\underset{\sim}{k}\omega')}{|\varepsilon(\underset{\sim}{k}\omega')|^2}\right.$$
$$\left.\delta_{+}\ (\omega-\omega')\right\}$$
$$P(\underset{\sim}{k}\omega) = P^*(\underset{\sim}{k}\omega) \tag{53}$$

Since

$$P(\underset{\sim}{k}\omega) = 2\ \mathrm{Re}\ P_{+}(\underset{\sim}{k}\omega)$$
$$= +2\frac{e^2}{\phi(k)} \quad \frac{\eta'(\underset{\sim}{k}\omega)}{|\varepsilon(\underset{\sim}{k}\omega)|^2} \tag{54}$$

$$= 2\pi \quad e^2 \quad \int F(\underset{\sim}{v}) \quad \frac{\delta(\omega-\underset{\sim}{k}\cdot\underset{\sim}{v})}{|\varepsilon(\underset{\sim}{k}\omega)|^2} \underset{\sim}{dv} \tag{55}$$

This is the desired result, which is formally identical to (27), but it is valid (i) for non-equilibrium distribution functions and (ii) separately for each species. The latter feature can be exploited to express the correlation of electron density fluctuations which is of interest in connection with e.m. scattering problems. To derive the two-component result we note that for a multispecies system $V$ and $W$, in addition to their velocity arguments, carry a species index. It is easy to show that

$$W_{AB}\ (\underset{\sim}{k}\omega;\underset{\sim}{v}\underset{\sim}{v}') = \frac{i}{\omega-\underset{\sim}{k}\cdot\underset{\sim}{v}_A} \left\{\delta(\underset{\sim}{v}_A-\underset{\sim}{v}_A') \quad \delta_{AB}\right.$$
$$\left. - \frac{\phi(k)}{\varepsilon(\underset{\sim}{k}\omega)} \quad \underset{\sim}{k}\cdot\frac{\partial F_A}{\partial \underset{\sim}{v}_A} \quad \frac{1}{\omega-\underset{\sim}{k}\cdot\underset{\sim}{v}_B'}\right\} \qquad A,B = e,i \tag{56}$$
$$\int W_{AB}(\omega\underset{\sim}{k};\underset{\sim}{v}\underset{\sim}{v}')\underset{\sim}{dv} = (\delta_{AB} - \frac{\alpha_A}{\varepsilon})\frac{i}{\omega-\underset{\sim}{k}\cdot\underset{\sim}{v}_B'}$$

(49) now becomes

$$\mathcal{G}_{AB} = \sum_{C,D} \int \frac{d\omega}{2\pi} \; W_{AC} \; W^{*}_{BD} \; I_{CD}$$

$$I_{AB} = i\phi(k) \; \underset{\sim}{k} \cdot \left( \frac{\partial F_A}{\partial \underset{\sim}{v}_{A1}} F_B - \frac{\partial F_B}{\partial \underset{\sim}{v}_{B2}} F_A \right) \tag{57}$$

while (47) can obviously be re-written as

$$\langle \nu_e \nu_e^{*} \rangle = F_e \; \delta(\underset{\sim}{v}_e - \underset{\sim}{v}_e{'}) + \mathcal{G}_{ee} \tag{58}$$

This implies that

$$\Pi_{+}^{(ee)} = W_{ee} \; F_e \; \delta(\underset{\sim}{v}_e - \underset{\sim}{v}_e{'}) + \int d\omega' \sum_{A,B} W_{eA} \; W^{*}_{eB} \; I_{AB} \; \delta_{+}(\omega - \omega') \tag{59}$$

and therefore

$$P_{+}^{(ee)}(\underset{\sim}{k}\omega) = \frac{e^2}{\phi(k)} \left\{ \eta_e \frac{1+\alpha_i}{\varepsilon} - \int d\omega' \; \delta_{+}(\omega-\omega') \sum_{A,B} (\delta_{eA} - \frac{\alpha^{(e)}}{\varepsilon})(\delta_{eB} - \frac{\alpha^{(e)*}}{\varepsilon^{*}}) \; (\eta_A \alpha_B{}^{*} + \eta_B{}^{*} \alpha_A) \right\} \tag{60}$$

The evaluation of $P^{(ee)}$ yields

$$P^{(ee)}(\underset{\sim}{k}\omega) = 2\pi e^2 \left\{ \int F_e(\underset{\sim}{v}) \; \delta(\omega - \underset{\sim}{k} \cdot \underset{\sim}{v}) \frac{|1+\alpha_i(\underset{\sim}{k}\omega)|^2}{|\varepsilon(\underset{\sim}{k}\omega)|^2} \, d\underset{\sim}{v} + \int F_i(\underset{\sim}{v}) \; \delta(\omega - \underset{\sim}{k} \cdot \underset{\sim}{v}) \frac{|\alpha_e(\underset{\sim}{k}\omega)|^2}{|\varepsilon(\underset{\sim}{k}\omega)|^2} \, d\underset{\sim}{v} \right\} \tag{61}$$

Both (55) and (61) can be interpreted in terms of fluctuations of uncorrelated dressed particles [25].

## III. NON-LINEAR FLUCTUATION DISSIPATION THEOREM

It is expected that higher order conductivities or response functions will be related to n-point correlations ($n > 2$). This chapter is devoted to the demonstration of such a relation for $n = 3$. The relation, although not unexpected, is far from being trivial and the derivation is much more involved than the similar derivation for the linear FDT. The difficulties arise on three different levels:

(i) $\Omega^{(2)}$ is *not* expressible in terms of time-displaced currents only;
(ii) the relation between response function and correlations appears as an integral equation, rather than an algebraic equation;
(iii) the final relation connects a combination of response functions with different arguments with the correlation function.

In contrast to the case of the linear FDT, no non-equilibrium equivalent of the non-linear FDT is known, although there is little doubt that such a relation exists. This problem is under study; its clarification would be of importance for non-linear scattering problems.

Due to the complexities of the algebra, only the main steps of the derivation are given in what follows. For details, the reader is referred to the original publication [7].

Only longitudinal (Coulomb) perturbation and response will be considered. We further specialize to an electron plasma (stationary ions) although an extension to electron-ion plasma will be included later.

### III-1. *Dynamical FDT*

We take now the hamiltonian in the simple form

$$H^o = \sum_i \frac{p_i^2}{2m} + \frac{1}{2} \sum_{ij} \phi(|\underset{\sim}{x}_i - \underset{\sim}{x}_j|) \tag{1}$$

The perturbation can be represented by

$$= -e \sum_{\underset{\sim}{k}} \sum_i \hat{\phi}_{\underset{\sim}{k}}(t)\, e^{i\underset{\sim}{k}\cdot\underset{\sim}{x}_i} = -e \sum_{\underset{\sim}{k}} \hat{\phi}_{\underset{\sim}{k}}(t)\, \rho_{-\underset{\sim}{k}} \tag{2}$$

With the aid of the previously introduced time evolution operator, $U(\tau) = \exp(-iL^0\tau)$, the perturbed distribution function, up to second order, can be written down immediately:

$$\Omega^{(1)}(t) = -i\int_{-\infty}^{t} dt_1 \; U(t-t_1) \; L^1(t_1)\,\Omega^0$$

$$\Omega^{(2)}(t) = -\int_{-\infty}^{t} dt_1 \int_{-\infty}^{t_1} dt_2\, U(t-t_1) \quad (3)$$

$$L^1(t_1)\; U(t_1-t_2)L^1(t_2)\Omega^0$$

The latter can be re-written with a repeated application of the time evolution operator

$$\Omega^{(2)}(t) = +\int_{-\infty}^{t} dt \int_{-\infty}^{t_1} dt_2 \sum_{\underset{\sim}{k}_1\underset{\sim}{k}_2} \frac{1}{k_1k_2}\,\hat{E}_{\underset{\sim}{k}_1}(t_1)\hat{E}_{\underset{\sim}{k}_2}(t_2)$$

$$U(t-t_1)\left\{\rho_{\underset{\sim}{k}_1}, U(t_1-t_2)\,[\rho_{\underset{\sim}{k}_2}, \Omega^0]\right]$$

$$= -\Omega^0 \int_{-\infty}^{t} dt_1 \int_{-\infty}^{t_1} dt_2 \sum_{\underset{\sim}{k}_1\underset{\sim}{k}_2} \hat{E}_{\underset{\sim}{k}_1}(t_1)\hat{E}_{\underset{\sim}{k}_2}(t_2) \quad (4)$$

$$\Big\{\beta^2 j_{-\underset{\sim}{k}_1}(t_1) - j_{-\underset{\sim}{k}_2}(t_2)$$

$$\frac{i\beta}{k_1}\,[\rho_{-\underset{\sim}{k}_1}(t_1), j_{-\underset{\sim}{k}_2}(t_2)]\Big\}$$

where the j-s now represent longitudinal currents.

The second order average longitudinal current evaluated with this distribution function takes the form

$$\langle j_{\underset{\sim}{k}} \rangle^{(2)} = - \int_{-\infty}^{t} dt_1 \int_{-\infty}^{t} dt_2 \sum_{\underset{\sim}{k}_1 \underset{\sim}{k}_2} \hat{E}_{k_1}(t_1)\, \hat{E}_{\underset{\sim}{k}_2}(t_2)$$

$$\Big\{ \beta^2\, j_{\underset{\sim}{k}}(t)\, j_{-\underset{\sim}{k}_1}(t_1)\, j_{-\underset{\sim}{k}_2}(t_2) - \frac{i\beta}{k_1} \langle j_{\underset{\sim}{k}}(t)\, [\rho_{-\underset{\sim}{k}_1}(t_1),\, j_{-\underset{\sim}{k}_2}(t_2)] \rangle \Big\} \tag{5}$$

$\underset{\sim}{k}_2 = \underset{\sim}{k} - \underset{\sim}{k}_1$

The right-hand-side of (5) can be recognized as the quadratic longitudinal condictivity. Thus, by symmetrizing and introducing the $\tau_1 = t - t_1$ and $\tau_2 = t - t_2$ relative time variables and the notation

$$\langle j_{-\underset{\sim}{k}_1}(-\tau_1)\, j_{-\underset{\sim}{k}_2}(-\tau_2)\, j_{\underset{\sim}{k}}(0) \rangle = Q(\underset{\sim}{k}_1\ \tau_1;\ \underset{\sim}{k}_2\ \tau_2) \equiv Q(120)$$

$$\langle \rho_{-\underset{\sim}{k}_1}(-\tau_1)\, \rho_{-\underset{\sim}{k}_2}(-\tau_2)\, \rho_{\underset{\sim}{k}}(0) \rangle = P(\underset{\sim}{k}_1 . \tau_1;\ \underset{\sim}{k}_2\ \tau_2) \equiv P(120) \tag{6}$$

(the latter to be used later), and

$$- \frac{i}{k_1} \langle [\rho_{-\underset{\sim}{k}_1}(-\tau_1),\, j_{-\underset{\sim}{k}_2}(-\tau_{02})]\, j_{\underset{\sim}{k}}(0) \rangle = S(\underset{\sim}{k}_1 \tau_1;\ \underset{\sim}{k}_2 \tau_2) \equiv S(120)$$

one can write

$$\hat{\sigma}(\underset{\sim}{k}\ \tau_1,\ \underset{\sim}{k}_2 \tau_2) = - \frac{\beta}{2}\, \Theta(\tau_1)\Theta(\tau_1) \Big\{ \beta Q(120) + \Theta(\tau_2 - \tau_1)\, S(120) + \Theta(\tau_1 - \tau_2)\, S(210) \Big\} \tag{7}$$

The first term in (7) is the expected triplet current correlation; the difficulty lies in the elimination of the second and third unwiedly Poisson bracket terms.

The elimination of these terms can be accomplished in t the following steps:

(i) One demonstrates the validity of the identity

$$S(120) + S(102) = -\beta Q(120) \tag{8}$$

(ii) With the aid of (8), (7) can be re-stated as follows:

$$\sigma(\underset{\sim}{k}_1\tau_1;\underset{\sim}{k}_2\tau_2) = -\frac{\beta}{2}\Theta(\tau_1)\Theta(\tau_2)\cdot\{\Theta(\tau_2-\tau_1)S(102) + \Theta(\tau_1-\tau_2)S(201)\} \tag{9}$$

(iii) Changing the arguments of (9) and constructing a symmetrized combination one finds (exploring the time inversion and reflection symmetry of S):

$$\hat{\sigma}(-\underset{\sim}{k}_1\tau_2-\tau_1;\ \underset{\sim}{k}\,\tau_2) + \hat{\sigma}(\underset{\sim}{k}\,\tau_1;-\underset{\sim}{k}_2\,\tau_1-\tau_2)$$

$$\begin{aligned} = -\frac{\beta}{2}\Big\{ &\Theta(\tau_1)\Theta(\tau_2)\Theta(\tau_2-\tau_1)S(\underset{\sim}{k}_1\,\tau_1;\ \underset{\sim}{k}_2\,\tau_2) \\ &+ \Theta(\tau_1)\ \Theta(\tau_2)\ \Theta(\tau_1-\tau_2)S(\underset{\sim}{k}_2\,\tau_2;\ \underset{\sim}{k}_1\,\tau_1) \\ &+ \Theta(-\tau_1)\ \Theta(\tau_2)\ \Theta(\tau_1-\tau_2)S(-\underset{\sim}{k}\,0;\ \underset{\sim}{k}_2\,\tau_2) \\ &+ \Theta(\tau_1)\ \Theta(-\tau_2)\ \Theta(\tau_1-\tau_2)S(-\underset{\sim}{k}\,0;\ \underset{\sim}{k}_1\,\tau_1)\Big\} \end{aligned} \tag{10}$$

(iv) For positive time differences ($\tau_1 > 0$, $\tau_2 > 0$) which are of interest here, only the first two terms in (10) contribute; thus multiplication by $\Theta(\tau_1)\Theta(\tau_2)$ eliminates the third and fourth terms.

The combination of the remaining two terms with (7) cancels the two S-terms there, leading to the final expression

$$\Psi(\underset{\sim}{k}_1\omega_1,\underset{\sim}{k}\,\omega_2) = -\frac{\beta^2}{2}\int_{-\infty}^{+\infty}d\mu\int_{-\infty}^{+\infty}d\nu\ \delta_+(\omega_1-\mu)\ \delta_+(\omega_2-\nu)\ Q(\underset{\sim}{k}_1\,\mu_1;\ \underset{\sim}{k}_2\nu) \tag{11}$$

with

$$\Psi(\underset{\sim}{k}_1\omega_1;\underset{\sim}{k}_2\omega_2) \equiv \hat{\sigma}(\underset{\sim}{k}_1\omega_1;\ \underset{\sim}{k}_2\omega_2) - \int_{-\infty}^{+\infty} d\mu\, \delta_+(\omega_1-\mu)\ \hat{\sigma}(-\underset{\sim}{k}_1,-\mu;\ \underset{\sim}{k}_2\omega_2+\mu) - \int_{-\infty}^{+\infty} d\mu\, \delta_+(\omega_2-\mu)\ \hat{\sigma}(\underset{\sim}{k}_1\omega_1+\mu;\ \underset{\sim}{k}_2\ -\mu) \tag{12}$$

Eq. (11) and (12) constitute the primitive form of the quadratic FDT. (Cf. II.63 for its linear version.) The real part of both sides of Eq. (11) can now be projected out providing a more useful form of the FDT:

$$\mathrm{Re}\ \Psi(\underset{\sim}{k}_1\omega_1;\underset{\sim}{k}_2\omega_2) = -\frac{\beta}{8}\,Q(\underset{\sim}{k}_1\omega_1;\underset{\sim}{k}_2\omega_2) + \frac{\beta^2}{8\pi^2}\,PP\int_{-\infty}^{+\infty} d\mu \int_{-\infty}^{+\infty} d\nu\ \frac{Q(\underset{\sim}{k}_1\mu;\underset{\sim}{k}_2\nu)}{(\omega_1-\mu)\ (\omega_2-\nu)} \tag{13}$$

The connection between the Re $\psi$ and the real and imaginary parts of $\hat{\sigma}$ can easily be shown to be

$$\begin{aligned} \mathrm{Re}\ \Psi\ (\underset{\sim}{k}_1\omega_1;\underset{\sim}{k}_2\omega_2) &= \hat{\sigma}'(\underset{\sim}{k}_1\omega_1;\underset{\sim}{k}_2\omega_2) \\ &- \frac{1}{2}\ \hat{\sigma}'(-\underset{\sim}{k}_1-\omega_1;\underset{\sim}{k}\omega) - \frac{1}{2}\ \hat{\sigma}'(\underset{\sim}{k}\omega;\ -\underset{\sim}{k}_2-\omega_2) \\ &+ \frac{1}{2\pi}\ P\int_{-\infty}^{+\infty} d\mu\ \frac{1}{\omega_1-\mu}\ \hat{\sigma}''(-\underset{\sim}{k}_1-\mu;\underset{\sim}{k}\ \omega_2+\mu) \\ &+ \frac{1}{2\pi}\ P\int_{-\infty}^{+\infty} d\mu\ \frac{1}{\omega_2-\mu}\ \hat{\sigma}''(\underset{\sim}{k}\ \ \omega_1+\mu;-\underset{\sim}{k}_2-\mu) \end{aligned} \tag{14}$$

Equation (13) can be regarded as the equivalent of (II.65); it however is an *integral-equation*, whose solution is required for obtaining an explicit relation between Q and $\hat{\sigma}$.

To solve the integral equation we note that Q should be linear functional of $\hat{\sigma}$; moreover, the general solution of the integral equation

$$\mathrm{Re}\ \Psi = -\frac{\beta^2}{8}\{1-HH\}\ Q \tag{15}$$

(where the H-s stand for the respective Hilbert transforms) can be written down as

$$-\frac{\beta}{8}Q=\mathrm{Re}\left[\frac{1}{2}\ \Psi + \sum_i a_i X_i\ \{\hat{\sigma}\}\ \right] \tag{16}$$

where the $X_i \equiv X_i(\omega_1,\omega_2)$ functions are plus-functions of one, and minus functions of their other frequency arguments (while $\Psi = \psi(\omega_1\omega_2)$ is a plus-function of both of its frequency arguments). This is indeed so because

$$(1-HH)\ \psi = 2\psi \tag{17}$$

and

$$(1-HH)X_i = 0 \tag{18}$$

and Q is real.

The required plus-minus functions can be built up from $\hat{\sigma}$ by constructing

$$X_1 = \int_{-\infty}^{+\infty} d\mu\ \delta_-\ (\omega_1-\mu)\ \hat{\sigma}(-\underset{\sim}{k}_1-\mu;\underset{\sim}{k}\ \omega_2-\mu) \tag{19}$$

and

$$X_2 = \int_{-\infty}^{+\infty} d\mu\ \delta_-\ (\omega_2-\mu)\ \hat{\sigma}(\underset{\sim}{k}\ \ \omega_1-\mu;\ -\underset{\sim}{k}_2-\mu) \tag{20}$$

The obvious "triangle-symmetry" requirement that Q has to satisfy

$$Q(\kappa_1;\kappa_2) = Q(\kappa_1;-\kappa) = Q(-\kappa;\kappa_2)$$
$$\kappa = (\underset{\sim}{k}\omega) \tag{21}$$

limits the choice to (19) and (20) and provides the constants. The final result is

$$Q(\underset{\sim}{k}_1\omega_1; \underset{\sim}{k}_2\omega_2) = -4\beta^2 \{\hat{\sigma}'(\underset{\sim}{k}_1\omega_1; \underset{\sim}{k}_2\omega_2) - -\hat{\sigma}'(-\underset{\sim}{k}_1-\omega_1;\underset{\sim}{k}\omega) - \hat{\sigma}(\underset{\sim}{k}\omega;-\underset{\sim}{k}_2-\omega_2)\} \tag{22}$$

or, in terms of the triplet charge density correlations and polarizabilities

$$P(\underset{\sim}{k}_1\omega_1; \underset{\sim}{k}_2\omega_2) = \frac{1}{\pi\beta^2}\frac{1}{\varepsilon(\underset{\sim}{k}\omega)} \; k\, k_1 k_2 \; x$$
$$\mathrm{Im} \; \frac{\alpha(\underset{\sim}{k}_1\omega_1 \; ; \; \underset{\sim}{k}_2\omega_2)}{\omega_1\omega_2 \; \varepsilon(\underset{\sim}{k}_1\omega_1) \; \varepsilon(\underset{\sim}{k}_2\omega_2)} \tag{23}$$
$$- \frac{\alpha(-\underset{\sim}{k}_1-\omega_1;\underset{\sim}{k}\omega)}{\omega_1\omega\varepsilon^*(\underset{\sim}{k}_1\omega_1)\varepsilon(\underset{\sim}{k}_2\omega_2)} - \frac{\alpha(\underset{\sim}{k}\omega; \; - \underset{\sim}{k}_2 \; -\omega_2)}{\omega\omega_2\varepsilon(\underset{\sim}{k}_1\omega_1)\varepsilon^*(\underset{\sim}{k}_2\omega_2)}$$

(23) is the closest analogue to the linear dynamical FDT (Eq.II.67) expressing the correlation of three density fluctuations in terms of response functions. It is obvious that peaks in the fluctuation spectrum emerge in the vicinity of the frequencies where the dispersion relation for the pairs $(\underset{\sim}{k}\omega)$, $(\underset{\sim}{k}_1\omega_1)$ $(\underset{\sim}{k}_2\omega_2)$ are satisfied.

III-2. *Static FDT*

The static FDT is derived from (23) by setting the time differences $\tau_1$ and $\tau_2$ equal to zero and by relating the static P to the triplet correlation function.

The triplet equilibrium correlation function h (123) can be defined by (H(123) is the three particle distribution function)

$$H(123) = F(1)F(2)F(3) \left\{1+g(12) + g(23) + g(31) + h(123)\right\} \tag{24}$$

Thus the relation between the three point equal time charge density correlation and the triplet correlation function is

$$P(\underset{\sim}{k}_1, \tau_1 = 0;\ \underset{\sim}{k}_2, \tau_2 = 0) = n^2 e^3 \Big\{ g(\underset{\sim}{k}) + g(\underset{\sim}{k}_1) + g(\underset{\sim}{k}_2) + n h(\underset{\sim}{k}_1 \underset{\sim}{k}_2) \Big\} \tag{25}$$

$$\sum_{\underset{\sim}{k}_1 \underset{\sim}{k}_2} h(\underset{\sim}{k}_1, \underset{\sim}{k}_2)\, e^{i[\underset{\sim}{k}_1 \cdot (\underset{\sim}{x}_1 - \underset{\sim}{x}_3) + \underset{\sim}{k}_2 \cdot (\underset{\sim}{x}_2 - \underset{\sim}{x}_3)]} = h(123)$$

In order to reduce the left hand side of (23) to its static limit, integration over $\omega_1$ and $\omega_2$ has to be performed. This can be done through the repeated application of the K-K relations and the Poincare-Bertrand theorem. The details of this are given in [7]. The result is

$$\int_{-\infty}^{+\infty} \frac{d\omega_1}{2\pi} \int_{-\infty}^{+\infty} \frac{d\omega_2}{2\pi} \left[ \frac{\hat{\alpha}''(\underset{\sim}{k}\ \omega_1; \underset{\sim}{k}_2 \omega_2)}{\omega_1 \omega_2} - \frac{\hat{\alpha}''(-\underset{\sim}{k}_1 - \omega_1; \underset{\sim}{k}\omega)}{\omega_1 \omega} - \frac{\hat{\alpha}''(\underset{\sim}{k}\omega; -\underset{\sim}{k}_2 - \omega_2)}{\omega \omega_2} \right] = -\frac{1}{2}\, \alpha''(\underset{\sim}{k}_1 0; \underset{\sim}{k}_2 0) \tag{26}$$

The combination of (25), (2.73) and (26) transform (23) into

$$n^2 h(\underset{\sim}{k}_1 \underset{\sim}{k}_2) = \frac{1}{n\beta^2 \phi(k_1 k_2)} \operatorname{Im} \frac{\alpha(\underset{\sim}{k}_1 \underset{\sim}{k}_2)}{\varepsilon(\underset{\sim}{k}_1)\varepsilon(\underset{\sim}{k}_2)\varepsilon(\underset{\sim}{k})} - \sum_{\underset{\sim}{q} = \underset{\sim}{k}_1 \underset{\sim}{k}_2 \underset{\sim}{k}} \frac{1}{n\beta\phi(q)} \operatorname{Re} \frac{\alpha(\underset{\sim}{q})}{\varepsilon(\underset{\sim}{q})} + 2 \tag{27}$$

$$\phi(k_1 k_2) = 2\pi \frac{e^3}{k_1 k_2 k}\ ;\ \alpha(\underset{\sim}{k}_1 \underset{\sim}{k}_2) = \alpha(\underset{\sim}{k}_1 0; \underset{\sim}{k}_2 0),\ \text{etc.}$$

The importance of this result can be stated as follows:

(a) it provides an exact relation for the triplet equilibrium correlation function;

(b) it has the "order-raising" property, manifested by the fact that response functions calculated to order n (in the plasma parameter) yield h to order n + 2;

(c) it establishes the connection between non-linear response functions and higher order correlation functions.

To conclude this section we demonstrate how the non-linear FDT reproduces the lowest order ($e^4$) contribution to h(123) from the response functions calculated in the Vlasov approximation.

The linear polarizability in the Vlasov approximation has the well known form

$$\alpha(\underset{\sim}{k}) = \frac{\kappa^2}{k^2} \qquad \kappa^2 = 4\pi e^2 n\beta$$

while the quadratic polarizability can easily be shown [10] to become

$$\alpha(\underset{\sim}{k}_1\underset{\sim}{k}_2) = i\frac{2\pi e^3 n\beta^2}{k\, k_1 k_2} \tag{28}$$

Moreover, from the linear FDT we have shown that

$$\frac{1}{\varepsilon(\underset{\sim}{k})} = 1 + n\, g(k) \tag{29}$$

Substitution of (27), (28), (29) into (26) yields

$$n^2 h(\underset{\sim}{k}_1\underset{\sim}{k}_2) = \prod_{\underset{\sim}{q}=\underset{\sim}{k}_1\underset{\sim}{k}_2\underset{\sim}{k}} \left\{1+ng(q))\right\} - \sum_{\underset{\sim}{q}=\underset{\sim}{k}_1\underset{\sim}{k}_2\underset{\sim}{k}} \left\{1+ng(q)\right\} + 2 \tag{30}$$

or

$$h(\underset{\sim}{k}_1\underset{\sim}{k}_2) = g(k_1)g(k_2) + g(k_2)\,g(k)+g(k)g(k_1) + ng(k_1)g(k_2)g(k) \tag{31}$$

This result has been derived by Salpeter [11] through equilibrium methods and by Rostoker and O'Neil [12] through the BBGKY hierarchy. The ease of the present derivation is, however, noteworthy.

## III-3. *Electron-Ion Plasma*

The non-linear FDT can be generalized without any serious difficulty to two-component systems. The usefulness of such an expression is however even more limited than that of the similar linear one, because two response functions have to be related to four triplet correlation functions (eee, eei, eii, iii). We quote again the static theorem only:

$$n^2 \sum_{\substack{B,C\\=e,i}} \frac{e_A e_B e_C}{e^3} h_{ABC}(\underset{\sim}{k}_1 \underset{\sim}{k}_2) \tag{32}$$

$$= -\frac{1}{n\beta^2 \phi(k_1 k_2)} \operatorname{Im} \frac{\alpha^{(A)}(\underset{\sim}{k}_1 \underset{\sim}{k}_2)}{\varepsilon(\underset{\sim}{k}_1)\varepsilon(\underset{\sim}{k}_2)\varepsilon(\underset{\sim}{k})}$$

$$-\sum_{\underset{\sim}{q}=\underset{\sim}{k}_1 \underset{\sim}{k}_2 \underset{\sim}{k}} \operatorname{Re} \frac{1}{n\beta\phi(q)} Z_A e_A \frac{\alpha^{(A)}(q)}{\varepsilon(\underset{\sim}{q})} + 2 Z_A^2 e_A$$

## IV. DIELECTRIC DESCRIPTION OF DENSE PLASMAS

The actual calculation of the dielectric function (or the polarizabilities) has to be performed in some approximation. In plasma problems the approximation can be characterized by the ordering according to the "plasma parameter" $\gamma = \kappa^3 n^{-1}$ (inverse of the number of particles within the Debye sphere*). In most of the cases the Vlasov approximation ($\gamma^0$) is quite satisfactory since $\gamma << 1$. Corrections to the Vlasov polarizability to order $\gamma$ (correlational corrections) can and have been calculated by a systematic expansion of the BBGKY hierarchy [13]. (In Appendix B we discuss the behavior of the Vlasov dielectric function and that of the

---

*The relation between $\gamma$, the Brush-Sahlin-Teller [18] parameter $\Gamma$ and the expansion parameter $r_s$ in the theory of the electron-gas is given by Table I.

TABLE I.

| | | |
|---|---|---|
| $\gamma = \kappa^3 n^{-1}$ | $\Gamma = \left(\frac{4\pi}{3}\right)^{1/3}$ | $r_s = \left(\frac{3}{4\pi}\right)^{1/3}$ |
| $= (4\pi)^{3/2}\beta^{3/2}\ e^3 n^{1/2}$ | $n^{1/3}\beta e^2$ | $\frac{1}{n^{1/3}}\ \frac{e^2 m}{h^2}$ |
| $\gamma =$ | $21.76\ \Gamma^{3/2}$ | $16 r_s{}^{3/2}$ |
| $\Gamma = 0.120\ \gamma^{2/3}$ | | $0.763 r_s$ |
| $r_s = 0.157\ \gamma^{2/3}$ | $1.308\ \Gamma$ | |

Note: $\kappa^2 = \frac{6\pi n e^2}{E_F}$, $E_F = \frac{p_F^2}{2m}$, $n = \frac{p_F^3}{3\pi^2 h^3}$ for a degenerate electrongas.

correlational corrections). In certain physical systems however (electron gas in solids, plasmas in stellar interiors, laser compressed plasmas - see Table II) $\gamma > 1$ and therefore a simple expansion technique fails. However, the FDT which furnishes an exact relation (independent of the value $\gamma$) for the dielectric response function (in terms of the correlation function g) can be exploited to provide a *non-perturbative* way to calculate $\varepsilon$ and g. In order to establish a self-consistent scheme, another link between $\varepsilon$ and g is required, which is established *via* the BBGKY hierarchy. This approach which was originally proposed by Hubbard [14] has been developed by Singwi and his collaborators [15,16] and also by Ichimaru [17]. The present chapter is devoted to the discussion of this approach and some of its results.

## IV-1. *Dielectric Function in Terms of the Pair Correlation Function: Hubbard Approximation*

The first equation of the BBGKY hierarchy, in the presence of an external field, can be written as follows:

$$\left\{\frac{\partial}{\partial t} + \underset{\sim}{v}_1\cdot\frac{\partial}{\partial \underset{\sim}{x}_1}\right\} F(1) = -\frac{\partial}{\partial \underset{\sim}{v}_1}\cdot \left\{-\frac{e}{m}\ \hat{\underset{\sim}{E}}(1)\ F(1) + \frac{1}{m}\int \underset{\sim}{K}\ (12)\ \ G(12)\ d2\right\} \tag{1}$$

TABLE II.

| | $n/cm^3$ | $T(E_F)/eV$ | $\kappa^{-1}/cm$ | $\gamma$ | $\Gamma$ |
|---|---|---|---|---|---|
| Metals | $(5 \sim 10)$ $\times 10^{22}$ | $4 \sim 7$ | $(5 \sim 6)$ $10^{-9}$ | $60 \sim 90$ | $1.8 \sim 2.4$ |
| Stellar interiors | $(0.1 \sim 1)$ $\times 10^{26}$ | $(0.7 \sim 1.5)$ $\times 10^3$ | $(2 \sim 6)$ $\times 10^{-9}$ | $0.5 \sim 2$ | $0.08 \sim 0.2$ |
| Laser compressed plasma | $10^{27}$ | $(0.1 \sim 1)$ $\times 10^4$ | $(1 \sim 3)$ $\times 10^{-9}$ | $0.12 \sim 3.6$ | $0.03 \sim 0.3$ |

where $\underset{\sim}{K}(12)$ is the elementary force originating from particle 2 and acting on particle 1. For Coulomb interaction*

$$\underset{\sim}{K}(12) = -i\sum_{\underset{\sim}{k}} \underset{\sim}{k}\phi(k)\, e^{-i\underset{\sim}{k}\cdot(\underset{\sim}{x}_1 - \underset{\sim}{x}_2)} \tag{2}$$

The two particle distribution function G(12) can be written as

$$G(12) = F(1)F(2)\left\{1 + g(12)\right\} \tag{3}$$

If the g(12) term is ignored, (1) reduces to the familiar Vlasov equation. Our purpose here, however, is to include the correlation terms, without making an ixpansion in the plasma parameter.

In equilibrium $g(12) = g^\circ(|\underset{\sim}{x}_1-\underset{\sim}{x}_2|)$, the equilibrium pair correlation function, and $F(1) = F^\circ(v\ )$ the Maxwell distribution. When the equilibrium is perturbed by an external field $\hat{E}$, both F(1) and G(12) become different from their equilibrium values. To first order in the perturbation

$$\begin{aligned} F(1) &= F^{o}(v_1) + F'(1) \\ g(12) &= g^{o}(|\underset{\sim}{x}_1-\underset{\sim}{x}_2|) + g'(12) \end{aligned} \tag{4}$$

* "Unit volume" is assumed throughout this chapter.

and

$$G(12) = F^{o}(v_1)F^{o}(v_2) \quad \{1 + g^{o}(|\underset{\sim}{x}_1-\underset{\sim}{x}_2|)\}$$
$$+ F^{o}(v_1)F'(2) \quad \{1 + g^{o}(|\underset{\sim}{x}_1-\underset{\sim}{x}_2|)\} \tag{5}$$
$$+ F'(1)F^{o}(v_2) \quad \{1 + g^{o}(|\underset{\sim}{x}_1-\underset{\sim}{x}_2|)\}$$
$$+ F^{o}(v_1)F^{o}(v_2) \; g'(12)$$

In the linearized version of (3) for F' the first term in (5) is of 0 order and thus does not contribute; the third term integrates out to zero; the fourth term (the change of correlations under the effect of the perturbation) is difficult to estimate or evaluate: Hubbard's approximation consists of ignoring this term and replacing the full correlation by its equilibrium value. Thus the Fourier transform of Eq. (3) becomes

$$-i(\omega-\underset{\sim}{k}\cdot\underset{\sim}{v}) \; F'(\underset{\sim}{k}\omega,\underset{\sim}{v}) - \frac{i}{m}\frac{\partial F^{o}}{\partial \underset{\sim}{v}}\{\underset{\sim}{k}\phi(k) + \sum_{\underset{\sim}{q}} \underset{\sim}{q} \; \phi(q)$$
$$g^{o}(\underset{\sim}{k}-\underset{\sim}{q})\}\int F'(\underset{\sim}{k}\omega;\underset{\sim}{u})d\underset{\sim}{u} = \frac{e}{m} \; \hat{\underset{\sim}{E}}_{\underset{\sim}{k}} \cdot \frac{\partial}{\partial \underset{\sim}{v}} \; F^{o}(\underset{\sim}{v}) \tag{6}$$

A glance at (6) shows that the only difference between this equation and the customary Vlasov equation is that $\phi(k)$ is replaced by an "effective interaction" $\psi(k)$

$$\psi(k) = \phi(k) + \sum_{\underset{\sim}{q}} \frac{\underset{\sim}{k}\cdot\underset{\sim}{q}}{k^2}\phi(q) \quad g^{o}(\underset{\sim}{k}-\underset{\sim}{q})$$
$$= \phi(k) \; \{1 + \sum_{\underset{\sim}{q}} \frac{\underset{\sim}{k}\cdot\underset{\sim}{q}}{q^2} \; g^{o}(\underset{\sim}{k}-\underset{\sim}{q})\} \tag{7}$$
$$= \phi(k) \quad \{1+u(k)\}$$

with

$$u(k) = \sum_{\underset{\sim}{q}} \frac{\underset{\sim}{k}\cdot\underset{\sim}{q}}{q^2} \quad g^{o}(\underset{\sim}{k}-\underset{\sim}{q}) \tag{8}$$

(The second line follows from the isotropy of the system, which requires that $\sum_{\underset{\sim}{q}} \underset{\sim}{q}\phi(q)g^{\circ}(\underset{\sim}{k}-\underset{\sim}{q})$ point along $\underset{\sim}{k}$).

We recall that the Vlasov $\varepsilon_V$ can be written as

$$\varepsilon_V(\underset{\sim}{k}\omega) = 1 + \frac{1}{m}\phi(k)\int \frac{\underset{\sim}{k}\cdot\frac{\partial F^o}{\partial \underset{\sim}{v}}}{\omega-\underset{\sim}{k}\cdot\underset{\sim}{v}}\, d\underset{\sim}{v} \tag{9}$$

$$\equiv 1 + \alpha_o(\underset{\sim}{k}\underset{\sim}{\omega})$$

which identifies $\alpha_0(\underset{\sim}{k}\omega)$. The replacement in (6) of $\phi(k)$ by $\psi(k)$ suggests that an improved $\varepsilon$ can be constructed by a similar replacement in (9). This is $\varepsilon_H$(Hubbard):

$$\varepsilon_H(\underset{\sim}{k}\omega) = 1 + \{1+u(k)\}\,\alpha_o(\underset{\sim}{k}\omega) \tag{10}$$

Although this construction is actually not correct, it is instructive to see whether $\varepsilon_H$ satisfies obvious limiting behavior (and thus sum rule) conditions.

$$\omega \to \infty \qquad \alpha_o \to -\frac{\omega_o^2}{\omega^2}$$

$$\lim_{\omega\to\infty} \varepsilon_H(\underset{\sim}{k}\omega) = 1 - \{1+u(k)\}\frac{\omega_o^2}{\omega^2} \tag{11}$$

which is obviously in conflict with the conductivity sum rule requirements.

The $\omega \to 0$ limit also follows immediately, since $\alpha_o(\underset{\sim}{k}0) = \frac{\kappa^2}{k^2}$ :

$$\varepsilon_H(\underset{\sim}{k}0) = 1 + \{1+u(k)\}\frac{\kappa^2}{k^2} \tag{12}$$

In order to see whether this limit satisfies the compressibility sum rule requirements, the evaluation of u(k) for $k\to 0$ is necessary. This can be done as follows (hereafter we drop the superscript $^0$ on g).

$$u(k) = \sum_{\underset{\sim}{q}} \frac{\underset{\sim}{k}\cdot\underset{\sim}{q}}{q^2} \, g(\underset{\sim}{k}-\underset{\sim}{q})$$

$$= \frac{1}{4\pi^2} \int_0^\infty dq q^2 \int_{-1}^{+1} d\mu \, \frac{kq\mu}{q^2} \, g(|\underset{\sim}{k}-\underset{\sim}{q}|) \tag{13}$$

$$\simeq \frac{k}{4\pi^2} \int_0^\infty dqq \int_{-1}^{+1} d\mu \, \mu \{g(q)-g'(q)\mu k\} =$$

$$= -\frac{k^2}{6\pi^2} \int_0^\infty dqq \, g'(q) = +\frac{k^2}{6\pi^2} \int_0^\infty dq \, g(q)$$

The last integral is bounded and expected to be negative. Thus it is useful to reexpress (13) as

$$u(k) \simeq -Z \frac{k^2}{\kappa^2} + \ldots$$

$$Z = \frac{\kappa^2}{6\pi^2} \int_0^\infty dq \, g(q) \tag{14}$$

Returning to (12), we have

$$\varepsilon_H(\underset{\sim}{k}0) = 1 + \{1 - Z \frac{k^2}{\kappa^2}\} \frac{\kappa^2}{k^2} = (1-Z) + \frac{\kappa^2}{k^2} \tag{15}$$

We see that this behavior violates the one required by the compressibility sum rule.

Ignoring these difficulties, Eq. (12) which expresses now as a functional of g, can be substituted into the FDT expression (II.30) which connects g with ε:

$$g(k) = \frac{1}{n} \left\{ \frac{k^2}{\kappa^2} [1 - \frac{1}{1 + [1+u(k)] \frac{\kappa^2}{k^2}} ] - 1 \right\}$$
$$= - \frac{1}{n} \frac{[1+u(k)]\kappa^2 - u(k) k^2}{[1+u(k)]\kappa^2 + k^2} \tag{16}$$

A further check on the consistency of $\varepsilon_H$ can be obtained from (16).

If g is expanded in powers of $\gamma$

$$g = g_1 + g_2 + \ldots \ldots \tag{17}$$

(we use subscripts in order to distinguish this expansion from the one in powers of $\hat{E}$) then (16) provides an iteration scheme to obtain higher order contributions from the lower order ones:

$$g_1(k) = - \frac{1}{n} \frac{\kappa^2}{\kappa^2+k^2}$$
$$g_2(k) = + \frac{1}{n} \frac{\kappa^4}{(\kappa^2+k^2)^2} \sum_{\underset{\sim}{q}} \frac{\underset{\sim}{q} \cdot \underset{\sim}{k}}{q^2} g_1(|\underset{\sim}{k}-\underset{\sim}{q}|) \tag{18}$$

This latter expression can be compared with the exact second order $g_2(k)$ calculated by O'Neil and Rostoker [12] from the BBGKY hierarchy with the aid of the ternary correlation function $h_2$, calculated from (3.31);

$$g_2(k) = -\frac{1}{n} \frac{\kappa^2 k^2}{(\kappa^2 + k^2)^2} \sum_{\underset{\sim}{q}} \frac{\underset{\sim}{q} \cdot \underset{\sim}{k}}{\kappa^2 + q^2} g_1(\underset{\sim}{k}-\underset{\sim}{q}) \tag{19}$$

The discrepancy between (18) and (19) is obvious. Eq. (16) can be solved numerically by iteration for g(k) and then for $\varepsilon(k\omega)$. This scheme has been followed by Hubbard [14] who obtained a g(k) which exhibited improved features over the Debye g(k). In view, however, of the difficulties associated with $\varepsilon_H$, we do not discuss this and in the next section we consider a more consistent approach for the determination of $\varepsilon$.

IV-2. *Dielectric Function in Terms of the Pair Correlation Function: Improved Construction*

We return now to Eq. (6) in order to determine $\varepsilon$ without any additional assumption. The result is the improved $\varepsilon_S$ (Singwi *et al.* [15]).

The solution of (6) can be written down immediately* as

$$F'(\underset{\sim}{k}\omega,\underset{\sim}{v}) = \frac{1}{\omega-\underset{\sim}{k}\cdot\underset{\sim}{v}}\frac{1}{m}\left[-\underset{\sim}{k}\psi(k)\int F'(\underset{\sim}{k}\omega,\underset{\sim}{u})\,d\underset{\sim}{u} + ie\underset{\sim}{E}_k\cdot\right]\cdot\frac{\partial}{\partial\underset{\sim}{v}}F^o(v) \tag{20}$$

which, when integrated over $\underset{\sim}{v}$, yields an expression for the average particle density $n' = \int F'd\underset{\sim}{v}$

$$i\underset{\sim}{k}\,\phi\,(k)n' = \frac{-\frac{\phi(k)}{m}\underset{\sim}{k}\cdot\int\frac{\partial F^o}{\partial\underset{\sim}{v}}\frac{1}{\omega-\underset{\sim}{k}\cdot\underset{\sim}{v}}\,d\underset{\sim}{v}}{1+\frac{\psi(k)}{m}\underset{\sim}{k}\cdot\int\frac{\partial F^o}{\partial\underset{\sim}{v}}\frac{1}{\omega-\underset{\sim}{k}\cdot\underset{\sim}{v}}\,d\underset{\sim}{v}}\,e\hat{\underset{\sim}{E}}_k \tag{21}$$

This can be compared with $\underset{\sim}{E} = -\frac{\alpha}{\varepsilon}\hat{\underset{\sim}{E}}$ from which one infers the following relation:

$$-\frac{\alpha}{\varepsilon} = -\frac{\alpha_o}{1+(1+u)\alpha_o}\,. \tag{22}$$

This then leads to the expression for $\varepsilon_S$:

$$\varepsilon_S(\underset{\sim}{k}\omega) = 1 + \frac{\alpha_o(\underset{\sim}{k}\omega)}{1+u(k)\alpha_o(\underset{\sim}{k}\omega)} \tag{23}$$

---

*It could be argued that the correlation function should be a functional of the *total* (plasma and external) fields and therefore $\underline{u}$ should - in contrast to (20) - multiply both of them; this, however, leads to a construction similar to Hubbard's, with all the previously mentioned difficulties.

The difference between $\varepsilon_H$ and $\varepsilon_S$ is evident; indeed,

$$\varepsilon_S = \frac{\varepsilon_H}{1+u\alpha_o} \tag{24}$$

The consistency tests for $\varepsilon_S$ yield a much more satisfactory result. For

$$\lim_{\omega \to \infty} \varepsilon_S(\underset{\sim}{k}\omega) = 1 - \frac{\omega_o^2}{\omega^2} \tag{25}$$

and for $\omega \to 0$

$$\varepsilon_S(\underset{\sim}{k}0) = 1 + \frac{\kappa^2}{k^2 + \kappa^2 u(k)} \tag{26}$$

which, in view of (14), gives

$$\lim_{k \to 0} \varepsilon_S(k0) = 1 + \frac{\kappa^2}{1-Z} \frac{1}{k^2} \tag{27}$$

Both of these results are in agreement with the sum rule requirements. When $\varepsilon_S$ is substituted into the FDT relation, the integral equation for g(k) becomes

$$g(k) = \frac{1}{n} \left\{ \frac{k^2}{\kappa 2} \frac{\alpha_o(\underset{\sim}{k}0)}{1+[1+u(k)]\alpha_o(\underset{\sim}{k}0)} - 1 \right\}$$

$$= - \frac{1}{n} \frac{[1+u(k)]\kappa^2}{k^2+[1+u(k)]\kappa^2} \tag{28}$$

Then, expansion in $\gamma$ yields

$$g_1(k) = -\frac{1}{n}\frac{\kappa^2}{k^2+\kappa^2}$$
$$g_2(k) = -\frac{1}{n}\frac{\kappa^2 k^2}{(\kappa^2+k^2)^2}\sum_{\underset{\sim}{q}}\frac{\underset{\sim}{k}\cdot\underset{\sim}{q}}{q^2}\,g_1(|\underset{\sim}{k}-\underset{\sim}{q}|) \tag{29}$$

This is now, apart from the $q^2$ denominator, in good agreement with the O'Neil-Rostoker [12] result.

IV-3. *Some Results*

The integral equation (28) when solved for g(k) provides, through (8) and (23) an expression for $\varepsilon_S(\underset{\sim}{k}\omega)$. No detailed study of the $\varepsilon_S(\underset{\sim}{k}\omega)$ so derived is, however, available. Some simple, though fairly significant results can be stated without a detailed investigation of $\varepsilon$ [15, 16].

(a) First the modification of the plasma oscillation dispersion relation can easily be derived by considering $\varepsilon_S(\underset{\sim}{k} \to 0,\ \omega)$:

$$\varepsilon_S(\underset{\sim}{k} \to 0,\omega) = 1+\frac{\alpha_o(k\to 0,\omega)}{1-Z\,\frac{k^2}{\kappa^2}\,\alpha_o(\underset{\sim}{k}\to 0,\omega)} \tag{30}$$

Since

$$\alpha_o(\underset{\sim}{k}\to 0\ \omega) = -\frac{\omega^2}{\omega^2}\left(1+a^2\frac{k^2}{\omega^2}\right) \simeq \frac{\omega_o}{\omega^2}\left(1+a^2\frac{k^2}{\omega_o{}^2}\right)$$
$$a = \sqrt{\frac{3}{\beta m}} \tag{31}$$

one immediately evaluates the solution of the dispersion relation $\varepsilon(\underset{\sim}{k}\omega) = 0$:

$$\omega^2 = \omega_o^2 + \left(a^2 - Z\frac{\omega_o^2}{\kappa^2}\right)k^2 = \omega_o^2 + \left(1-\frac{Z}{3}\right)a^2k^2 \tag{32}$$

This result indicates that for sufficiently high positive Z the slope of the plasma dispersion curve can change sign and become negative.

(b) If the static dispersion relation

$$\varepsilon_S\ (\underset{\sim}{k}\ 0) = 0 \tag{33}$$

can be satisfied for a *real k*, it can be inferred that the system will undergo phase transition to a state characterized by a periodicity corresponding to the value of this particular $\underset{\sim}{k}$ vector (cf. 18]):

$$1 + [1+u(k)]\ \frac{\kappa^2}{k^2} = 0 \tag{34}$$

or for small k

$$1 + \{1 - Z\frac{k^2}{\kappa^2}\}\ \frac{\kappa^2}{k^2} = 0$$

$$k^2 = \frac{\kappa^2}{Z-1} \tag{35}$$

Evidently, the validity of (35) is assured for fairly large Z only; this, however, casts some doubt on the feasibility of replacing u(k) by $-Z(k^2/\kappa^2)$.

(c) A further condition for the instability of the plasma state is the compressibility $K \rightarrow \infty$ or becoming negative. Since, from (I.33)

$$K = K_o\ \frac{k^2}{\kappa^2}\ \ \alpha(\underset{\sim}{k} \longrightarrow 0, 0)$$

we find

$$K = \frac{K_o}{1-Z} \tag{36}$$

The requirement for the phase transition is again $Z \geq 1$. Since Z plays an important role in determining the behaviour of the system, it is instructive to estimate its dependence on. To do this, we replace g(q) in (14) by $g_1(q)$, to find

$$Z = +\frac{\kappa^2}{6\pi^2} \quad \frac{1}{n} \int_0^\infty \frac{\kappa^2}{\kappa^2+q^2} \, dq$$

$$= \frac{\kappa^3}{6\pi^2 n} \int_0^\infty \frac{1}{1+x^2} \, dx \tag{37}$$

$$= \frac{\gamma}{12\pi}$$

Detailed numerical work on the solution of the integral Equation (28) for a classical plasma has been performed by Berggren [16] for $0.05 < \Gamma < 2.5$. His result can be summarized as follows:

(i) The self-consistent pair correlation function almost completely eliminates the pathological large negative region for r (particle-separation) $\to 0$, that emerges from the Debye-Hückel approximation.

(ii) At $\Gamma \sim 2$ oscillations appear in g(r), indicating the onset of a short range order.

(iii) The decrease in the steepness of the dispersion curve is established, but no change in the sign of the slope is found. (However, such a change has been found for a degenerate electrongas by Singwi *et al.* [15] for $r_s > 6$.

(iv) The extrapolated inverse compressibility goes to 0 for $\Gamma \simeq 3$.

## IV-4. *Some Speculations*

Further improvements on the construction of u(k) have recently been suggested by Singwi and his collaborators [19] and by Ichimaru [17]. Singwi *et al.* [19] suggest the following expression to replace u(k):

$$u(\underset{\sim}{k}\omega) = \frac{1}{e^2 n^2} \sum_{\underset{\sim}{q}} \int \frac{d\nu}{2\pi} \frac{\underset{\sim}{k}\cdot\underset{\sim}{q}}{q^2} \frac{1}{\varepsilon(\underset{\sim}{q}\ \nu)} P(\underset{\sim}{k}-\underset{\sim}{q},\omega-\nu) \tag{38}$$

where P is the dynamical charge density correlation (cf. Eq. (II.49)). In the limit when the frequency dependence in (38) is ignored one finds

$$u(k) = \sum_{\underset{\sim}{q}} \frac{\underset{\sim}{k}\cdot \underset{\sim}{q}}{q^2 \varepsilon(\underset{\sim}{q}\, 0)} \; g\,(\underset{\sim}{k}-\underset{\sim}{q}) \tag{39}$$

which to second order in γ exactly reproduces the O'Neil-Rostoker result [12]. The numerical work performed by Berggren [16] does not indicate any drastic modification of the previously quoted result by the inclusion of the screening factor $\varepsilon(\underset{\sim}{q}0)$ in (39).

Ichimaru's [17] suggestion is to use

$$u(k) = \frac{1}{n^2} \sum_{\underset{\sim}{q}} \frac{\underset{\sim}{k}\cdot \underset{\sim}{q}}{q^2} \{1+ng(\underset{\sim}{k}-\underset{\sim}{q})\}\ \{1+ng(q)\} \tag{40}$$

which also conforms with the O'Neil-Rostoker limit [12].

No systematic evaluation of the ignored g' (12) term in (5) is available at the present time, although it should be possible to evaluate $g^\circ(12)$* (and possibly of $h^\circ(123)$) and incorporate this in the self-consistent scheme for ε and $g^\circ$.

A more ambitious improvement on the self consistent scheme could be feasible by exploiting *both* the linear and non-linear FDT. Should one pursue the calculation of g'(12), ε instead of being truncated as a functional of $g^\circ$, could be constructed as a functional of binary and ternary equilibrium correlations functions (by replacing only h by its equilibrium value):

$$\varepsilon\ \{g^o\} \rightarrow \varepsilon\{g^o, h^o\} \tag{41}$$

Similarly, the quadratic polarizability could be expressed as $\alpha_2\{g^\circ,h^\circ\}$. On the other hand, the linear FDT provides $g^\circ\{\varepsilon\}$ while the non-linear FDT complements this with $h^\circ\{\varepsilon,\alpha_2,\}$, thus completing the self-consistent scheme. If one assumes that the higher order correlations are less affected by the external perturbation than the lower order ones (which is a reasonable conjecture) then such a scheme would provide a much improved calculation for dense plasmas.

---

*To first order in γ this has been done by Coste [20] and Guernsey [21]. Their result is discussed in Appendix A.

APPENDIX IV-A: *Correlations in a Homogeneous, Stationary System*

The binary correlation function correct to the lowest order in the plasma parameter $\gamma$ can be determined for a stationary, homogeneous system without great difficulty [23, 24]. Thus the result includes - but is not restricted to - equilibrium; the way the equilibrium form of the correlation function follows form the more general result will be indicated in some detail.

Apart from its interest on its own merits, the present derivation serves as the paradigm for the inhomogeneous, nonstationary case, which is the basis of the calculation of the higher order corrections for $\varepsilon$ and which will be discussed in the subsequent Appendix.

The second equation of the BBGKY hierarchy can be Fourier transformed with respect to $\underset{\sim}{x}_1 - \underset{\sim}{x}_2$ and t to yield

$$-i\{\omega - V(\underset{\sim}{k};1) - V(-\underset{\sim}{k};2)\} \quad G(\underset{\sim}{k};\underset{\sim}{v}_1\underset{\sim}{v}_2)$$

$$= i\frac{\phi(k)}{m}\underset{\sim}{k} \cdot \left\{\frac{\partial}{\partial\underset{\sim}{v}_1}F(\underset{\sim}{v}_1)\ F(\underset{\sim}{v}_2) - \frac{\partial}{\partial\underset{\sim}{v}_2}F(\underset{\sim}{v}_2)\ F(\underset{\sim}{v}_1)\right\} \cdot 2\pi\delta(\omega)$$

$$+i\frac{\phi(k)}{m}\underset{\sim}{k} \cdot \left(\frac{\partial}{\partial\underset{\sim}{v}_1} - \frac{\partial}{\partial\underset{\sim}{v}_2}\right) G(\underset{\sim}{k};\underset{\sim}{v}_1\underset{\sim}{v}_2) \tag{42}$$

$$+\sum_{\underset{\sim}{q}}\int\left\{i\phi(q)\ \underset{\sim}{q}\cdot\left(\frac{\partial}{\partial\underset{\sim}{v}_1} - \frac{\partial}{\partial\underset{\sim}{v}_3}\right) H(\underset{\sim}{k}-\underset{\sim}{q}\ ,\ -\underset{\sim}{k}\ ,\ \underset{\sim}{q}\ ;\ \underset{\sim}{v}_1\underset{\sim}{v}_2\underset{\sim}{v}_3)\right.$$

$$\left.-i\phi(q)\ \underset{\sim}{q}\cdot\left(\frac{\partial}{\partial\underset{\sim}{v}_2} - \frac{\partial}{\partial\underset{\sim}{v}_3}\right) H(\underset{\sim}{k}\ ,\ -\underset{\sim}{k}+\underset{\sim}{q}\ ,\ -\underset{\sim}{q}\ ;\ \underset{\sim}{v}_1\underset{\sim}{v}_2\underset{\sim}{v}_3)\right\} d\underset{\sim}{v}_3$$

In (42) $V(k;\,1) \equiv V(k;\, v_1 v_1{}')$ etc. are the Vlasov-Dupree operators discussed earlier (Appendix 2-A); $G$ and $H$ are defined

$$G\ (12) = F(1)\ F(2) + \mathcal{G}(12)$$

$$H\ (123) = F(1)\ F(2)\ F(3) + F(1)\mathcal{G}(23) + F(2)\mathcal{G}(31) + F(3)\mathcal{G}(12) + \mathcal{H}(123) \qquad (43)$$

The presence of the $\delta(\omega)$ term is a consequence of the stationarity of F.

To the first order in the plasma parameter all but the first term on the right hand side (which we denote by I)

$$I(\underset{\sim}{k};\underset{\sim}{v}_1\underset{\sim}{v}_2) = i\phi(k)\underset{\sim}{k}\cdot\left(\frac{\partial}{\partial \underset{\sim}{v}_1} - \frac{\partial}{\partial \underset{\sim}{v}_2}\right) F(\underset{\sim}{v}_1)\ F(\underset{\sim}{v}_2) \qquad (44)$$

are negligible. The formal solution* of (44)

$$\mathcal{G}\ (\underset{\sim}{k};\underset{\sim}{v}_1\underset{\sim}{v}_2) = \frac{i}{-\mathcal{V}(\underset{\sim}{k};1) - \mathcal{V}(-\underset{\sim}{k};2)}\ I(\underset{\sim}{k},\underset{\sim}{v}_1\underset{\sim}{v}_2)\ 2\pi\delta(\omega)$$

$$= i\int\frac{d\nu}{2\pi}\ \mathcal{W}(\nu,\underset{\sim}{k};\underset{\sim}{v}_1\underset{\sim}{v}_1')\ \mathcal{W}(\omega-\nu,-\underset{\sim}{k};,\underset{\sim}{v}_2\underset{\sim}{v}_2') \qquad (45)$$

$$I(\underset{\sim}{k};\underset{\sim}{v}_1'\underset{\sim}{v}_2')\ 2\pi\delta(\omega)$$

$$= i\int\frac{d\nu}{2}\ \mathcal{W}(\nu\underset{\sim}{k};\underset{\sim}{v}_1 v_1')\ \mathcal{W}^*(\nu\underset{\sim}{k};\underset{\sim}{v}_2\underset{\sim}{v}_2')$$

$$I(\underset{\sim}{k};\underset{\sim}{v}_1'\underset{\sim}{v}_2')\ 2\pi\delta(\omega)$$

We have ignored the possible presence of an eventual initial value term $\mathcal{G}(t{=}0)$; thus $\mathcal{G}$ is now completely time independent and the trivial $2\pi\delta(\omega)$ coefficient can be ignored. Recalling the explicit representation of $\mathcal{W}$, one is left with the evaluation of

---

*We recall that integration over the repeated velocity arguments is understood. Also, $\mathcal{W}$ is defined such a way that, $\mathcal{W}(t)f(t') = 0$ for $t < t'$.

$$G(\underset{\sim}{k};\underset{\sim}{v}_1\underset{\sim}{v}_2) = i\int\frac{d\nu}{2\pi}\,\frac{i}{\nu-\underset{\sim}{k}\cdot\underset{\sim}{v}_1}$$

$$\left\{\delta(\underset{\sim}{v}_1-\underset{\sim}{v}_1') - \underset{\sim}{k}\cdot\underset{\sim}{\partial}F_1\frac{\phi(k)}{\varepsilon(\underset{\sim}{k}\nu)}\,\frac{1}{\nu-\underset{\sim}{k}\cdot\underset{\sim}{v}'}\right\}$$

$$\frac{i}{-\nu+\underset{\sim}{k}\cdot\underset{\sim}{v}_2}\left\{\delta(\underset{\sim}{v}_2-\underset{\sim}{v}_2') + \underset{\sim}{k}\cdot\underset{\sim}{\partial}F_2\frac{\phi(k)}{\varepsilon^*(\underset{\sim}{k}\nu)}\,\frac{1}{-\nu+\underset{\sim}{k}\cdot\underset{\sim}{v}_2'}\right\} \tag{46}$$

$$\phi(k)\left\{\underset{\sim}{k}\cdot\underset{\sim}{\partial}F_{1'}F_{2'} - \underset{\sim}{k}\cdot\underset{\sim}{\partial}F_{2'}F_{1'}\right\}$$

where

$$F_i \equiv F(\underset{\sim}{v}_i)$$

$$\underset{\sim}{\partial}F_i \equiv \frac{\partial}{\partial\underset{\sim}{v}_i}F(\underset{\sim}{v}_i)$$

and singular denominators (in the form presented) are understood to carry a + io(o → 0) term. A little algebra leads to the sequence of steps listed below.

$$G(\underset{\sim}{k};\underset{\sim}{v}_1\underset{\sim}{v}_2) = -\frac{\phi(k)}{\underset{\sim}{k}\cdot(\underset{\sim}{v}_2-\underset{\sim}{v}_1)}\left\{\underset{\sim}{k}\cdot\underset{\sim}{\partial}F_1F_2 - \underset{\sim}{k}\cdot\underset{\sim}{\partial}F_2F_1\right.$$

$$-\left[\frac{\alpha(\underset{\sim}{k}\underset{\sim}{k}\cdot\underset{\sim}{v}_2)}{\varepsilon(\underset{\sim}{k}\underset{\sim}{k}\cdot\underset{\sim}{v}_2)}\,F_2\underset{\sim}{k}\cdot\underset{\sim}{\partial}F_1 + i\,\frac{\eta(\underset{\sim}{k}\underset{\sim}{k}\cdot\underset{\sim}{v}_2)}{\varepsilon(\underset{\sim}{k}\underset{\sim}{k}\cdot\underset{\sim}{v}_2)}\,\underset{\sim}{k}\cdot\underset{\sim}{\partial}F_1\underset{\sim}{k}\cdot\underset{\sim}{\partial}F_2\right]$$

$$+\left[\frac{\alpha^*(\underset{\sim}{k}\underset{\sim}{k}\cdot\underset{\sim}{v}_1)}{\varepsilon^*(\underset{\sim}{k}\underset{\sim}{k}\cdot\underset{\sim}{v}_1)}\,F_1\underset{\sim}{k}\cdot\underset{\sim}{\partial}F_2 - i\,\frac{\eta^*(\underset{\sim}{k}\underset{\sim}{k}\cdot\underset{\sim}{v}_1)}{\varepsilon^*(\underset{\sim}{k}\underset{\sim}{k}\cdot\underset{\sim}{v}_1)}\,\underset{\sim}{k}\cdot\underset{\sim}{\partial}F_1\;\underset{\sim}{k}\cdot\underset{\sim}{\partial}F_2\right] \tag{47}$$

$$+\phi(k)\int\frac{d\omega}{2\pi}\,\frac{1}{\omega-\underset{\sim}{k}\cdot\underset{\sim}{v}_1}\,\frac{1}{\underset{\sim}{k}\cdot\underset{\sim}{v}_2-\omega}\,\frac{1}{|\varepsilon(\underset{\sim}{k}\omega)|^2}$$

$$\left.\underset{\sim}{k}\cdot\partial F_1 \underset{\sim}{k}\cdot\partial F_2 \; [\alpha(\underset{\sim}{k}\omega)\eta^*(\underset{\sim}{k}\omega) + \alpha^*(\underset{\sim}{k}\omega)\eta(\underset{\sim}{k}\omega)]\right\}$$

$$= -\frac{\phi(k)}{\underset{\sim}{k}\cdot(\underset{\sim}{v}_2-\underset{\sim}{v}_1)}\left\{\frac{1}{\varepsilon(\underset{\sim}{k}\underset{\sim}{k}\cdot\underset{\sim}{v}_1)} F_2\underset{\sim}{k}\cdot\partial F_1\right.$$

$$- \frac{1}{\varepsilon^*(\underset{\sim}{k}\underset{\sim}{k}\cdot\underset{\sim}{v}_1)} F_1\underset{\sim}{k}\cdot\partial F_2 \tag{47}$$

$$-i\left[\frac{\eta(\underset{\sim}{k}\underset{\sim}{k}\cdot\underset{\sim}{v}_2)}{\varepsilon(\underset{\sim}{k}\underset{\sim}{k}\cdot\underset{\sim}{v}_2)} + \frac{\eta^*(\underset{\sim}{k}\underset{\sim}{k}\cdot\underset{\sim}{v}_1)}{\varepsilon^*(\underset{\sim}{k}\underset{\sim}{k}\cdot\underset{\sim}{v}_1)}\right] \underset{\sim}{k}\cdot\partial F_1\underset{\sim}{k}\cdot\partial F_2$$

$$\left.+i\underset{\sim}{k}\cdot\partial F_1\underset{\sim}{k}\cdot\partial_2 F\left[\frac{\eta(\underset{\sim}{k}\underset{\sim}{k}\cdot\underset{\sim}{v}_2)}{\varepsilon(\underset{\sim}{k}\underset{\sim}{k}\cdot\underset{\sim}{v}_2)} + \frac{\eta^*(\underset{\sim}{k}\underset{\sim}{k}\cdot\underset{\sim}{v}_1)}{\varepsilon^*(\underset{\sim}{k}\underset{\sim}{k}\cdot\underset{\sim}{v}_1)}\right]\right\}$$

$$-\phi(k)\int\frac{d\omega}{2\pi}\,\frac{1}{\omega-\underset{\sim}{k}\cdot\underset{\sim}{v}_1}\,\frac{1}{\underset{\sim}{k}\cdot\underset{\sim}{v}_2-\omega}\,\frac{1}{|\varepsilon(\underset{\sim}{k}\omega)|^2}$$

$$\left[\eta(\underset{\sim}{k}\omega) + \eta^*(\underset{\sim}{k}\omega)\right] \underset{\sim}{k}\cdot\partial F_1\underset{\sim}{k}\cdot\partial F_2$$

with

$$\alpha(\underset{\sim}{k}\omega) = \frac{\phi(k)}{m}\int\frac{\underset{\sim}{k}\cdot\partial F(\underset{\sim}{v})}{\omega-\underset{\sim}{k}\cdot\underset{\sim}{v}}\,d\underset{\sim}{v}$$

$$\eta(\underset{\sim}{k}\omega) = i\phi(k)\int\frac{F(\underset{\sim}{v})}{\omega-\underset{\sim}{k}\cdot\underset{\sim}{v}}\,d\underset{\sim}{v} \tag{48}$$

$$\varepsilon = 1 + \alpha$$

Recalling that the correlation of density fluctuations is (cf. Appendix II-A).

$$P(\underset{\sim}{k}\omega) = 2\pi e^2\int\frac{\delta(\omega-\underset{\sim}{k}\cdot\underset{\sim}{v})}{|\varepsilon(\underset{\sim}{k}\omega)|^2}F(\underset{\sim}{v})\,d\underset{\sim}{v}$$

$$= \frac{k^2}{2\pi}\;\frac{\eta'(\underset{\sim}{k}\omega)}{|\varepsilon(\underset{\sim}{k}\omega)|^2}$$

(47) can be written as

$$G(\underset{\sim}{k};\underset{\sim}{v}_1\underset{\sim}{v}_2) = -\frac{\phi(k)}{\underset{\sim}{k}\cdot(\underset{\sim}{v}_2-\underset{\sim}{v}_1)}\Bigg\{\Big[\frac{1}{\varepsilon(\underset{\sim}{k},\underset{\sim}{k}\cdot\underset{\sim}{v}_2)}F_2\underset{\sim}{k}\cdot\underset{\sim}{\partial}F_1 - \frac{1}{\varepsilon(\underset{\sim}{k},\underset{\sim}{k}\cdot\underset{\sim}{v}_1)}F_1\underset{\sim}{k}\cdot\underset{\sim}{\partial}F_2\Big] + \frac{4\pi i}{k^2}[P_+(\underset{\sim}{k}\ \underset{\sim}{k}\cdot\underset{\sim}{v}_2)+P_-(\underset{\sim}{k}\ \underset{\sim}{k}\cdot\underset{\sim}{v}_1)]\ \underset{\sim}{k}\cdot\underset{\sim}{\partial}F_1\underset{\sim}{k}\cdot\underset{\sim}{\partial}F_2\Bigg\} \tag{49}$$

The rather obvious interpretation of (49) is that the first two terms in the first bracket represent the screened electric force generated by particle 2(1) and modifying the distribution of particle 1(2), while the second two terms in the second bracket represent the forces associated with the density fluctuations created by particles 1 and 2 and reacting on their respective distribution functions.

(49) is correct for any (quasi-) stationary homogeneous distribution function $F(\underset{\sim}{v})$. In general, it is *not* factorizable into velocity - and $\underset{\sim}{k}$ - dependent parts. In equilibrium, however, it reduces to $G^\circ(\underset{\sim}{k};\underset{\sim}{v}_1\underset{\sim}{v}_2) = g^\circ(k)F^\circ(v_1)F^\circ(v_2)$

$$g^o(k) = -\frac{1}{n}\frac{\kappa^2}{k^2+\kappa^2} \tag{50}$$

$F^\circ(v)$ is maxwellian, implying*

$$\frac{\partial F^o}{\partial \underset{\sim}{v}} = -\beta\underset{\sim}{v}F^o \tag{51}$$

The demonstration [24] of (50) relies upon the relation that exists in equilibrium between $\eta$ and $\alpha$, in view of (51):

---

*We set m = 1 in the sequel.

$$\begin{aligned}\eta(\underset{\sim}{k}\omega) &= i\phi(k)\int \frac{F^{o}(\underset{\sim}{v})}{\omega-\underset{\sim}{k}\cdot\underset{\sim}{v}}\, d\underset{\sim}{v} \\ &= i\frac{\phi(k)}{\omega}\int\left(1 + \frac{k\cdot v}{\omega-\underset{\sim}{k}\cdot\underset{\sim}{v}}\right) d\underset{\sim}{v} \\ &= \frac{i}{\beta\omega}\ \{\beta\phi(k)n - \alpha(\underset{\sim}{k}\omega)\}\end{aligned} \tag{52}$$

The substitution of (52) into (69) yields

$$G^{o}(\underset{\sim}{k};\underset{\sim}{v}_1\underset{\sim}{v}_2) = \frac{\phi(k)}{\underset{\sim}{k}\cdot(\underset{\sim}{v}_2-\underset{\sim}{v}_1)}\ \beta F_1F_2$$

$$\Bigg\{\left[\frac{\underset{\sim}{k}\cdot\underset{\sim}{v}_1}{\varepsilon(\underset{\sim}{k}\underset{\sim}{k}\cdot\underset{\sim}{v}_2)} - \frac{\underset{\sim}{k}\cdot\underset{\sim}{v}_2}{\varepsilon^{*}(\underset{\sim}{k}\underset{\sim}{k}\cdot\underset{\sim}{v}_1)}\right]$$

$$+\, i\int\Bigg[\frac{d\omega}{2\pi}\quad \frac{1}{\omega+io}\ \frac{1}{\varepsilon^{*}(\underset{\sim}{k}\omega)} - \frac{1}{\omega-io}\ \frac{1}{\varepsilon^{*}(\underset{\sim}{k}\omega)} \tag{53}$$

$$+\, 2\pi i\delta(\omega)\quad \frac{1+\beta\phi(k)n}{|\varepsilon(\underset{\sim}{k}\omega)|^{2}}\Bigg]\quad \underset{\sim}{k}\cdot\underset{\sim}{v}_1\underset{\sim}{k}\cdot\underset{\sim}{v}_2$$

$$\frac{1}{\omega-\underset{\sim}{k}\cdot\underset{\sim}{v}_1+io}\quad \frac{1}{\underset{\sim}{k}\cdot\underset{\sim}{v}_2-\omega+io}$$

The correct consideration of the plus and minus character of the singular denominators is of crucial importance here and therefore has explicitly been indicated. The integration over $\omega$ now can easily be performed by rearranging the products of plus- and minus- functions in such a way that only one plus- (minus-) function appears in a particular term, and then closing the contour over the lowest (upper) large semicircle. Detailing this procedure one finds

$$G^{o}(\underset{\sim}{k};\underset{\sim}{v}_1\underset{\sim}{v}_2) = \phi(k)\beta F_1F_2$$

$$\Bigg\{\frac{1}{\underset{\sim}{k}\cdot(\underset{\sim}{v}_2-\underset{\sim}{v}_1)}\ \left[\left[\frac{\underset{\sim}{k}\cdot\underset{\sim}{v}_1}{\varepsilon(\underset{\sim}{k}\underset{\sim}{k}\cdot\underset{\sim}{v}_2)} - \frac{\underset{\sim}{k}\cdot\underset{\sim}{v}_2}{\varepsilon^{*}(\underset{\sim}{k}\underset{\sim}{k}\cdot\underset{\sim}{v}_1)}\right]\right.$$

$$+ \underset{\sim}{k}\cdot\underset{\sim}{v}_1\ \underset{\sim}{k}\cdot\underset{\sim}{v}_2 \quad \frac{1}{\underset{\sim}{k}\cdot\underset{\sim}{v}_1\varepsilon^*(\underset{\sim}{k}\underset{\sim}{k}\cdot\underset{\sim}{v}_1)} - \frac{1}{\underset{\sim}{k}\cdot\underset{\sim}{v}_2\varepsilon(\underset{\sim}{k}\underset{\sim}{k}\cdot\underset{\sim}{v}_1)}$$

$$+ \frac{2}{\varepsilon(\underset{\sim}{k}0)}\left(\frac{1}{\underset{\sim}{k}\cdot\underset{\sim}{v}_1} - \frac{1}{\underset{\sim}{k}\cdot\underset{\sim}{v}_2}\right)\Bigg]\Bigg] + \frac{\underset{\sim}{k}\cdot\underset{\sim}{v}_1\underset{\sim}{k}\cdot\underset{\sim}{v}_2}{|\varepsilon(\underset{\sim}{k}o)|^2}\ \frac{1+\beta\phi(k)n}{\underset{\sim}{k}\cdot\underset{\sim}{v}_1\underset{\sim}{k}\cdot\underset{\sim}{v}_2}\Bigg\}$$

Since $\varepsilon(\underset{\sim}{k}o) = 1 + \beta\phi(k)n = 1 + \kappa^2/k^2$, (54) simplifies to

$$G^{o}(\underset{\sim}{k};\underset{\sim}{v}_1\,\underset{\sim}{v}_2) = -\frac{1}{n}\ \frac{\kappa^2}{k^2+\kappa^2}\ F_1F_2 \tag{54}$$

Appendix IV-B. *Correlational Contribution to ε: Perturbation Calculation*

In this Appendix we derive the exact first order (in $\gamma$) correction to $\varepsilon(\underset{\sim}{k}\omega)$. The original resolution of this question in its generality is due to Coste [20]; the $k \to 0$ limit was evaluated earlier by Oberman, Ron and Dawson [13].

The derivation rests upon the evaluation of the binary correlation function $G^1(12)$ in the presence of an external perturbation, to first order on the perturbation, and to first order in $\gamma$.* Earlier in this Chapter $G^1(12)$ (to arbitrary order in $\gamma$) was approximated by $\{F^\circ(1)f^1(2) + f^1(1)F^\circ(2)\}$ $g^\circ(12)$. The approximate character of this assumption, however, has been emphasized and should be obvious. The second equation of the BBGKY hierarchy for a homogeneous system has been given in Appendix A, Eq. (42). The equivalent equation can easily be established for an inhomogeneous system, but will not be needed in its general form. Since we are interested in $G_1^1$ only, the linearized version of the equation is sufficient. Writing down this equation is basically a bookkeeping problem, and therefore it is useful to employ diagrams** as a memotechnique device.

The way the diagrams are constructed and are to be read should be rather self-explanatory. We note that the left-hand side and diagrams (7) and (8) can be combined into

---

* Superscripts indicate the order in the external perturbation, subscripts in $\gamma$.

**These diagrams are modified [26] Balescu-Prigogine diagrams [4].

$V(q,\omega;v_1v_1')$ and $V(k-q;v_2v_2')$ and therefore the full equation for an inhomogeneous system, completed with the term associated with the external driving field (which appears in conjunction with the plasma field)

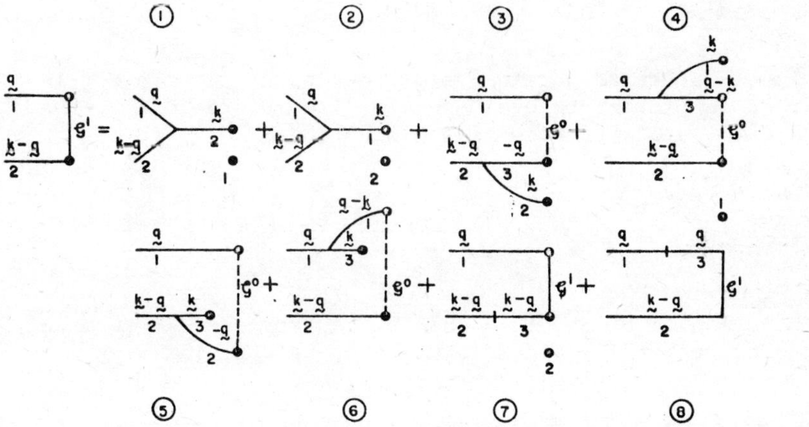

Figure 1. Diagrammatic representation of the equation for $G_1^1(\underset{\sim}{q},\underset{\sim}{k}-\underset{\sim}{q};\underset{\sim}{v}_1\underset{\sim}{v}_2)$. Lines are labeled by a number referring to the particle and the wave-vector associated with that particle. Particle 3 is integrated over. Wave vector conservation is satisfied in vertices. A vertex for the particle pair i, j is associated with $i\phi(\Delta\underset{\sim}{k}_i)\ \Delta\underset{\sim}{k}_i\cdot(\frac{\partial}{\partial v_i} - \frac{\partial}{\partial v_j})$ where $\Delta\underset{\sim}{k}_i - \Delta\underset{\sim}{k}_j$ is the wave vector change for particle i in that vertex.

can be exhibited as follows:

$$
\begin{aligned}
&-i\{\omega - V(\underset{\sim}{q},\omega;\underset{\sim}{v}_1\underset{\sim}{v}_1') - V(\underset{\sim}{k}-\underset{\sim}{q},\omega;\underset{\sim}{v}_2\underset{\sim}{v}_2')\}\, G(\underset{\sim}{q},\underset{\sim}{k}-\underset{\sim}{q};\underset{\sim}{v}_1\underset{\sim}{v}_2')\\
&= i\phi(q)\ \underset{\sim}{q}\cdot\{\partial F(\underset{\sim}{v}_1) f(\underset{\sim}{k}\omega;\underset{\sim}{v}_2) - F(\underset{\sim}{v}_1)\partial f(\underset{\sim}{k}\omega;\underset{\sim}{v}_2)\\
&- i\phi(q)\ \underset{\sim}{q}\cdot\underset{\sim}{\partial} f(\underset{\sim}{k}\omega;\underset{\sim}{v}_2)F(\underset{\sim}{v}_1)n\ g^{O}(q) \qquad (55)\\
&+ i\phi(k)\ \underset{\sim}{k}\cdot\underset{\sim}{\partial}\ F(\underset{\sim}{v}_2)\ F(\underset{\sim}{v}_1)g^{O}(q)\ \{n(\underset{\sim}{k}\omega) + \hat{n}(\underset{\sim}{k}\omega)\}\}\\
&+ [1 \rightarrow 2,\ 2 \rightarrow 1;\ \underset{\sim}{q} \rightarrow \underset{\sim}{k}-\underset{\sim}{q}]
\end{aligned}
$$

where

$$n(\underset{\sim}{k}\omega) = \int f(\underset{\sim}{k}\omega;\underset{\sim}{v})d\underset{\sim}{v} \tag{56}$$

and $\hat{n}(k\omega)$ is the "external" density associated with the perturbing field,

$$i\phi(k) \cdot \underset{\sim}{k}\, \hat{n}(\underset{\sim}{k}\omega) = e\hat{\underset{\sim}{E}}(\underset{\sim}{k}\omega). \tag{57}$$

In order to obtain $G = G_1^1$ correct to first order in $\gamma$, it is sufficient to evaluate $f = F^1$ to 0 order in $\gamma$, that is, to express it from the Vlasov equation:

$$f_o(\underset{\sim}{k}\omega;\underset{\sim}{v}) = \frac{1}{\omega-\underset{\sim}{k}\cdot\underset{\sim}{v}}\,\phi(k)\underset{\sim}{k}\cdot\frac{\partial F}{\partial\underset{\sim}{v}}\{n_o(\underset{\sim}{k}\omega) + \hat{n}(\underset{\sim}{k}\omega)\} \tag{58}$$

The solution of Eq. (55) can now be expressed in terms of the corresponding time evolution operators, yielding

$$G(\underset{\sim}{q},\underset{\sim}{k}-\underset{\sim}{q};\underset{\sim}{v}_1\underset{\sim}{v}_2) = \int\frac{d\nu}{2\pi}\,W(\underset{\sim}{q}\nu;\underset{\sim}{v}_1\underset{\sim}{v}_1')\quad W(\underset{\sim}{k}-\underset{\sim}{q}\;\omega-\nu;\;\underset{\sim}{v}_2\underset{\sim}{v}_2')$$
$$L(\underset{\sim}{k}\omega;\underset{\sim}{v}_1'\underset{\sim}{v}_2') \tag{59}$$

while L can be evaluated from (55) with a little algebra:

$$L(\underset{\sim}{k}\omega,\underset{\sim}{q}\nu;\underset{\sim}{v}_1\underset{\sim}{v}_2)$$

$$= i\beta\phi\;(k)\left\{\frac{1}{\omega-\underset{\sim}{k}\cdot\underset{\sim}{v}_2}\;[\psi(q)(\beta\;\underset{\sim}{k}\cdot\underset{\sim}{v}_2\;\;[\;\omega-\underset{\sim}{k}\cdot\underset{\sim}{v}_2+\underset{\sim}{q}\cdot\underset{\sim}{v}_2]\;-\right.$$

$$-\;\underset{\sim}{q}\cdot\underset{\sim}{k}\;\;\frac{\omega}{\omega-\underset{\sim}{k}\cdot\underset{\sim}{v}_2})-\beta\phi(q)\;\;\underset{\sim}{k}\cdot\underset{\sim}{v}_2\;\;\underset{\sim}{q}\cdot\underset{\sim}{v}_1]$$

$$\left.+[1\rightarrow 2,2\rightarrow 1\;,\underset{\sim}{q}\rightarrow\underset{\sim}{k}-\underset{\sim}{q}]\right\}$$

$$F(v_1)\ F(v_2)\quad \{n(\underset{\sim}{k}\omega) + \hat{n}(\underset{\sim}{k}\omega)\}$$

$$\psi(k) = \frac{4\pi e^2}{k^2+\kappa^2} \tag{60}$$

In deriving (60) we have used the fact F is the equilibrium (maxwellian) distribution function.

In order to find $\varepsilon(k\omega)$ we now consider the first equation of the hierarchy for $\tilde{f}_1$:

$$-i(\omega-V)f_1(\underset{\sim}{k}\omega;\underset{\sim}{v}_1) = i\sum_{\underset{\sim}{q}} \phi(|\underset{\sim}{k}-\underset{\sim}{q}|)(\underset{\sim}{k}-\underset{\sim}{q})$$

$$\cdot\frac{\partial}{\partial v_1}\int G\,\frac{1}{1}(\underset{\sim}{q},\underset{\sim}{k}-\underset{\sim}{q};\underset{\sim}{v}_1\underset{\sim}{v}_2) \tag{61}$$

It is evident that (59) and (60) can be combined in the concise expression

$$G(\underset{\sim}{q},\underset{\sim}{k}-\underset{\sim}{q};\underset{\sim}{v}_1\underset{\sim}{v}_2) = J(\underset{\sim}{q},\underset{\sim}{k}-\underset{\sim}{q};\underset{\sim}{v}_1\underset{\sim}{v}_2)\ \frac{\hat{n}(k\omega)}{\varepsilon(\underset{\sim}{k}\omega)} \tag{62}$$

this latter relation providing the definition of J. $f_1$ and subsequently $n_1$ can be expressed from (62) leading to

$$n_1(\underset{\sim}{k}\omega) = -\sum_{\underset{\sim}{q}}\int d\underset{\sim}{v}_1\,d\underset{\sim}{v}_2\ \frac{1}{\omega-\underset{\sim}{k}\cdot\underset{\sim}{v}_1}\ \frac{1}{\varepsilon(\underset{\sim}{k}\omega)}$$

$$\phi(|\underset{\sim}{k}-\underset{\sim}{q}|)\ (\underset{\sim}{k}-\underset{\sim}{q})\cdot\frac{\partial}{\partial\underset{\sim}{v}_1}\ J(\underset{\sim}{q},\underset{\sim}{k}-\underset{\sim}{q};\ \underset{\sim}{v}_1\underset{\sim}{v}_2)\ \frac{\hat{n}(\underset{\sim}{k}\omega)}{\varepsilon(\underset{\sim}{k}\omega)} \tag{63}$$

Comparing this result with the general expression

$$n \simeq n_o+n_1 = -\frac{\alpha}{1+\alpha}\,\hat{n} \simeq$$

$$-\left\{\frac{\alpha_o}{1+\alpha_o} + \frac{\alpha_1}{1+\alpha_o} - \frac{\alpha_o\alpha_1}{(1+\alpha_o)^2}\right\}\hat{n} \tag{64}$$

which implies

$$n_1 = -\frac{\alpha_1}{\varepsilon^2}\,\hat{n} \tag{65}$$

one concludes that

$$\alpha_I(\underset{\sim}{k}\omega) = \sum_{\underset{\sim}{q}} \int d\underset{\sim}{v}_1 d\underset{\sim}{v}_2 \frac{1}{\omega - \underset{\sim}{k}\cdot\underset{\sim}{v}_1} \phi(|\underset{\sim}{k}-\underset{\sim}{q}|)\ (\underset{\sim}{k}-\underset{\sim}{q})\cdot$$

$$\frac{\partial}{\partial \underset{\sim}{v}_1} J(\underset{\sim}{q},\underset{\sim}{k}-\underset{\sim}{q};\ \underset{\sim}{v}_1\underset{\sim}{v}_2)$$

The integral over $\underset{\sim}{v}_2$ can easily be performed since it requires

$$\int d\underset{\sim}{v}_2\ W(\underset{\sim}{k}-\underset{\sim}{q},\omega-\nu;\ \underset{\sim}{v}_2\underset{\sim}{v}_2') = \frac{i}{\varepsilon(\underset{\sim}{k}-\underset{\sim}{q}\omega-\nu)} \frac{1}{\omega-\nu-(\underset{\sim}{k}-\underset{\sim}{q})\cdot\underset{\sim}{v}_2'} \tag{67}$$

only. The integral over $\underset{\sim}{v}_1$ is somewhat more involved:

$$\phi(|\underset{\sim}{k}-\underset{\sim}{q}|) \int d\underset{\sim}{v}_1 \frac{1}{\omega-\underset{\sim}{k}\cdot\underset{\sim}{v}_1} (\underset{\sim}{k}-\underset{\sim}{q}) \cdot \frac{\partial}{\partial\underset{\sim}{v}_1} W(\underset{\sim}{q}\nu;\ \underset{\sim}{v}_1\underset{\sim}{v}_1')$$

$$= -i\phi(|\underset{\sim}{k}-\underset{\sim}{q}|) \frac{\underset{\sim}{k}\cdot(\underset{\sim}{k}-\underset{\sim}{q})}{(\omega-\underset{\sim}{k}\cdot\underset{\sim}{v}_1')^2} \frac{1}{\nu-\underset{\sim}{q}\cdot\underset{\sim}{v}_1'} \tag{68}$$

$$+ \frac{i}{\nu-\underset{\sim}{q}\cdot\underset{\sim}{v}_1'} \frac{a(\underset{\sim}{k}\omega;\ \underset{\sim}{q}\nu)}{\varepsilon(\underset{\sim}{q}\ \nu)}$$

$$a(\underset{\sim}{k}\omega;\ \underset{\sim}{q}\nu)$$

$$= \phi(|\underset{\sim}{k}-\underset{\sim}{q}|)\ \phi(q) \int \frac{\underset{\sim}{k}\cdot(\underset{\sim}{k}-\underset{\sim}{q})}{(\omega-\underset{\sim}{k}\cdot\underset{\sim}{v})^2}\ \frac{1}{\nu-\underset{\sim}{q}\cdot\underset{\sim}{v}}\ \underset{\sim}{q}\cdot\frac{\partial F}{\partial\underset{\sim}{v}}\ d\underset{\sim}{v}$$

The latter is related to the nonlinear polarizability.

Thus the final result can be written down as follows:

$$\alpha_1(\underset{\sim}{k}\omega) = -i\beta\ \phi(k)\sum_{\underset{\sim}{q}}$$

$$\int\frac{d\nu}{2\pi}\int d\underset{\sim}{v}_1\, d\underset{\sim}{v}_2\ \frac{1}{\nu-\underset{\sim}{q}\cdot\underset{\sim}{v}_1}\ \frac{1}{\omega-\nu-(\underset{\sim}{k}-\underset{\sim}{q})\cdot\underset{\sim}{v}_2}$$

$$\left\{\frac{1}{\varepsilon(\underset{\sim}{k}-\underset{\sim}{q},\ \omega-\nu)}\ \phi(\underset{\sim}{k}-\underset{\sim}{q})\ \frac{\underset{\sim}{k}\cdot(\underset{\sim}{k}-\underset{\sim}{q})}{(\omega-\underset{\sim}{k}\cdot\underset{\sim}{v}_1)^2} - \frac{a(\underset{\sim}{q}\nu,\ \underset{\sim}{k}\omega)}{\varepsilon(\underset{\sim}{q}\nu)\,\varepsilon(\underset{\sim}{k}-\underset{\sim}{q},\ \omega-\nu)}\right\}$$

$$\left\{\frac{1}{\omega-\underset{\sim}{k}\cdot\underset{\sim}{v}_2}\left[\beta\psi(q)\ \underset{\sim}{k}\cdot\underset{\sim}{v}_2\ (\omega-[\underset{\sim}{k}-\underset{\sim}{q}]\cdot\underset{\sim}{v}_2) - \underset{\sim}{q}\cdot\underset{\sim}{k}\ \psi(q)\right.\right.$$

$$\left.\frac{\omega}{\omega-\underset{\sim}{k}\cdot\underset{\sim}{v}_2} - \beta\phi(q)\ \underset{\sim}{k}\cdot\underset{\sim}{v}_2\ \underset{\sim}{q}\cdot\underset{\sim}{v}_1\right] + \frac{1}{\omega-\underset{\sim}{k}\cdot\underset{\sim}{v}_1}\left[\beta\psi(\underset{\sim}{k}-\underset{\sim}{q})\right.$$

$$\underset{\sim}{k}\cdot\underset{\sim}{v}_1\ (\omega-\underset{\sim}{q}\cdot\underset{\sim}{v}_1) - (\underset{\sim}{k}-\underset{\sim}{q})\cdot\underset{\sim}{k}\ \psi(\underset{\sim}{k}-\underset{\sim}{q})\ \frac{\omega}{\omega-\underset{\sim}{k}\cdot\underset{\sim}{v}_1} - \beta\phi(\underset{\sim}{k}-\underset{\sim}{q})$$

$$\left.\left.\underset{\sim}{k}\cdot\underset{\sim}{v}_1\ (\underset{\sim}{k}-\underset{\sim}{q})\cdot\underset{\sim}{v}_2\right]\right\}\ F(\underset{\sim}{v}_1)\ F(\underset{\sim}{v}_2) \tag{69}$$

We now evaluate the static ($\omega = 0$) limit of (67):

$$\alpha_1(\underset{\sim}{k}o) = -i \lim_{\omega\to o} \beta^2 \phi(k) \sum_{\underset{\sim}{q}} \int \frac{d\nu}{2\pi} \int d\underset{\sim}{v}_1 \; d\underset{\sim}{v}_2 \; F(\underset{\sim}{v}_1)\, F(\underset{\sim}{v}_2)$$

$$\frac{1}{\nu-\underset{\sim}{q}\cdot\underset{\sim}{v}_1} \; \frac{1}{\omega-\nu-(\underset{\sim}{k}-\underset{\sim}{q})\cdot\underset{\sim}{v}_2} \left\{ \frac{1}{\varepsilon(\underset{\sim}{k}-\underset{\sim}{q},\ \omega-\nu)} \; \phi(\underset{\sim}{k}-\underset{\sim}{q}) \right.$$

$$\left. \frac{\underset{\sim}{k}\cdot(\underset{\sim}{k}-\underset{\sim}{q})}{(\omega-\underset{\sim}{k}\cdot\underset{\sim}{v}_1)^2} - \frac{a(\underset{\sim}{q}\nu;\ \underset{\sim}{k}o)}{\varepsilon(\underset{\sim}{q}\nu)\ \varepsilon(\underset{\sim}{k}-\underset{\sim}{q},\ \omega-\nu)} \right\}$$

$$\left\{ \left[\phi(q) + \psi(\underset{\sim}{k}-\underset{\sim}{q})\right] \underset{\sim}{q}\cdot\underset{\sim}{v}_1 + \left[\phi(\underset{\sim}{k}-\underset{\sim}{q}) + \psi(q)\right] (\underset{\sim}{k}-\underset{\sim}{q})\cdot\underset{\sim}{v}_2 \right.$$

$$\equiv [11] + [21] + [12] + [22] \tag{70}$$

The algebra at this point becomes indeed tedious, but it is useful to display the steps in detail. Calculation of the four contributions designated by [11] etc. yields the results as follows:

[11] =

$$- i \lim_{\omega\to o} \sum_{\underset{\sim}{q}} \int \frac{d\nu}{2\pi} \int \beta\phi(q)\,\phi(\underset{\sim}{k}-\underset{\sim}{q}) \; \frac{\underset{\sim}{q}\cdot\underset{\sim}{v}_1}{\nu-\underset{\sim}{q}\cdot\underset{\sim}{v}_1}$$

$$\cdot \frac{\underset{\sim}{k}\cdot(\underset{\sim}{k}-\underset{\sim}{q})}{(\omega-\underset{\sim}{k}\cdot\underset{\sim}{v}_1)^2} \; F(\underset{\sim}{v}_1)\, d\underset{\sim}{v}_1$$

$$\int \beta\phi(\underset{\sim}{k}-\underset{\sim}{q}) \; \frac{1}{\omega-\nu} \left(1 + \frac{(\underset{\sim}{k}-\underset{\sim}{q})\cdot\underset{\sim}{v}_2}{\omega-\nu-(\underset{\sim}{k}-\underset{\sim}{q})\cdot\underset{\sim}{v}_2}\right)$$

$$\cdot\, F(\underset{\sim}{v}_2)\, d\underset{\sim}{v}_2 \; \xi(\underset{\sim}{k}-\underset{\sim}{q}) \; \frac{1}{\varepsilon(\underset{\sim}{k}-\underset{\sim}{q},\omega-\nu)}$$

$$= + i \lim_{\omega\to o} \sum_{\underset{\sim}{q}} \int \frac{d\nu}{2\pi} \; \frac{a(\underset{\sim}{q}\nu;\underset{\sim}{k}\omega)}{\varepsilon(\underset{\sim}{k}-\underset{\sim}{q},\omega-\nu)} \; \frac{1}{\omega-\nu} \{\alpha(\underset{\sim}{k}-\underset{\sim}{q},0)$$

$$- \alpha(\underset{\sim}{k}-\underset{\sim}{q},\omega-\nu)\}\ \xi(\underset{\sim}{k}-\underset{\sim}{q}) \tag{71}$$

$$[21] = + i \lim_{\omega \to 0} \sum_{\underset{\sim}{q}} \int \frac{d\nu}{2\pi} \int \beta\phi(q) \frac{\underset{\sim}{q}\cdot\underset{\sim}{v}_1}{\nu - \underset{\sim}{q}\cdot\underset{\sim}{v}_1} F(\underset{\sim}{v}_1)\, d\underset{\sim}{v}_1$$

$$\cdot \int \beta\, \phi\, (\underset{\sim}{k}-\underset{\sim}{q}) \frac{1}{\omega-\nu} \left(1 + \frac{(\underset{\sim}{k}-\underset{\sim}{q})\cdot \underset{\sim}{v}_2}{\omega-\nu-(\underset{\sim}{k}-\underset{\sim}{q})\cdot \underset{\sim}{v}_2}\right) F(\underset{\sim}{v}_2)\, d\underset{\sim}{v}_2$$

$$\cdot\ \xi(\underset{\sim}{k}-\underset{\sim}{q}) \frac{a(\underset{\sim}{q}\nu;\, \underset{\sim}{k}\omega)}{\varepsilon(\underset{\sim}{q}\nu)\ \varepsilon(\underset{\sim}{k}-\underset{\sim}{q},\omega-\nu)}$$

$$= - i \lim_{\omega \to 0} \sum_{\underset{\sim}{q}} \int \frac{d\nu}{2\pi} \frac{a(\underset{\sim}{q}\nu;\underset{\sim}{k}\omega)}{\varepsilon(\underset{\sim}{q}\nu)\ \varepsilon(\underset{\sim}{k}-\underset{\sim}{q},\omega-\nu)} \alpha(\underset{\sim}{q}\nu)$$

$$\cdot\ \{\alpha(\underset{\sim}{k}-\underset{\sim}{q},0) - \alpha(\underset{\sim}{k}-\underset{\sim}{q},\omega-\nu)\}\quad \xi(\underset{\sim}{k}-\underset{\sim}{q}) \tag{72}$$

$$[12] = - i \lim_{\omega \to 0} \sum_{\underset{\sim}{q}} \int \frac{d\nu}{2\pi} \int \beta\phi(q)\ \phi(\underset{\sim}{k}-\underset{\sim}{q}) \frac{1}{\nu}\left(1 + \frac{\underset{\sim}{q}\cdot\underset{\sim}{v}_1}{\nu-\underset{\sim}{q}\cdot\underset{\sim}{v}_1}\right)$$

$$\frac{\underset{\sim}{k}\cdot(\underset{\sim}{k}-\underset{\sim}{q})}{(\omega-\underset{\sim}{k}\cdot\underset{\sim}{v}_1)^2} F(\underset{\sim}{v}_1)\ d\underset{\sim}{v}_1 \int \beta\phi(\underset{\sim}{k}-\underset{\sim}{q}) \frac{(\underset{\sim}{k}-\underset{\sim}{q})\cdot\underset{\sim}{v}_2}{\omega-\nu-(\underset{\sim}{k}-\underset{\sim}{q})\cdot\underset{\sim}{v}_2}$$

$$F(\underset{\sim}{v}_2)\ d\underset{\sim}{v}_2 \quad \xi(\underset{\sim}{q}) \frac{1}{\varepsilon(\underset{\sim}{k}-\underset{\sim}{q},\ \omega-\nu)}$$

$$= - i \lim_{\omega \to 0} \sum_{\underset{\sim}{q}} \int \frac{d\nu}{2\pi} \frac{1}{\nu} \frac{\alpha(\underset{\sim}{k}-\underset{\sim}{q},\omega-\nu)}{\varepsilon(\underset{\sim}{k}-\underset{\sim}{q},\omega-\nu)}$$

$$\{\ a(\underset{\sim}{q}\nu;\underset{\sim}{k}\omega) - p(\underset{\sim}{k},\underset{\sim}{q})\quad \alpha(\underset{\sim}{k}\omega)\} \cdot \xi(\underset{\sim}{q}) \tag{73}$$

$$[22] = - i \lim_{\omega \to 0} \sum_{\underset{\sim}{q}} \int \frac{d\nu}{2\pi} \int \beta\phi(q) \frac{1}{\nu} \left(1 + \frac{\underset{\sim}{q}\cdot\underset{\sim}{v}_1}{\nu-\underset{\sim}{q}\cdot\underset{\sim}{v}_1}\right)$$

$$\cdot F(\underset{\sim}{v}_1)\ d\underset{\sim}{v}_1 \int \beta\phi\ (\underset{\sim}{k}-\underset{\sim}{q}) \frac{(\underset{\sim}{k}-\underset{\sim}{q})\cdot\underset{\sim}{v}_2}{\omega-\nu-(\underset{\sim}{k}-\underset{\sim}{q})\cdot\underset{\sim}{v}_2} F(\underset{\sim}{v}_2)\, d\underset{\sim}{v}_2$$

$$\cdot\ \xi(\underset{\sim}{q}) \frac{a(\underset{\sim}{q}\nu\ ;\ \underset{\sim}{k}\omega)}{\varepsilon(\underset{\sim}{q}\nu)\,\varepsilon(\underset{\sim}{k}-\underset{\sim}{q},\ \omega-\nu)} \tag{74}$$

$$= - i \lim_{\omega \to 0} \sum_{\tilde{q}} \int \frac{d\nu}{2\pi} \frac{1}{\nu} \frac{\alpha(\tilde{k}-\tilde{q},\omega-\nu)}{\varepsilon(\tilde{q}\nu)\varepsilon(\tilde{k}-\tilde{q},\ \omega-\nu)} \{\alpha(\tilde{q}\ 0)-\alpha(\tilde{q}\nu)$$

$$\cdot\ \xi(\tilde{q})\ a(\tilde{q}\nu;\ \tilde{k}\omega) \tag{74}$$

with

$$\xi(\tilde{q}) = \frac{\phi(k)}{\phi(q)\ \phi(|\tilde{k}-\tilde{q}|)} \left[ \phi(|\tilde{k}-\tilde{q}|) + \psi(q) \right]$$

$$p(\tilde{k},\tilde{q}) = \beta \frac{\tilde{k}\cdot(\tilde{k}-\tilde{q})}{|\tilde{k}-\tilde{q}|^2} \phi(q) \tag{75}$$

Combining these results and exploiting the fact that $\varepsilon(\nu)$, $a(\nu,\omega)$ are plus-functions and $\varepsilon(\omega-\nu)$ is a minus-function of $\nu$ which permits performing the integration over $\nu$ of pieces which contain one plus or minus function only, one obtains:

$$[11] + [12] = -i \lim_{\omega \to 0} \sum_{\tilde{q}} \int \frac{d\nu}{2\pi} \frac{1}{\omega-\nu} \frac{a(\tilde{q}\nu;\ \tilde{k}\omega)}{\varepsilon(\tilde{k}-\tilde{q}\ \ \omega-\nu)}$$

$$\left(1 - \frac{\alpha(\tilde{q}\nu)}{\varepsilon(\tilde{q}\nu)}\right) \left(\varepsilon(\tilde{k}-\tilde{q}\ \ \omega-\nu) - \varepsilon(\tilde{k}-\tilde{q}\ 0)\right)$$

$$\xi(\tilde{k}-\tilde{q})$$

$$= - \sum_{\tilde{q}} \left\{ \frac{a(q\ 0,\ \tilde{k}\ 0)}{\varepsilon(q\ 0)} \ i\varepsilon(\tilde{k}-\tilde{q},0) \lim_{\omega \to 0} \int \frac{d\nu}{2\pi} \frac{1}{\omega-\nu} \right.$$

$$\left. \frac{a(\tilde{q}\nu\ ;\tilde{k}\omega)}{\varepsilon(\tilde{q}\nu)\varepsilon(\tilde{k}-\tilde{q}\ \omega-\nu)} \right\} \xi(\tilde{k}-\tilde{q})$$

$$[12] + [22] = \sum_{\underset{\sim}{q}} \left\{ \frac{\alpha(\underset{\sim}{k}-\underset{\sim}{q},0)\ \alpha(\underset{\sim}{k}\ 0)}{\varepsilon(\underset{\sim}{k}-\underset{\sim}{q}\ 0)}\ p(\underset{\sim}{k},\underset{\sim}{q}) + i\varepsilon\ (\underset{\sim}{q}0) \lim_{\omega\to 0} \int \frac{d\nu}{2\pi}\ \frac{1}{\nu}\ \frac{a(\underset{\sim}{q}\nu;\ \underset{\sim}{k}\omega)}{\varepsilon(\underset{\sim}{q}\nu)\ \varepsilon(\underset{\sim}{k}-\underset{\sim}{q}\ \ \omega-\nu)} \right\} \xi(\underset{\sim}{q}) \tag{76}$$

Since

$$\xi(\underset{\sim}{k}-\underset{\sim}{q})\ \ \varepsilon(\underset{\sim}{k}-\underset{\sim}{q}\ 0) = \xi(\underset{\sim}{q})\ \ \varepsilon(\underset{\sim}{q}\ 0) \tag{77}$$

the principal value terms in the two remaining integrals cancel each other and only the $\delta$-function terms survive, leading to the complete expression

$$\alpha_1(\underset{\sim}{k}\ 0) = -\sum_{\underset{\sim}{q}} \frac{1}{2} a(\underset{\sim}{q}\ 0,\ \underset{\sim}{k}\ 0) \left\{ \frac{\xi(\underset{\sim}{k}-\underset{\sim}{q})}{\varepsilon(\underset{\sim}{q}\ 0)} - \frac{\xi(\underset{\sim}{q})}{\varepsilon(\underset{\sim}{k}-\underset{\sim}{q},0)} \right\} + \sum_{\underset{\sim}{q}} \frac{\alpha(\underset{\sim}{k}-\underset{\sim}{q}\ 0)\,\alpha\ (\underset{\sim}{k}\ 0)}{\varepsilon\ (\underset{\sim}{k}-\underset{\sim}{q},0)}\ p(\underset{\sim}{k},\underset{\sim}{q})\ \ \xi(\underset{\sim}{q}) \tag{78}$$

The first term is zero in virtue of (77). The second term becomes using (66) and (69)

$$\begin{aligned}
\alpha_1(\underset{\sim}{k}\ 0) &= \sum_{\underset{\sim}{q}} \beta^3 n^2 \frac{\underset{\sim}{k}\cdot(\underset{\sim}{k}-\underset{\sim}{q})}{k^2}\ \phi(k)\ \ \psi(|\underset{\sim}{k}-\underset{\sim}{q}|)\{\psi(q) + \phi(\underset{\sim}{k}-\underset{\sim}{q})\\
&= \sum_{\underset{\sim}{q}} \beta^3 n^2 \frac{\underset{\sim}{k}\cdot\underset{\sim}{q}}{k^2}\ \ \phi(k)\ \psi(q)\{\psi(\underset{\sim}{k}-\underset{\sim}{q}) + \phi(q)\}\\
&= \sum_{q} \beta^3 n^2\ \frac{\underset{\sim}{k}\cdot\underset{\sim}{q}}{k^2}\ \ \phi(k)\psi(q)\psi(\underset{\sim}{k}-\underset{\sim}{q})\\
&= \sum_{\underset{\sim}{q}} \frac{1}{2}\ \beta^3 n^2\ \phi(k)\psi(q)\psi(|\underset{\sim}{k}-\underset{\sim}{q}|)
\end{aligned} \tag{79}$$

The second step in (79) follows by shifting the summation variable, the third by virtue of the oddness of the second term, and the last through the combination of the first and second lines.

Eq. (79) provides the exact result (to first order in $\gamma$) for the dielectric function. When substituted in the FDT (II.30), the resulting $g_2$ is identical to the O'Neil-Rostoker [12] expression (19). We note that the integral over $\underset{\sim}{q}$ in (77) can be evaluated [20] as

$$\frac{1}{2}\,\beta^3 n^2 \phi(k) \sum_{\underset{\sim}{q}} \psi(q)\ \psi(|\underset{\sim}{k}-\underset{\sim}{q}|)$$

$$= \frac{\gamma}{8\pi}\, y^{-3} \tan^{-1} \frac{y}{2} \tag{80}$$

$$y = k/\kappa$$

The moral of the foregoing lengthy derivation can be summarized as follows:

(i) Eq. (39) gives the correct expression for $\varepsilon(\underset{\sim}{k}0)$ to lowest order.

(ii) As far as $\varepsilon(\underset{\sim}{k}\omega)$ is concerned, the dynamical result (69) is much more complex than suggested by (38) or any of the simpler schemes.

(iii) In a consistent derivation g is a functional of the *total* field only and does *not* depend on the plasma and external fields separately (cf. footnote in section IV-2).

ACKNOWLEDGEMENT

The difficult text of these notes has been typed by Miss Genevieve Sullivan and Miss Susan Le Vangie to whom I wish to express my sincere appreciation.

## REFERENCES

1. GOLDEN, K.I. and KALMAN, G., *J. Stat. Phys. 1*, 415 (1969). See this paper also for an extensive list of further references.

2. KUBO, R., *J. Phys. Soc. Jap.*, *12*, 570 (1957). SITENKO, A.G., *Electromagnetic Fluctuations in Plasmas*, Academic Press, New York, 1967.

3. See for example, LANDAU, L.D. and LIFSCHITZ, E.M., *Electrodynamics of Continuous Media*, Pergamon Press, N.Y., 1960.

4. BALESCU, R., *Statistical Mechanics of Charged Particles*, Interscience, New York, 1963.

5. For similar relations in solid state physics see, for example, PINES, D. and NOZIERES, P., *Theory of Quantum Liquids*, W.A. Benjamin Inc., N.Y., 1966.

6. For some related discussions see, for example, BLOEMBERGEN, *Nonlinear Optics*, W.A. Benjamin, Inc., N.Y., 1962.

7. GOLDEN, K.I., KALMAN, G., SILEVITCH, M.B., *J. Stat. Phys.*, *6*, 87 (1972).

8. See, for example, GOLDMAN, M.V., in *Nonlinear Effects in Plasmas*, KALMAN, G. and FEIX, M. (eds.), Gordon and Breach, 1969, pp. 335-361. DUBOIS, D.F. and GOLDMAN, M.V., *Phys. Rev. Let.*, *28*, 218 (1972), JACKSON, E.A., *Phys. Rev.*, *153*, 235 (1966).

9. ICHIMARU, S., *Ann. Phys.*, *20*, 78 (1962), N.Y. KALMAN, G., Lecture Notes, Université de Paris, Faculté des Sciences, Centre d'Orsay, Laboratoire des Hautes Energies, TN-5 (1963).

10. KALMAN, G. and POMEAU, *Proc. VIIth Int. Conf. on Phenomena in Ionized Gases*, Belgrade, 1966, v. 2, p. 91.

11. SALPETER, E.E., *Ann. Phys.*, *5*, 183 (1958), N.Y.

12. O'NEIL, T. and ROSTOKER, N., *Phys. Fluids*, *8*, 1109 (1965).

13. OBERMAN, C., RON, A. and DAWSON, J., *Phys. Fluids, 5* 1514 (1962). PEREL, V.I. and ELIASHBERG, G.M., *J. Exptl. Theoret. Phys. (U.S.S.R.)* (1962), *Soviet Phys.-JETP, 15* 633 (1962)]. BERK, H.L., *Phys. Fluids, 7*, 257 (1964). COSTE, J., *Nuclear Fusion, 5*, 284, 293 (1965).

14. HUBBARD, J., *Phys. Let., 25A*, 709 (1967).

15. SINGWI, K.S., TOSI, M.P., LAND, R.H. and SJOLANDER, A., *Phys. Rev., 176*, 589 (1968).

16. BERGGREN, K.F., *Phys. Rev., A1*, 1783 (1969).

17. ICHIMARU, S., *Phys. Rev., A2*, 494 (1970).

18. BRUSH, S.G., SAHLIN, H.L. and TELLER, E., *J. Chem. Phys., 45*, 2192 (1966), KALMAN, G., *Phys. Rev., 161*, 156 (1967).

19. SINGWI, K.S., SJOLANDER, A., TOSI, M.P. and LAND, R.H., *Solid State Comm., 7*, 1503 (1969). SINGWI, K.S., SJOLANDER, A., TOSI, M.P. and LAND, R.H., *Phys. Rev., B1*, 1044 (1970).

20. See Ref. 13. COSTE.

21. GUERNSEY, R.L., *Phys. Fluids, 5*, 322 (1962).

22. ROSTOKER, M.N. and ROSENBLUTH, N., *Phys. Fluids, 5*, 776 (1962).

23. KLIMONTOVICH, Yu.L. and SILIN, V.P., *J. Exptl. Theoret. Phys. (U.S.S.R.) 42*, 268 (1962). [*Soviet Phys.-JETP, 15*, 199, 1962].

24. BALESCU, R. and TAYLOR, H.S., *Phys. Fluids, 4*, 85 (1961).

25. ROSTOKER, N., *Phys. Fluids, 7*, 491 (1964).

26. KALMAN, G., *Phys. Fluids, 7*, 910 (1964).

# STRONGLY MAGNETIZED CLASSICAL PLASMA MODELS

DAVID MONTGOMERY

*Department of Physics and Astronomy, University of Iowa,*
*Iowa City, Iowa 52242, U.S.A.*

CONTENTS

## INTRODUCTION

Every beginning course in classical mechanics has as an exercise the derivation of the Rutherford scattering cross section for two classical Coulomb point charges. Many people are surprised to learn that this is still apparently an unsolved problem if a uniform dc magnetic field is assumed to be present. Since the interaction of classical point charges is fundamental for all discrete-particle processes that go on in most plasmas, this simple example should make clear that there are many things we do not know about such processes. The situation becomes worse if we add in the complexities associated with the long range of the Coulomb force, and the statistical mechanical side of the problem.

There is a seriously underpopulated border region of plasma physics that lies between controlled thermonuclear research and the older, less volatile society of classical statistical mechanics. This territory was initially explored in classic papers of Landau [1,2], Vlasov, [3], and Bogolyubov [4]. Ten or twelve years ago a few permanent settlements appeared to have been established there. Life on this rugged frontier proved to be harsh, however, and many of these settlements have now been abandoned. One hope of these lectures will be to recruit some vigorous pioneers who will be willing to try again at the neglected task of developing this important area.

Our subject matter may be roughly defined as that class of plasma processes for which the so-called "Vlasov approximation" is inadequate [5]. Such phenomena include: equations of state and other equilibrium thermodynamic relations, thermal relaxation phenomena, transport properties such as diffusion and electrical conduction, and microscopic statistical fluctuations in such quantities as the electric field and the charge density. All these may be loosely called "discrete particle" processes.

Convincing calculations of these quantities have been given in some cases for the case of a plasma in the absence of an external magnetic field. Calculations in the presence of a strong magnetic field are usually absent or less than convincing, though progress has been made within the last year or two. Reliable laboratory measurements of these quantities, that would pass muster in other branches of physics, are usually lacking, both with and without the external magnetic field.

Results in this area come slowly, and it is rather easy to make embarrassing conceptual mistakes, even though it is only classical physics we are dealing with. There is also great economic pressure for plasma physics to show results in a hurry, more rapidly than it is realistic to expect these rather deep fundamental problems to sort themselves out. To yield to this economic pressure is to gamble upon achieving practical success in plasma situations *without needing* to understand these phenomena. Such a gamble may be justified for some people, but I think it would be unwise for *all* of us to take it.

The question of why practitioners of orthodox statistical mechanics have largely stayed away from these problems is more difficult to answer. Their training and extensive experience with similar problems in neutral particle systems would appear to be an excellent preparation for plasmas. Perhaps it has to do with their perceptions of the sometimes desparate professional style of the plasma physics community. But for whatever reason, the classical statistical mechanics experts have largely remained aloof from considerations of systems with long range forces, one suspects much to the detriment of both subjects.

The subject of discrete-particle processes is much better understood in the absence of an externally-imposed dc magnetic field than in the presence of such a field. Many, if not most, of the important laboratory plasmas involve external magnetic fields in a fundamental way. The emphasis in this article will be almost exclusively on the situation in which a strong external magnetic field is present. A recent volume by the author [6] surveys the theory of the unmagnetized, or field-free plasma, and many other good survey volumes exist.

The following material is divided into four sections. In Section I, a number of results from the equilibrium statistical mechanics of two-dimensional plasmas are derived. The corresponding results for three dimensions are already well known. These results are independent of the presence of an external dc magnetic field, and are of interest in their own right as well as being essential to what follows. Section II is concerned with the non-equilibrium statistical mechanics of the electrostatic guiding-center plasma, a two-dimensional plasma model of great power and simplicity, recently introduced by Taylor and McNamara [7]. Section III concerns the generalization of this model to three dimensions. Section IV returns to two dimensions and relaxes the guiding-center model to include finite Larmor radius effects.

## REFERENCES FOR THE INTRODUCTION

1. LANDAU, L.D., *Physik. Z. Sowjetunion 10,* 154 (1936).

2. LANDAU, L.D., *J. Physics (U.S.S.R.) 10,* 25 (1946).

3. VLASOV, A.A., *J. Physics (U.S.S.R.) 9,* 25 (1945).

4. BOGOLYUBOV, N.N., *J. Physics (U.S.S.R.) 10,* 257 (1946).

5. The "Vlasov approximation" is a limit in which the charges e of mass m are supposed to be infinitely subdivided in such a way that $e \to 0$, $m \to 0$, but e/m and $n_0 e$ stay finite. $n_0$ is the average number density. See, e.g. N. ROSTOKER and M.N. ROSENBLUTH, *Phys. Fluids 3,* 1 (1960), or the following reference.

6. MONTGOMERY, D., *Theory of the Unmagnetized Plasma,* New York, Gordon and Breach,(1971).

7. TAYLOR, J.B. and McNAMARA, B., *Phys. Fluids, 14,* 1492 (1971). This paper can be said to have originated the subject we are going to be developing, though it has many antecedents.

## I. EQUILIBRIUM ENSEMBLE AVERAGES; STATIONARY CORRELATION FUNCTIONS

### *(A) Phase Space and the Gibbs Canonical Ensemble*

As is the case with any system with a very large number of degrees of freedom, our description is necessarily probabilistic. The dynamical state of the system can be specified by identifying it with a point of the *phase space* of the system. The coordinates of any such point (X, say) are the positions and velocities of every particle. The dynamical state of the system is then equivalent to a random variable defined over the phase space. Equilibrium statistical mechanics [1] demonstrates that the normalized probability distribution of this random variable appropriate to a system in thermal equilibrium with a temperature $\theta = KT$ (in energy units) is

$$D_{eq.} \equiv \frac{\exp(-\mathcal{E}/\theta)}{\int dX \exp(-\mathcal{E}/\theta)} \tag{1}$$

This is, of course, the canonical distribution of Gibbs; $\mathcal{E}$ is the total energy, uniquely defined for each point X of the phase space, and dX is the volume element.

Thermodynamic quantities and other macroscopic observables are interpretable as "ensemble averages", or expectation values computed with respect to $D_{eq}$. Thus, if A(X) is any dynamical variable, its expectation value will be written as

$$\langle A \rangle = \int dX \, A(X) \, D_{eq.} \tag{2}$$

where the integration runs over the entire phase space accessible to the system.

Thermodynamic functions are derivable from the *partition function*, which, apart from quantum mechanical factors which will not be of interest to us here, is

$$Z = \int dX \exp(-\mathcal{E}/\theta) \tag{3}$$

In particular, the equation of state for the pressure [2] is

$$p = \theta \frac{\partial}{\partial V} \ln Z = \frac{\theta}{Z} \frac{\partial Z}{\partial V} \tag{4}$$

where V is the spatial volume in which the system is confined. If one wants to calculate extensive quantities (such as the Helmholtz free energy or the entropy) accurately, one has to put the additional factors into Eq. (3) correctly, but since they are volume-independent, they cancel out of Eq. (4).

For definiteness, we will usually consider N positive charges +e and N negative ones -e. For $\varepsilon$ we have the expression

$$\mathcal{E} = T + U \tag{5}$$

$$T = \sum_i \frac{1}{2} m_i \vec{v}_i^2 \tag{6}$$

$$U = \sum_{i<j} \varphi_{ij} \tag{7}$$

T is the total kinetic energy and U is the total potential (interaction) energy. The $\Sigma_i$ in (6) runs over all particles of both signs, and $m_i$ and $\vec{v}_i$ are the mass and velocity of the ith particle. In (7), $\phi_{ij}$ is the two-body Coulomb potential between charged particles i and j. The sum runs over all pairs. We make the convenient and important restriction to *electrostatic interactions*, ignoring all retardation effects and magnetic interactions as being down by a factor of $O(v^2/c^2)$ from the ones we shall include. (This approximation is often called, somewhat incorrectly, the "low beta" approximation by plasma physicists. "Beta" is a rather important plasma quantity which is essentially the ratio of thermal energy density to magnetic energy density.) It is also important to note that the things that have been said so far are unchanged by the addition of an external, constant, and spatially uniform magnetic field of arbitrary strength, since $\mathcal{E}$ is unchanged by the addition of such a field. Thus, the calculation of such ensemble averages as Eq. (2) are *independent* of $\vec{B}$, the external field, so long as A is a time-independent function of X.

*(B) Reduced Probability Distributions* [2-6]

It is convenient to project the probability distribution (1) onto a subspace which is just the phase of s particles, where s is some small integer (1,2,3,...). Let $X_j \equiv (\vec{x}_j, \vec{v}_j)$ be the position and velocity of the jth charge. Then

$$\frac{f_s^{eq.}(X_1 \ldots X_s)}{V^s} \equiv \int dX_{s+1}\, dX_{s+2} \cdots D_{eq.} \tag{8}$$

is the *reduced probability distribution* in the phase space of s particles. The volume factor $V^s$ is for later convenience, and the integration is over all the coordinates of all the remaining charges. To discuss two charged species, it is convenient to partition s into $s_i$ "ions" (positive charges) and $s_e$ "electrons" ($s_e + s_i = s$) and define

$$f^{eq.}_{s_i,s_e} = f^{eq.}_{s_i,s_e}(X^i_1 \ldots X^i_{s_i} ; X^e_1 \ldots X^e_{s_e})$$

$$= V^{s_i+s_e} \int dX^i_{s_i+1} \ldots dX^i_N \, dX^e_{s_e+1} \ldots dX^e_N \, D_{eq.} \tag{9}$$

where $X^i_j = (\vec{x}^i_j, \vec{v}^i_j)$ are the phase space coordinates of the jth ion; a similar expression applies for electrons. The superscripts will indicate the species of the charged particles. Note that all the $f_{s_i,s_e}$ are symmetric under the interchange of like particle coordinates.

It is immediately obvious that the velocity-space parts of the $f_{s_i,s_e}$ are trivial and may be separated from the configuration-space parts:

$$f^{eq.}_{s_i,s_e} = \left\{ \prod_{\ell=1}^{s_i} \frac{\exp(-m_i \vec{v}^{i\,2}_\ell / 2\theta)}{(2\pi\theta/m_i)^{d/2}} \right\} \tag{10}$$

$$\cdot \left\{ \prod_{j=1}^{s_e} \frac{\exp(-m_e \vec{v}^{e\,2}_j / 2\theta)}{(2\pi\theta/m_e)^{d/2}} \right\} n_{s_i,s_e}$$

where d is the dimensionality of the system, and

$$n_{s_i,s_e} \equiv \frac{V^{s_i+s_e}\int d\vec{x}^i_{s_i+1}\ldots d\vec{x}^i_N\, d\vec{x}^e_{s_e+1}\ldots d\vec{x}^e_N \exp(-U/\theta)}{\int d\vec{x}^i_1\ldots d\vec{x}^i_N\, d\vec{x}^e_1\ldots d\vec{x}^e_N \exp(-U/\theta)} \tag{11}$$

is the configuration-space probability distribution of the subsystem.

It will be generally understood that we are dealing throughout with very large systems: $N \gg 1$, and $N/V \equiv n_0$, the number density. The "thermodynamic limit" of $N \to \infty$ and $V \to \infty$ may or may not be well defined for such expressions as Eqs. (3) and (4). It is not an entirely academic question as to whether such limits exist. Some rigorous work has been done in one dimension [7,8] and in three dimensions [9], and more is badly needed (in two dimensions, especially). We shall generally assume that the limit does exist; but that V and N are just very large but finite quantities is perhaps a slighly more acceptable way for us to look at the problem. The large-system limit is desirable in that we may hope that removing the system's boundaries to infinity will leave us with a "bulk" limit in which the surface contributions to extensive quantities will be negligible compared to the volume contributions, and that nothing that happens near the boundaries will affect the intensive quantities. (This may appear to be obvious, but it is not.) Making this assumption, it is easy to show that $n_{1,0} = n_{0,1} = \text{const.} = 1$, and that the higher $n_{s_i,s_e}$ depend only on the separations of the particle coordinates and not on their absolute spatial locations.

### *(C) The Equilibrium BBGKY Hierarchy*

Most of the macroscopic functions of interest can be expressed as integrals of $n_{2,0}$, $n_{0,2}$, and $n_{1,1}$, as it turns out. Any approximate method of calculating these quantities is therefore of some interest. A convenient starting place is the well-knwon *equilibrium BBGKY* [2-5] *hierarchy*, which can be readily obtained by taking gradients of Eq. (11). It is left as an exercise to show that, assuming periodic boundary conditions over a very large volume V, for any j between 1 and $s_i$,

$$\frac{\partial n_{s_i,s_e}}{\partial \vec{x}^i_j} + \frac{n_{s_i,s_e}}{\theta} \frac{\partial \Phi(s_i,s_e)}{\partial \vec{x}^i_j}$$

$$= - \frac{N - s_i}{\theta V} \int d\vec{x}^i_{s_i+1} \frac{\partial \varphi^{ii}\left(\vec{x}^i_j - \vec{x}^i_{s_i+1}\right)}{\partial \vec{x}^i_j} n_{s_i+1,s_e} \quad (12)$$

$$- \frac{N - s_e}{\theta V} \int d\vec{x}^e_{s_e+1} \frac{\partial \varphi^{ie}\left(\vec{x}^i_j - \vec{x}^e_{s_e+1}\right)}{\partial \vec{x}^i_j} n_{s_i,s_e+1}$$

The notation used is the following: $\phi^{ii}$ is the Coulomb potential interaction of the two ions, and $\phi^{ee}$ is that of two electrons. $\phi^{ii}(\vec{x}) = \phi^{ee}(\vec{x})$. $\phi^{ie} = \phi^{ei} = -\phi^{ee}$ is the Coulomb potential between an electron and an ion. $\Phi(s_i,s_e)$ is the total Coulomb potential energy of the first $s_i$ ions and $s_e$ electrons If $s_e$ and $s_i$ are not large, we can replace $(N - s_i)/V$ and $(N - s_e)/V$ by the average number density $n_o$, on the right-hand side of (12). A similar relation comes from taking gradients of Eq. (11) with respect to the electron coordinates:

$$\frac{\partial n_{s_i,s_e}}{\partial \vec{x}^e_j} + \frac{n_{s_i,s_e}}{\theta} \frac{\partial \Phi(s_i,s_e)}{\partial \vec{x}^e_j}$$

$$= - \frac{n_o}{\theta} \int d\vec{x}^i_{s_i+1} \frac{\partial \varphi^{ei}\left(\vec{x}^e_j - \vec{x}^i_{s_i+1}\right)}{\partial \vec{x}^e_j} n_{s_i+1,s_e}$$

$$- \frac{n_o}{\theta} \int d\vec{x}^e_{s_e+1} \frac{\partial \varphi^{ee}\left(\vec{x}^e_j - \vec{x}^e_{s_e+1}\right)}{\partial \vec{x}^e_j} n_{s_i,s_e+1} \tag{13}$$

where we have assumed $N >> s_i$ or $s_e$. The normalization conditions to be obeyed by $n_{s_i,s_e}$ are, from (11),

$$\int d\vec{x}^i_1 \ldots d\vec{x}^i_{s_i} \, d\vec{x}^e_1 \ldots d\vec{x}^e_{s_e} \, n_{s_i,s_e} = V^{s_i+s_e} \tag{14}$$

and $n_{0,0} = n_{1,0} = n_{0,1} = 1$.

*(D) Expansion in the "Plasma Parameter"* $\varepsilon$

For the entirely general case, there is no particular advantage in writing the thermal equilibrium relations in the form of Eqs. (12) and (13). Rather, their advantage lies in the ease with which perturbation theory can be done on them. It turns out that for a wide range of parameters, the potential energy of two charges in a plasma which lie at a distance equal to the nearest-neighbor separation is much less than the average kinetic energy per particle, $\theta$. This means that the interactions are mostly *weak* in the sense that $\phi^{ee}/\theta$, $\phi^{ii}/\theta$ and $\phi^{ie}/\theta$ can all be considered to be << 1 over most of the phase space. We can thus do a weak-coupling expansion in $\langle\phi\rangle_{avg.}/\theta$, (where $\langle\phi\rangle_{avg.}$ is some average interaction potential). The expansion is complicated slightly by the long range of the potential, which raises some of the integral terms in Eqs. (12), (13) by one order in $\langle\phi\rangle_{avg.}/\theta$.

The details of this expansion for the three-dimensional case have been adequately described several times (see e.g., Guernsey [10] or Montgomery [11]), and there is no purpose in doing so again here. We shall outline the development, however, for the two-dimensional [12] case, since so much of the later theory to be developed involves two-dimensional models. Equations (12) and (13) apply to one, two, or three dimensions, with appropriate modifications of the Coulomb potential $\phi^{ii} = \phi^{ee} = -\phi^{ei} = -\phi^{ie}$. In two dimensions, the charges are imagined as very long parallel "rods" of length $\ell$ and charge e, and the potential between two rods of positive charge +e each is, for example,

$$\varphi^{ii}(\vec{x}_\ell^i - \vec{x}_j^i) = -\frac{2e^2}{\ell}\, \ln |\vec{x}_\ell^i - \vec{x}_j^i| \tag{15}$$

with similar expressions for the other interactions.

The zeroth order (i.e., non-interacting) values of the $n_{s_i,s_e}$ are seen to be those solutions of Eqs. (12) and (13) with all the potential energy parts neglected and which also satisfy (14). These are, obviously,

$$n^{(0)}_{s_i,s_e} = 1, \qquad \text{all } s_i \text{ and } s_e \tag{16}$$

The superscript zero means of zeroth order in powers of the interaction potential, and we imagine (16) as the first term of the series.

$$n_{s_i,s_e} = n^{(0)}_{s_i,s_e} + n^{(1)}_{s_i,s_e} + \dots \tag{17}$$

We expect the higher-order terms of this series, the "correlation functions", all to vanish as the particle separations become very large, and $n_{s_i,s_e} \to n^{(0)}_{s_i,s_e}$ then.

The lowest order there is, according to (16), no correlation between the positions of the particles: what one would expect from non-interacting particles. We proceed to (12) and (13) to obtain the next terms in the series. However, the long range of the Coulomb potential means that one has to keep the next order terms in the integrals in Eq. (12), above those a purely formal estimate would indicate.

A self-consistent expansion procedure is readily found in which the correlations between larger numbers of particles is progressively weaker in powers of $\langle\phi\rangle/\theta$. The next order can be found in terms of *pair correlations*:

$$n^{(1)}_{s_i,s_e} = \sum_{k<\ell=1}^{s_i} p^{ii}(\vec{x}_k^i - \vec{x}_\ell^i)$$

$$+ \sum_{k<\ell=1}^{s_e} p^{ee}(\vec{x}_k^e - \vec{x}_\ell^e)$$

$$+ \sum_{k=1}^{s_i} \sum_{\ell=1}^{s_e} p^{ei}(\vec{x}_\ell^e - \vec{x}_k^i) \tag{18}$$

where $p^{ii} = p^{ee} = -p^{ei}$.

With the assumptions, it can be shown that the first-order part of *all* of Eqs. (12) and (13) is satisfied if one sets $p^{ee}(\vec{x}) = e^2\mathcal{P}(x)$, $\phi^{ee}(\vec{x}) = e^2\psi(x)$, and

$$\frac{\partial \mathcal{P}(x_{12})}{\partial \vec{x}_{12}} + \frac{1}{\theta} \frac{\partial \psi(x_{12})}{\partial \vec{x}_{12}}$$

$$= - \frac{n_o e^2}{\theta} \int d\vec{x}_3 \frac{\partial \psi(x_{13})}{\partial \vec{x}_{13}} \mathcal{P}(x_{23}) \tag{19}$$

$$- \frac{n_o e^2}{\theta} \int d\vec{x}_3 \frac{\partial \psi(x_{13})}{\partial \vec{x}_{13}} \mathcal{P}(x_{23})$$

In Eq. (19), $x_{12} \equiv |\vec{x}_1 - \vec{x}_2|$, and so on. It is a differentio-integral equation of the convolution type, and can be solved by the Fourier transformation. Setting

$$\mathcal{O}(x) = \int d\vec{k}\ \mathcal{O}(\vec{k})\ e^{i\vec{k}\cdot\vec{x}}$$

$$\psi(x) = \int d\vec{k}\ \psi(\vec{k})\ e^{i\vec{k}\cdot\vec{x}} \tag{20}$$

we find

$$\mathcal{O}(\vec{k}) = \frac{-\psi(\vec{k})/\theta}{1 + \frac{2n_o e^2}{\theta}(2\pi)^2\ \psi(\vec{k})} \tag{21}$$

or

$$p^{ee}(\vec{k}) = -\frac{\varphi^{ee}(\vec{k})/\theta}{1 + (2n_o/\theta)\ \varphi^{ee}(\vec{k})(2\pi)^2} \tag{22}$$

Poisson's equation, Fourier-transformed in two dimensions, gives $\phi^{ee}(\vec{k}) = e^2/\pi\ell k^2$, so that we may finally write (22) as the classical *Debye-Hückel expression*:

$$p^{ee}(\vec{k}) = -\left(\frac{1}{8\pi^2 n_o}\right)\left(\frac{1}{1 + k^2\lambda_D^2}\right) \tag{23}$$

Here, we have introduced the very important *Debye length* into the problem, namely

$$\lambda_D^2 \equiv \frac{\ell\theta}{8\pi n_o e^2} \equiv K_D^{-2} \tag{24}$$

If we Fourier transform Eq. (23) back to $\vec{x}$-space, we get

$$p^{ee}(x) = -\frac{2e^2}{\ell\theta}\ K_o(x/\lambda_D) \tag{25}$$

where $K_o$ is a Bessel function of imaginary argument, and has the asymptotic form

$$p^{ee}(x) \xrightarrow[x \gg \lambda_D]{} -\frac{2e^2}{\ell\theta}\sqrt{\frac{\pi\lambda_D}{2x}}\exp(-x/\lambda_D) \qquad (26)$$

We see that $\lambda_D$ has the important physical significance of being the distance over which pair correlations persist, just as in three dimensions, though the functional form is somewhat different. Since $p^{ii} = p^{ee} = -p^{ei}$, we now have all the first-order solution.

The calculation of the correlation functions to higher order becomes quite messy, but as long as $p^{ee}(x)$ is $\ll 1$, Eqs. (17), (18), and (25) are a satisfactory approximation. It is readily seen that the condition for validity is thus $2e^2/\ell\theta \ll 1$, or what is equivalent,

$$\epsilon \equiv (4\pi n_o \lambda_D^2)^{-1} \ll 1 \qquad (27)$$

This condition is essentially that the number of charged rods per Debye square is very large. The well-known analogue [13] of this condition in three dimensions is that the number of point charges per Debye cube is very large.

*(E) Further Comments on the Validity of the Expansion*

The failure of the weak-interaction approximation over small regions of the phase space (at small particle separations) is indicated by the fact that the short-range form of Eq. (25), $p_{ee}(x \ll \lambda_D) \approx -(2e^2/\ell\theta)\ln(x/\lambda_D)$. This ceases to be $\ll 1$ at a distance $x_{min.}$ given by

$$-\frac{2e^2}{\ell}\ ln(x_{min.}/\lambda_D) \approx \theta$$

or

$$x_{min.} \cong \lambda_D\, e^{-1/\epsilon} \qquad (28)$$

Compared with the nearest-neighbor separation $n_o^{-\frac{1}{2}}$, this gives

$$\frac{x_{min.}}{n_o^{-\frac{1}{2}}} = \frac{\sqrt{\epsilon}}{4\pi} \exp\,(-1/\epsilon) \tag{29}$$

which is smaller than any power of $\varepsilon$, but is not zero. Because the separation of two charges is so seldom less than $x_{min.}$, it is reasonable to hope that the important features of the theory will be unchanged by ignoring the short range of $\phi^{ee}(x)$. To render certain integrals convergent, it is often convenient to work not with $\phi^{ee}(x)$, but with another function which differs from it only for $x \lesssim x_{min}$, namely

$$\varphi_{eff.}(x) = \int_0^{k_{max}} k\,dk \int_0^{2\pi} d\theta \, \frac{e^2 e^{ikx\cos\theta}}{\pi \ell k^2} \tag{30}$$

where $k_{max} = 1/x_{min}$. The length $x_{min}$ has the physical interpretation in the unmagnetized case of being the distance of closest approach of two colliding thermal particles with zero impact parameter. To remove this somewhat unsatisfactory state of affairs at short range requires quantum mechanical modifications in three dimensions, and has not been carried out in two dimensions. It has been concluded by everyone who has seriously looked at the three-dimensional calculations that the dominant effects are correctly obtained from the modification of $\phi^{ee}$ of Eq. (3) at short distances. We shall have to hope that the same thing is true in two dimensions. The effect of this modification of the interaction on the pair correlations is just to cut off the Fourier integrals of such expressions as (23) at $|\vec{k}| = 1/x_{min}$

*(F) Ensemble Averages of Macroscopic Functions* [14]

To summarize our results to this point, we have calculated the four pair correlations in this problem in terms of a single function, which in Fourier transform language reads

$$p^{ee}(\vec{k}) = p^{ii}(\vec{k}) = -p^{ei}(\vec{k}) = -p^{ie}(\vec{k})$$

$$= -\frac{(8\pi^2 n_o)^{-1}}{1 + k^2\lambda_D^2} \tag{31}$$

This is not an exact result, and is valid to the extent that the inequality (27) holds. We have immediately from it expressions for all the various $n_{s_i,s_e}$, from (17) and (18), to the same accuracy. We can calculate, in terms of it, ad-tional quantities of physical interest. For example, the electric field

$$\vec{E}(\vec{x}) = -\frac{\partial}{\partial\vec{x}}\left\{\sum_{j=1}^{N}\left(-\frac{2e}{\ell}\ln|\vec{x}-\vec{x}_j^i|\right)\right. \qquad (32)$$

$$\left.+\sum_{k=1}^{N}\left(\frac{2e}{\ell}\ln|\vec{x}-\vec{x}_k^e|\right)\right\}$$

can be readily shown to have zero ensemble average

$$\langle\vec{E}(\vec{x})\rangle = \int D_{eq.}\,\vec{E}\,dX = 0 \qquad (33)$$

as can the net electrical charge density

$$\rho(\vec{x}) = \frac{e}{\ell}\sum_{j=1}^{N}\left[\delta(\vec{x}-\vec{x}_j^i) - \delta(\vec{x}-\vec{x}_j^e)\right] \qquad (34)$$

For, using our previous expressions for $p^{ee}$, $p^{ii}$, $p^{ie}$, we can also show

$$\langle\rho(\vec{x})\rangle = 0 \qquad (35)$$

However, the vanishing of the ensemble averages of $\vec{E}$ and $\rho$ should not be construed to mean they are zero, because, for example, $\langle\vec{E}^2\rangle \neq 0$. Let us calculate the *autocorrelation* of the charge density,

$$\langle\rho(\vec{x})\,\rho(\vec{x}+\vec{r})\rangle = \int D_{eq.}\,\rho(\vec{x})\,\rho(\vec{x}+\vec{r})\,dX \qquad (36)$$

For this purpose it is convenient to insert (35), (9), and (10) into (36), getting:

$$\langle \rho(\vec{x})\ \rho(\vec{x}+\vec{r}) \rangle$$

$$= \int dX\ \frac{e^2}{\ell^2} \sum_{j=1}^{N} \sum_{q=1}^{N} \left\{ \delta(\vec{x}-\vec{x}^{\,i}_j) - \delta(\vec{x}-\vec{x}^{\,e}_j) \right\}$$

$$\left\{ \delta(\vec{x}+\vec{r}-\vec{x}^{\,i}_q) - \delta(\vec{x}+\vec{r}-\vec{x}^{\,e}_q) \right\} D_{eq.}$$

$$= \frac{e^2}{\ell^2} \frac{N(N-1)}{V^2} \left\{ n_{2,0}(\vec{x},\vec{x}+\vec{r}) + n_{0,2}(\vec{x},\vec{x}+\vec{r}) \right\} \quad (37)$$

$$- \frac{e^2}{\ell^2} \frac{N^2}{V^2} \left\{ n_{1,1}(\vec{x},\vec{x}+\vec{r}) + n_{1,1}(\vec{x},\vec{x}+\vec{r}) \right\}$$

$$+ \frac{e^2}{\ell^2} \frac{N}{V} \left\{ n_{1,0}(\vec{x})\ \delta(\vec{r}) + n_{0,1}(\vec{x})\ \delta(\vec{r}) \right\}$$

Using $n_{1,0} = n_{0,1} = 1$, and passing to the limit of N,V very large, we use (18) and our previously derived expressions for the pair correlations to write:

$$\langle \rho(\vec{x})\ \rho(\vec{x}+\vec{r}) \rangle = 4\ \frac{n_o^2 e^2}{\ell^2}\ p^{ee}(\vec{r}) + \frac{2 n_o e^2}{\ell^2}\ \delta(\vec{r}) \quad (38)$$

which depends only upon the separation of the points $\vec{r}$, as we know it should.

It is often useful to decompose expressions like Eq. (38) into a Fourier sum or integral, and write them in terms of their *spectral densities*. Thus,

$$\langle \rho(\vec{x})\ \rho(\vec{x}+\vec{r})\rangle = \int d\vec{k}\ S_\rho(\vec{k})\ e^{i\vec{k}\cdot\vec{r}} \tag{39}$$

where Fourier transformation of Eq. (38) and use of (23) shows that the spectral density $S_\rho(\vec{k})$ is

$$S_\rho(\vec{k}) = \frac{n_o e^2}{2\pi^2 \ell^2}\left(\frac{k^2\lambda_D^2}{1+k^2\lambda_D^2}\right) \tag{40}$$

Of considerable interest also will be the spectral density of the auto-correlation tensor of the electric field, $\vec{\vec{S}}_E(\vec{k})$, defined by

$$\langle \vec{E}(\vec{x})\ \vec{E}(\vec{x}+\vec{r})\rangle = \int d\vec{k}\ \vec{\vec{S}}_E(\vec{k})\ e^{i\vec{k}\cdot\vec{r}} \tag{41}$$

Two applications of Poisson's equation shows that $\vec{\vec{S}}_E$ and $S_\rho$ are related by

$$\vec{\vec{S}}_E(\vec{k}) = \frac{16\pi^2\vec{k}\vec{k}}{k^4}\ S_\rho(\vec{k}) \quad , \tag{42}$$

so that

$$\vec{\vec{S}}_E(\vec{k}) = \frac{\theta}{\ell\pi}\ \frac{\vec{k}\vec{k}}{k^2}\ \frac{1}{1+k^2\lambda_D^2} \ . \tag{43}$$

It will be instructive for the reader to compute other auto-correlations as exercices, such as the current-current auto-correlation, $\langle\vec{j}(\vec{x})\ \vec{j}(\vec{x}+\vec{r})\rangle$

*(G) Field Energies, Thermal Energies, and a Warning*

It is interesting to calculate the rms electric field strength

$$\langle \vec{E}^2 \rangle = \text{Trace} \int d\vec{k}\, \vec{\vec{S}}_E(\vec{k}) = \frac{2\theta}{\ell} \int_0^\infty \frac{k\, dk}{1 + k^2 \lambda_D^2} \tag{44}$$

Equation (44) contains a logarithmic divergence at large k, which is to be associated with small particle separations; it can be rendered finite by the cutoff (30). However, this is misleading, for (44) contains in it a term proportional to the infinite electrostatic self-energies of all the point charges, which originates in the $\delta(\vec{r})$ term in Eq. (38). Subtracting this off will give a more revealing quantity. The self energy of one charge is proportional to $E^2_{avg} \equiv (8\pi e^2/\ell^2 V)$ $\int dk/k$, and the self field-energy for 2N of them is an additive quantity, so

$$\langle \vec{E}^2 \rangle_{self} = \frac{16\pi n_o e^2}{\ell^2} \int \frac{dk}{k}$$

and

$$\langle \vec{E}^2 \rangle - \langle \vec{E}^2 \rangle_{self} = -\frac{2\theta}{\ell \lambda_D^2} \int_0^\infty \frac{dk}{k(1 + k^2 \lambda_D^2)} \tag{45}$$

Note that Eq. (45) has now a small $\vec{k}$ divergence, which must be associated with *large* spatial separations. This has no analogy in the three-dimensional theory, even though (23) is essential the same as the three-dimensional expression. This pathology has its origin in the fact that the Coulomb potential (15) has a divergence at infinity in two dimensions, but the three-dimensional expression does not. The divergence strongly suggests that we may anticipate other unexpected small-$\vec{k}$ divergences, and that at times we may want to limit the discussion to a large but finite plasma volume $V = L^2$, to provide an effective cutoff in k space at $k \sim 2\pi/L$.

An interesting ratio is the ratio of the electric field energy density to the thermal energy density $2n_o\theta/\ell$. Dividing (44) by $16\pi n_o\theta/\ell$ and cutting the integral off at $x_{min.} = \lambda_D e^{-1}$

$$\frac{\langle \vec{E}^2 \rangle}{16\pi n_o \theta/\ell} = \frac{1}{8\pi n_o \lambda_D^2} \ln(\lambda_D/x_{min.})$$

$$= \frac{1}{2} \varepsilon \ln e^{1/\varepsilon} = \frac{1}{2}$$

This is not small small compared to unity for any $\varepsilon$. However, if we compute the comparable ratio for the part of $\langle \vec{E}^2 \rangle$ that is to be associated with interaction energy,

$$\frac{\langle \vec{E}^2 \rangle - \langle \vec{E}^2 \rangle_{self}}{16\pi n_o \theta/\ell} = -\frac{\varepsilon}{2} \ln(L/2\pi\lambda_D) \tag{46}$$

As long as the right-hand side of (46) is small compared to unity it makes sense to speak about the field fluctuation energy as being small relative to the kinetic energy, which is the essential approximation that has been made. But it is clear that such formulas as (46) lead to grave difficulties if one tries to pass to the limit $L \to \infty$. Whether these are difficulties with the perturbation expansion (17), or correspond to more basic questions connected with the existence of the thermodynamic limit in two dimensions is something that at the time of this writing, apparently nobody knows. Definitive answers to these questions will involve investigations of a considerably higher degree of rigor than the one presented here. In the meantime, one wants to be especially cautious of attempts to evaluate extensive quantities, such as the total energy, by integrating intensive ones such as $\langle \vec{E}^2 \rangle$. The circumstance may arise (it does in one dimension [15]) in which small terms of $O(1/V)$ may be missed in the perturbation theory, which give no contribution to $\langle \vec{E}^2 \rangle$ in the limit, but which may nonetheless give finite or even divergent contributions to integrals over the entire volume.

*(H) Equation of State, Collapse at Low Temperatures*

As a next comment on the equilibrium theory, let us remark on one rather remarkable property of this system, namely that its equation of state is exactly calculable. This has apparently been discovered at least twice [16,17]. It is not a complicated manipulation [2] to reduce Eq. (4) to

$$p = 2\,\frac{N}{V}\,\theta - \frac{1}{2V^3}\sum_{i<j}\int d\vec{x}_i\, d\vec{x}_j\, \varphi'_{ij}(x_{ij})\, x_{ij}\, n_2(\vec{x}_i,\vec{x}_j) \tag{47}$$

where

$$n_2(\vec{x}_i,\vec{x}_j) \equiv \frac{V^2 \int d^{8N-4} X\, D_{eq.}}{\int dX\, D_{eq.}} \tag{48}$$

In (48), $\int d^{8N-4} X$ means the integral over all phase space coordinates except for $\vec{x}_i$ and $\vec{x}_j$. The $\sum_{i<j}$ is over all pairs of both signs.

Equation (47) applies to any force law $\phi_{ij}$. The unique feature of the two-dimensional Coulomb potential is that $x_{ij}\phi'(x_{ij}) = -2e_i e_j/\ell$, independently of $\vec{x}_i$ and $\vec{x}_j$. What remains is just the normalization integral for the two-body distributions, and the problem has been reduced to counting particle pairs. This gives, after a little algebra

$$\frac{p}{2\,n_o\,\theta} = 1 - \frac{\varepsilon}{4} \tag{49}$$

The equation of state, while becoming that of an ideal gas for high temperatures ($\varepsilon << 1$), has the feature of predicting a *collapse* ($p < 0$) at low temperatures, corresponding to $\varepsilon = 4$. This is nothing remarkable, since it has its origin in the fact that the ion-electron interaction (and thus the total energy) is unbounded from below. In fact, one sees from (3) that when $\vec{x}_i$ is close to $\vec{x}_j$, Z contains a factor

$$\sim \int d\vec{x}_{ij}\, \exp\left(\frac{2\,e_i\,e_j}{\ell\theta}\, \ln x_{ij}\right) \sim \int x_{ij}\, dx_{ij}\, x_{ij}^{\;2e_i e_j/\ell\theta}$$

If $e_i$ and $e_j$ are of unlike sign, this will diverge when $-(2^2/\ell\theta + 1 < -1$, or $\varepsilon > 2$. If more than two particles are considered, the divergence may even occur at higher $\theta$ (see Knorr [18]). Some lower bound on the interaction would be required to eliminate this collapse, as quantum mechanics is required in three dimensions.

*(I) Summary*

Starting from the Gibbs distribution for an equilibrium system of electrostatically interacting charged rods aligned parallel, *approximate* expressions for the phase space probability distributions have been derived by an expansion in the plasma parameter (number of rods per Debye square). The program has been carried out in parallel to the familiar [10,13] expansion in three dimensions. These approximate probability distributions permit the calculation of ensemble averages of stationary fluctuation quantities and their spectral densities; these are independent of external magnetic field strength. The thermodynamic equation of state has been noted to be exact. Anomalies which may possibly result from taking the "thermodynamic" limit ($N \to \infty$ $V \to \infty$) have been noted. Up to this point the theory is the same with or without an external magnetic field.

## II. THE TWO-DIMENSIONAL GUIDING-CENTER PLASMA

### *(A) Some General Remarks on Transport Processes*

A central concern in statistical physics has always been the matter of *transport properties* [19]. That is, gradients in the macroscopic hydrodynamic parameters will result in the net transport of mass, momentum, energy, electric current, etc., across an imaginary surface in a gas, liquid, or plasma. This transport comes about as a direct consequence of the microscopic interactions among the particles, and is inevitably a problem in non-equilibrium statistical mechanics.

This problem has been of acute concern for strongly magnetized plasmas, because many controlled fusion schemes propose to confine a very hot plasma away from material walls by means of externally-imposed dc magnetic fields. Often, actual machines appear to confine plasmas less well than they are supposed to, and this "anomalous" transport of plasma has been a matter of concern for many years. It was addressed early by Bohm [20], disappeared from sight for several years, re-appeared about 1960 [21,22] when controlled fusion research became public knowledge, and has been the motivation for many of the exotic instability and "weak turbulence" calculations which were a major pre-occupation of plasma theory in the 1960's. The subject is far from exhausted, though it has become difficult of access by virtue of being surrounded with hundreds of rather diffuse calculations.

Actually, the transport may not be so "anomalous" as it might appear at first sight. Most of the more sober calculations [23] have started from a *kinetic equation*. By kinetic equation, one means a differentio-integral equation such as Boltzmann's equation, which will advance the time-dependent analogues of $f_{1,0}$ and $f_{0,1}$ of Eqs. (9) in time. A crucial ingredient of any kinetic equation is a *collision term* in which the irreversibility originates, which is a functional of the one-body distribution ($f_{1,0}$ or $f_{0,1}$), and into the derivation of which goes some microscopic model of the particle-particle interaction process. Different forms of the collision term arise from different microscopic models [24]. Thus uncorrelated, strong, two-body collisions lead to the Boltzmann equation, and weak two-body collisions lead to the Fokker-Planck equation. In the plasma case, however, most of the microscopic models used to describe the interactions of charged particles have *not* included the effects of an external magnetic field on the particle-particle interaction process. Putting in an external magnetic field renders the strongly-interacting two-body problem insoluble, so there is at present no analogue of Boltzmann's equation for the case in which an intense, external magnetic field is present. In a weak-interaction model of a plasma, a kinetic equation results [25-27] but is so complicated that even the derivation of it is quite strenuous, and heroic efforts will be required to extract many predictions about transport properties from it.

From the foregoing remarks we may conclude that the subject of transport properties in a strongly-magnetized plasma is still very much an open one. One expects significant gains in understanding only to the extent that clean and bold simplifying approximations can be found. A recent breakthroug in this direction was due to Taylor and McNamara [28]. They introduced a two-dimensional model of a plasma consisting of charged rods aligned parallel which are free to move perpendicularly to a strong dc magnetic field. They further simplified the dynamics by introducing the *guiding-center approximation* for the particle motion. Discussion of the various properties of this model will be the subject-matter of this chapte Various features of the model can be generalized to three dimensions, but discussion of these is better deferred until after the two-dimensional model has been presented.

*(B) The Guiding-Center Plasma ("Zero Gyro-Radius" Plasma)*

The physical system is the same as that described in Section I: a collection of N very long rods of charge +e, length $\ell$, mass $m_i$, and N more of charge −e, length $\ell$, mass $m_e$ aligned parallel to the z-axis of coordinates, and interacting through the Coulomb potential, with an interaction potential given by $\phi_{\ell m} = -(2e_\ell e_m/\ell)\ln|\vec{x}_\ell - \vec{x}_m|$ for any two charges.

The equation of motion for the velocity of a given charged rod of charge-to-mass ratio $e_j/m_j$ is

$$\frac{d\vec{v}_j}{dt} = \frac{e_j}{m_j}\left(\vec{E} + \frac{\vec{v}_j}{c} \times \vec{B}\right) \tag{50}$$

where $\vec{v}_j$, $\vec{E}$ are two-dimensional vectors lying in the xy plane, $\vec{B} = B\hat{e}_z$, with B = a constant, and $\vec{E}$ is evaluated at the location of the particle. $\vec{E} = -\nabla\phi$ obeys the two-dimensional Poisson equation,

$$\nabla^2\phi = -4\pi\sum_j \frac{e_j}{\ell}\,\delta(\vec{x} - \vec{x}_j) \tag{51}$$

where the sum runs over all charges of both signs. Equations (50) and (51), together with $d\vec{x}_j/dt = \vec{v}_j$ and the appropriate boundary conditions, constitute a complete dynamical description of the system. This dynamical description is still quite complicated, and we shall have more to say about it later. At present, we wish to discuss an even more simplified version of the dynamics, namely, the *guiding-center approximation.*

The reader is referred to Northrop [29] for a discussion of the guiding center approximation; in practice, it amounts to neglecting the left-hand side of Eq. (50), so that $\vec{E} + (\vec{v}_j/c) \times \vec{B} \cong 0$, or in this two-dimensional geometry,

$$\vec{v}_j = c\,\vec{E}(\vec{x}_j,t) \times \vec{B}/B^2 \tag{52}$$

The $\vec{v}_j$ in (52) is to be distinguished from the $\vec{v}_j$ in (50); (52) describes the motion of the *guiding center* of the particle. This guiding center motion has superimposed on it the "fast" gyration of the particle with characteristic gyro-frequency $\Omega_j = e_jB/m_jc$ and characteristic radius of the order of $r_j = \sqrt{\theta/m_j}/\Omega_j$, for a particle of thermal energy $\theta$. The "fast" motion is averaged out in writing down Eq. (52). The conditions for the validity of Eq. (52) as a satisfactory approximation are that, most importantly, the characteristic time $\tau$ and length $\lambda$ over which $\vec{E}(\vec{x},t)$ varies significantly must satisfy the inequalities

$$\lambda \gg r_j$$
$$\Omega_j \tau \gg 1 \qquad (53)$$

The actual microscopic electric field in the plasma will be de-composable into components which do and do not satisfy the inequalities (53). Only the former will be treated accurately by (52). At the present time, no satisfactory quantitative estimate exists as to the errors which result from this approximation.

*(C) Connection with Ideal Hydrodynamics*

The replacement of the differential equation (50) by the algebraic equation (52) is a considerable simplification. It also has the unexpected benefit of leaving us with a Hamiltonia system, the Hamiltonian of which is essentially the same as the potential energy of the original system! To see this, we digress briefly on the hydrodynamics of incompressible, invisci fluids in two dimensions. In the theory of parallel "line" vortex motion [30], it is shown that the fluid velocity field which results from a collection of line vortices of vortex strength $\vec{K}_j = K_j\hat{e}_z$ located at positions $\vec{x}_j$ is

$$\vec{v}(\vec{x}) = \sum_j \frac{\vec{K}_j}{2\pi} \times \frac{\vec{x} - \vec{x}_j}{|\vec{x} - \vec{x}_j|^2} \qquad (54)$$

The vortices move along streamlines, so in particular the velocity of the ith vortex is

$$\vec{v}_i = \sum_{j \neq i} \frac{\vec{K}_j}{2\pi} \times \frac{\vec{x}_{ij}}{x_{ij}^2} \qquad (55)$$

Now consider (50), (51), and (52). The electric field seen by charge i is

$$\vec{E}(\vec{x}_i) = \sum_{j \neq i} \frac{2e_j}{\ell} \frac{\vec{x}_{ij}}{x_{ij}^2} \tag{56}$$

so its velocity is

$$\vec{v}_i = \sum_{j \neq i} \frac{2ce_j}{\ell B} \frac{\vec{x}_{ij}}{x_{ij}^2} \times \frac{\vec{B}}{B} \tag{57}$$

If we identify $(2\pi)^{-1}$ times the vortex strength, $K_j/2\pi$, with $-2ce_j/\ell B$, then the two mathematical descriptions become *identical*. We may use language appropriate to the guiding center plasma interchangably with language appropriate to line vortex motion, providing that

$$K_j = -4\pi c e_j/\ell B \tag{58}$$

That the system is Hamiltonian with Hamiltonian

$$H = -\frac{1}{2\pi} \sum_{i<j} K_i K_j \ln |\vec{x}_i - \vec{x}_j| \tag{59}$$

can readily be seen by picking canonical coordinates $[\vec{x}_j \equiv (x_j, y_j)]$ $q_j, p_j$:

$$(q_j, p_j) = |K_j|^{\frac{1}{2}} (x_j, y_j \operatorname{sgn} K_j) \tag{60}$$

for each vortex (or charge). It is readily verified that (55) is the same as

$$\dot{p}_i = -\frac{\partial H}{\partial q_i}, \qquad \dot{q}_i = \frac{\partial H}{\partial p_i} \tag{61}$$

Thus the canonical coordinates and momenta are essentially just the Cartesian spatial coordinates $x_i, y_i$ of the charges (or vortices), with a change in sign of $y_i$ for the negative vortices.

Since H is not explicitly time dependent, it is a constant of the motion. The canonical distribution is again given by

Eq. (1), since H is up to a constant factor, just the interaction energy $\sum_{i<j} \phi_{ij}$ of all the pairs of charges. All the stationary thermal equilibrium correlation functions and spectral densities calculated in Section I are identical to the corresponding quantities for the guiding center plasma. Of course time dependent quantities will evolve much differently, and we pass now to a consideration of some of these.

*(D) Formulation of the Problem of Spatial Diffusion in the Green-Kubo Picture*

A measure of the rate of spatial transport can be obtained in the following simple situations. Suppose a very tenuous distribution of labeled particles, so rarified that their interaction with each other is negligible, is released near $\vec{x} = 0$. As time goes on, due to the fluctuating microscopic electric field, the charges (they will hereafter be called "test" particles) will be moved away from the origin. The position after time t for one of them will be

$$\vec{x} = \int_0^t \vec{v}(\tau)\, d\tau \tag{62}$$

where $\vec{v}(t)$ is the velocity. By (52), this is

$$\vec{x} = c \int_0^t \frac{\vec{E}(\tau) \times \vec{B}}{B^2}\, d\tau \tag{63}$$

where $\vec{E}(\tau) = \vec{E}[\vec{x}(\tau),\tau]$ is the electric field evaluated at the instantaneous test particle position.

The path (63) is erratic and complicated, and to evaluate it would involve knowing, among other things, the exact microscopic electric field $\vec{E}(\vec{x},t)$. An ensemble average prediction is the best we can hope for. So we imagine a set of similar plasmas, distributed according to the Gibbs distribution (1), and release a collection of test particles near $\vec{x} = 0$ at $t = 0$ in each one. Hereafter we address ourselves to the question of calculating ensemble averages over this ensemble [31].

Since $\langle\vec{E}\rangle = 0$, by Eq. (33), we have at once that $\langle\vec{x}\rangle = 0$, which simply says that particle transport is isotropic. A measure of the amount of diffusion is $\langle\vec{x}^2\rangle \neq 0$; thus

$$\langle\vec{x}^2\rangle = \frac{c^2}{B^2} \int_0^t d\tau_1 \int_0^t d\tau_2\, \langle\vec{E}(\tau_1) \cdot \vec{E}(\tau_2)\rangle \tag{64}$$

*Auto-correlations* like $\langle\vec{E}(\tau_1) \cdot \vec{E}(\tau_2)\rangle$ play a central role in what follows. If the ensemble is stationary, they will depend upon $\tau_2 - \tau_1$ only, so that $\langle\vec{E}(\tau_1) \cdot \vec{E}(\tau_2)\rangle = \langle\vec{E}(0) \cdot \vec{E}(\tau_2 - \tau_1)\rangle$, and it is expected that for all physical systems, $\langle\vec{E}(0) \vec{E}(\tau)\rangle$ will $\to 0$ as $\tau \to \pm\infty$. It is also to be expected that $\langle\vec{E}(0) \vec{E}(\tau)\rangle$ is an *even* function of $\tau$. We shall assume so.

Defining $\tau_2 - \tau_1 \equiv \tau$, $(\tau_2 + \tau_1)/2 \equiv T$, the integral (64) can be rewritten as

$$\langle\vec{x}^2\rangle = \frac{c^2}{B^2} \int dT \int d\tau \, \langle\vec{E}(0) \cdot \vec{E}(\tau)\rangle \tag{65}$$

The region of integration in (65) is as shown below (Fig. II-1).

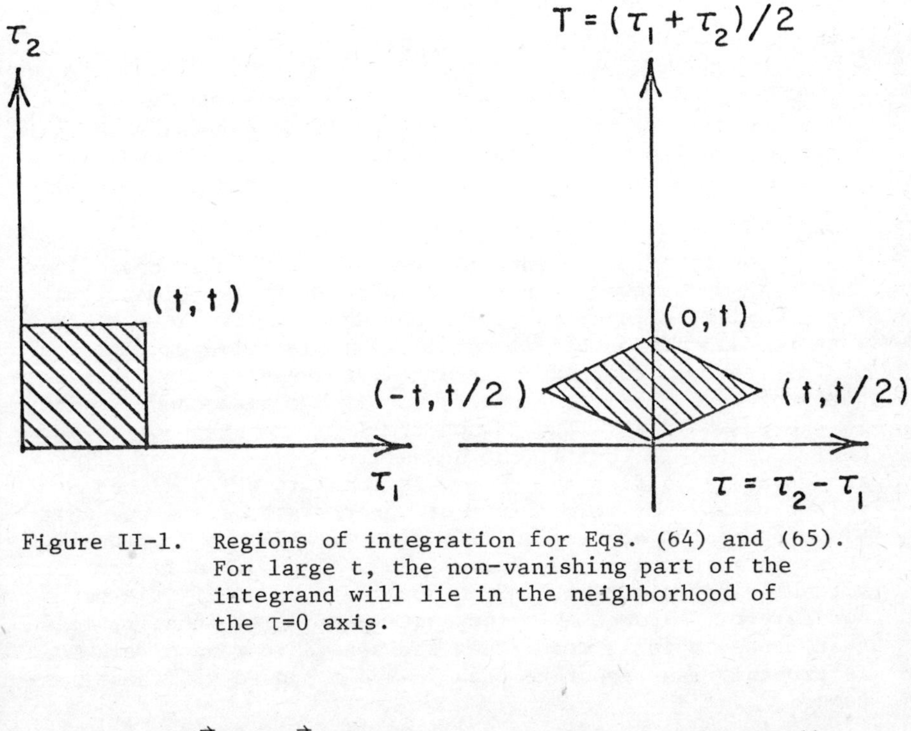


Figure II-1. Regions of integration for Eqs. (64) and (65). For large t, the non-vanishing part of the integrand will lie in the neighborhood of the $\tau$=0 axis.

Now *if* $\langle\vec{E}(0) \cdot \vec{E}(\tau)\rangle$ goes to zero sufficiently rapidly (more rapidly than $1/\tau$) for large $\tau$, we may consider the case when t is >> the time in which $\langle\vec{E}(0) \cdot \vec{E}(\tau)\rangle$ goes to zero, and write, from (65),

$$\langle \vec{x}^2 \rangle \cong \frac{c^2}{B^2} \int_0^t dT \int_{-\infty}^{\infty} d\tau \ \langle \vec{E}(0) \cdot \vec{E}(\tau) \rangle$$

$$= t \left\{ \frac{2c^2}{B^2} \int_0^{\infty} \langle \vec{E}(0) \cdot \vec{E}(\tau) \rangle \, d\tau \right\} \tag{66}$$

The ratio $\langle \vec{x}^2 \rangle / 2t$ thus becomes time-independent for long times and this ratio D,

$$D = \frac{c^2}{B^2} \int_0^{\infty} \langle \vec{E}(0) \cdot \vec{E}(\tau) \rangle \, d\tau \tag{67}$$

is called the *diffusion coefficient*. It is generally thought of as a good measure of the diffusiveness of a medium. (It is also frequently defined by $\langle \vec{x}^2 \rangle / 2td$, where d is the dimensionality of the system).

Recently a lot of interest has arisen in the possibility that $\langle \vec{x}^2 \rangle$ may grow *faster* than t as $t \to \infty$ in two dimensions for a variety of systems [32]. For example, if $\langle \vec{E}(0) \cdot \vec{E}(\tau) \rangle$ were to fall off as $O(1/\tau)$ for large $\tau$, detailed examination of (65) reveals that $\langle \vec{x}^2 \rangle \sim t \ln t$ for large t. This has been hard for many people to accept, but it has also shown up now in the plasma case. The reason the $\langle \vec{x}^2 \rangle \sim t$ behavior is so well accepted is that it can be derived *regardless of dimensionality* [31], from a random-walk model in which a random walker takes a large number n of uncorrelated steps per unit time, each step of very small length $\sigma$, with $n\langle \sigma^2 \rangle$ = const.. (D is in fact, just proportional to $n\langle \sigma^2 \rangle$.) This is true for virtually any distribution law for the steps. If $\langle \vec{x}^2 \rangle$ turns out to grow *faster* than t for systems of dimensionality less than three, then it means that the venerable random walk model is probably less accurate than we had supposed for these systems.

In any case, we will be concerned in the following pages with attempts to calculate the function

$$D(t) \equiv \frac{c^2}{B^2} \int_0^t \langle \vec{E}(0) \cdot \vec{E}(\tau) \rangle \, d\tau \tag{68}$$

for the guiding-center plasma whose dynamics are given by (51) and (52). *If* the integral converges, it will be called the "diffusion coefficient",

$$D = D(\infty) \qquad \text{(if it exists!)} \ . \tag{69}$$

But if the integral D(t) fails to approach a limit, it still is an interesting measure of the way the test particles diffuse.

### (E) *Evaluation of* $\int_0^t \langle \vec{E}(0) \cdot \vec{E}(\tau)\rangle \, d\tau$ *and D*

It is convenient to evaluate (68) in a large but finite volume $V = L^2$. $\vec{E}(\tau) = \vec{E}[\vec{x}(\tau),\tau]$, and the exact electric field due to the 2N charges is

$$\vec{E}(\vec{x},t) = -\sum_{\vec{k}} \frac{4\pi i\vec{k}}{k^2} \rho_{\vec{k}}(t) \, e^{i\vec{k}\cdot\vec{x}} \tag{70}$$

where the Fourier transform of the charge density is

$$\rho_{\vec{k}}(t) = \frac{1}{V} \int d\vec{x} \; e^{-i\vec{k}\cdot\vec{x}} \rho(\vec{x},t) \tag{71}$$

Using (34), and inserting the result into (70) gives

$$\vec{E}(\vec{x},t) = -\sum_{\vec{k}} \sum_{j} \frac{e_j}{\ell} \frac{4\pi i\vec{k}}{k^2 V} e^{i\vec{k}\cdot[\vec{x}-\vec{x}_j(t)]} \tag{72}$$

where the sum is over all charges of both signs, and the values of $\vec{k}$ have components which are integer multiples of $2\pi/L$ (we impose periodic boundary conditions).

$\vec{E}(\tau) = \vec{E}[\vec{x}(\tau), \tau]$ is then

$$\vec{E}(\tau) = \sum_{\vec{k},j} \frac{4\pi e_j \vec{k}}{ik^2 \ell V} e^{i\vec{k}\cdot[\vec{x}(\tau)-\vec{x}_j(\tau)]} \tag{73}$$

where $\vec{x}(\tau)$ is the trajectory of a particle which starts at the origin: $\vec{x}(0) = 0$. For shorthand, (73) will also be written as

$$\vec{E}(\tau) = \sum_{\vec{k}} \vec{E}_{\vec{k}}(\tau)\, e^{i\vec{k}\cdot\vec{x}(\tau)} \tag{74}$$

in an obvious notation.

The problem is then one of evaluating

$$\langle \vec{E}(0) \cdot \vec{E}(\tau) \rangle = \sum_{\vec{k}_1,\vec{k}_2} \left\langle \vec{E}_{\vec{k}_1}(0) \cdot \vec{E}_{\vec{k}_2}(\tau) \exp\{i\vec{k}_2 \cdot \vec{x}(\tau)\} \right\rangle \tag{75}$$

This result is exact, and we cannot go further without an approximation or an hypothesis. One which can be made [28] but which some day will have to be given careful scrutiny is to neglect the correlations between the positions of the test particles and those of the background plasma, the $\vec{x}_j(\tau)$. That assumption reduces (75) to

$$\langle \vec{E}(0) \cdot \vec{E}(\tau) \rangle \cong \sum_{\vec{k}_1,\vec{k}_2} \left\langle \vec{E}_{\vec{k}_1}(0) \cdot \vec{E}_{\vec{k}_2}(\tau) \right\rangle \left\langle e^{i\vec{k}_2\cdot\vec{x}(\tau)} \right\rangle \tag{76}$$

Further, noting that $\left\langle E_{\vec{k}_1}(t_1)\, E_{\vec{k}_2}(t_2) \right\rangle = 0$ unless $\vec{k}_1 + \vec{k}_2 = 0$ for a spatially-uniform ensemble, and that $\vec{E}_{-\vec{k}}(t) = \vec{E}^*_{\vec{k}}(t)$ since $\vec{E}$ is real, we have, from (76),

$$\langle \vec{E}(0) \cdot \vec{E}(\tau) \rangle = \sum_{\vec{k}} \left\langle \vec{E}^*_{\vec{k}}(0) \cdot \vec{E}_{\vec{k}}(\tau) \right\rangle \left\langle e^{i\vec{k}\cdot\vec{x}(\tau)} \right\rangle \tag{77}$$

The next step necessarily involves expressing the two terms on the right-hand side of (77) in an approximate way in terms of $\langle \vec{E}(0) \cdot \vec{E}(\tau) \rangle$, to provide an equation to be solved for this quantity. The last term can be written as

$\langle \exp[-i\vec{k} \cdot \vec{x}(\tau)] \rangle$

$$= \left\langle \exp\left[ - \frac{ic\vec{k}}{B^2} \cdot \int_0^\tau d\xi \; \vec{E}(\xi) \times \vec{B} \right] \right\rangle$$

$$= \lim_{M\to\infty} \left\langle \exp\left[ \frac{ic\vec{k} \times \vec{B}}{B^2} \; \Delta t \cdot \sum_{m=1}^{M} \vec{E}(m) \right] \right\rangle \tag{78}$$

where $\vec{E}(\vec{m})$ is the electric field between times $(m - 1)\tau/M$ and $m\tau/M$. $\Delta t$ is $\tau/M$.

The computation of the expectation value in (78) involves not just the probability distribution of the electric field seen by the particle, but the *joint* probability distribution of the electric field at a whole sequence of times, $\vec{E}(1)$, $\vec{E}(2)$, $\vec{E}(3)$. . . If we call this joint probability distribution $P_M[\vec{E}(1), \vec{E}(2), \ldots \vec{E}(M)]$, then formally

$$\langle \exp[-i\vec{k} \cdot \vec{x}(\tau)] \rangle$$

$$= \lim_{M\to\infty} \int d\vec{E}(1) \ldots d\vec{E}(M) \; P[\vec{E}(1) \ldots \vec{E}(M)] \tag{79}$$

$$\exp\left[ \frac{ic\vec{k} \times \vec{B}}{B^2} \; \Delta t \cdot \sum_{m=1}^{M} \vec{E}(m) \right]$$

It has not been possible to establish an expression for $P[\vec{E}(1) \ldots \vec{E}(M)]$ from first principles[33]. A reasonable conjecture is that it is a *jointly normal probability distribution* [34,35]; that is,

$$P[\vec{E}(1) \ldots \vec{E}(M)] = \eta \exp\left[ - \sum_{i,j=1}^{M} \vec{\vec{a}}_{ij} : \vec{E}(i) \, \vec{E}(j) \right] \tag{80}$$

where $\eta$ is a normalizing constant and the $\vec{\vec{a}}_{ij}$ are constant $2 \times 2$ positive-definite dyadics. A generalization of the central limit theorem, for example, makes (80) look extremely plausible [35]. In the present context, the theorem says the following. Consider $\vec{E}(1), \vec{E}(2), \ldots$ to be the sum $\Sigma_j [\vec{E}_j(1), \vec{E}_j(2), \ldots]$.

Here $\vec{E}_j(n)$ is the electric field at the test particle produced by particle j in time interval n, and the $\Sigma_j$ runs over all the plasma charges. Then, under very weak conditions on the moments of the probability distributions of the individual $\vec{E}_j(n)$, and the additional assumption that the $\vec{E}_j(n)$ are uncorrelated with each other for different j, the probability distribution of the $\Sigma_j$ (where j goes from 1 to 2N) converges to (80) for $N \to \infty$. This neglect of correlation between the trajectories of the plasma particles is the key approximation, and is made quite plausible by the weakness of the pair correlation [Eqs. (21) to (25)] for small plasma parameter.

It is shown in Appendix I that the result of substituting (80) into (79) is

$$\langle \exp[-i\vec{k} \cdot \vec{x}(\tau)] \rangle$$

$$\lim_{M\to\infty} \exp\left[ -\frac{c^2}{2} \sum_{m_1=1}^{M} \sum_{m_2=1}^{M} \langle \vec{E}(m_1)\, \vec{E}(m_2) \rangle : \frac{(\vec{k} \times \vec{B})(\vec{k} \times \vec{B})}{B^4} \Delta t\, \Delta t \right]$$

$$= \exp\left[ -\frac{c^2 k^2}{2B^2} \int_0^\tau d\tau_1 \int_0^\tau d\tau_2\, \langle \vec{E}(\tau_1)\, \vec{E}(\tau_2) \rangle : (\hat{k} \times \hat{B})(\hat{k} \times \hat{B}) \right]$$

$$= \exp\left[ -\frac{c^2 k^2}{4B^2} \int_0^\tau d\tau_1 \int_0^\tau d\tau_2\, \langle \vec{E}(\tau_1) \cdot \vec{E}(\tau_2) \rangle \right] \qquad (81)$$

Inserting (81) into (77) yields the result that

$$\langle \vec{E}(0) \cdot \vec{E}(\tau) \rangle$$

$$\sum_{\vec{k}} \left\langle \vec{E}_{\vec{k}}^{\,*}(0) \cdot \vec{E}_{\vec{k}}(\tau) \right\rangle \exp\left\{ -\frac{c^2 k^2}{4B^2} \int_0^\tau d\tau_1 \int_0^\tau d\tau_2\, \langle \vec{E}(\tau_1) \cdot \vec{E}(\tau_2) \rangle \right\} \qquad (82)$$

Equation (82) is a satisfactory starting point for calculation of $\langle \vec{E}(0) \cdot \vec{E}(\tau) \rangle$ as soon as we have an acceptable expression for $\langle \vec{E}^*_{\vec{k}}(0) \cdot \vec{E}_{\vec{k}}(\tau) \rangle$. First, note that

$$\langle \vec{E}_{\vec{k}}^*(0) \cdot \vec{E}_{\vec{k}}(\tau) \rangle$$

$$= \sum_{j,\ell} \frac{16\pi^2 e_j e_\ell}{\ell^2 k^2 V^2} \left\langle e^{i\vec{k}\cdot[\vec{x}_j(0)-\vec{x}_\ell(\tau)]} \right\rangle \tag{83}$$

$$= \sum_{j,\ell} \frac{16\pi^2 e_j e_\ell}{\ell^2 k^2 V^2} \left\langle \exp\left\{ i\vec{k} \cdot \left[ \vec{x}_j(0) - \vec{x}_\ell(0) - c\int_0^\tau d\xi \, \frac{\vec{E}_\ell(\xi) \times \vec{B}}{B^2} \right] \right\} \right\rangle$$

By $\vec{E}_\ell(\tau)$, we mean the electric field seen by the $\ell$th particle, of charge $e_\ell$. In (83), we can treat the $\ell$th particle in the same approximations we treated the test particles in Eqs. (76) through (82), getting

$$\langle \vec{E}_{\vec{k}}^*(0) \cdot \vec{E}_{\vec{k}}(\tau) \rangle$$

$$\cong \sum_{j,\ell} \frac{16\pi^2 e_j e_\ell}{\ell^2 k^2 V} \left\langle \exp i\vec{k} \cdot [\vec{x}_j(0) - \vec{x}_\ell(0)] \right\rangle \cdot \left\langle \exp \frac{ic\vec{k} \times \vec{B}}{B^2} \cdot \int_0^\tau \vec{E}_\ell(\xi)\, d\xi \right\rangle$$

$$= \left\langle \vec{E}_{\vec{k}}^*(0) \cdot \vec{E}_{\vec{k}}(0) \right\rangle \exp\left\{ -\frac{c^2 k^2}{4B^2} \int_0^\tau d\tau_1 \int_0^\tau d\tau_2 \, \langle \vec{E}(\tau_1) \cdot \vec{E}(\tau_2) \rangle \right\} \tag{84}$$

Since the $\vec{E} \times \vec{B}$ drift is independent of both the mass and sign of the charge, the statistical properties of $\vec{E}_\ell(\xi)$ will be the same for all $\ell$.

Inserting (84) into (82) gives

$$\langle \vec{E}(0) \cdot \vec{E}(\tau) \rangle \tag{85}$$

$$= \sum_{\vec{k}} \langle |\vec{E}_{\vec{k}}|^2 \rangle \exp \left\{ - \frac{c^2 k^2}{2 B^2} \int_0^\tau d\tau_1 \int_0^\tau d\tau_2 \langle \vec{E}(\tau_1) \cdot \vec{E}(\tau_2) \rangle \right\}$$

Since $<|\vec{E}_{\vec{k}}|^2>$ is a stationary quantity obtainable from the methods of Section I, Eq. (85) is now a closed equation that can be used to determine $<\vec{E}(0) \cdot \vec{E}(\tau)>$. Defining $R(\tau) \equiv (c^2/2B^2)$.

$\int_0^\tau d\tau_1 \int_0^\tau d\tau_2 \, Q(\tau_2 - \tau_1)$, where $Q(\tau) \equiv \frac{1}{2} \langle \vec{E}(0) \cdot \vec{E}(\tau) \rangle$

$dR(\tau)/d\tau = (c^2/B^2) \int_0^\tau Q(\xi)\, d\xi$, and $d^2R/d\tau^2 = c^2\, Q(\tau)/B^2$, whic[h] means that (85) can be expressed as

$$2 \frac{B^2}{c^2} \frac{d^2 R}{d\tau^2} = \sum_{\vec{k}} \langle |\vec{E}_{\vec{k}}|^2 \rangle \exp\left[-2 k^2 R(\tau)\right] \tag{86}$$

From (65), R(t) has the physical interpretation of being $<\vec{x}^2>/4$, where $<\vec{x}^2>$ is the mean square spread after time t of the distribution of test particles.

Equation (86) has a first integral

$$\frac{1}{2}\left(\frac{dR}{d\tau}\right)^2 - \frac{c^2}{2 B^2} \sum_{\vec{k}} \langle |\vec{E}_{\vec{k}}|^2 \rangle \left\{ \frac{1 - \exp(-2k^2 R)}{2k^2} \right\} = 0 \tag{87}$$

where the fact that $dR(0)/d\tau = 0$ has been used to evaluate the constant of integration. To proceed further with Eq. (87) requires numerical integration, if $R(\tau)$ is desired. This integration is discussed in a later section. We can extract the long time behavior analytically, however, by noting that $R(\tau) \xrightarrow[\tau\to\infty]{} +\infty$, so that

$$\left(\frac{dR}{d\tau}\right)^2 \xrightarrow[\tau\to\infty]{} \frac{c^2}{2B^2} \sum_{\vec{k}} \frac{\langle|\vec{E}_{\vec{k}}|^2\rangle}{k^2} \tag{88}$$

or

$$D(t) = \frac{2c^2}{B^2} \int_0^t Q(\tau)\, d\tau = 2\,\frac{dR(t)}{dt}$$

approaches the limit

$$D(\infty) = 2\left[\frac{c^2}{2B^2} \sum_{\vec{k}} \frac{\langle|\vec{E}_{\vec{k}}|^2\rangle}{k^2}\right]^{\frac{1}{2}} \tag{89}$$

Up to a factor of $\sqrt{2}$, this is the result of Taylor and McNamara [28].

Several remarks are in order concerning (89) before we pass to the details of its evaluation. First, it varies as $1/B$, confirming the elusive conjecture of Bohm [20]. Secondly, it will contain a factor $\sqrt{\varepsilon}$, since it contains $\sum_{\vec{k}} \langle|\vec{E}_{\vec{k}}|^2\rangle/k^2$, and will vanish in the Vlasov limit ($\varepsilon \to 0$). Finally, the expression will *diverge* in the infinite volume limit, since the minimum value of $|\vec{k}|$ is $2\pi/L$, and a factor $1/k^2$ occurs in the $\Sigma_{\vec{k}}$. This is a reflection of the fact that the long-wavelength, small -k contributions to (89) are the dominant terms, in contrast to "classical" diffusion models based on random two-body encounters.

We may estimate (89) for a large volume $V = L^2$ by replacing the sum over discrete values of k by an integral over continuous k according to the prescription

$$\sum_{\vec{k}} \langle |\vec{E}_{\vec{k}}|^2 \rangle \longrightarrow \int d\vec{k} \ \text{Trace} \ \overleftrightarrow{S}_E(\vec{k})$$

$$= 2\pi \int k \, dk \, \frac{\theta}{\ell\pi} \frac{1}{1 + k^2 \lambda_D^2} \tag{90}$$

where use has been made of (43). The integral in (90) is to be cut off at $k_{min.} = 2\pi/L$ from below. The result of using (90) in (89) for the diffusion coefficient is

$$D(\infty) = \frac{2c}{B} \left[ \frac{\theta}{\ell} \int_{k_{min}}^{\infty} \frac{dk}{k(1 + k^2 \lambda_D^2)} \right]^{\frac{1}{2}}$$

$$\cong \frac{2c}{B} \left(\frac{\theta}{\ell}\right)^{\frac{1}{2}} \ln (L/2\pi \lambda_D)^{\frac{1}{2}} \tag{91}$$

$$= \sqrt{2} \frac{c\theta}{eB} \epsilon^{\frac{1}{2}} [\ln (L/2\pi \lambda_D)]^{\frac{1}{2}}$$

The Bohm result, never justified in detail, was $D(\infty) = \frac{1}{16} \frac{c\theta}{eB}$.

Notice, however, the following limitation in the derivation of Eqs. (89) and (91). It was arrived at by assuming that the terms involving $\exp(-2k^2R)$ in (87) had become << 1. Since $k_{min.} = 2\pi/L$ and $R = \langle \vec{x}^2 \rangle/4$, this is equivalent to

$$2 \frac{\langle \vec{x}^2 \rangle}{4} \frac{4\pi^2}{L^2} >> 1$$

or $\langle \vec{x}^2 \rangle >> L^2/2\pi^2 \cong L^2/20$. However, when $\langle \vec{x}^2 \rangle$ becomes $\gtrsim L^2$, the test particles will have encountered the boundary once, and in view of the periodic boundary conditions, the diffusion model with its monotonically increasing $\langle \vec{x}^2 \rangle$ will have broken down. Thus there is a relatively narrow temporal interval,

$$\frac{L^2}{20} \ll 2t\, D(\infty) \ll L^2 \tag{92}$$

in which (89) and (91) may be expected to apply. Both before and after this interval, the development of $\langle \vec{x}^2 \rangle$ will be more complicated.

Finally, we remark that if one attempts to apply (86) to *unbounded* two-dimensional plasmas, a variation [36] of $\langle \vec{x}^2 \rangle \sim t\sqrt{\ln t}$ occurs for large values of t.

### (F) *Liouville Equation, BBGKY Hierarchy, and the Time-Dependent "Vlasov" Limit* [12]

The diffusion calculation could proceed in a relatively unsystematic way, with the assumption (80) as the key ingredient For other purposes, a more systematic formalism for the kinetic theory of the two-dimensional guiding center plasma is desirable. Proceeding in parallel to the well-known program of the non-equilibrium BBGKY [2-6,11] hierarchy, it is useful to construct a description based on the *Liouville equation*.

The phase-space probability distribution of any Hamiltonian system obeys a Liouville equation, which simply expresses the constancy of that probability distribution as seen by an observer who moves along any phase-space trajectory traced out by the system as it moves through its phase space. Since we have seen the system to be Hamiltonian in Eqs. (59) to (61), we can write down the Liouville equation immediately as

$$\frac{dD}{dt} = \left( \frac{\partial}{\partial t} + \sum_i \frac{d\vec{x}_i(t)}{dt} \cdot \frac{\partial}{\partial \vec{x}_i} \right) D = 0 \tag{93}$$

where $D = D(\vec{x}_1, \vec{x}_2, \ldots, t)$ is the probability distribution of all the particles in their phase space. Noting that $d\vec{x}_i(t)/dt = c\,\vec{E}(\vec{x}_i,t) \times \vec{B}/B^2$, (93) becomes

$$\left( \frac{\partial}{\partial t} + \sum_i \frac{c\vec{E}(\vec{x}_i,t) \times \vec{B}}{B^2} \cdot \frac{\partial}{\partial \vec{x}_i} \right) D = 0, \tag{94}$$

where $\vec{E}(\vec{x}_i,t) = + \frac{\partial}{\partial \vec{x}_i} \sum_{j\neq i} \frac{2e_j}{\ell} \ln|\vec{x}_i - \vec{x}_j|$, and the summations run over all charges of both signs. The single most important solution of (94) is, of course, $D_{eq.} = \exp(-\sum_{i<j} \varphi_{ij}/\theta)/\int dX \exp(-\sum_{i<j} \varphi_{ij}/\theta)$, though for a non-equilibrium situation, D will *not* be $D_{eq.}$. We shall generally assume, however, that the various correlation functions have the same *orders of magnitude* as the thermal equilibrium correlations of Section I. We also assume D to be symmetric under interchange of like particle coordinates. The reduced probability distributions are defined in analogy to Eqs. (9) as

$$f_{s_i,s_e} = V^{s_i+s_e} \int d\vec{x}^i_{s_i+1} \ldots d\vec{x}^i_N \, d\vec{x}^e_{s_e+1} \ldots d\vec{x}^e_N \, D \quad (95)$$

where it is to be noted that

$$f_{s_i,s_e} = f_{s_i,s_e}(\vec{x}^i_1 \ldots \vec{x}^i_{s_i}, \vec{x}^e_1 \ldots \vec{x}^e_{s_e}, t)$$

is a function of *position space coordinates* only (the dimensionality of the phase space is now 4N).

Integrating (94) over a big box of volume $V = L^2$ gives:

$$\left\{ \frac{\partial}{\partial t} + \sum_{j=1}^{s_i} \frac{c\,\vec{E}_j^{(s_i,s_e)} \times \vec{B}}{B^2} \cdot \frac{\partial}{\partial \vec{x}^i_j} \right.$$

$$\left. + \sum_{k=1}^{s_e} \frac{c\,\vec{E}_k^{(s_i,s_e)} \times \vec{B}}{B^2} \cdot \frac{\partial}{\partial \vec{x}^e_k} \right\} f_{s_i,s_e}$$

$$= n_0 \sum_{j=1}^{s_i} \int d\vec{x}^i_{s_i+1} \left[ \frac{\partial \varphi^{ii}\left(\vec{x}^i_j - \vec{x}^i_{s_i+1}\right)}{\partial \vec{x}^i_j} \times \frac{\vec{B}}{B^2} \right] \cdot \frac{\partial f_{s_i+1}}{\partial \vec{x}_j}$$

$$+ n_o \sum_{k=1}^{s_e} \int d\vec{x}^i_{s_i+1} \left[ \frac{\partial \varphi^{ie}\left(\vec{x}^e_k - \vec{x}^i_{s_i+1}\right)}{\partial \vec{x}^e_k} \times \frac{\vec{B}}{\vec{B}^2} \right] \cdot \frac{\partial f_{s_i+1,s_e}}{\partial \vec{x}^e_k}$$

$$+ n_o \sum_{j=1}^{s_e} \int d\vec{x}^e_{s_e+1} \left[ \frac{\partial \varphi^{ei}\left(\vec{x}^i_j - \vec{x}^e_{s_e+1}\right)}{\partial \vec{x}^i_j} \times \frac{\vec{B}}{B^2} \right] \cdot \frac{\partial f_{s_i,s_e+1}}{\partial \vec{x}^i_j}$$

$$+ n_o \sum_{k=1}^{s_e} \int d\vec{x}^e_{s_e+1} \left[ \frac{\partial \varphi^{ee}\left(\vec{x}^e_k - \vec{x}^e_{s_e+1}\right)}{\partial \vec{x}^e_k} \times \frac{\vec{B}}{B^2} \right] \cdot \frac{\partial f_{s_i,s_e+1}}{\partial \vec{x}^e_k}$$

(96)

This is the *non-equilibrium* BBGKY hierarchy for the system, and is the analogue of (13). By $\vec{E}_J^{(s_i,s_e)}$, we mean $-\frac{\partial}{\partial \vec{x}^i_j} \cdot \Phi(s_i,s_e)$, where $\Phi(s_i,s_j)$ is the $\sum_{i<j} \phi_{ij}$ for the first $s_i$ ions and the first $s_e$ electrons. Similarly $\vec{E}_k^{(s_i,s_e)} = -\partial\, \Phi(s_i,s_e)/\partial \vec{x}^e_k$.

Equations (94) and (96) are not as complicated as they look, and we can profitably introduce abbreviations for the linear differential and integral operators which appear in them. Thus, (94) will sometimes be written as $\left(\frac{\partial}{\partial t} + H_o\right)D = 0$, and (96) as

$$\left[\frac{\partial}{\partial t} + H(s_i,s_e)\right] f_{s_i,s_e} = \left(L^{ii}_{s_i s_e} + L^{ie}_{s_i s_e}\right) f_{s_i+1,s_e}$$

$$+ \left(L^{ei}_{s_i s_e} + L^{ee}_{s_i s_e}\right) f_{s_i,s_e+1}$$ in an obvious notation. $H_o$ is the

Liouville operator for all 2N particles, and $H(s_i, s_e)$ is the Liouville operator for $s_i$ ions and $s_e$ electrons.

The BBGKY hierarchy (96) is analogous to the well-known hierarchy for ordinary plasmas but it has important differences which are worth mentioning. It is interesting to consider the "Vlasov limit" in which the correlations vanish; the corresponding thermal equilibrium limit (see Section I) is that in which $\varepsilon \to 0$. Thus we write

$$f_{s_i, s_e} = \left[ \prod_{j=1}^{s_i} f_{1,0}(\vec{x}^i_j, t) \right] \left[ \prod_{k=1}^{s_e} f_{0,1}(\vec{x}^e_j, t) \right] \text{ and all of Eqs. (96)}$$

become satisfied if

$$\frac{\partial f_{1,0}}{\partial t}(\vec{x}^i_1, t) + \frac{c\,\vec{E}(\vec{x}^i_1, t) \times \vec{B}}{B^2} \cdot \frac{\partial f_{1,0}(\vec{x}^i_1, t)}{\partial \vec{x}^i_1} = 0$$

$$\frac{\partial f_{0,1}}{\partial t}(\vec{x}^e_1, t) + \frac{c\,\vec{E}(\vec{x}^e_1, t) \times \vec{B}}{B^2} \cdot \frac{\partial f_{0,1}(\vec{x}^e_1, t)}{\partial \vec{x}^e_1} = 0 \tag{97}$$

where

$$\vec{E}(\vec{x}, t) = + \frac{\partial}{\partial \vec{x}} \, n_o \int \left[ f_{1,0}(\vec{x}', t) - f_{0,1}(\vec{x}', t) \right] \frac{2e}{\ell} \ln |\vec{x} - \vec{x}'| \, d\vec{x} \tag{98}$$

Equations (97) and (98) are the analogues of the classical Vlasov-Poisson system for the two-dimensional guiding-center plasma. The spatially-uniform state, $f_{1,0} = f_{0,1} = 1$ and $\vec{E} = 0$, is obviously a solution. However, the analogue of the Landau problem, in which one linearizes about the spatially uniform state, is apparently insoluble; for the first non-vanishing terms in (97) are intrinsically of second order in powers of the departure from spatial uniformity.

Equations (97), (98) are completely equivalent to the equations of motion for incompressible, inviscid flow. They can be made to look simpler by subtracting Eqs. (97) from each other to get

$$\frac{\partial \rho}{\partial t} + \frac{c\,\vec{E} \times \vec{B}}{B^2} \cdot \frac{\partial \rho}{\partial \vec{x}} = 0 \tag{99}$$

where

$$\frac{\partial}{\partial \vec{x}} \cdot \vec{E} = 4\pi\rho \tag{100}$$

Very little information exists on the solutions of the system (99), (100), although a number of numerical simulations exist which are relevant [37-39]. Taylor and Thompson [40] have argued for the existence of a large class of *oscillatory* solutions of this system with a frequency which vanishes as the amplitude of $|\vec{E}|^2$ vanishes. This derivation, however, is not wholly transparent. Any oscillations which do exist *will* be intrinsically nonlinear.

Another peculiar feature of the hierarchy (96) is the apparent absence of any tendency toward thermal equilibrium. For the conventional hierarchy for a normal plasma, assuming an initially spatially uniform state leads to rapidly-developing pair correlations. These in turn relax to functionals of $f_{1,0}$ and $f_{0,1}$ which, when substituted into the equations for $\partial f_{1,0}/\partial t$ and $\partial f_{0,1}/\partial t$, lead to the Balescu-Lenard equation [7,41-43]. From this, an approach to thermal equilibrium can be proved. The situation is quite different for the guiding-center plasma. It is clear by inspection that the completely uncorrelated state, in which $f_{s_i,s_e} = \prod_{j=1}^{s_i} \prod_{k=1}^{s_e} f_{1,0}(\vec{x}_j^i)\, f_{0,1}(\vec{x}_k^e)$ with $f_{1,0} = f_{0,1} = 1$, is a *time-independent* solution of (96)! It is not known whether there is *any* sense in which the solutions of (96) relax to the thermal equilibrium values (11). This is one of the really mysterious parts of the theory. A number of time independent solutions with non-thermal equilibrium correlations for (96) can be constructed, but there is no particular reason for preferring one over the other.

*(G) Calculation of the Electrical Conductivity by the Kubo Method* [44,45]

A calculation of a second transport coefficient, the electrical conductivity, proceeds most naturally from the Liouville equation, Eq. (94) will be abbreviated as

$$\left(\frac{\partial}{\partial t} + H_o\right) D = 0 \tag{94a}$$

The addition of a spatially uniform electric field $\vec{E}_o e^{-i\omega t}$ modifies (94a) to read

$$\left(\frac{\partial}{\partial t} + H_o\right) D = - \frac{c\,\vec{E}_o \exp(-i\omega t) \times \vec{B}}{B^2}$$

$$\cdot \left\{ \sum_{j=1}^{N} \left( \frac{\partial D}{\partial \vec{x}^{\,i}_j} + \frac{\partial D}{\partial \vec{x}^{\,e}_j} \right) \right\} \equiv - \exp(-i\omega t)\, H_1 D \tag{101}$$

where the right-hand side of Eq. (101) defines the linear operator $H_1$. We now regard $\vec{E}_o$ as *weak*, and seek a perturbation solution to (101) in the form $D = D^{(0)} + D^{(1)} + \ldots$, where the superscript indicates the order in $H_1$. Thus $D^{(0)}$ obeys

$$\left(\frac{\partial}{\partial t} + H_o\right) D^{(0)} = 0 \tag{102}$$

and $D^{(1)}$ obeys

$$\left(\frac{\partial}{\partial t} + H_o\right) D^{(1)} = - \exp(-i\omega t)\; H_1 D^{(0)} \tag{103}$$

The appropriate solution to (102) is the Gibbs distribution

$$D^{(0)} = \eta \exp\Big(- \sum_{i<j} \varphi_{ij}/\theta\Big) \tag{104}$$

and the $\sum_{i<j} \phi_{ij}$ runs over all pairs of particles of both signs. As usual, $\eta$ is chosen so that $\int D^{(0)}\, dX = 1$. Since (102) is simply a statement of the constancy of $D^{(0)}$ along a trajectory in the total phase space, it is just a statement that the formal solution

$$D^{(0)}(t) = e^{-tH_o} D^{(0)}(0) \tag{105}$$

is just a function of the constants of the motion [6]. $\sum_{i<j} \phi_{ij}$ has already been noted to be such a constant, and (104) is the solution appropriate to thermal equilibrium.

We imagine the external uniform field $\vec{E}_o$ to be turned on abruptly at $t = 0$. The solution to (103) can be found by integrating the equivalent equation

$$\frac{\partial}{\partial t}\left(e^{tH_o} D^{(1)}\right) = - e^{tH_o} e^{-i\omega t} H_1 e^{-tH_o} D^{(0)} \qquad (106)$$

between $t = 0$ and $t = t$. Setting $D^{(1)} = 0$ at $t = 0$, noting that $e^{-tH_o} D^{(0)} = D^{(0)}$, and rearranging the integral gives

$$D^{(1)} = - e^{i\omega t} \int_o^t d\tau \, e^{i\omega\tau} e^{-\tau H_o} H_1 D^{(0)} \qquad (107)$$

Now for any dynamical variable whose value is determined by the phase space coordinates (A, say), the expectation value $\langle A \rangle = \int A \, D \, dX$. For variables which have expectation values which vanish in thermal equilibrium, we have the result that to first order in $E_o$, $\langle A \rangle = \int A \, D^{(1)} \, dX$. Introducing the explicit form of $H_1$, we have

$$H_1 D^{(0)} = + \frac{c \vec{E}_o \times \vec{B}}{B^2} \cdot \sum_j \left[\frac{e \vec{E}(\vec{x}_j^i)}{\theta} - \frac{e \vec{E}(\vec{x}_j^e)}{\theta}\right] D^{(0)} \qquad (108)$$

$$= - \frac{1}{\theta} \vec{E}_o \cdot \sum_j e(\vec{v}_j^i - \vec{v}_j^e) \, D^{(0)}$$

We now note that the effect of the operator $e^{-\tau H_o}$ applied to any function of the coordinates obeying Liouville's equation is to trace those coordinates back $\tau$ units along the phase space trajectory and assign them their values there. Thus, for instance $e^{-\tau H_o} \vec{v}_j^i = \vec{v}_j^i(-\tau)$, where $\vec{v}_j^i(-\tau)$ is the velocity that the jth ion would have had $\tau$ seconds ago if it now has velocity $\vec{v}_j^{\,i} = \vec{v}_j^{\,i}(0)$. Therefore

$$\exp(-\tau H_o)\, H_1\, D^{(0)} = -\frac{1}{\theta}\, \vec{E}_o \cdot \sum_j e\left[\vec{v}_j^{\,i}(-\tau) - \vec{v}_j^{\,e}(-\tau)\right] D^{(0)} \tag{109}$$

since $D^{(0)}$ is constant along the phase space trajectory. Inserting (109) into (107) and the result into the expression for $\langle A\rangle$ gives

$$\langle A\rangle = \frac{\exp(-i\omega t)}{\theta} \int_0^\infty d\tau\, \exp(i\omega\tau) \tag{110a}$$

$$\langle A \sum_j e\, \vec{E}_o \cdot [\vec{v}_j^{\,i}(-\tau) - \vec{v}_j^{\,e}(-\tau)]\rangle$$

We have not yet committed ourselves to an explicit expression for the dynamical variable A. To calculate the electric current density produced in response to the electric field $\vec{E}_o$, we pick A a function which makes $\langle A\rangle$ equal to the volume current density $\langle\vec{j}\rangle$. We must remember to apply this procedure only far away from the plasma boundary. Thus

$$A \rightarrow \vec{A} = \sum_j e(\vec{v}_j^{\,i} - \vec{v}_j^{\,e})/\ell L^2$$

Noting that $\langle\vec{v}(0) \cdot \vec{v}(\tau)\rangle$ will be an even function of $\tau$ for all particles of both signs,

$$\langle\vec{A}\rangle = \langle\vec{j}\rangle_\omega\, e^{-i\omega t}$$

$$= \frac{e^2 \exp(-i\omega t)}{\ell L^2 \theta} \int_0^\infty d\tau\, e^{i\omega\tau}$$

$$\sum_{j,\ell} \langle [\vec{v}_j^{\,i}(0) - \vec{v}_j^{\,e}(0)]\, [\vec{v}_\ell^{\,i}(\tau) - \vec{v}_\ell^{\,e}(\tau)]\rangle \cdot \vec{E}_o \qquad (110b)$$

The coefficient of $\vec{E}_o$ exp $(-i\omega t)$ in (110b) can be identified as the *conductivity tensor* $\vec{\vec{\sigma}}(\omega)$:

$$\vec{\vec{\sigma}}(\omega) = \frac{e^2}{\ell\,\theta\,L^2} \int_0^\infty d\tau\, e^{i\omega\tau} \sum_{j,\ell} \langle [\vec{v}_j^{\,i}(0) - \vec{v}_j^{\,e}(0)]\, [\vec{v}_\ell^{\,i}(\tau) - \vec{v}_\ell^{\,e}(\tau)]\rangle \qquad (111)$$

The terms in (111) with $j \neq \ell$ and not of the same species can be shown to give a contribution which is down by a factor of $O(1/N)$ from those *with* $j = \ell$. Therefore $\vec{\vec{\sigma}}(\omega)$ becomes, since all the $j = \ell$ terms of the same species are equal.

$$\vec{\vec{\sigma}}(\omega) = \frac{2Ne^2}{\ell L^2 \theta} \int_0^\infty d\tau\, e^{i\omega\tau} \langle \vec{v}(0)\, \vec{v}(\tau)\rangle$$

$$= \frac{2n_o e^2 c^2}{\ell\,\theta\,B^4}\, \vec{B} \times \int_0^\infty d\tau\, e^{i\omega\tau} \langle \vec{E}(0)\, \vec{E}(\tau)\rangle \times \vec{B} \qquad (112)$$

The $\langle\vec{E}(0)\, \vec{E}(\tau)\rangle$ is the same quantity whose trace appears in Eq. (64), and the same methods can be used to calculate it. Under the apparently straightforward assumption that $\langle\vec{E}(0)\, \vec{E}(\tau)\rangle$ is diagonal (for it to be otherwise would imply a preferred direction in space), we have

$$\vec{\vec{\sigma}}(\omega) = \vec{\vec{1}}\, \sigma(\omega)$$

where

$$\sigma(\omega) = \frac{2n_o e^2 c^2}{\ell\,\theta\,B^2} \int_0^\infty Q(\tau)\, e^{i\omega\tau}\, d\tau \qquad (113)$$

with $Q(\tau)$ given by $\frac{1}{2} \langle \vec{E}(0) \cdot \vec{E}(\tau) \rangle$, as before.

In the dc limit, $\omega = 0$,

$$\sigma(0) = \frac{2 n_o e^2 c^2}{\ell \, \theta \, B^2} \int_0^\infty Q(\tau) \, d\tau$$

$$= \frac{n_o e^2}{\ell \, \theta} D(\infty) \tag{114}$$

showing a direct proportionality with the diffusion coefficient $D(\infty)$ of Eq. (89). Such a proportionality between the conductivity and diffusion coefficient was called by Kubo [44] a "generalized Einstein relation", for systems whose dynamics are governed by Newton's laws. But it was far from obvious that such a proportionality would exist for the guiding-center plasma.

The ac conductivity requires a Fourier transform of $Q(\tau)$, which is not obtainable analytically in closed form. For that, one needs to solve Eq. (87) numerically. This has been done, and the details are reported by Montgomery and Tappert [45]. Typical results are shown, in dimensionless units whose interpretation we can ignore here, in Figs. II-2 and II-3. The same remarks about divergences for $L \to \infty$ apply as were applicable in the discussion of the diffusion coefficient.

(Note added in proof: Recent numerical simulations by G. Joyce and the author have indicated that the conductive behavior of the guiding center plasma may be somewhat more complicated than the present development would indicate. A transverse electric field induces large scale vortex motion of a non-local character in a wide variety of simulations. We are also indebted to Dr. Leaf Turner for pointing out to us certain shortcomings of the present derivation of a local conductivity.)

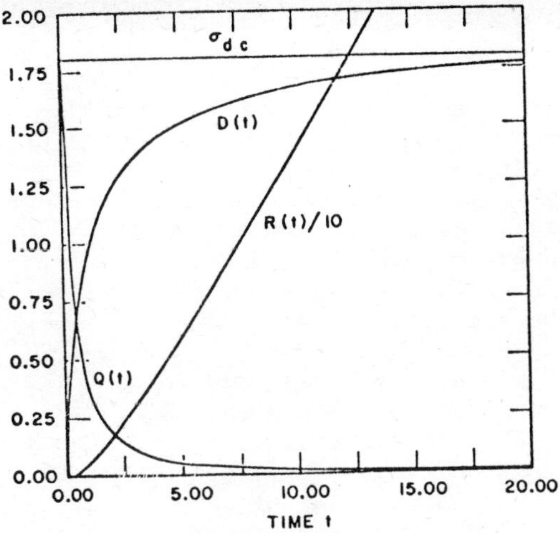


Figure II-2. Numerical solution of Eq. (87), from Montgomery and Tappert [45]. Dimensionless units are: $k/K_D$, $R/2K_D^2$, $t[8\pi n_0 c e \varepsilon^{\frac{1}{2}}]/\ell B$. Notice that R increases linearly with t almost from the beginning.

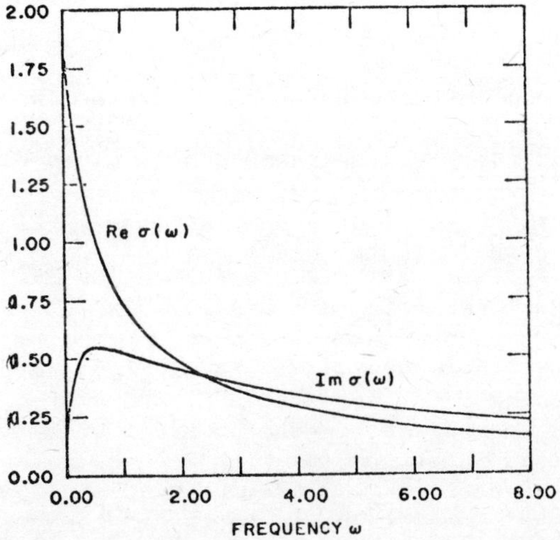


Figure II-3. Real and imaginary parts of the conductivity from Eq. (113). The units of $\sigma$ are $n_0 e c \varepsilon^{\frac{1}{2}}/\ell B$.

### (H) *The "Negative Temperature" Instability for the Guiding-Center Plasma*

A rather unusual feature of the guiding-center plasma is that for interaction energies $\sum_{i<j} \phi_{ij}$ greater than a certain amount, statistical mechanics predicts no stable, thermal equilibrium, spatially uniform state. In the discussion immediately after Eq. (49), the absence of such a state for sufficiently *low* energies was noted. It was noted some time ago by Onsager [46] that, for different reasons, no equilibrium states exist for *high* enough energies.

Onsager's prediction was made in the context of the hydrodynamic formulation of the problem, Eqs. (59) through (61). We can understand it in the following way. Equation (3) for the partition function can be rewritten as (again omitting factors which are not of interest here)

$$Z = \int dX \exp(-\mathcal{E}/\theta)$$
$$= \int d\mathcal{E}\, \Omega(\mathcal{E}) \exp(-\mathcal{E}/\theta) \tag{115}$$

where $\varepsilon$ is now just the numerical value of the Hamiltonian (59), $d\mathcal{E} = \prod_{i=1}^{2N} (dq_i\, dp_i)$ is the phase space volume element for all the vortices (charges) of both signs, and $\Omega(\mathcal{E})$ is the *structure function*, or phase space volume per unit $\mathcal{E}$. $\Omega(\mathcal{E}) = d\Phi(\mathcal{E})/d\mathcal{E}$, where $\Phi(\mathcal{E})$ is the total phase space volume with values of H less than $\mathcal{E}$: $\Phi(\mathcal{E}) = \int_{-\infty}^{\mathcal{E}} d\mathcal{E}'\, \Omega(\mathcal{E}')$.

For most thermodynamic systems, $\Omega(\mathcal{E})$ increases rapidly with increasing $\mathcal{E}$ for very large N, and $\exp(-\mathcal{E}/\theta)$ decreases rapidly for very large N. The competition between the two means that the integrand in (115) will be rather sharply peaked about some value $\mathcal{E} = \mathcal{E}_o$. Indeed, this is nothing but the statement that the fluctuations in energy in the ensemble are small, and that most of the systems are sharply concentrated in energy near the mean value, which is the only reason statistical mechanics works at all. Thus we can write

$$Z = \int d\mathcal{E} \exp[\psi(\mathcal{E}) - \mathcal{E}/\theta] \tag{116}$$

where $\psi(\mathcal{E}) = \ln\Omega(\mathcal{E})$. The integrand in (116) can be well approximated near $\mathcal{E}_0$ by $\exp[\psi(\mathcal{E}_0) - \mathcal{E}_0/\theta + \psi''(\mathcal{E}_0)(\mathcal{E} - \mathcal{E}_0)^2/2]$, and is negligibly small away from $\mathcal{E}_0$. The integral can be performed to give

$$Z = \exp\,[\psi(\mathcal{E}_0) - \mathcal{E}_0/\theta]\ |\psi''(\mathcal{E}_0)|^{-\frac{1}{2}}\ \sqrt{2\pi} \tag{117}$$

but the theory of the canonical ensemble also shows

$$Z = \exp\left[\frac{S(\mathcal{E}_0)}{K} - \mathcal{E}_0/\theta + \text{const.}\right] \tag{118}$$

where $\mathcal{E}_0$ is the thermodynamic internal energy of the system being represented and $S(\mathcal{E}_0)$ is the corresponding entropy. The maximum occurs at $1/\theta = \psi'(\mathcal{E}_0) = \Omega'(\mathcal{E}_0)/\Omega(\mathcal{E}_0)$. Noting that $\psi$ and S are both O(N) quantities, comparing (118) and (117) gives for the entropy

$$S(\mathcal{E}_0) = K \ln \Omega(\mathcal{E}_0) + \text{const.} \tag{119}$$

plus terms negligible compared to O(N), and for the temperature,

$$\frac{1}{\theta} = \frac{\Omega'(\mathcal{E}_0)}{\Omega(\mathcal{E}_0)} \tag{120}$$

Equations (119) and (120) are standard formulas but they have some peculiar implications for the system described by Eqs. (59)-(61). For notice that each one of the 2N particles (vortices) has available to it a maximum phase space volume $L^2|K|$. All 2N of them have a phase-space volume $L^{4N}|K|^{2N}$. This means that $\Phi(\mathcal{E})$, which is a non-decreasing function of $\mathcal{E}$ by definition, increases monotonically from 0 to $L^{4N}|K|^{2N}$ as $\mathcal{E}$ goes from $-\infty$ to $+\infty$. This further implies a maximum (at $\mathcal{E} = \mathcal{E}_m$, say) in $\Omega(\mathcal{E})$, where $\Omega'(\mathcal{E})$ goes from positive to negative. Equation (120) then implies that if $\mathcal{E}_0 > \mathcal{E}_m$, the temperature $\theta$ is *negative*.

Landau and Lifshitz [47] show quite generally that a stationary thermal equilibrium state is not to be expected for negative temperatures, provided the component parts of the system are free to move with respect to each other. That condition is fulfilled for the present system, so that no stationary state is expected when the value of H exceeds the quantity $\mathcal{E}_m$. One can get a qualitative picture of what the states of the system are for $H = \mathcal{E} > \mathcal{E}_m$ by considering the limit of very large values of H. These clearly correspond to states in which the charges (vortices) of unlike sign as they can get. The macroscopic flow pattern is that of a pair of large counter-rotating vortices, spatially separated in different parts of the volume.

Recently, such a flow pattern emerged [37] rather unexpectedly in a numerical solution to the two-dimensional equations of incompressible fluid flow at high Reynolds number (essentially Eqs. (99) and (100) with a small viscous damping). An "ergodic boundary" separating those initial conditions where the large pair of vortices is formed from those where they are not was proposed by Deem and Zabusky. A possible explanation in terms of the "negative temperature" instability was offered by Montgomery [48], who also observed that the prediction was not totally applicable to the Deem-Zabusky simulation in view of the lack of a distinction between "self" and "interaction" energies in the continuum representation (Onsager's prediction applies only to a "line" vortex model, and $H = \mathcal{E}$ is the *interaction* energy). Cook and Taylor [49] had previously offered an explanation of the Deem-Zabusky results in terms of the impossibility of relaxation to the thermal equilibrium energy spectrum [essentially Eq. (44)], as compatible with conservation of *enstrophy*, for certain sets of initial conditions. [Enstrophy is the volume integral of the square of the vorticity, or for the plasma interpretation, the square of the charge density. It is one among many constants of the motion for the ideal incompressible fluid, where the integral of any power of the vorticity is also such a constant. Due to the non-square-integrability of the delta function, it is ill-defined for the discrete-charge or discrete-vortex system].

Joyce and Montgomery [50] performed a simulation of the discrete vortex situation in which the equations of motion of individual vortices were advanced numerically in time. At high initial interaction energies (energies >> the self energies, in the particle-in-cell representation), the Deem-Zabusky phenomenon reappeared: formation of a pair of large vortices composed of many of the smaller discrete ones.

A calculation due to Taylor [51] attempts to determine the threshold for the negative temperature instability by evaluation of the stationary-phase approximation to $\frac{d}{d\varepsilon_o} \ln\Omega(\varepsilon_o)$. The rather remarkable result emerges that, when the infinite self-energy is taken into account, $\mathcal{E}_m = 0$. [there is also a somewhat confusing attempt to connect the result with the previous Cook-Taylor [49] calculation for *continuous* fluids. We are unable to appreciate this argument, and shall remark no more on it here]. An improved calculation of $\mathcal{E}_m$ is due to C.E. Seyler, Jr., in Physical Review Letters, 32 515 (1974).

The result $\mathcal{E}_m = 0$ would be a remarkable conclusion for the following reason. Since the plasma parameter does not enter the calculation, we may assume that the result holds down to $\mathcal{E} = 0$. But $\mathcal{E} = 0$ is the "Vlasov" limit, or for the fluid case, the limit of ideal, incompressible, inviscid hydrodynamics. The spatially-uniform, field-free, quiescent state would be by this criterion *unstable*. The interaction energy per particle in the Vlasov limit is zero. Therefore, *any* perturbation on the spatially-uniform state would have the effect of adding a positive interaction energy per particle. (The energy $\int d\vec{x} \langle \vec{E}^2 \rangle / 8\pi$ is a positive-definite functional of any charge density perturbation). Thus *any* perturbation from uniformity would put the system on the negative-temperature side of the boundary.

Finally, it may be speculated that the energy at which p becomes < 0 according to (49) might lie *above* $\mathcal{E}_m$. This eventuality, which seems unlikely, would deny the existence in principle of a stable spatially-uniform thermodynamic equilibrium state of the guiding-center plasma in two dimensions for *any* energy.

From these various calculations, it is to be concluded that the guiding center plasma in two dimensions is an extraordinarily rich statistical-mechanical system, one about which much remains to be discovered. It is also one whose pathologies and singular features should make the investigator wary until we have developed a better intuitive understanding of the system than now exists.

## III. GUIDING-CENTER PLASMAS IN THREE DIMENSIONS

### *(A) Dynamical Description, Canonical Ensemble, and Time-Independent Fluctuations*

The two-dimensional guiding-center plasma model developed in Sections I and II is of intrinsic interest for statistical physics, but any practical significance it may have depends in large part upon finding a way to generalize it to three dimensions. This program is still in its infancy, and though some of the results which have been achieved will now be summarized, they will very probably have been considerably enlarged upon by the time this material appears in print.

The natural generalization to three dimensions involves allowing point charges to move with the $c\vec{E} \times \vec{B}/B^2$ drift perpendicularly to a constant uniform magnetic field $\vec{B}$, but to move according to Newton's laws in the parallel direction. If the velocity of the jth charge is $\vec{v}_j$, we decompose it (and other vectors) into parallel and perpendicular parts,

$$\vec{v}_j = v_{j\parallel} \hat{b} + \vec{v}_{j\perp} \tag{121}$$

where $\hat{b} \equiv \vec{B}/B$ is in the z-direction, say, and $\hat{b} \cdot \vec{v}_{j\perp} = 0$. For $\vec{v}_{j\perp}$, we have

$$\vec{v}_{j\perp} = c\vec{E}(\vec{x}_j,t) \times \vec{B}/B^2 \tag{122}$$

where $\vec{E}(\vec{x}_j,t)$ is the electric field evaluated at the instantaneous position $\vec{x}_j$ of the jth charged particle ($d\vec{x}_j/dt = \vec{v}_j$). For the parallel motion, we have

$$m_j \frac{d v_{j\parallel}}{dt} = e_j \hat{b} \cdot \vec{E}(\vec{x}_j,t) \tag{123}$$

Once again, we use the electrostatic approximation (more for convenience than anything else) and determine $\vec{E}$ by Poisson's equation, or equation, or equivalently, by its solution

$$\vec{E}(\vec{x},t) = \sum_j e_j \frac{(\vec{x} - \vec{x}_j)}{|\vec{x} - \vec{x}_j|^3} \tag{124}$$

where the sum runs over all charges of both signs. (Again we take N positive ions and N electrons inside a large volume V. Now, $V = L^3$.)

Equations (121)-(124) form a closed system. The dynamics they express can be given a Hamiltonian form; this will guarantee, among other things, the existence of a Liouville equation and a BBGKY hierarchy, with their attendant possibilities for doing a systematic perturbation theory. Letting the jth particle have coordinates $\vec{x}_j = (x_j, y_j, z_j)$, the Hamiltonian function can be chosen to be

$$H = \sum_{i<j} e_i e_j \left[\left(\sqrt{\frac{c}{B|e_i|}}\, q_i - \sqrt{\frac{c}{B|e_j|}}\, q_j\right)^2\right.$$

$$+\left(\sqrt{\frac{c}{B|e_i|}}\,p_i - \sqrt{\frac{c}{B|e_j|}}\,p_j\right)^2 + (z_i - z_j)^2\Bigg]^{-\frac{1}{2}} \tag{125}$$

$$+\sum_j p_{zj}^2/2m_j$$

where the summations run over all charges of both signs, and the canonically conjugate Hamiltonian variables for the perpendicular motion are

$$q_j = \sqrt{\frac{B|e_j|}{c}}\,x_j$$

$$p_j = \sqrt{\frac{B|e_j|}{c}}\,y_j \text{ sgn } e_j \tag{126}$$

For the parallel motion, the canonical variables are just $z_j$ and $p_{zj}$. It is straightforward to show that Hamiltonian's equations $\dot{q}_i = \partial H/\partial p_i$, $\dot{p}_i = -\partial H/\partial q_i$ reproduce (122), while $\dot{z}_j = \partial H/\partial p_{zj}$, $\dot{p}_{zj} = -\partial H/\partial z_j$ reproduce (123) and the definition of $v_{j\parallel} \equiv v_{jz}$.

Expressed in terms of the positions and velocities, H becomes

$$H = \sum_{i<j} \varphi_{ij} + \sum_j m_j v_{jz}^2/2 \tag{127}$$

where $\phi_{ij}$ is the two-body Coulomb potential $e_i e_j/|\vec{x}_i - \vec{x}_j|$. The canonical distribution of Gibbs will be

$$D_{eq.} = \frac{e^{-H/\theta}}{\int dX\, e^{-H/\theta}} \tag{128}$$

where, if H is expressed in the form (127), $\int dX$ means an integral over the $2N \cdot 4 = 8N$ coordinates$(x_j, y_j, z_j, v_{jz})$, where j runs over all charges of both signs.

That the equilibrium phase space probability density has the form that it does [Eq. (128)] means that, in three dimensions as in two, the thermal equilibrium configuration-space probability density (using superscripts to label particle species, now) has the form:

$$n_{N,N}(\vec{x}_1^{\,i} \ldots \vec{x}_N^{\,i}; \vec{x}_1^{\,e} \ldots \vec{x}_N^{\,e})/V^{2N}$$

$$= \frac{\exp[-\Phi(N,N)/\theta]}{\int d\vec{x}_1^{\,i} \ldots d\vec{x}_N^{\,i}\; d\vec{x}_1^{\,e} \ldots d\vec{x}_N^{\,e}\; \exp[-\Phi(N,N)/\theta]}$$

$$\equiv \exp[-\Phi(N,N)/\theta]/Q_{2N} \tag{129}$$

Equation (129) is the same as for the ordinary plasma in three dimensions, with or without an external magnetic field. Here, $\Phi(N,N) = \sum_{i<j} e_i\, e_j/x_{ij}$ is the total potential energy of N ions and N electrons. All the reduced probability distributions

$$n_{s_i, s_e} = V^{s_i + s_e} \int d\vec{x}^{\,i}_{s_i+1} \cdots d\vec{x}^{\,i}_N \quad d\vec{x}^{\,e}_{s_e+1} \cdots d\vec{x}^{\,e}_N \quad \frac{\exp[-\Phi(N,N)/}{Q_{2N}}$$

will be the same, too, as will all single-time thermal equilibrium ensemble averages. This simplifies matters considerably, since these quantities are familiar ones and have been for several years (see, for a review, reference 13). They are, as before, independent of B. Without proof we shall list a number of readily-proved results obtained from $n_{2,0}$, $n_{0,2}$, and $n_{1,1}$.

Interest frequently focuses on the case $n_0\lambda_D^3 \gg 1$, where now $n_0 = N/V$ is the average density (particles/cm$^3$) for both species. The Debye length $\lambda_D$ in three dimensions is defined by $\lambda_D^2 = \theta/8\pi n_0 e^2$. For $\varepsilon = (4\pi n_0 \lambda_D^3)^{-1} << 1$, we have $n_{1,0} = n_{0,1} = 1$ exactly, and to $O(\varepsilon)$,

$$
\begin{aligned}
n_{2,0} &= 1 + p^{ii}(\vec{x}_1^i - \vec{x}_2^i) \\
n_{0,2} &= 1 + p^{ee}(\vec{x}_1^e - \vec{x}_2^e) \qquad (130) \\
n_{1,1} &= 1 + p^{ie}(\vec{x}_1^i - \vec{x}_2^e)
\end{aligned}
$$

Here, the pair correlation $p^{ii} = p^{ee} = -p^{ei} = -p^{ie}$ is given by

$$
\begin{aligned}
p^{ee}(\vec{x}) &\cong -\frac{e^2}{\theta} \frac{e^{-|\vec{x}|/\lambda_D}}{|\vec{x}|} \\
&= \int d\vec{k}\, e^{i\vec{k}\cdot\vec{x}}\, p^{ee}(\vec{k}) \qquad (131a)
\end{aligned}
$$

where

$$p^{ee}(\vec{k}) = -\frac{1}{16\pi^3 n_o} \frac{1}{1 + k^2 \lambda_D^2} \qquad (131b)$$

From such expressions, other quantities such as $\langle \vec{E}(\vec{x})\, \vec{E}(\vec{x}')\rangle$ may be similarly calculated. Thus, the autocorrelation of the electrical charge density $\rho(\vec{x})$ is

$$
\begin{aligned}
\langle \rho(\vec{x})\, \rho(\vec{x}')\rangle &= 4 e^2 n_o^2\, p^{ee}(\vec{x} - \vec{x}') \\
&\quad + 2 n_o e^2\, \delta(\vec{x} - \vec{x}')
\end{aligned}
$$

$$= \int S_\rho(\vec{k})\, e^{i\vec{k}\cdot\vec{x}}\, d\vec{k} \tag{132}$$

where

$$S_\rho(\vec{k}) = \frac{n_o e^2}{4\pi^3} \frac{k^2 \lambda_D^2}{1 + k^2 \lambda_D^2} \tag{133}$$

Also

$$\langle \vec{E}(\vec{x})\, \vec{E}(\vec{x}') \rangle = \int \vec{\vec{S}}_E(\vec{k})\, e^{i\vec{k}\cdot(\vec{x}'-\vec{x})}\, d\vec{k}$$

where

$$\vec{\vec{S}}_E(\vec{k}) = \frac{4 n_o e^2}{\pi} \frac{\vec{k}\vec{k}}{k^2} \frac{\lambda_D^2}{1 + k^2 \lambda_D^2} \tag{134}$$

from which

$$\frac{\langle \vec{E}^2 \rangle}{8\pi} = \frac{\theta}{2} \int \frac{d\vec{k}}{(2\pi)^3} \frac{1}{1 + k^2 \lambda_D^2} \tag{135}$$

Equation (135) contains a divergence at large $\vec{k}$ which, as in two dimensions, may be associated with the infinite self-energies of the 2N particles

$$\frac{\langle \vec{E}^2 \rangle_{self}}{8\pi} \equiv \frac{n_o e^2}{2\pi^2} \int \frac{d\vec{k}}{k^2} = \frac{\theta}{2} \int \frac{d\vec{k}}{(2\pi)^3} \frac{1}{k^2 \lambda_D^2} \tag{136}$$

The *interaction energy* can then be measured by

$$\frac{\langle \vec{E}^2 \rangle - \langle \vec{E}^2 \rangle_{self}}{8\pi} = -\frac{\theta}{2} \int \frac{d\vec{k}}{(2\pi)^3} \frac{1}{k^2 \lambda_D^2 (1 + k^2 \lambda_D^2)} \tag{137}$$

so that the ratio of interaction energy to thermal energy is approximately

$$\frac{[\langle\vec{E}^2\rangle - \langle\vec{E}^2\rangle_{self}]/8\pi}{3n_o\theta} = -\frac{1}{24\pi n_o \lambda_D^3} = -\frac{\epsilon}{6} \quad (138)$$

The smallness of this quantity is the basis of the weak-interaction approximation in which the kinetic theory of a plasma is usually treated. The limit $\varepsilon \to 0$ for the non-thermal equilibrium problem (where $n_o$ and $\theta/m$ are defined in terms of an average number density and average thermal velocity) is usually called the "Vlasov limit". The non-equilibrium properties of this limit have occupied the attentions of more plasma theoriests than any other single topic.

To render $\langle\vec{E}^2\rangle$ finite, we may cut offthe radial k-integration in (135) in the usual way at the inverse distance of closest approach of a thermal particle, $k_{max.} = \theta/e^2$. This gives $\langle\vec{E}^2/8\pi\rangle/3n_o\theta = 2\pi/3$, which, it should be noted, is *not* small [unlike (138)] for any value of $\varepsilon$. The failure to distinguish between interaction energy and the spurious self-energy of point Coulomb charges can lead to serious conceptual errors.

*(B) Transverse Diffusion in Three Dimensions* [52]

The coefficient of transverse spatial diffusion, for the three-dimensional guiding-center plasma, can be written in the same way as (67), in terms of the part of $\vec{E}$ that is perpendicular to $\vec{B}$:

$$D_\perp = \frac{c^2}{B^2}\int_0^\infty \langle\vec{E}_\perp(0)\cdot\vec{E}_\perp(\tau)\rangle\, d\tau \quad (139)$$

Again, the problem is one of evaluating the electric field auto-correlation $\langle\vec{E}(0)\,\vec{E}(\tau)\rangle$, where $\vec{E}(\tau)$ is the electric field seen at time $\tau$ by a "test" ion which is located at the z-axis at $\tau = 0$. $\vec{E}(\tau)$ is related to the Eulerian electric field $\vec{E}(\vec{x},t)$ by

$$\vec{E}(\tau) = \vec{E}[\vec{x}(\tau),\tau] \quad (140)$$

and

$$\vec{E}(\vec{x},t) = \sum_{\vec{k}} \sum_{j} \frac{4\pi\vec{k}}{i\,k^2} \frac{e_j}{V} \exp i\vec{k}\cdot[\vec{x} - \vec{x}_j(t)]$$

where $\vec{x}(\tau)$ is the orbit of the test particle. $\vec{x}_j(t)$ is the location of the jth charge in the plasma at time t. We have toFourier-analyze $\vec{E}(x,t)$ in a large volume V, rather than an infinite volume, for reasons which will become clear as we go on. Equation (141) will also be written as

$$\vec{E}(\vec{x},t) = \sum_{\vec{k}} \vec{E}_{\vec{k}}(t)\, e^{i\vec{k}\cdot\vec{x}} \quad , \tag{142}$$

so that (139) becomes

$$D_{\perp} = \frac{c^2}{B^2} \int_0^{\infty} \sum_{\vec{k}} \langle \vec{E}^*_{\perp\vec{k}}(0) \cdot \vec{E}_{\perp\vec{k}}(\tau)\, e^{i\vec{k}\cdot\vec{x}(\tau)} \rangle \, d\tau \qquad (143$$

since $<\vec{E}_{\vec{k}_1} \vec{E}_{\vec{k}_2}> = 0$ for $\vec{k}_1 \neq -\vec{k}_2$ for a spatially uniform ensemble

If we once again ignore the correlation between the spatial locations of the test ions and the plasma particles [it will be small if it is of the same order as (131)], we have a relation analogous to (77), and

$$D_{\perp} = \frac{c^2}{B^2} \int_0^{\infty} \sum_{\vec{k}} \langle \vec{E}^*_{\perp\vec{k}}(0) \cdot \vec{E}_{\perp\vec{k}}(\tau) \rangle \langle e^{i\vec{k}\cdot\vec{x}(\tau)} \rangle \, d\tau \tag{144}$$

Once again, we can evaluate $D_\perp$ if we can evaluate $<\exp i\vec{k} \cdot x(\tau)>$. The same methods will be applicable in evaluating

$$\langle \vec{E}^*_{\perp\vec{k}}(0) \cdot \vec{E}_{\perp\vec{k}}(\tau) \rangle$$

$$= \sum_{ij} \frac{16\pi^2 e_i e_j}{k^2 V^2} \frac{k_\perp^2}{k^2} \langle \exp i\vec{k} \cdot [\vec{x}_i(\tau) - \vec{x}_j(0)] \rangle \tag{145}$$

The program to be followed closely parallels that for two dimensions. For $\vec{x}(\tau)$, we have

$$\vec{x}(\tau) = \vec{x}(0) + c\int_0^{\tau} \frac{\vec{E}(\tau') \times \vec{B}\, d\tau'}{B^2}$$

$$+ \hat{b}\, v_\parallel\, \tau + \int_0^{\tau} d\tau' \int_0^{\tau'} d\tau'' \frac{e_i\, \hat{b} \cdot \vec{E}(\tau'')\, \hat{b}}{m_i} \tag{146}$$

In (146), the "test" particle has been assumed to be an ion with charge-to-mass ratio $e_i/m_i$. Equation (146) follows from integrating (122) and (123). $\vec{x}(0)$ is the initial position and $v_\parallel$ the initial velocity of the test ion. $\vec{x}(0)$ can be set equal to zero with no loss of generality. $v_\parallel$ is a statistically distributed quantity which may be assumed to obey the Maxwell-Boltzmann distribution.

Evaluating $\langle \exp i\vec{k} \cdot \vec{x}(\tau)$ is no simple matter, and Eq. (146) is considerably more complicated than (78). About the best that has been done so far in evaluating it is the following. Because $\vec{x}(\tau)$ is the position of a random variable which is initially localized near $\vec{x}(0) = 0$, and because its probability distribution $P[\vec{x}(\tau),\tau]$ is expected to spread out with time,

$$\langle \exp i\vec{k} \cdot \vec{x}(\tau) \rangle = \int d\vec{x}(\tau)\, e^{i\vec{k}\cdot\vec{x}(\tau)}\, P[\vec{x}(\tau),\tau]$$

will damp with increasing $\tau$. $\langle \exp i\vec{k} \cdot \vec{x}(\tau)\rangle$ is the "characteristic function" of the random variable $\vec{x}(\tau)$, in the language of probability. Roughly speaking, the damping will occur for two reasons: motion parallel to the field lines, and motion perpendicular to them. Only the latter mechanism was operative in the two-dimensional case. The latter is inhibited by increasing B, but the former is not. The damping due to the parallel motion will be more extreme than that due to the perpendicular motion. Therefore, for very strong magnetic fields, it is useful to forget the perpendicular drift *except* when $\vec{k}$ is normal to $\vec{B}$. This is tantamount to assuming

$$\langle \exp i\vec{k} \cdot \vec{x}(\tau) \rangle$$

$$= \begin{cases} \left\langle \exp \left( \dfrac{-i\, c\vec{k} \times \vec{B}}{B^2} \cdot \displaystyle\int_0^\tau \vec{E}(\tau')\, d\tau' \right) \right\rangle & \text{if } \vec{k} \cdot \vec{B} = 0 \\[2ex] \left\langle \exp \left\{ i\, \vec{k} \cdot \left[ \hat{b}\, v_\parallel\, \tau + \displaystyle\int_0^\tau d\tau' \int_0^{\tau'} d\tau'' \, \dfrac{e_i\, \hat{b}\hat{b} \cdot \vec{E}(\tau'')}{m_i} \right] \right\} \right\rangle & \text{if } \vec{k} \cdot \vec{B} \neq 0 \end{cases} \tag{147}$$

The first of these two expressions can be approximated by the assumption which led to Eq. (81), but the second is more complicated.

That the statistical properties of the parallel electric field seen by a particle can be rather different from the properties of the perpendicular field can be seen as follows. Even if the parallel electric field obeys some jointly normal distribution such as (80), there is the important difference that it can *accelerate* a particle, whereas the perpendicular components cannot. While the perpendicular motion can be visualized as a limit of small increments in position space, the parallel motions are a sum of small increments in *velocity* space. This means that eventually some of the particles will be moving very fast. Such fast-moving particles will, as is well known, eventually begin losing energy due to the radiation of plasma oscillations, and due to the relatively infrequent close collisions with other particles. Therefore, the electric field seen in the parallel direction can*not* in general be well represented by an expression such as (80) in which the probability of the electric field seen by a particle is given by a velocity-independent expression.

The parallel motion is closer in some ways to the conventional [31] Langevin approximation of the theory of Brownian motion in one dimension. There, a test particle in one dimension obeys the equation

$$\frac{dv}{dt} = -F(v) + A(t) \tag{148}$$

with a frictional "drag" coefficient F(v) and a stationary random force field A [analogous to our $\hat{b}\hat{b} \cdot \vec{E}(t)$]. The statistics of A can be given by some such expression as (80). For $F(v) = \beta v$, where $\beta^{-1}$ is a constant long compared to the time necessary for $\langle A(0)\, A(t)\rangle$ to go to zero, the stochastic solutions to (148) are tractable [31]. However, we know that for the plasma case, F(v) is a considerably more complicated nonlinear function of v. Methods apparently do not currently exist for the calculation of such quantities as $\langle \exp[ik \int_0^t v(t')dt']\rangle$ when (148) governs v(t), with a more elaborate nonlinear form for F(v). Therefore we must admit an inability to treat the second expressions in (147) properly over the whole range of $\tau$, and seek an approximation which will render it tractable, even if the approximation is not entirely satisfactory.

One possible procedure is to neglect the contribution of the parallel electric field in the second of Eqs. (147). We are treating only the low-frequency, long-wavelength parts of the electric field spectrum correctly anyway. For this reason, one may expect that the mechanism of *free streaming* of particles down the field lines will be dominant in destroying the correlations between the initial electric field seen by the particle and that field at a later time. The parallel motion of a particle will endure for something of the order of a mean free path, and we may imagine that the auto-correlation length for the electric field will be $\lesssim$ this mean free path. If we ignore $E_{\cdot}$, we get for $\vec{k} \cdot \vec{B} \neq 0$,

$$\langle \exp i\vec{k} \cdot \vec{x}(\tau)\rangle_{k_\parallel \neq 0} \cong \langle \exp(i\vec{k} \cdot \hat{b}\, v_\parallel\, \tau)\rangle$$

(149)

$$= \int_{-\infty}^{\infty} dv_\parallel \left(\frac{m_i}{2\pi\theta}\right)^{\frac{1}{2}} \exp\left(\frac{-m_i v_\parallel^2}{2\theta}\right) \exp(i\, k\, v_\parallel\, \tau)$$

$$= \exp(-k_\parallel^2 \theta \tau^2 / 2m_i) = \exp(-k_\parallel^2 V_{ion}^2 \tau^2 / 2),$$

where we have assumed the test ions to be distributed with a thermal equilibrium velocity distribution. $V_{ion}^2 \equiv \theta / m_i$

We shall make the approximation (149) and explore its consequences, both for the test particles and the plasma particles. Later we shall remark upon the possibilities of relaxing it. For the case of purely perpendicular $\vec{k}$'s, a development parallel to that of

$$\langle \exp i\vec{k} \cdot \vec{x}(\tau)\rangle_{k_\parallel = 0}$$

(150)

$$= \exp\left(\frac{-c^2 k_\perp^2}{4 B^2} \int_0^\tau d\tau_1 \int_0^\tau d\tau_2 \langle \vec{E}_\perp(\tau_1) \cdot \vec{E}_\perp(\tau_2)\rangle\right)$$

Before substituting (150) and (149) into (144), we need expressions for the $\langle \vec{E}^*_{\perp\vec{k}}(0) \cdot \vec{E}_{\perp\vec{k}}(\tau)\rangle$. Making the same approximations that led to (149) and (150) on the plasma charges in (145) gives

$$\langle \vec{E}^{*}_{\perp\vec{k}}(0) \cdot \vec{E}_{\perp\vec{k}}(\tau)\rangle \cong$$

$$\begin{cases} \displaystyle\sum_{ij} \frac{16\pi^2 e_i e_j}{k^2 V^2} \frac{k_\perp^2}{k^2} \langle \exp i\vec{k}\cdot[\vec{x}_i(0) - \vec{x}_j(0)]\rangle \cdot \exp\left(\frac{-k_\parallel^2 V_{oj}^2 \tau^2}{2}\right) & \text{for } k_\parallel \neq 0 \\ \displaystyle\sum_{ij} \frac{16\pi^2 e_i e_j}{k_\perp^2 V^2} \exp\left(\frac{-c^2 k_\perp^2}{4B^2} \int_0^\tau d\tau_1 \int_0^\tau d\tau_2 \langle \vec{E}_\perp(\tau_1)\cdot\vec{E}_\perp(\tau_2)\rangle\right) \cdot \langle \exp i\vec{k}\cdot[\vec{x}_i(0) - \vec{x}_j(0)]\rangle & \text{for } k_\parallel = 0 \end{cases}$$

(151)

In (151), $V_{oj}^2 \equiv \theta/m_j$ is the square of the thermal speed for the species to which charge j belongs. We have also implicitly assumed that the jointly normal probability distribution of the perpendicular electric field is the same for all the particles.

Finally, we need $\langle \exp i\vec{k}\cdot[\vec{x}_i(0) - \vec{x}_j(0)]\rangle$, which is $\int d\vec{x}_1\, d\vec{x}_2\, n_2(\vec{x}_1,\vec{x}_2)\, e^{-i\vec{k}\cdot\vec{x}_{12}}/V^2$, where $n_2$ is the probability distribution (130) of whatever two particles are involved. If i and j are the same particle, the < > is clearly unity. If they are different, $i \neq j$, then

$$\langle \exp i\vec{k} \cdot [\vec{x}_i(0) - \vec{x}_j(0)]\rangle$$

$$= \delta_{Kr}(\vec{k}) - \frac{1}{2n_o V}\left(\frac{K_D^2}{k^2 + K_D^2}\right) \quad \text{sgn } e_i \text{ sgn } e_j \tag{152}$$

(152) is just (129)-(131), expressed in a finite-volume representation.

Inserting (152) and (151) into the relation

$$\langle \vec{E}_\perp(0) \cdot \vec{E}_\perp(\tau)\rangle = \sum_{\vec{k}} \langle \vec{E}^*_{\perp\vec{k}}(0) \cdot \vec{E}_{\perp\vec{k}}(\tau)\rangle\langle e^{i\vec{k}\cdot\vec{x}(\tau)}\rangle \tag{153}$$

we have

$$\langle \vec{E}_\perp(0) \cdot \vec{E}_\perp(\tau)\rangle$$

$$= \sum_{i=j} \sum_{k_\parallel \neq 0, \vec{k}} \frac{16\pi^2 e_i e_j}{k^2 V^2} \frac{k_\perp^2}{k^2} \exp\left(\frac{-k_\parallel^2 V_{oj}^2 \tau^2}{2}\right) \exp\left(\frac{-k_\parallel^2 V_{ion}^2 \tau^2}{2}\right)$$

$$+ \sum_{i=j} \sum_{k_\parallel = 0, \vec{k}} \frac{16\pi^2 e_i e_j}{k^2 V^2} \exp\left(\frac{-c^2 k_\perp^2}{2B^2} \int_0^\tau d\tau_1 \int_0^\tau d\tau_2 \langle \vec{E}_\perp(\tau_1) \cdot \vec{E}_\perp(\tau_2)\rangle\right)$$

$$+\sum_{i\neq j}\sum_{\substack{\vec{k}\\ k_\parallel \neq 0}} \frac{16\pi^2 e_i e_j}{k^2 V^2} \frac{k_\perp^2}{k^2} \exp\left(\frac{-k_\parallel^2 V_{oj}^2 \tau^2}{2}\right)$$

$$\left\{\delta_{Kr}(\vec{k}) - \frac{1}{2 n_o V} \frac{K_D^2}{k^2 + K_D^2} \operatorname{sgn} e_i e_j\right\} \exp \frac{-k_\parallel^2 V_{ion}^2 \tau^2}{2}$$

$$+\sum_{i\neq j}\sum_{\substack{\vec{k}\\ k_\parallel = 0}} \frac{16\pi^2 e_i e_j}{k^2 V^2} \exp\left(\frac{-c^2 k_\perp^2}{2B^2}\int_0^\tau d\tau_1 \int_0^\tau d\tau_2 \langle \vec{E}_\perp(\tau_1)\cdot\vec{E}_\perp(\tau_2)\rangle\right)$$

$$\cdot\left\{\delta_{Kr}(\vec{k}) - \frac{1}{2 n_o V} \frac{K_D^2}{k^2 + K_D^2} \operatorname{sgn} e_i e_j\right\} \qquad (154)$$

This is an integral equation for $<\vec{E}_\perp(0) \cdot \vec{E}_\perp(\tau)>$. If we had defined the Fourier transform of the Coulomb potential $\phi^{ee}(\vec{k}) = \lim_{\lambda\to 0} e^2/2\,\pi^2(k^2+\lambda^2)$ more carefully, the $\delta_{Kr}(\vec{k})$ terms would not have been present in $<\vec{E}^*_{\vec{k}}(0) \cdot \vec{E}_{\vec{k}}(\tau)>$, and they will not contribute to the summation. Hereafter they are dropped. (This is only one of the places where one has to be alert for the pathologies that follow from the fact that the Coulomb potential is neither square integrable nor absolutely integrable, and so has no Fourier transform in a rigorous sense, except as such a limit as the one just described.) The $\Sigma_{ij}$ can be performed in (154) and the result is:

$$\langle \vec{E}_\perp(0) \cdot \vec{E}_\perp(\tau)\rangle$$

$$= \sum_{\substack{\vec{k} \\ k_\parallel \neq 0}} \frac{16\pi^2 e^2 n_o k_\perp^2}{V k^2 (k^2 + K_D^2)} \exp\left(\frac{-k_\parallel^2 V_{ion}^2 \tau^2}{2}\right) \tag{155}$$

$$\cdot \left[\exp\left(\frac{-k_\parallel^2 V_{ion}^2 \tau^2}{2}\right) + \exp\left(\frac{-k_\parallel^2 V_{oe}^2 \tau^2}{2}\right)\right]$$

$$+ \sum_{k_\parallel = 0} \frac{32\pi^2 e^2 n_o}{V(k^2 + K_D^2)} \exp\left(\frac{-c^2 k_\perp^2}{2B^2} \int_o^\tau d\tau_1 \int_o^\tau d\tau_2 \langle \vec{E}_\perp(\tau_1) \cdot \vec{E}_\perp(\tau$$

Equation (155) can be simplified by defining, as in the two-dimensional case,

$$\frac{1}{2} \langle \vec{E}_\perp(0) \cdot \vec{E}_\perp(\tau)\rangle \equiv Q_\perp(\tau)$$

$$R_\perp(\tau) \equiv \frac{c^2}{2B^2} \int_o^\tau d\tau_1 \int_o^\tau d\tau_2 \, Q_\perp(\tau_2 - \tau_1) \tag{156}$$

$$\frac{d^2 R_\perp(\tau)}{d\tau^2} = \frac{c^2 Q_\perp(\tau)}{B^2}$$

These definitions can be used in (155) to give

$$2Q_{\perp}(\tau) = \frac{16\pi^2 e^2 n_o}{V} \sum_{\substack{\vec{k} \\ k_{\parallel}\neq 0}} \frac{k_{\perp}^2/k^2}{(k^2 + K_D^2)}$$

$$\cdot \left[\exp\left(\frac{-k_{\parallel}^2 V_{ion}^2 \tau^2}{2}\right) + \exp\left(\frac{-k_{\parallel}^2 V_{oe}^2 \tau^2}{2}\right)\right] \exp\left(\frac{-k_{\parallel}^2 V_{ion}^2 \tau^2}{2}\right) \tag{157}$$

$$+ \frac{32\pi^2 e^2 n_o}{V} \sum_{\substack{\vec{k} \\ k_{\parallel}=0}} \frac{1}{k_{\perp}^2 + K_D^2} \exp\left[-2k_{\perp}^2 R_{\perp}(\tau)\right]$$

Defining $\epsilon_b \equiv 16\pi^2 e^2 c^2/B^2 V K_D^2$, this becomes, replacing $\tau$ by $t$,

$$\frac{d^2 R_{\perp}(t)}{dt^2} = \epsilon_b \sum_{\substack{\vec{k} \\ k_{\parallel}=0}} \frac{\exp[-2k_{\perp}^2 R_{\perp}(t)]}{1 + k_{\perp}^2 \lambda_D^2} + \epsilon_b \sum_{\substack{\vec{k} \\ k_{\parallel}\neq 0}} f(\vec{k},t) \tag{158}$$

where $f(\vec{k},t)$ is a known function of $t$:

$$f(\vec{k},t) \equiv \frac{1}{2} \frac{k_{\perp}^2}{k^2} \frac{\exp -k_{\parallel}^2 V_{ion}^2 t^2/2}{1 + k^2 \lambda_D^2}$$

$$\cdot \left[\exp\left(\frac{-k_{\parallel}^2 V_{ion}^2 t^2}{2}\right) + \exp\left(\frac{-k_{\parallel}^2 V_{oe}^2 t^2}{2}\right)\right] \tag{159}$$

Equation (158) is the analogue of Eq. (86) for three dimensions. It is considerably more complicated because of the last term, however, and so far has not been solved in the general case. It can be integrated once, noting that $dR_{\perp}(0)/dt = 0$ to yield:

$$\frac{dR_{\perp}(t)}{dt} = \epsilon_b \sum_{\substack{\vec{k} \\ k_{\parallel}=0}} \int_0^t d\tau \, \frac{\exp[-2k_{\perp}^2 R_{\perp}(\tau)]}{1 + k_{\perp}^2 \lambda_D^2}$$

$$+ \epsilon_b \sum_{\substack{\vec{k} \\ k_{\parallel}\neq 0}} \int_0^t f(\vec{k},t)\, d\tau \tag{160}$$

and we may show how to extract the leading term in powers of 1/B from Eq. (160).

The only place the magnetic field enters (158) or (160) is in the denominator of the small quantity $\varepsilon_b$. We seek the leading term, in powers of $\varepsilon_b$, of the solution to (158) and (160). We shall see presently that for large t, $R_{\perp}(t)$ varies proportionally to $\varepsilon_b^{1/2}$ t, which makes the first term on the right of (160) an $O(\varepsilon_b^{1/2})$ term. Since the second term on the right of (160) is $O(\varepsilon_b)$, the result follows that it can be neglected to lowest significant order, either in (160) or in (158).

If we neglect $f(\vec{k},t)$ in (158), we get

$$\frac{d^2 R_{\perp}(t)}{dt^2} \cong \epsilon_b \sum_{\substack{\vec{k} \\ k_{\parallel}=0}} \frac{\exp[-2k_{\perp}^2 R_{\perp}(t)]}{1 + k_{\perp}^2 \lambda_D^2}$$

or equivalently,

$$\frac{1}{2}\left(\frac{dR_{\perp}(t)}{dt}\right)^2 = \frac{1}{2}\epsilon_b \sum_{\vec{k}_{\perp}} \frac{[1 - \exp(-2k_{\perp}^2 R_{\perp})]}{k_{\perp}^2(1 + k_{\perp}^2 \lambda_D^2)} \tag{161}$$

Since $R_\perp$ and its first two derivatives are always positive for $t > 0$, last term in (161) becomes negligible as $t \to \infty$, and

$$\frac{1}{2}\left(\frac{d\,R_\perp(\infty)}{dt}\right)^2 = \frac{\epsilon_b}{2}\sum_{\vec{k}_\perp}\frac{1}{k_\perp^2(1 + k_\perp^2\,\lambda_D^2)} \tag{162}$$

But $D_\perp/2$ is just $d\,R_\perp(\infty)/dt$, so

$$D_\perp = 2\,\epsilon_b^{\frac{1}{2}}\left[\sum_{\vec{k}_\perp}\frac{1}{k_\perp^2(1 + k_\perp^2\,\lambda_D^2)}\right]^{\frac{1}{2}} + \text{terms of } O(\epsilon_b) \tag{163}$$

Equation (163) is the three-dimensional analogue of (89). To approximate it by an integral for a large volume, we make the replacement [as in (90)]:

$$\sum_{\vec{k}_\perp} \longrightarrow \frac{V^{\frac{2}{3}}}{2\pi}\int_{k_{min}}^{\infty} k_\perp\,dk_\perp$$

where $k_{min} = 2\pi/L = 2\pi/V^{\frac{1}{3}}$ is the lower limit of $k_\perp$ integration which results from the finite box size.

Carrying out the integration and introducing the definition of $\epsilon_b$, (163) becomes

$$D_\perp = \frac{c\theta}{eB}\left(\frac{1}{2\pi n_o\,\lambda_D^3}\right)^{\frac{1}{2}}\left(\frac{\lambda_D}{L}\,\ln\,\frac{L}{2\pi\lambda_D}\right)^{\frac{1}{2}} \tag{164}$$

Equation (164) is strikingly similar to (91). It contains the factors $c\theta/eB$, the square root of the plasma parameter, and a volume-dependent factor. Only the volume-dependent factor is different: it diverges slowly as $L \to \infty$ in two dimensions, but it approaches zero slowly in three dimensions. This slow approach to zero puts a constraint on the volume in order that the second term of (160) really be negligible compared to the first. Replacing the second $\Sigma_{\vec{k}}$ in (160) by an integral (which must be cut off at $k_{max.} = \theta/e^2$ to avoid an unphysical short-range divergence) the condition that the term involving $f(\vec{k},t)$ be negligible is that

$$\frac{B^2}{n_o m_i c^2} \gg \frac{1}{32 n_o \lambda_D^3} \frac{L/2\pi\lambda_D}{\ln(L/2\pi\lambda_D)} \left[\ln\left(\frac{L\,k_{max}}{2\pi}\right)\right]^4 \tag{165}$$

This can be regarded as a constraint on either B or L, that (164) shall adequately represent the coefficient of transverse diffusion. The limits of large B and large V are not interchangeable. At fixed volume, however large, (164) becomes accurate as $B \to \infty$. But at fixed B, however large, (164) ceases to be accurate as $V \to \infty$, because (165) ceases to hold.

Arguments which led to the inequality (92) on the time over which true diffusion-like behavior can be observed likewise complicate the picture for the solutions of (161).

*(C) Problems Associated with a Convergent Infinite-Volume Theory*

The form of Eq. (164) leaves little doubt of its inadequacy if we are interested in passing to the rigorous limit of an infinite volume, $V \to \infty$. One might at first guess that the way to obtain the infinite-volume limit would be to drop the *first* term in (160) and simply write

$$D_\perp(\infty) = 2 \lim_{t\to\infty} \frac{d\,R_\perp(t)}{d\,t}$$

$$= 2\,\epsilon_b\,V \int \frac{d\,\vec{k}}{(2\pi)^3} \int_0^\infty f(\vec{k},\tau)\,d\tau \tag{166}$$

However, upon substituting (159) into (166), one finds that the integral diverges, and so what superficially appears to be a nice expression with the correct "classical" dependence $\sim B^{-2}$ is actually useless.

The reasons for this are not difficult to see, but they are not easy to correct. The reasons lie in the separations, represented by Eqs. (147), (149), and (150), of the damping of $\langle \exp i\vec{k} \cdot \vec{x}(\tau) \rangle$ into purely perpendicular and purely parallel contributions. The point is that as V increases and the allowed values of $\vec{k}$ become more and more dense, this separation will become less and less realistic for very small but non-zero values of $k_{\parallel}$. The damping due to the $\vec{E} \times \vec{B}$ drift will compete with the free-streaming damping below a value of $k_{\parallel}$ that can be estimated by setting $k_{\parallel}^2 V_{ion}^2 t^2 \cong k^2 D_{\perp} t$ at a time when both are of order unity; i.e., below $k_{\parallel} \approx k_{\perp}^2 D_{\perp}/V_{ion}$. Similarly the parallel electric field which has been unceremoniously dropped up to now from the second of Eqs. (147) can become important in the small $k_{\parallel}$ region. It has been pointed out recently by Vahala [53] that mild assumptions on the statistics of the parallel electric field can lead to a term in (147) which has the essential temporal behavior $\sim \exp\{-k_{\parallel}^2 t^3$ times a quantity which $\to 0$ as $\varepsilon \to 0\}$. This mechanism, essentially a parallel collision damping, is a third approximately equal competitor in the role of damping $f(\vec{k},\tau)$ to zero for large $\tau$ at small $k_{\parallel}$.

It should be apparent that a rather complex limiting process, involving the three small parameters $\varepsilon$, $\varepsilon_b$, and 1/V is involved; a simple formula which covers all cases is too much to hope for in the near future. But formula (164), supplemented by the inequality (165), may be supposed to be a good measure of the regime in which the guiding-center plasma exhibits Bohm-like diffusion in thermal equilibrium. What are not yet clear are the infinite volume or finite 1/B limits in which "classical" diffusion may be expected. The "classical" diffusion formulas, though very familiar, lack convincing derivations in a strongly magnetized plasma. One may expect this part of the picture to clear up before long, but jumping to premature conclusions is to be avoided.

### *(D) A Systematic Kinetic Theory of the Three-Dimensional Guiding-Center Plasma*

Since the system is Hamiltonian [Eqs. (125)-(127)], it will obey a Liouville equation. From this a BBGKY hierarchy can be derived, and provides a possibility of doing a systematic kinetic theory via an expansion in the plasma parameter $\varepsilon$. This in some ways is similar to the theory for a one-dimensional plasma, but the correspondence is not perfect. We shall adopt a compact notation, dispensing with the superscripts to identify charged species, as in the previous two sections.

For the phase space coordinates of the jth particle, we write $\underset{\sim}{X}_j \equiv (x_j, y_j, z_j, p_{zj}) \equiv (\vec{x}_j, p_{zj})$, where $p_{zj} = m_j v_{zj}$. The Liouville equation then reads

$$\left( \frac{\partial}{\partial t} + \sum_j \frac{d\underset{\sim}{X}_j}{dt} \cdot \frac{\partial}{\partial \underset{\sim}{X}_j} \right) D = 0 \tag{167}$$

where the probability distribution D is a function of all the $\underset{\sim}{X}_j$'s for all particles of both species, and is assumed to be symmetric under the interchange of like particle coordinates. Written out in detail, (167) is

$$\left\{ \frac{\partial}{\partial t} + \sum_j \left( \frac{c\,\vec{E}_j \times \vec{B}}{B^2} \cdot \frac{\partial}{\partial \vec{x}_j} + v_{zj} \frac{\partial}{\partial z_j} + \frac{e_j}{m_j} E_{j\parallel} \frac{\partial}{\partial v_{zj}} \right) \right\} D = 0 \tag{168}$$

Here, $\vec{E}_j = \Sigma_{\ell \neq j}\, e_\ell (\vec{x}_j - \vec{x}_\ell)/|\vec{x}_j - \vec{x}_\ell|^3$, and $E_{j\parallel} = \hat{b} \cdot \vec{E}_j$ is the z-component of the electric field at the location of particle j. Summations run over all particles of both signs.

The reduced probability distributions are

$$f_s \equiv V_s \int D\, d\underset{\sim}{X}_{s+1}\, d\underset{\sim}{X}_{s+2} \ldots d\underset{\sim}{X}_{2N} \tag{169}$$

and integrating (168) yields

$$\frac{\partial f_s}{\partial t} + \sum_{j=1}^{s} \left( \frac{c\,\vec{E}^s_j \times \vec{B}}{B^2} \cdot \frac{\partial}{\partial \vec{x}_j} + v_{zj} \frac{\partial}{\partial z_j} + \frac{e_j}{m_j} E^s_{j\parallel} \frac{\partial}{\partial v_{zj}} \right) f_s$$

$$= -c\,n_o \sum_{\text{species}} \sum_{i=1}^{s} \int d\underset{\sim}{X}_{s+1}$$

(170)

$$\left[\left(\frac{\vec{B}}{B^2} \times \frac{\partial \tilde{\varphi}_{i,s+1}}{\partial \vec{x}_i}\right) \cdot \frac{\partial f_{s+1}}{\partial \vec{x}_i} + \frac{e_i}{m_i c} \frac{\partial \tilde{\varphi}_{i,s+1}}{\partial z_i} \cdot \frac{\partial f_{s+1}}{\partial v_{zi}}\right]$$

Here,

$$\vec{E}_j^s = \sum_{\substack{\ell=1 \\ \ell \neq j}}^{s} \frac{e_\ell(\vec{x}_j - \vec{x}_\ell)}{|\vec{x}_j - \vec{x}_\ell|^3}$$

$$\tilde{\varphi}_{i,s+1} = e_{s+1}/|\vec{x}_i - \vec{x}_{s+1}|$$

$$n_o = N/V$$

and the $\sum_{\text{species}}$ means to sum over the (s+1) st particle, first treating it as an ion and then as an electron. Equation (170) is a generalization of (96), but because of the velocity dependence, greatly exceeds (96) in dynamical content.

The "Vlasov approximation" amounts to ignoring all the correlations and writing $f_s = \prod_{i=1}^{s} f_1(i)$, as usual. This gives, for the jth species,

$$\frac{\partial f_1}{\partial t} + v_z \frac{\partial f_1}{\partial z} + \frac{e_j E_z}{m_j} \frac{\partial f_1}{\partial v_z} + \frac{c\,\vec{E} \times \vec{B}}{B^2} \cdot \frac{\partial f_1}{\partial \vec{x}} = 0 \tag{171}$$

where

$$\vec{E}(\vec{x},t) = -\frac{\partial}{\partial \vec{x}} \sum_{\text{species}} n_o e \int \frac{f_1(\vec{x}', v_z', t)}{|\vec{x} - \vec{x}'|} d\vec{x}' dv_z' \tag{172}$$

This is the limit of the *Vlasov-Poisson* system in a very strong magnetic field. An equivalent limit was treated some time ago by Harris [54] and Rosenbluth [55], although only after the solution to the full linearized Vlasov-Poisson system had been carried out.

Standard methods* [13] can be applied to solve the linearized version of (171)-(172). That is, we keep only linear terms in the departure from the spatially-uniform, field-free state. Linearized, Eq. (171) takes the form

$$\frac{\partial f_{i,e}^{(1)}}{\partial t} + v_z \frac{\partial f_{i,e}^{(1)}}{\partial z} \pm \frac{e E_z}{m_{i,e}} \frac{\partial f_{i,e}^{(0)}}{\partial v_z} = 0 \tag{173}$$

where i,e stands for ions and electrons. Poisson's equation (172) takes the form

$$\vec{E}(\vec{x},t) = -\frac{\partial}{\partial \vec{x}} \sum_{\text{i and e}} (\pm e n_o) \int \frac{f_{i,e}^{(1)}(\vec{x}', v_z', t)\, d\vec{x}'\, dv_z'}{|\vec{x} - \vec{x}'|} \tag{174}$$

$f_{i,e}^{(0)} = f_{i,e}^{(0)}(v_z)$ only are the spatially uniform equilibria we are perturbing about, and $f_{i,e}^{(1)}(\vec{x}, v_z, t)$ are the perturbations. Assuming $\vec{E}, f_{i,e}^{(1)} \sim \exp(i\vec{k} \cdot \vec{x})$ and that $f_{i,e}^{(1)}(\vec{x}, v_z, 0) = g_{i,e}(v_z)\, e^{i\vec{k}\cdot\vec{x}}$, the solution to (173) is (dropping $e^{i\vec{k}\cdot\vec{x}}$):

* L. D. Landau, *J. Phys. (U.S.S.R) 10*, 25 (1946).

$$f^{(1)}_{i,e}(v_z,t) = g_{i,e}(v_z)\, e^{-ik_z v_z t}$$

$$\mp \frac{e}{m_{i,e}} \frac{\partial f^{(0)}_{i,e}(v_z)}{\partial v_z} \int_0^t d\tau\, E_z(\tau)\, e^{ik_z v_z(\tau - t)} \tag{175}$$

Substituting (175) into (174) gives

$$i\vec{k}\cdot\vec{E}(t) = i\,k\,E(t) = 4\pi e n_o \int_{-\infty}^{\infty} d\,v_z$$

$$\left\{ [g_i(v_z) - g_e(v_z)]\, e^{-ik_z v_z t} - \frac{e}{m}\left( \frac{\partial f_i^{(0)}}{\partial v_z} + \frac{\partial f_e^{(0)}}{\partial v_z} \right) \right. \tag{176}$$

$$\left. \cdot \int_0^t d\,\tau\; E_z(\tau)\, e^{-ik_z v_z (t-\tau)} \right\}$$

Since $E_z(t) = (k_z/k)\,E(t)$, (176) is an integral equation of the convolution type for $E(t)$ which can be solved by standard techniques [13]. The result is conveniently obtained as a Laplace transform

$$E(\tau) = \int_{\sigma - i\infty}^{\sigma + i\infty} \frac{d\,s}{2\pi i}\, E(s)\, e^{st} \tag{177}$$

where s is the complex Laplace transform variable, and

$$E(s) = \frac{- \int \frac{d\, v_z}{v_z - is/k_z} \, \frac{4\pi e n_o}{k\, k_z} \, [g_e(v_z) - g_i(v_z)]}{D(\vec{k},s)} \tag{178}$$

$D(\vec{k},s)$ is the dielectric function, defined by (the usual [13] remarks about the contours of s and $v_z$ integration apply):

$$D(\vec{k},s) = 1 - i \sum_j \frac{\omega_{pj}^2}{k^2} \, k_z \int \frac{dv_z \, \partial f_j^{(0)}(v_z)/\partial v_z}{s + i k_z v_z} \tag{179}$$

The $\Sigma_j$ in (179) is a sum over species (two terms) and $\omega_{pj}^2 \equiv 4\pi n_o e^2/m_j$, $j$ = i or e. The vanishing of $D(\vec{k},s)$ at $s = - i\omega(\vec{k}) + \gamma(\vec{k})$ determines the angular frequencies $\omega(\vec{k})$ at which the medium can support electrostatic oscillations; the corresponding Landau damping decrement is $-\gamma(\vec{k})$. Antecedents of Eq. (179) were first derived by Harris [54] and Rosenbluth [55].

Equation (179) becomes identical with that for an unmagnetized plasma upon replacing $k_z$ by k, $v_z$ by the component of $\vec{v}$ along $\vec{k}$, and letting $f_j^{(0)}(v_z)$ be replaced by the velocity distribution function of that component of $\vec{v}$. It follows that from all the previous detailed work on the solutions of $D(\vec{k},s)$ = we can immediately infer the character of the possible modes of oscillation of the 3-dimensional guiding center plasma.

In summary, there are two important branches of the $\omega$ versus $\vec{k}$ curve for Maxwellian distributions, one associated with electron oscillations and one associated with ion acoustic waves. For $k \ll K_D$, the electron branch has $\omega \approx \omega_{pe} k_z/k$, and goes over into ordinary electron plasma oscillations for parallel propagation ($k_z = k$). The ion acoustic branch is heavily damped unless $T_e \gg T_i$, in which case its dispersion relation

becomes $\omega/k \simeq (\omega_{pi}(k_z/k))/(k^2 + K_{De}^2)^{1/2}$, where $K_{De}^2 \equiv 4\pi n_0 e^2/\theta_e$, and this also goes over into it unmagnetized value at $k_z = k$. The damping decrements depend linearly, for the case $\gamma^2 \ll \omega^2$, on the values of $\partial f_{i,e}^{(0)}/\partial v_z(v_z = \omega/k_z)$, and are large when these quantities are large. The most important difference between the solution of $D(\vec{k},s) = 0$ here and in the unmagnetized case is perhaps the possibility that undamped electrostatic oscillations at large angles ($k_z/k \ll 1$) can exist at considerably lower values than the electron plasma frequency: $\omega \approx \omega_{pe} k_z/k$. Since $\omega/k_z \approx \omega_{pe}/k$, these are still essentially undamped as long as $\omega_{pe}/k$ is greater than an electron thermal speed. For sufficiently large wavelength (small k) and large angles, these can lie below the ion gyrofrequency in a real plasma and can render the drift approximation more satisfactory than it would otherwise be.

Other kinetic theory problems can be approached through (170). For example, it may be expected that *kinetic equations* for the evolution of non-Maxwellian one-body distributions, such as the Balescu-Lenard equation, might be extracted from an expansion in $\varepsilon$ for the spatially uniform state. The strong similarity to the one-dimensional case suggests, however, that a messy, $O(\varepsilon^2)$, calculation might be required in order to predict relaxation to thermal equilibrium. A calculation of the spectral density of the fluctuations in the next sub-section points to a possible pathological feature of any such expansion, and re-emphasizes the very special role played in the guiding-center plasma by wave numbers with $k_z/k \ll 1$.

*(E) Fluctuations for the Three-Dimensional Guiding-Center Plasma*

It is possible, for the spatially-uniform three-dimensional guiding-center plasma, to calculate the auto-correlation function of the electric field in the Eurlerian representation:

$$\langle \vec{E}(\vec{x},t)\, \vec{E}(\vec{x}+\vec{r},t+\tau)\rangle = \int d\vec{k}\, d\omega\, \vec{\vec{S}}_E(\vec{k},\omega)\quad e^{i(\vec{k}\cdot\vec{r}+\omega\tau)} \tag{180}$$

The calculation can be carried out one of two ways: either by the Rostoker method of super-posed "dressed" test particles [14] or through the two-time BBGKY hierarchy [13], which can be shown to lead to the same results. Both methods are by now well known, and we shall use the former since it is simpler.

The calculation is instructive because it illustrates clearly the singular behavior associated with those components of $\vec{k}$ which lie nearly in the perpendicular direction to $\vec{B}$. The unusual physical behavior associated with $k \ll K_D$ and $\vec{k} \cdot \vec{B} = 0$ has already been shown to give the dominant contribution to the "anomalous" diffusion coefficient (164). A calculation of $\vec{\vec{S}}_E(\vec{k},\omega)$ from the conventional expansion in the strength of the interaction sheds additional light on the unusual features of this small, crucial region of $\vec{k}$-space.

"Test particles", in the sense they were introduced into kinetic theory by Rostoker [14] and Thompson and Hubbard [56], are uncorrelated, freely-streaming Coulomb charges carrying a modified potential. The modification of the potential is achieved by Fourier transforming it in space and time, then dividing the transformed potential $\phi(\vec{k},\omega)$ by $D(\vec{k},i\omega)$. Such fluctuation quantities as $S_E(\vec{k},\omega)$ are then obtained by superposing *uncorrelated* particles with this potential. The foundations of the method are discussed, e.g., in detail in Ref. 13, where it is shown to be connected with the more rigorous derivations proceeding from the two-time BBGKY hierarchy.

For particles of species j, the exact number density for non-interacting charges is:

$$n_j(\vec{x},t) = \sum_{i=1}^{N} \delta(\vec{x} - \vec{x}^j_{io} - \vec{v}^j_{io}\, t) \quad .$$

For the 3-dimensional guiding-center plasma, $\vec{v}^j_{io} = \hat{b}\, v^j_{io}$, but $\vec{x}^j_{io}$ can be anything. The Fourier-transformed charge density

$$n_j(\vec{k},t) = \sum_{i=1}^{N} (2\pi)^{-3} \exp -i\vec{k} \cdot (\vec{x}^j_{io} + \vec{v}^j_{io} t) \tag{181}$$

will give

$$\langle n_j(\vec{k},t)\, n_j(\vec{k}',t')\rangle$$

$$\left\langle \sum_{i,\ell=1}^{N} (2\pi)^{-6} \exp[-i\vec{k} \cdot (\vec{x}^j_{io} + \vec{v}^j_{io} t) - i\vec{k}' \cdot (\vec{x}^j_{\ell o} + \vec{v}_{\ell o}\, t')] \right\rangle \tag{182}$$

If the various charges are uncorrelated, the only terms which will contribute to the electric field spectrum for large t, t' are those with $i = \ell$, so

$$\langle n_j(\vec{k},t)\, n_j(\vec{k}',t+\tau)\rangle$$

$$= \frac{N_j}{V}\,\frac{\delta(\vec{k}+\vec{k}')}{(2\pi)^3}\int d\vec{v}^{\,j}_{io}\, e^{-i\vec{k}\cdot\vec{v}^{\,j}_{io}\tau}\, f_j(\vec{v}^{\,j}_{io}) \tag{183}$$

plus terms which do not contribute to $\overrightarrow{\overrightarrow{S}}_E$. The velocity integral for the 3D guiding-center plasma is a single integration, so we can write, for the guiding center plasma

$$\langle n_j(\vec{x},t)\, n_j(\vec{x}+\vec{r},\, t+\tau)\rangle$$

$$= \int d\vec{k}\, d\omega\, S_{jj}(\vec{k},\omega)\exp[i\,\vec{k}\cdot\vec{r} + i\,\omega\tau] \tag{184}$$

where

$$S_{jj}(\vec{k},\omega) = \frac{n_{oj}}{(2\pi)^3}\,\frac{f_j(-\omega/k_z)}{|k_z|} \tag{185}$$

plus terms which do not contribute to $\overrightarrow{\overrightarrow{S}}_E$. $n_{oj} \equiv N_j/V = n_o$ for both ions and electrons. Summing (185) over species and applying Poisson's equation, we have the spectral density of the electric field fluctuations for "bare" (i.e., uncorrelated and non-interacting) particles:

$$\overleftrightarrow{S}_E(\vec{k},\omega) = \sum_j \frac{2 n_{oj} e_j^2}{\pi} \frac{\vec{k}\vec{k}}{k^4} \frac{f_j(-\omega/k_z)}{|k_z|} \tag{186}$$

("bare" particles).

The canonical recipe now requires division of (186) by $|D(\vec{k},i\omega)|^2$ to give:

$$\overleftrightarrow{S}_E(\vec{k},\omega) = \sum_j \frac{2 n_{oj} e_j^2}{\pi} \frac{\vec{k}.\vec{k}}{k^4} \frac{1}{|k_z|} \frac{f_j(-\omega/k_z)}{|D(\vec{k},i\omega)|^2} \tag{187}$$

The corresponding expression for the three-dimensional *un*magnetized plasma is [13,14]

$$\overleftrightarrow{S}_E(\vec{k},\omega)\Big|_{B=0} = \sum_j \frac{2 n_{oj} e_j^2}{\pi} \frac{\vec{k}\vec{k}}{k^4} \frac{1}{k} \frac{F_j(-\omega/k)}{|D(\vec{k},i\omega)|^2} \tag{188}$$

where $F_j$ is the velocity distribution function for the component of velocity along $\vec{k}$.

Equations (187) and (188) are similar, but there is a

striking difference. Namely, for $\omega/k_z$ = const., k finite, and $k_z \to 0$, (187) diverges, whereas (188) does not. In other words, (187) predicts that $<|E(\vec{k},\omega)|^2>$ will be in no sense small for $\vec{k} \perp \vec{B}$, even though the plasma parameter may be small.

Superficially more complicated, but similar, divergences plague the full three-dimensional plasma in a very strong magnetic field [57]. There exists at the present time no satisfactory calculation of $\overleftrightarrow{S}_E(\vec{k},\omega)$ for those components of $\vec{k}$ nearly perpendicular to $\vec{B}$. The weak-coupling procedures simply will not work. In view of the already displayed importance of these components for plasma transport properties, the problem of calculating them accurately appears to be of the utmost importance. Because of the discussion following Eq. (166), it is not reasonable to expect this calculation to be easy.

### (F) *Summary*

We have outlined a basis for the systematic calculation of statistical mechanical quantities for a plasma in which the plasma particles $\vec{E} \times \vec{B}$ drift transverse to a strong dc magnetic field, but respond according to Newton's laws in the parallel direction. The leading term in the coefficient of transverse spatial diffusion for a large but finite plasma has been calculated, and has been shown to fall off as O(1/B), even in thermal equilibrium. Spectral density calculations for the electric field fluctuations based on the more conventional weak-coupling expansion are shown to lead to an unphysical divergence for those $\vec{k}$ vectors nearly perpendicular to $\vec{B}$. It is exactly these components which are involved in "anomalous" transport properties. The conventional expansion in the plasma parameter is therefore inadequate in such calculations.

## IV. FINITE GYRORADIUS EFFECTS AND THERMAL RELAXATION IN TWO DIMENSIONS

### (A) *Preliminary Considerations; Liouville Equation; Hierarchy*

Though the guiding-center approximation in two dimensions has greatly facilitated calculation of thermal equilibrium transport coefficients, certain other processes associated with finite values of the plasma parameter apparently lie outside its scope. Mot importantly, there is apparently no tendency toward thermal equilibrium in the conventional sense. Any uncorrelated spatially uniform state is an exact solution of (96):

setting $f_{s_i,s_e} = \prod_{i=1}^{s_i} \prod_{e=1}^{s_e} f_{1,0}(i)\, f_{0,1}(e)$ gives $\partial/\partial t \equiv 0$ in (96)

if $f_{1,0} = f_{0,1} = \text{const.} = 1$. There is no tendency, as there is for other Hamiltonian systems that obey a hierarchy, for the pair correlations to develop and go to their thermal equilibrium values [such as (25)]. Moreover, any state with vanishing triplet correlations and with two-particle distributions which depend only upon the magnitudes of the separations of the particl is *also* a time independent solution of the equations from (96) for which $(s_i, s_e) = (1,1)$, $(2,0)$, and $(0,2)$. Therefore from the *dynamical* point of view, there is no particular reason to prefer (43) over any other spectral density. For example, the purely random distribution, with all pair correlations $\equiv 0$ and $\langle \vec{E}^2 \rangle \equiv \langle \vec{E}^2 \rangle_{self}$ has been studied in some detail numerically by Taylor and McNamara [28].

Clearly what is required in order to study thermalization processes in a two-dimensional magnetized plasma is a relaxation of the guiding-center approximation. By averaging over the "fast" gyromotion, one arrives at a system of dynamical equations which, among other things, makes the potential energy a constant of the motion. Since total energy is also conserved this makes it impossible for a ratio of potential to total energy which is different from that demanded by thermal equilibrium to adjust itself. This is an additional proof of the impossibility of complete thermalization within the framework of guiding-center theory.

In this section we shall examine the possibility of predicting thermal relaxation for a plasma made up of two-dimensio charges whose equation of motion is

$$\frac{d\vec{v}_j}{dt} = \frac{e_j}{m_j} \left[ \vec{E}(\vec{x}_j, t) + \frac{\vec{v}_j}{c} \times \vec{B} \right]$$

instead of (52). To keep the description simple, we shall also make the popular approximation of a uniform immobile ion background, so that only the negative charges are involved in the dynamics: all $e_j = -e$. The Liouville equation is simple to write down, and the BBGKY hierarchy which results from integrating over the phase space of the last N-s charges is

$$\left\{ \frac{\partial}{\partial t} + \sum_{j=1}^{s} \left( \vec{v}_j \cdot \frac{\partial}{\partial \vec{x}_j} + \frac{e_j}{m_j} \frac{\vec{v}_j}{c} \times \vec{B} \cdot \frac{\partial}{\partial \vec{v}_j} \right) - \frac{1}{m} \sum_{i<j=1}^{s} \frac{\partial \varphi_{ij}}{\partial \vec{x}_{ij}} \cdot \left( \frac{\partial}{\partial \vec{v}_i} - \frac{\partial}{\partial \vec{v}_j} \right) \right\} f_s \quad (189)$$

$$= \frac{n_o}{m} \sum_{i=1}^{s} \int d\vec{x}_{s+1}\, d\vec{v}_{s+1} \frac{\partial \varphi_{i,s+1}}{\partial \vec{x}_i} \cdot \frac{\partial f_{s+1}}{\partial \vec{v}_i}$$

where $\phi_{ij}$ is given by (15). $f_s = f_s(\vec{x}_1\ \vec{v}_1 \ldots \vec{x}_s\ \vec{v}_s;t)$ is a function of the 4s phase space variables necessary to describe the first s particles, and is normalized so that $\int f_s\, d\vec{x}_1\, d\vec{v}_1 \ldots d\vec{x}_s\, d\vec{v}_s = V^s = L^{2s}$.

Equation (189) looks much more like the familiar hierarchies that have been investigated than does (96), but the magnetic field term introduces some non-trivial complications.

It is realistic to confine our attention to the spatially uniform and gyrotropic limit, so that $f_1 = f_1(v_1^2/2,t)$ only, and $f_2 = f_2\ (\vec{x}_1 - \vec{x}_2, \vec{v}_1, \vec{v}_2, t)$ only. The s = 1 and s = 2 equations become, from (189),

$$\frac{\partial f_1}{\partial t} = \frac{n_o}{m} \int d\vec{x}_2\, d\vec{v}_2 \frac{\partial \varphi_{12}}{\partial \vec{x}_1} \cdot \frac{\partial f_2}{\partial \vec{v}_1} \quad (190)$$

$$\left\{\frac{\partial}{\partial t} + \vec{v}_1 \cdot \frac{\partial}{\partial \vec{x}_1} + \vec{v}_2 \cdot \frac{\partial}{\partial \vec{x}_2} - \frac{e}{m}\frac{\vec{v}_1 \times \vec{B}}{c} \cdot \frac{\partial}{\partial \vec{v}_1}\right.$$

$$\left. - \frac{e}{m}\frac{\vec{v}_2 \times \vec{B}}{c} \cdot \frac{\partial}{\partial \vec{v}_2} - \frac{1}{m}\frac{\partial \varphi_{12}}{\partial \vec{x}_{12}} \cdot \left(\frac{\partial}{\partial \vec{v}_1} - \frac{\partial}{\partial \vec{v}_2}\right)\right\}$$

$$= \frac{n_o}{m}\int d\vec{x}_3\, d\vec{v}_3 \left(\frac{\partial \varphi_{13}}{\partial \vec{x}_1} \cdot \frac{\partial}{\partial \vec{v}_1} + \frac{\partial \varphi_{23}}{\partial \vec{x}_2} \cdot \frac{\partial}{\partial \vec{v}_2}\right) f_3 \tag{191}$$

Another useful way of writing (191) compactly is

$$\left(\frac{\partial}{\partial t} + H_2\right) f_2 = L_2 f_3 \tag{192}$$

where the linear operator $H_2$ is the *Liouville operator for two particles*, and $L_2$ is defined by the right-hand side of (191).

### *(B) Boltzmann's Equation for the Two-Dimensional Strongly-Magnetized Plasma*

Equations (190) and (191) are general for any two-body potential $\phi_{12}$. Their analogues in three dimensions (usually minus the magnetic field) are the starting point for the most satisfactory existing derivations of *kinetic equations*: approximate equations of the form $\partial f_1/\partial t = \{$a functional of $f_1$ only$\}$ which predict relaxation to a Maxwellian $f_1$. Expansion in the density, for example, leads to the Boltzmann equation; expansion in the potential leads to the Fokker-Planck equation; and expansion in the plasma parameter leads to the Balescu-Lenard equation. It is natural to attempt these expansions on (190) and (191).

The perturbation theory is not entirely straightforward in the three-dimensional case, and is less so for (190) and (191). Normally the complications result from the presence of disparate time scales (first articulated by Bogolyubov [2]) in the underlying physical processes. This must be incorporated into the mathematics either formally [2,58], or by a fortunate insight [59]. The complication of more than one time scale is present in (190)-(191) also, but another difficulty is more fundamental. The conventional kinetic equations have in common that the two-body distribution $f_2$ evolves in terms of $f_1$ and what is essentially a two-particle dynamical process that goes to completion in a time short compared with $f_1/\partial f_1/\partial t$. For instance, in the derivation of Boltzmann's equation [60,61], the two-body process is a collision in which two particles approach each other, collide, and separate in well-defined stages during the duration of which $f_1$ is sensibly constant. For two colliding charges in two dimensions, the situation is much different in the presence of a strong magnetic field, as we shall now show.

Two equal charges in two dimensions obey the equations

$$\frac{d\vec{v}_1}{dt} = -\frac{e}{m}\left(\vec{E}_{12} + \frac{\vec{v}_1}{c} \times \vec{B}\right)$$

$$\frac{d\vec{v}_2}{dt} = -\frac{e}{m}\left(\vec{E}_{21} + \frac{\vec{v}_2}{c} \times \vec{B}\right) \tag{193}$$

where $\vec{E}_{12} = -(2e/\ell)\,\vec{x}_{12}/x_{12}^2$, and $\vec{x}_{12} = \vec{x}_1 - \vec{x}_2$. Adding Eqs. (193) gives, since $\vec{E}_{12} = -\vec{E}_{21}$

$$\frac{d\vec{V}}{dt} = -\frac{e}{mc}\,\vec{V} \times \vec{B} \tag{194}$$

where $\vec{V} \equiv (\vec{v}_1 + \vec{v}_2)/2$ is the center-of-mass velocity of the two charges. Equation (194) is the equation of motion of a free particle in a uniform magnetic field and so the solution for the center-of-mass motion is immediate.

Subtracting Eqs. (193) gives

$$\frac{d\vec{v}_{12}}{dt} = -\frac{2e}{m}\vec{E}_{12} - \frac{e\,\vec{v}_{12}}{mc} \times \vec{B}$$

(195)

$$\doteq \frac{4e^2}{\ell m}\frac{\vec{x}_{12}}{x_{12}^2} - \frac{e\,\vec{v}_{12}}{mc} \times \vec{B}$$

where $\vec{v}_{12} \equiv \vec{v}_1 - \vec{v}_2$ is the relative velocity of the two charges. Two constants of the motion of (195) are (as can be readily verified)

$$\frac{m\,\vec{v}_{12}^{\,2}}{2} - \frac{4e^2}{\ell}\ln|\vec{x}_{12}| = \Sigma = \text{const.}$$

(196)

$$m\hat{b}\cdot(\vec{x}_{12} \times \vec{v}_{12}) - \frac{eB}{2c}x_{12}^2 = P_\theta = \text{const.}$$

Expressing $\vec{x}_{12} = (r\cos\theta,\ r\sin\theta)$ in terms of polar coordinates and eliminating $\dot{\theta}$, we have

$$\frac{m}{2}\dot{r}^2 + \text{"}V(r)\text{"} = \mathcal{E} = \text{const.} \qquad (197)$$

where the "effective potential" is

$$\text{"}V(r)\text{"} \equiv \frac{P_\theta^2}{2mr^2} + \frac{e^2B^2r^2}{8mc^2} - \frac{4e^2}{\ell}\ln r\ . \qquad (198)$$

The solution to (197) is

$$\pm\int\frac{dr}{\sqrt{\mathcal{E} - \text{"}V(r)\text{"}}} = \sqrt{\frac{2}{m}}\int dt$$

and since "V(r)" has a single minimum and $\to\infty$ as $r \to 0$ or $r \to \infty$, the r motion is always bounded and periodic, with limits defined by the intersection of "V(r)" with $\mathcal{E}$.

Since $\dot{\theta} = P_\theta/mr^2 + eB/2mc$, $\dot{\theta}$ is periodic with the same period as r, and $\theta$ can be written as a constant times t plus a periodic function of t with the same period as the r motion and with average value zero. The motion may or may not encircle the origin, but there is always a maximum radius r which is reached periodically.

This shows that a significant difference exists between the basic two-body interaction in a magnetized two-dimensional plasma and a three-dimensional one (or a two-dimensional one with $\vec{B} = 0$). There is no time before which one can confidently assume that the spatial correlations of two colliding charges vanish. This renders suspect many of the assumptions of the BBGKY kinetic theory. Nevertheless, alternative equally-plausible assumptions have yet to be suggested, so it is still of interest to take the kinetic theory as far as possible in parallel to its usual form.

The easiest expansion to perform on Eqs. (190)-(192) is in powers of the density. To lowest order in the density, (192) is just the two-particle Liouville equation

$$\left(\frac{\partial}{\partial t} + H_2\right) f_2 \approx 0 \tag{199}$$

whose solution is any function of the constants of the motion of the two-body problem defined by (193). If we go ahead and ignore the pair correlation at $t = 0$, $f_2 \approx f_1 f_1$ then, the solution to (199)

$$f_2(\vec{x}_1,\vec{v}_1,\vec{x}_2,\vec{v}_2,\tau) = f_1[\vec{v}_1^{\,2}(-\tau)/2,0]\ f_1[\vec{v}_2^{\,2}(-\tau)/2,0] \tag{200}$$

where $\vec{v}_1(-\tau)$, $\vec{v}_2(-\tau)$ are the solutions to (193) that lead to two particles, being at $\vec{x}_1,\vec{v}_1,\vec{x}_2,\vec{v}_2$, at time $\tau$. Substitution of (200) into (190) would lead to secularity for reasons which are well-known [11,61,62]. This is avoided by the method of Dupree [59] or its more systematic formulation in the method of multiple time scales [11,58,61].

One notes that an equally accurate solution to (199) is obtained by allowing $f_1$ to have a slow dependence on the time. Equation (190) predicts $\frac{\partial f_1}{\partial t} = O(n_o)$, and so a slowly-varying $f_1$ results for low densities. If we write a formal dimensionless expansion parameter $\tilde{\varepsilon}$ in front of $n_0$ to remind us that it is treated as very small, (200) can to equal accuracy be replaced by

$$f_2(\vec{x}_1,\vec{v}_1,\vec{x}_2,\vec{v}_2,\tau) \cong f_1[\vec{v}_1^2(-\tau)/2,\tilde{\varepsilon}\tau]\ f_1[\vec{v}_2^2(-\tau)/2,\tilde{\varepsilon}\tau] \quad (201)$$

since when substituted into (199), the only additional terms generated are of $O(\tilde{\varepsilon})$. At the end of the calculation, we can then let $\tilde{\varepsilon} \to 1$.

The canonical recipe would be to substitute (201) into (190) after passing to the limit $\tau \to \infty$, $\tilde{\varepsilon}\tau$ finite. That is, one would write

$$\frac{\partial f_1(\vec{v}_1^2/2,\tilde{\varepsilon}t)}{\partial(\tilde{\varepsilon}t)} = \lim_{\tau\to\infty} \frac{n_0}{m}\int d\vec{x}_2\, d\vec{v}_2\, \frac{\partial \varphi_{12}}{\partial \vec{x}_{12}} \cdot \frac{\partial}{\partial \vec{v}_1} f_1\left[\frac{\vec{v}_1^2(-\tau)}{2},\tilde{\varepsilon}t\right] f_1\left[\frac{\vec{v}_2^2(-\tau)}{2},\tilde{\varepsilon}t\right] \quad (202)$$

In the $\vec{B} = 0$ case, (202) is nothing but Boltzmann's equation [2], and only a few additional manipulations are required [6] to bring it into the form given by Boltzmann. For the case under discussion, however, it is not so simple. The limit of the integrand does not exist; it is an oscillatory function of $\tau$. The period of oscillation does vary from point to point in the two-particle phase space so that the limit of the integral may exist even though the limit of the integrand may not. But this has not been proved. An additional *ad hoc* step which makes the limit of (202) well defined is that of *time averaging*, reasonable in view of the success of the Bogolyubov "method of averaging" [63,64] in other situations. It consists of replacing (202) by its time average

$$\frac{\partial f_1(v_1^2/2, \tilde{\epsilon}t)}{\partial(\tilde{\epsilon}t)} = \lim_{T\to\infty} \frac{1}{T} \int_0^T d\tau \frac{n_o}{m}$$

$$\int d\vec{x}_2\, d\vec{v}_2 \frac{\partial \varphi_{12}}{\partial \vec{x}_{12}} \cdot \frac{\partial}{\partial \vec{v}_1} f_1[\vec{v}_1^2(-\tau)/2, \tilde{\epsilon}t]\, f_1[\vec{v}_2^2(-\tau)/2, \tilde{\epsilon}t] \tag{203}$$

If the limit in (202) exists, it is identical to (203).

We propose (203) [or (202)] as the analogue of Boltzmann's equation for the two-dimensional magnetized plasma [12]. We suspect that it obeys an H-theorem and conservation laws, but have not been able to prove any of these.

*(C) The Weak-Coupling Approximation; The Fokker-Planck Equation*

A second expansion of (191) which lies in the mainstream of kinetic theory is an expansion in powers of the interaction. That is, we shall treat the terms containing a two-body potential as of one order higher in a formal expansion parameter $\tilde{\varepsilon}$ than the terms which do not. Corresponding to this (and gleaning an insight from the equilibrium theory), we expect the correlation functions to get successively smaller in powers of $\tilde{\varepsilon}$ as well. Since $\frac{\partial f_1}{\partial t} = O(\tilde{\epsilon}^2)$,

$$f_1(1) = f_1(\vec{v}_1^2/2, \tilde{\varepsilon}^2 t)$$

$$f_2(1,2) = f_1(\vec{v}_1^2/2, \tilde{\varepsilon}^2 t)\, f_1(\vec{v}_2^2/2, \tilde{\varepsilon}^2 t)$$

$$+ \tilde{\epsilon}\, P(\vec{x}_1,\vec{v}_1,\vec{x}_2,\vec{v}_2;t,\tilde{\epsilon}^2 t)$$

$$f_3(1,2,3) = f_1(\vec{v}_1^{\,2}/2,\tilde{\epsilon}^2 t)\, f_1(\vec{v}_2^{\,2}/2,\tilde{\epsilon}^2 t)\, f_1(\vec{v}_3^{\,2}/2,\tilde{\epsilon}^2 t) \tag{204}$$

$$+ \tilde{\epsilon}\{f_1(\vec{v}_1^{\,2}/2,\tilde{\epsilon}^2 t)\, P(\vec{x}_2,\vec{v}_2,\vec{x}_3,\vec{v}_3;t,\tilde{\epsilon}^2 t) + \text{cyclic permutations}\}$$

$$+ O(\tilde{\epsilon}^2)$$

and so on.

Dropping terms of $O(\tilde{\varepsilon}^2)$ from (191) after substituting in (204) gives

$$\left(\frac{\partial}{\partial t} + \vec{v}_1\cdot\frac{\partial}{\partial \vec{x}_1} + \vec{v}_2\cdot\frac{\partial}{\partial \vec{x}_2} - \frac{e}{mc}\vec{v}_1 \times B\cdot\frac{\partial}{\partial \vec{v}_1} - \frac{e}{mc}\vec{v}_2 \times \vec{B}\cdot\frac{\partial}{\partial \vec{v}_2}\right) P$$

$$= \frac{1}{m}\frac{\partial \varphi_{12}}{\partial \vec{x}_{12}}\cdot\left(\frac{\partial}{\partial \vec{v}_1} - \frac{\partial}{\partial \vec{v}_2}\right) f_1(\vec{v}_1^{\,2}/2,\tilde{\epsilon}^2 t)\; f_1(\vec{v}_2^{\,2}/2,\tilde{\epsilon}^2 t) \tag{205}$$

for the pair correlation P.

The left-hand side of (205) can be recognized as the total time derivative dP/dt *computed along the noninteracting trajectories in the two-particle phase space*. Integrating (205) from 0 to t and assuming that no pair correlations exist initially, we get

$$P = -\int_0^t d\tau \; \frac{2e^2}{\ell m} \frac{\vec{x}_{12}(\tau) \cdot \vec{v}_1(\tau)}{x_{12}^2(\tau)} \; f_1'(v_1^2/2, \tilde{\varepsilon}^2 t)\, f_1(v_2^2/2, \tilde{\varepsilon}^2 t)$$

$$+ (1 \leftrightarrow 2) \tag{206}$$

In (206), the prime means differentiation with respect to $v_1^2/2$, and $(1 \leftrightarrow 2)$ means the same term with coordinates of particles 1 and 2 interchanged.

Since $\vec{x}_{12}(\tau)$, $\vec{v}_1(\tau)$, $\vec{v}_2(\tau)$ are simple periodic functions of $\tau$, readily expressible in terms of $\vec{x}_1, \vec{v}_1, \vec{x}_2$, and $\vec{v}_2$, the integral

$$I(1) = \frac{2e^2}{\ell m} \int_0^t d\tau \; \frac{\vec{x}_{12}(\tau) \cdot \vec{v}_1(\tau)}{x_{12}^2(\tau)} \tag{207}$$

can be done. $I(1)$ has the physical interpretation of being $m^{-1}$ times the kinetic energy transferred, to lowest order in perturbation theory, by particle 2 to particle 1. As long as $I(1)$ remains $<< \vec{v}_1^2/2$, the perturbation theory remains valid.

The integral (207) is messy, but can be carried out in terms of elementary functions. It is performed in detail in Vahala's thesis [12]. Taken over an integral number of gyroperiods $T_o \equiv 2\pi mc/eB$, it is

$$I(1) = \begin{cases} 0 \, , & x_{12}^2 + \dfrac{2\,\vec{x}_{12} \cdot (\hat{b} \times \vec{v}_{12})}{\Omega} > 0 \\[2ex] \dfrac{t}{T_o} \dfrac{4\pi e^2}{\ell m} \dfrac{\vec{v}_1 \cdot [\vec{v}_{12} \times \hat{b}]}{v_{12}^2} \, , & x_{12}^2 + \dfrac{2\,\vec{x}_{12} \cdot (\hat{b} \times \vec{v}_{12})}{\Omega} < 0 \end{cases} \tag{208}$$

Here, $\Omega \equiv eB/mc$. The content of (208) is roughly that only particles initially close enough together contribute to P, and this contribution grows secularly with time.

We must confess lack of a systematic understanding of what the proper use of (208) in (190) is. The formal substitution of (208) into (190) leaves a secularly-growing term which is unphysical. It was previously dealt with [12] by the somewhat *ad hoc* procedure of cutting off the integration at an upper limit corresponding to $mI(1) \sim \theta$. Here we shall use a time averaging procedure motivated by "stroboscopic" perturbation theory [65] and obtain the same answer, up to a numerical factor.

The motivating physical inequality is that in one gyroperiod, $f_1$ should have changed by very little. Therefore, (190) will be replaced by

$$\frac{\partial f_1}{\partial t} = \tilde{\epsilon} \frac{n_o}{m} \int d\vec{x}_2 \, d\vec{v}_2 \, \frac{\partial \varphi_{12}}{\partial \vec{x}_1} \cdot \frac{\partial}{\partial \vec{v}_1} \bar{f}_2$$

$$= \tilde{\epsilon}^2 \frac{n_o}{m} \int d\vec{x}_2 \, d\vec{v}_2 \, \frac{\partial \varphi_{12}}{\partial \vec{x}_1} \cdot \frac{\partial}{\partial \vec{v}_1} \bar{P} \qquad (209)$$

where the bar over $f_2$ and P indicates an average over a gyroperiod. Since in the region where it is non-zero, $\overline{I(1)} = (2\pi e^2/\ell m) \cdot \vec{v}_1 \cdot (\vec{v}_{12} \times \vec{b})/v_{12}^2$, and since

$$P = -I(1)\, f_1'(v_1^2/2, \tilde{\epsilon}^2 t)\, f_1(v_2^2/2, \tilde{\epsilon}^2 t) + (1 \leftrightarrow 2) \qquad (210)$$

we may substitue (208) and (210) into (209) to get a kinetic equation

$$\frac{\partial f_1}{\partial(\tilde{\epsilon}^2 t)}(\vec{v}_1^2/2, \tilde{\epsilon}^2 t)$$

$$= \frac{n_o}{m}\int d\vec{x}_2\, d\vec{v}_2\, \frac{\partial \varphi_{12}}{\partial \vec{x}_1} \cdot \frac{\partial}{\partial \vec{v}_1}\left\{- \overline{I(1)}\, f_1'(v_1^2/2, \tilde{\epsilon}^2 t)\, f_1(v_2^2/2, \tilde{\epsilon}^2 t) + (1 \leftrightarrow 2)\right\} \tag{211}$$

with the region of integration to run over all the region defined by the second inequality in (208).

The integrals in (211) can be performed [12] and the result is

$$\frac{\partial f_1(v_1^2/2, \tilde{\epsilon}^2 t)}{\partial(\tilde{\epsilon}^2 t)} = -\frac{\partial}{\partial \vec{v}_1} \cdot \int d\vec{v}_2\, \vec{\vec{Q}}(\vec{v}_1, \vec{v}_2) \cdot \left(\frac{\partial}{\partial \vec{v}_1} - \frac{\partial}{\partial \vec{v}_2}\right) f_1(v_1^2/2, \tilde{\epsilon}^2 t)\, f_1(v_2^2/2, \tilde{\epsilon}^2 t) \tag{212}$$

where the dyadic $\vec{\vec{Q}}$ is given by

$$\vec{\vec{Q}} = -\frac{4\pi^2 n_o e^4}{m^2 \ell^2 \Omega}\, \frac{(\vec{v}_{12} \times \hat{b})(\vec{v}_{12} \times \hat{b})}{v_{12}^2}$$

$$= -\frac{4\pi^2 n_o e^4}{m^2 \ell^2 \Omega}\, \frac{v_{12}^2 \vec{\vec{I}} - \vec{v}_{12}\vec{v}_{12}}{v_{12}^2} \tag{213}$$

Equation (212) is a standard form in which the Fokker-Planck equation has been written before. It is easy to prove from it conservation of particles, momentum (zero in this case) and kinetic energy. It also obeys an H-theorem, and thus predicts an approach to a Maxwellian $f_1$ as $t \to \infty$:

$$f_1(v_1^2/2, \tilde{\epsilon}^2 t) \xrightarrow[\tilde{\epsilon}^2 t \to \infty]{} \frac{m}{2\pi\theta} \exp(-m\vec{v}_1^2/2\theta)$$

where $\theta n_0$ is the initial kinetic energy density. The relaxation that occurs is predicted to occur on a time scale $t_{rel.}$, where

$$\frac{1}{t_{rel.}} \approx \frac{1}{4 n_o \lambda_D^2} \frac{\omega_{pe}^2}{\Omega}$$

That $t_{rel.}$ shall be long compared to other time scales involved is a necessary condition for the applicability of the guiding-center model to situations other than thermal equilibrium, as done, for example, by Taylor and McNamara [28] for the case of a random initial distribution (P = 0).

Up to an arbitrary numerical factor associated with the cut-off of I(1), Eq. (212) is the same as that given by Vahala and Montgomery [12]. [See, however, reference [67].]

*(D) The "Weak, Long-Range", or Balescu-Lenard, Limit*

Note that no long-range cutoff of the potential was required to render (213) finite, unlike the corresponding situation in the unmagnetized case. Once can thus question the necessity for going to the "weak long-range" limit usually thought necessary to account properly for the long-range part of the Coulomb force. Equation (212) may thus be surmised to represent the "Balescu-Lenard limit" as well as the "Fokker-Planck limit". Nevertheless, the Balescu-Lenard limit was investigated in some detail in Vahala's thesis [12]. P was re-calculated by adding in terms

$$- \frac{n_o}{m} \frac{\partial f_1}{\partial \vec{v}_1} \cdot \int d\vec{x}_3\, d\vec{v}_3 \frac{\partial \varphi_{13}}{\partial \vec{x}_1} P(\vec{x}_2,\vec{v}_2,\vec{x}_3,\vec{v}_3,t)$$

$$+ (1 \leftrightarrow 2)$$

on the left-hand side of (205). The calculation is still manageable, though very lengthy. The rather surprising result is that following the standard procedures, $\partial f_1/\partial t = 0$, through terms of first order in the plasma parameter. This not entirely expected vanishing of the Balescu-Lenard collision term has as yet no satisfactory physical interpretation, and requires additional investigation.

*(E) The "Test Particle" Problem; Absence of Dynamic Screening on Vlasov Time Scales*

One of the most important physical characteristics of the unmagnetized Vlasov plasma in both two and three dimensions is the separation of "Vlasov" and "collisional" time scales. These can be roughly associated with processes which occur at finite rates even when the plasma parameter $\varepsilon$ goes to zero and those which do not. In the former category, we can put plasma oscillations, Landau damping, and relaxation to the "kinetic stage" of Bogolyubov, all of which occur on time scales measured by the inverse plasma frequency $\omega_p^{-1}$. In the latter class is relaxation to thermal equilibrium, which occurs on time scales $\sim \omega_p^{-1}\, \varepsilon^{-1}$ in two dimensions, and $\sim \omega_p^{-1}\, \varepsilon^{-1}/\ln\varepsilon^{-1}$ in three.

Many of the novel features of two-dimensional strongly magnetized plasmas (and the nearly perpendicular $\vec{k}$-vectors in three) are associated with the fact that these two time scales no longer enjoy a clear separation. There are no relaxation processes which occur for components of the fluctuation spectra with $\vec{k}$ perpendicular to $\vec{B}$ that occur on a time scale independent of $\varepsilon$.

This can be illustrated by adapting a very standard kind of calculation for three-dimensional plasma [56] to the two-dimensional magnetized plasma: the linear response of a Vlasov plasma to a "test" charge introduced at $t = 0$.

The Vlasov equation obtained from (189) by ignoring all correlations is

$$\left[\frac{\partial}{\partial t} + \vec{v} \cdot \frac{\partial}{\partial \vec{x}} - \frac{e}{m} \left(\vec{E} + \frac{\vec{v}}{c} \times \vec{B}\right) \cdot \frac{\partial}{\partial \vec{v}}\right] f_1 = 0 \tag{214}$$

where

$$\vec{E}(\vec{x},t) = \frac{2e}{\ell} n_o \int f(\vec{x}',\vec{v}',t) \frac{(\vec{x} - \vec{x}')}{|\vec{x} - \vec{x}'|^2} d\vec{x}' d\vec{v}' \tag{215}$$

It is soluble, as a linearized initial value problem, by standard methods. If the initial conditions are such that a test charge $e_t$ is introduced at $\vec{x}_o$ with velocity $\vec{v}_o = (v_o \cos\Theta, v \sin\Theta)$ at $t = 0$, the electric field which results for $t > 0$ is

$$\vec{E}(\vec{x},t) = \int d\vec{k}\; e^{i\vec{k}\cdot\vec{x}} \int_{\sigma-i\infty}^{\sigma+i\infty} \frac{ds}{2\pi i} e^{st} \vec{E}_{\vec{k}}(s) \tag{216}$$

where

$$\vec{E}(\vec{k},s)$$

$$= \frac{\left(\vec{k} i e_t/\pi k^2 \ell\right)}{D(\vec{k},s)} e^{-i\vec{k}\cdot\vec{x}_o + (ik v_o/\Omega_e)\sin(\Theta-\alpha)} \sum_{n=-\infty}^{\infty} \frac{J_n\left(\frac{k v_o}{\Omega_e}\right) e^{in(\Theta-}}{s - in\Omega_e} \tag{217}$$

The new symbols appearing in (217) are $\Omega_e \equiv |eB/mc|$, $\vec{k} = (k \cos\alpha, k \sin\alpha)$, and the dielectric function $D(\vec{k},s)$ is

$$D(\vec{k},s) = 1 - \frac{2\pi i\, \omega_{pe}^2 \Omega_e}{k^2} \sum_{n=-\infty}^{\infty} \frac{n}{s + in\Omega} \int_o^{\infty} v dv\, f_o'(v^2/2)\, J_n^2 \left(\frac{kv}{\Omega_e}\right) \tag{218}$$

$J_n$ is the Bessel function of its argument and $f_o = (2\pi\theta)^{-1}\cdot \exp(-mv^2/2\theta)$ is the two-dimensional Maxwellian; $\omega_{pe} \equiv 4\pi n_o e^2/\ell m$.

If the charge is stationary, $v_o = 0$, and $kV_e \ll \Omega_e$, $|s| \ll \Omega_e$, (217) simplifies considerably to

$$\vec{E}_{\vec{k}}(s) \cong \frac{i e_t \vec{k}}{\pi k^2 \ell} \frac{e^{-i\vec{k}\cdot\vec{x}_o}}{s\left(1 + \frac{\omega_{pe}^2}{\Omega_e^2}\right)} \tag{219}$$

When we Fourier-Laplace invert (219) we get, for the electric field produced by the test charge

$$\vec{E}(\vec{x},t) \cong \frac{2 e_t(\vec{x} - \vec{x}_o)}{\ell\,|\vec{x} - \vec{x}_o|^2} \frac{1}{1 + \frac{\omega_{pe}^2}{\Omega_e^2}}, \quad \text{all } t, \tag{220}$$

which is the vacuum value divided by the low-frequency limit of the dielectric function $1 + \omega_{pe}^2/\Omega_e^2$.

If, however, we had considered the limit $\Omega_e \to 0$ in (217) we would have gotten instead of (220),

$$\vec{E}(\vec{x},t) \cong \frac{2 e_t(\vec{x} - \vec{x}_o)}{\ell\,|\vec{x} - \vec{x}_o|} \sqrt{\frac{\pi K_D}{2\,|\vec{x} - \vec{x}_o|}}\; e^{-K_D|\vec{x}-\vec{x}_o|} \tag{221}$$

for $\omega_{pe}t \gg 1$ and $K_D|\vec{x} - \vec{x}_o| \gg 1$.

The differences between (220) and (221) are considerable. The latter is Debye shielded and has an effective range $\sim K_D^{-1} = \lambda_D$, while the former is not, and has an effective range of the order of the size of the system.

We know, of course, that the field of the test charge will eventually be shielded; the thermal equilibrium theory tells us that. The point is that it occurs on time scales which are not identifiable with Vlasov's equation, but with processes which slow down to stationary as $\varepsilon \to 0$.

*(F) Comments on the Problem of Spatial Diffusion for Finite Larmor Radii*

Dawson, Okuda, and Carlile have published a preliminary [66] account of a calculation of a two-dimensional coefficient of spatial diffusion which endeavors to include finite Larmor radius effects. The essentially new result, for which some evidence from numerical simulation is presented, is that the Taylor-McNamara result [Eqs. (89) and (91) of this article] appears divided by the square root of the dielectric function at zero frequency, $\left[1 + \sum_j \omega_{pj}^2/\Omega_j^2\right]^{\frac{1}{2}}$. The development, however, does not aspire to the deductive character of the Taylor-McNamara development and its generalizations, and a number of the steps are little more than order-of-magnitude estimates. These fortunately combine to produce exactly the right numerical coefficient to agree with Taylor-McNamara in the limit $B \to \infty$. A supplementary derivation based on a two-particle encounter model of diffusion is used to interpret the data for lower values of B and a "classical" [i.e., $O(1/B^2)$] diffusion coefficient is claimed for these lower values. This derivation omits contributions from the very-long-wavelength components of the fluctuation spectrum, however, and so contains no volume dependence. Since it is precisely these components that led to the volume divergence in Eq. (91) and in the calculations of Ref. 32, one suspects they may play a similar role in the magnetized finite-Larmor-radius plasma. This "classical" expression therefore probably needs, at best, an inequality on the plasma volume in order that it be in fact the dominant term.

One can hope for a clarification of these, and other, important unanswered questions connected with the finite gyroradius plasma in the next few years.

ACKNOWLEDGMENTS

This work was supported by NASA Grant NGL-16-001-043 at the University of Iowa.

Valuable discussions with the following individuals are gratefully acknowledged: Dr. George Vahala (in collaboration with whom many of these results were obtained), Dr. Frederick Tappert, Dr. Glenn Joyce, Dr. C. -S. Liu, and Dr. Bruce Langdon. An air-mail preprint of the important manuscript of Taylor and McNamara, shortly before the English postal strike of 1970-71, enabled us to begin work on these problems many months earlier than we would otherwise have been able to do. Mrs. Rozanne Huff skillfully performed the very demanding typing.

## APPENDIX I: DERIVATION OF EQ. (81)

We are interested in evaluating expressions of the type

$$I \equiv \int d\vec{E}_1\, d\vec{E}_2 \ldots d\vec{E}_M\, P(\vec{E}_1\, \vec{E}_2 \ldots \vec{E}_M) \exp\left(i \sum_{m=1}^{M} \vec{\lambda}_m \cdot \vec{E}_m\right) \tag{A1}$$

where

$$P(\vec{E}_1\, \vec{E}_2 \ldots \vec{E}_M) = \eta \exp\left(- \sum_{i,j=1}^{M} \vec{\vec{a}}_{ij} : \vec{E}_i\, \vec{E}_j\right) \tag{A2}$$

and $\eta$ is chosen so that

$$\int d\vec{E}_1\, d\vec{E}_2 \ldots P(\vec{E}_1\, \vec{E}_2 \ldots) = 1 \tag{A3}$$

The $\vec{\vec{a}}_{ij}$ are positive definite dyadics which satisfy $\vec{x} \cdot \vec{\vec{a}}_{ij} \cdot \vec{x} > 0$ for any non-zero $\vec{x}$.

The $(\vec{E}_1, \vec{E}_2, \ldots \vec{E}_M)$ can be thought of as a column vector whose ith element is $\mathcal{E}_i$, say. The $\vec{\vec{a}}_{ij}$ can be thought of as a real, symmetric, positive definite matrix whose ijth element is $A_{ij}$, say. Finally the $(\vec{\lambda}_1, \vec{\lambda}_2, \ldots \vec{\lambda}_M)$ can be thought of as a column vector whose jth element is $\ell_j$, say. We can then write the desired integral I as

$$I = \eta \int \left(\prod_j d\mathcal{E}_j\right) \exp\left\{- \sum_{ij} A_{ij}\, \mathcal{E}_i\, \mathcal{E}_j + i \sum_j \ell_j\, \mathcal{E}_j\right\} \tag{A4}$$

The exponentials in (A4) are scalars, and a unitary transformation of the $\mathcal{E}_i$'s always exists which will make $A_{ij}$ diagonal in the new representation. Call the unitary transformation $U_{ij}$. Thus

$$\varepsilon_i' = \sum_j U_{ij}\, \varepsilon_j$$

$$A_{ij}' = \sum_{kn} U_{ik}\, A_{kn} (U^{-1})_{nj} \tag{A5}$$

where $A_{ij}' = A_i' \delta_{ij}$, with $A_i' > 0$, all i.

The Jacobian of a unitary transformation is unity, so we can write (A4) as

$$I = \eta \int \left( \prod_j d\varepsilon_j' \right) \exp\left\{ - \sum_j A_j' \varepsilon_j'^2 + i \sum_j \ell_j' \varepsilon_j' \right\} \tag{A6}$$

where $\ell_j' = \sum_k U_{jk}\, \ell_k$. The integrals in (A6) are now elementar and

$$I = \eta \prod_j \left\{ \int_{-\infty}^{\infty} d\varepsilon_j'\, e^{-\left(\sqrt{A_j'}\, \varepsilon_j' - i \ell_j' \varepsilon_j' / 2\sqrt{A_j'}\right)^2} e^{-\ell_j'^2/4A_j'} \right\}$$

$$= \eta \prod_j \left( \frac{\pi^{\frac{1}{2}}}{\sqrt{A_j'}}\, e^{-\ell_j'^2/4A_j'} \right) \tag{A7}$$

Now $\eta \prod_j (\pi^{\frac{1}{2}}/\sqrt{A_j'}) = 1$, and

$$\prod_j \exp(-\ell_j'^2/4A_j') = \exp\left( - \sum_{ij} \ell_i (A^{-1})_{ij}\, \ell_j / 4 \right) \quad \text{, so}$$

$$I = \exp\left( - \sum_{ij} \ell_i (A^{-1})_{ij}\, \ell_j / 4 \right) \tag{A8}$$

We also note from (A4) that

$$- \left. \frac{\partial^2 I}{\partial \ell_i \, \partial \ell_j} \right|_{\text{all } \ell_k = 0} = \langle \mathcal{E}_i \, \mathcal{E}_j \rangle$$

so

$$\langle \mathcal{E}_i \, \mathcal{E}_j \rangle = (A^{-1})_{ij}/2$$

from (A8), and

$$I = \exp\left( - \frac{1}{2} \sum_{ij} \ell_i \langle \mathcal{E}_i \, \mathcal{E}_j \rangle \, \ell_j \right) \tag{A9}$$

If we now revert to the vector notation,

$$I = \exp\left( - \frac{1}{2} \sum_{i,j=1}^{M} \vec{\lambda}_i \, \vec{\lambda}_j : \langle \vec{E}_i \, \vec{E}_j \rangle \right) \tag{A10}$$

To derive Eq. (81) of the text from (A10), replace $\vec{\lambda}_i$ by $(c\vec{k} \times \vec{B}/B^2)\,\Delta t$, all for i.

Equation (A2) has the property that integrating it over all the $\vec{E}_i$ but one leaves a Gaussian distribution in that $\vec{E}_i$. However, the converse is not true: that the electric field have a Gaussian distribution at any instant is not a sufficient condition that the joint distribution have the form (A2).

Considerable use has been made of this theorem in statistical physics in recent years. See, e.g., the review article by A.J.F. Siegert, in *Statistical Mechanics at the Turn of the Decade*, E.G.D. Cohen, ed.; New York, Marcel Dekker, 1971.

REFERENCES

1. See, e.g. HILL, T.L., *Statistical Mechanics*, McGraw-Hill, New York, 1956.

2. BOGOLYUBOV, N.N., *Problems of a Dynamical Theory in Statistical Physics*, State Technical Press, Moscow, 1946. English trans. by E.K. GORA in *Studies in Statistical Mechanics*, Vol. 1, ed. by J. DE BOER and G.E. Uhlenbeck, North-Holland Publishing Company, Amsterdam, 1962.

3. YVON, J., *La Théorie des Fluides et l'Equation d'État: Actualités Scientifiques et Industrielles*, Hermann et Cie., Paris, 1935.

4. KIRKWOOD, J., *J. Chem. Phys.* *14*, 180 (1946); *15*, 72 (1947).

5. BORN, M. and GREEN, H.S., *A General Kinetic Theory of Liquids*, Cambridge Univ. Press, London, 1949.

6. UHLENBECK, G.E. and FORD, G.W., *Lectures in Statistical Mechanics*, American Mathematical Society, Providence (R.I.), 1963.

7. LENARD, A., *J. Math. Phys.* *2*, 682 (1961).

8. BAXTER, R.J., *Proc. Camb. Phil. Soc.* *59*, 779 (1963).

9. DYSON, F.J. and LENARD, A., *J. Math. Phys.* *8*, 423 (1967). LENARD, A. and DYSON, F.J., *J. Math. Phys.* *9*, 698 (1968).

10. GUERNSEY, R.L., *Phys. Fluids* *7*, 792 (1964).

11. MONTGOMERY, D. in *Kinetic Theory: Lectures in Theoretical Physics*, Vol. 9C; Ed. by W. BRITTIN, A. BARUT, and M. GUENIN, Gordon and Breach, New York, 1967.

12. VAHALA, G. and MONTGOMERY, D., *J. Plasma Phys.* *6*, 425 (1971), and G. VAHALA, Ph.D. Thesis, University of Iowa, 1971.

13. See, e.g., MONTGOMERY, D., *Theory of the Unmagnetized Plasma*, Gordon and Breach, New York, 1971.

14. ROSTOKER, N., *Nucl. Fusion* *1*, 101 (1961).

15. LENARD, A. (private communication).

16. SALZBERG, A. and PRAGER, S., *J. Chem. Phys.* *38*, 2587 (1963).

17. MAY, R.M., *Physics Letters* *25A*, 282 (1967).

18. KNORR, G., *Physics Letters* *28A*, 166 (1968).

19. See, e.g., CHAPMAN, S. and COWLING, T.G., *The Mathematical Theory of Non-uniform Gases*, 3rd Ed., Cambridge University Press, Cambridge, 1970.

20. BOHM, D. in *The Characteristics of Electrical Discharges in Magnetic Fields*, ed. by A. GUTHRIE and R.K. WAKERLING, McGraw-Hill, New York, 1949.

21. SPITZER, L. Jr., *Phys. Fluids* *3*, 659 (1960).

22. Ref. 14, Sec. 2.7.

23. See, e.g., the encyclopaedic work of I.P. SHKAROVSKY, T.W. JOHNSTON, and M. P. BACHYNSKI, *The Particle Kinetics of Plasmas*, Addison-Wesley, Reading, Mass., 1966.

24. For a review of the literature, see, e.g., Ref. 13.

25. ROSTOKER, N., *Phys. Fluids* *3*, 922 (1960).

26. HAGGERTY, M.J. and DE SOBRINO, L.G., *Canad. J. Phys.* *42*, 1969 (1964).

27. HAGGERTY, M.J., *Canad. J. Phys.* *43*, 122 and 1750 (1965).

28. TAYLOR, J.B. and McNAMARA, B., *Phys. Fluids* *14*, 1492 (1971).

29. NORTHROP, T.G., *The Adiabatic Motion of Charged Particles*, Interscience-Wiley, New York, 1963.

30. See, e.g., SOMMERFELD, A., *Mechanics of Deformable Bodies*, Academic Press, New York, 1950.

31. See, e.g., *Selected Papers on Noise and Stochastic Processes*, edited by N. WAX, Dover, New York, 1954; for background in the problems of spatial diffusion.

32. See, e.g., WAINWRIGHT, T.E., ALDER, B.J., and GASS, D.M., *Phys. Rev.* *A4*, 233 (1971), and references therein.

33. See BERAN, M.J., *Statistical Continuum Theories*, Interscience, New York, 1968, for a physicist's discussion of these problems.

34. PAPOULIS, A., *Probability, Random Variables and Stochastic Processes*, McGraw-Hill, New York, 1965, pp. 253-254. Some use had previously been made of this result in plasma theory by J. WEINSTOCK, *Phys. Fluids 11*, 354 (1968).

35. CRAMÉR, H., *Random Variables and Probability Distribitions*, Cambridge University Press, Cambridge, 1962, Chapter X.

36. This observation is due to F. Tappert and J. Doles III.

37. DEEM, G.S. and ZABUSKY, N.J., *Phys. Rev. Letters 27*, 396 (1971).

38. LILLY, D.K., *J. Fluid Mech. 45*, 395 (1971).

39. TAPPERT, F. and HARDIN, R. (unpublished).

40. TAYLOR, J.B. and THOMPSON, W.B., *Phys. Fluids 16*, 111 (197 See also: G. Joyce, D. Montgomery and M. Emery, *Phys. Fluid 17*, 110 (1974).

41. BALESCU, R., *Phys. Fluids 3*, 52 (1960).

42. GUERNSEY, R.L., Ph.D. Thesis (Michigan), 1960.

43. MONTGOMERY, D. and TIDMAN, D.A., *Plasma Kinetic Theory*, McGraw-Hill, New York, 1964, pp. 89-92.

44. KUBO, R., *J. Phys. Soc. (Japan) 12*, 570 (1957).

45. MONTGOMERY, D. and TAPPERT, F., *Phys. Fluids 15*, 683 (1972)

46. ONSAGER, L., *Nuovo Cimento Suppl. Series 9, 6*, 279 (1949).

47. LANDAU, L.D. and LIFSHITZ, E.M., *Statistical Physics*, Addison-Wesley, Reading, Mass., 1969 (2nd Ed.), pp. 35-36 and 211-214.

48. MONTGOMERY, D., *Physics Letters 39A*, 7 (1972).

49. COOK, I. and TAYLOR, J.B., *Phys. Rev. Letters 28*, 82 (1972)

50. JOYCE, G. and MONTGOMERY, D., Simulation of the 'Negative Temperature' Instability for Line Vortices, *Physics Letters 39A*, 371 (1972), and *J. Plasma Phys. 10*, 107 (1973).

51. TAYLOR, J.B., Negative Temperatures in Two-Dimensional Vortex Motion, *Physics Letters 40A*, 1 (1972).

52. MONTGOMERY, D., LIU, C.-S., and VAHALA, G., *Phys. Fluids 15*, 815 (1972). See also: G. Vahala, *Phys. Fluids 16*, 1876 (1973).

53. VAHALA, G., *Phys. Rev. Letters 29*, 93 (1972).

54. HARRIS, E.G., *Phys. Rev. Letters 2*, 34 (1959).

55. ROSENBLUTH, M.N., in *Plasma Physics*, ed. by A.M. HAMENDE, International Atomic Energy Agency, Vienna, 1965.

56. THOMPSON, W.B. and HUBBARD, J., *Revs. Mod. Phys. 32*, 714 (1960).

57. SALPETER, E.E., *Phys. Rev. 122*, 1663 (1961); and D.W. SWIFT, Univ. of Iowa Report 72-13 (1972).

58. FRIEMAN, E.A., *J. Math. Phys. 4*, 410 (1963).
SANDRI, G., *Ann. Phys. (N.Y.) 24*, 332 and 380 (1963).

59. DUPREE, T.H., *Phys. Fluids 4*, 696 (1961).

60. BOGOLYUBOV, N.N., *J. Phys. (U.S.S.R.) 10*, 265 (1946).

61. McCUNE, J.E., SANDRI, G., and FRIEMAN, E.A., Rarified Gas Dynamics, *Proc. 3rd Int. Symposium on Rarified Gas Dynamics*, Palais de l'UNESCO, Paris, 1962, ed. by J.A. LAURMANN, Academic Press, New York, 1963, Vol. 1, p. 102.

62. Ref. 6, Chapters 4, 5, and 6.

63. BOGOLYUBOV, N.N. and MITROPOLSKII, Y.A., *Asymptotic Methods in the Theory of Nonlinear Oscillations* (transl. from the Russian), Gordon and Breach, New York, 1961.

64. MORRISON, J.A., *S.I.A.M. Review 8*, 66 (1966).

65. MINORSKI, N., *Nonlinear Oscillations*, D. Van Nostrand, Princeton (N.J.), 1962, Ch. 16.

66. DAWSON, J.M., OKUDA, M., and CARLILE, R.N., *Phys. Rev. Letters 27*, 491 (1971).

67. J.-Y. Hsu, D. Montgomery, and G. Joyce *J. Plasma Phys.* (to be published, 1974).

# TABLE OF CONTENTS